Lecture Notes in Networks and Systems

Volume 182

The series "Lecture Notes in Networks and Systems" publishes the latest developments in Networks and Systems—quickly, informally and with high quality. Original research reported in proceedings and post-proceedings represents the core of LNNS.

Volumes published in LNNS embrace all aspects and subfields of, as well as new challenges in, Networks and Systems.

The series contains proceedings and edited volumes in systems and networks, spanning the areas of Cyber-Physical Systems, Autonomous Systems, Sensor Networks, Control Systems, Energy Systems, Automotive Systems, Biological Systems, Vehicular Networking and Connected Vehicles, Aerospace Systems, Automation, Manufacturing, Smart Grids, Nonlinear Systems, Power Systems, Robotics, Social Systems, Economic Systems and other. Of particular value to both the contributors and the readership are the short publication timeframe and the world-wide distribution and exposure which enable both a wide and rapid dissemination of research output.

The series covers the theory, applications, and perspectives on the state of the art and future developments relevant to systems and networks, decision making, control, complex processes and related areas, as embedded in the fields of interdisciplinary and applied sciences, engineering, computer science, physics, economics, social, and life sciences, as well as the paradigms and methodologies behind them.

Indexed by SCOPUS, INSPEC, WTI Frankfurt eG, zbMATH, SCImago.

All books published in the series are submitted for consideration in Web of Science.

More information about this series at http://www.springer.com/series/15179

Henry Selvaraj · Grzegorz Chmaj ·
Dawid Zydek
Editors

Proceedings of the 27th International Conference on Systems Engineering, ICSEng 2020

Springer

Editors
Henry Selvaraj
Department of Electrical and Computer Engineering
University of Nevada Las Vegas
Las Vegas, NV, USA

Grzegorz Chmaj
Department of Electrical and Computer Engineering
University of Nevada Las Vegas
Las Vegas, NV, USA

Dawid Zydek
Department of Electrical and Computer Engineering
University of Nevada Las Vegas
Las Vegas, NV, USA

ISSN 2367-3370 ISSN 2367-3389 (electronic)
Lecture Notes in Networks and Systems
ISBN 978-3-030-65795-6 ISBN 978-3-030-65796-3 (eBook)
https://doi.org/10.1007/978-3-030-65796-3

This Springer imprint is published by the registered company Springer Nature Switzerland AG
The registered company address is: Gewerbestrasse 11, 6330 Cham, Switzerland

Preface

ICSEng 2020

This important series of conferences arose from technical cooperation among Wrocław University of Science and Technology (Poland), Coventry Polytechnic (UK), Wright State University (USA) and the University of Nevada, Las Vegas (USA). Prior to 1980, a series of International Conferences on Systems Science had been held in Wroclaw, Poland. In 1980, it was decided that the conference would change venue on alternate years, be held in Coventry, UK, and, at the same time, change emphasis from science to engineering when hosted in UK. Consequently, the First and Second International Conferences on Systems Engineering were held in Coventry, UK, in 1980 and 1982. In 1982, it was decided that the engineering series of conferences would be held in the USA each third year. The Third and Fifth International Conferences on Systems Engineering were held in Dayton, Ohio, USA. The Seventh, Ninth, Eleventh, Thirteenth, Fifteenth, Seventeenth, Nineteenth, Twenty-First, Twenty-Third and Twenty-Fifth International Conferences on Systems Engineering were held in Las Vegas, Nevada, USA. In the year 2018, the twenty-sixth conference was held at the University of Technology Sydney, Australia. This year 2020, the Twenty-Seventh International Conference on Systems Engineering is held in Las Vegas, Nevada, USA. Due to unforeseen circumstances arising out of the COVID-19 outbreak, the conference is held remotely.

Research in the discipline of systems engineering is an important concept in the advancement of engineering and information sciences. Systems engineering attempts to integrate many of the traditional engineering disciplines to solve large complex functioning engineering systems, dependent on components from all the disciplines. The research papers contained in these proceedings reflect the state of the art in systems engineering from all over the world and should serve as vital references to researchers to follow.

We received 92 submissions. Each paper was reviewed by at least two independent reviewers. A total of 44 papers were selected as full papers that gives an acceptance rate of 48%. The University of Nevada, Las Vegas, is very pleased to host this Twenty-Seventh International Conference on Systems Engineering.

We would like to thank all the authors, reviewers, participants and student volunteers for making the conference a success. We wish all the participants a fruitful conference.

Henry Selvaraj
ICSEng 2020 General Chair

Grzegorz Chmaj
Dawid Zydek
ICSEng 2020 Organizing Committee Co-chairs

Conference Organization

ICSEng 2020

General Chair

H. Selvaraj	University of Nevada, Las Vegas, USA

Organizing Committee Co-chairs

G. Chmaj	University of Nevada, Las Vegas, USA
D. Zydek	University of Nevada, Las Vegas, USA

Steering Committee Chair

H. Selvaraj	University of Nevada, Las Vegas, USA

Steering Committee

H. Selvaraj	University of Nevada, Las Vegas, USA
G. Chmaj	University of Nevada, Las Vegas, USA
D. Zydek	University of Nevada, Las Vegas, USA
J. Swiatek	Wrocław University of Science and Technology, Poland
P. Swiatek	Wrocław University of Science and Technology, Poland
K. Burnham	University of Wolverhampton, UK

Organizing Committee

G. Chmaj	University of Nevada, Las Vegas, USA
D. Zydek	University of Nevada, Las Vegas, USA
B. Tiwari	University of Nevada, Las Vegas, USA
S. Sharma	University of Nevada, Las Vegas, USA

Program Committee

ICSEng 2020

R. J. Baker	University of Nevada, Las Vegas, USA
G. Borowik	Deep Dive Lab, Poland
A. P. F. Carvalho	University of Sao Paulo at Sao Carlos, Brazil
Z. Chaczko	University of Technology Sydney, Australia
G. Chmaj	University of Nevada, Las Vegas, USA
M. Ezra	Universiti Tunku Abdul Rahman, Malaysia
T. Fujimoto	Toyo University, Japan
L. Gewali	University of Nevada, Las Vegas, USA
T. F. Gonzalez	University of California at Santa Barbara, USA
T. Gorski	Polish Naval Academy, Poland
A. Iyer	Virginia Commonwealth University, USA
L. Jozwiak	Eindhoven University of Technology, Netherlands
J. Kacprzyk	Polish Academy of Sciences, Poland
A. Kasprzak	Wrocław University of Science and Technology, Poland
A. Khamis	Military Technical College, Cairo, Egypt
M. H. Kolekar	Indian Institute of Technology Patna, India
L. Koszalka	Wrocław University of Science and Technology, Poland
T. Luba	Warsaw University of Technology, Poland
H. Migliore	Portland State University, USA
S. Muthaly	Asia Pacific University of Technology & Innovation, Malaysia
I. Pozniak-Koszalka	Wrocław University of Science and Technology, Poland
J. Rozenblit	University of Arizona, USA
H. Selvaraj	University of Nevada, Las Vegas, USA
D. Selvathi	Mepco Schlenk Engineering College, India
S. Singh	University of Nevada, Las Vegas, USA
P. Stubberud	University of Nevada, Las Vegas, USA
S. Stubberud	Boeing Company, USA
M. Thoma	University of Hannover, Germany
A. Vadivel	SRM AP, Amaravati, India
R. Venkat	University of Nevada, Las Vegas, USA
L. Wang	Harbin Institute of Technology, China
M. Wozniak	Wrocław University of Science and Technology, Poland
D. Zydek	University of Nevada, Las Vegas, USA

Sponsors

ICSEng 2020

Contents

Intelligent Systems

Machine Learning and Analytics

Internet of Things

AeroSpace Systems

A Track Registration Association Technique – Feature Analysis

Stephen C. Stubberud[1] and Kathleen A. Kramer[2(✉)]

[1] General Atomics – EMS, San Diego, CA, USA
scstubberud@ieee.org
[2] Department of Electrical Engineering, University of San Diego, San Diego, CA, USA
kramer@sandiego.edu

Abstract. A recent track-to-track association technique was developed to associate tracks from multiple sensor platforms where position and heading errors were incorporated into the reports. When this issue, known as target registration, is present, standard association routines often result in misassociations or non-associations between the tracks of the same target. This new technique develops a track representation that is invariant in translation errors and insensitive to rotation errors. It has been demonstrated to be effective for registration problems with reasonable heading errors and any translation errors. In this work, the features used in the track representation are analyzed for their effectiveness and sensitivity for various registration error sizes, providing the foundation for the weighting of the features to enhance registration association.

Keywords: Target registration · Registration association · Target tracking · Translation error · Rotation error

1 Introduction

Target registration occurs when multiple target tracking platforms are reporting their tracks to other platforms, but bias errors in one or more readings causing the potential for misassociation. Standard association routines are based on nearest neighbor or the chi-squared-metric [1, 2].

Figures 1 and 2 illustrate the registration problem. In Fig. 1, two platforms are tracking the same target. If both platforms knew their local coordinates and headings exactly, their reported tracks would be as shown. Figure 2 illustrates the difference in reported tracks if one of them had a bias in its reported position, and an offset in its heading. Standard association routines in such as case can undergo gate out. A gate out occurs when the association score exceeds a threshold value that defines a score that indicates that the no association should be allowed.

In [3], an association approach for registration was presented. The approach was an extension of the association routine first developed in [4, 5]. This image-correlation technique for general track association gave the foundation for a track representation that is invariant when registration occurs. The problem with the technique is that when uncertainty in the track is present, the correlation falls apart.

H. Selvaraj et al. (Eds.): ICSEng 2020, LNNS 182, pp. 3–11, 2021.
https://doi.org/10.1007/978-3-030-65796-3_1

A recently developed registration-association technique [6] used the idea of from [3] of track representation that would be invariant under registration. This association scheme used the chi-squared metric as the foundation. The key difference between this technique and the standard application is the track vector representation. The core idea was to develop a registration-invariant representation of tracks. The results indicated that the technique is invariant for translation but not for rotation. It demonstrated insensitivity to rotation of the tracks but not a true invariance.

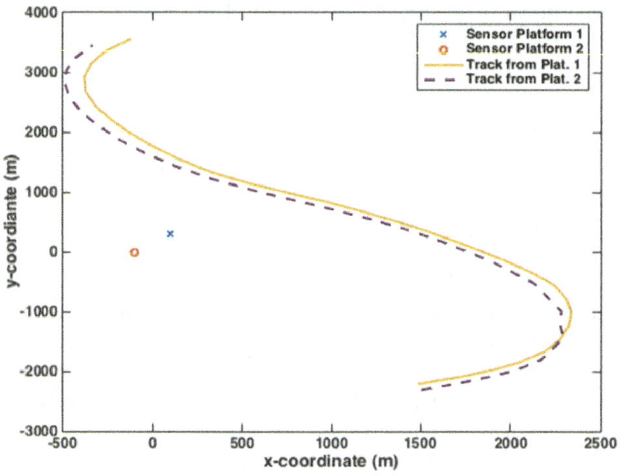

Fig. 1. Two sensor platforms provide two track reports of the same target.

In this paper, the individual features of the vector representing the tracks are analyzed. The impact on the association score for each feature is analyzed. The impact is then used to estimate weighting on the feature to determine is the insensitivity over different regions of rotation can be increased. Section 2 provides a summary overview the registration-association technique. In Sect. 3, the feature's effects on the association score under different rotation conditions are analyzed. This followed in Sect. 4 with proposed weighting schemes in the association routine.

2 Registration Association Technique

The standard tracking algorithm is the extended Kalman filter (EKF). The track state, usually referred to as $\boldsymbol{x}$, is a Gaussian random vector with an associated covariance, often denoted as $\mathbf{P}$. The standard association is referred to as the chi-squared metric. For standard track-to-track association, the metric is defined as

$$\chi^2 = \left(\boldsymbol{x}_{PlatA} - \boldsymbol{x}_{PlatB}\right)^T \left(\boldsymbol{P}_{PlatA} + \boldsymbol{P}_{PlatB}\right)^{-1} \cdot \left(\boldsymbol{x}_{PlatA} - \boldsymbol{x}_{PlatB}\right) \tag{1}$$

Since this is the multiplication of two Gaussian random variables, the metric has a chi-squared distribution. The vector of errors between the two tracks is weighted by the

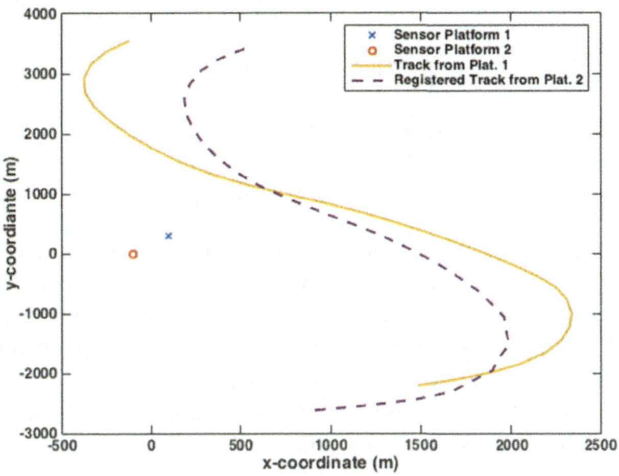

Fig. 2. Sensor Platform 2 has a registration error in reporting, both translation and rotation.

vector uncertainties. Thus, differences that are associated with large uncertainties are weighted less while small differences are amplified when associated with small uncertainties. An analogy would be close stars in astronomy as compared to missed putts in golf.

The standard vector used in track association is a vector comprised of position, velocity, and, sometimes, acceleration. The position-velocity vector is defined as

$$x^T = [\,x \quad \dot{x} \quad y \quad \dot{y}\,] \tag{2}$$

Each track-pair score is then entered into an assignment matrix after being processed as

$$e^{-\chi^2} \tag{3}$$

An assignment matrix that provides the maximum total score is produced by applying a version of the Munkres algorithm [7] upon the full set of track-pair scores. A new track-vector is defined in Eq. (4) as the distance traveled, the direction travelled, the change in speed, and the target heading for each time step. This vector represents the track differently than Eq. (2) which would be impacted significantly by translation and rotation errors.

$$\begin{bmatrix} \tilde{d}_n \\ \tilde{\theta}_n \\ \tilde{s}_n \\ \tilde{\varphi}_n \end{bmatrix} = \begin{bmatrix} (\tilde{x}_n^2 + \tilde{y}_n^2)^{\frac{1}{2}} \\ \left(\tan^{-1}\left(\frac{\tilde{y}_n}{\tilde{x}_n}\right)\right) - \left(\tan^{-1}\left(\frac{\tilde{y}_n - 1}{\tilde{x}_n - 1}\right)\right) \\ (\tilde{\dot{x}}_n^2 + \tilde{\dot{y}}_n^2)^{\frac{1}{2}} \\ \left(\tan^{-1}\left(\frac{\dot{y}_n}{\dot{x}_n}\right)\right) \end{bmatrix} \tag{4}$$

where

$$\begin{bmatrix} \tilde{x}_n \\ \tilde{y}_n \\ \tilde{\dot{x}}_n \\ \tilde{\dot{y}}_n \end{bmatrix} = \begin{bmatrix} x_n - x_{n-1} \\ y_n - y_{n-1} \\ \dot{x}_n - \dot{x}_{n-1} \\ \dot{y}_n - \dot{y}_{n-1} \end{bmatrix}. \tag{5}$$

It is important to note that the fourth element in Eq. (4) is only the current heading information provided by the track state. Unlike the first (distance) element and third (speed change) element, which were both based upon current and previous state elements, the heading angle change proved too noisy to be effective, even with smoothing over time, as used in the second element for the position angle.

The chi-squared metric to determine the association score is from Sensor Platform A to Sensor Platform B therefore generated as.

$$\chi^2_{Reg} = \left(\boldsymbol{V}_{Track\_pltA} - \boldsymbol{V}_{Track\_pltB}\right)^T \left(\boldsymbol{H}\boldsymbol{P}_{pltA}\boldsymbol{H}^T + \boldsymbol{H}\boldsymbol{P}_{pltB}\boldsymbol{H}^T\right)^{-1} \left(\boldsymbol{V}_{Track\_pltA} - \boldsymbol{V}_{Track\_pltB}\right), \tag{6}$$

where

$$\left(\boldsymbol{V}_{Track_{plt\,A}} - \boldsymbol{V}_{Track_{plt\,B}}\right) = \left(\begin{bmatrix} \tilde{\boldsymbol{d}}_n \\ \tilde{\theta}_n \\ \tilde{s}_n \\ \tilde{\varphi}_n \end{bmatrix}_{Track_{plt\,A}} - \begin{bmatrix} \tilde{\boldsymbol{d}}_n \\ \tilde{\theta}_n \\ \tilde{s}_n \\ \tilde{\varphi}_n \end{bmatrix}_{Track_{plt\,B}} \right) \tag{7}$$

and the matrix **P** is the associated standard track four-state position and velocity covariance matrix. The Jacobian is defined in Eq. (8) as.

$$\boldsymbol{H} = \begin{bmatrix} \frac{\partial h_d}{\partial x} & \frac{\partial h_\theta}{\partial x} & \frac{\partial h_s}{\partial x} & \frac{\partial h_\varphi}{\partial x} \\ \frac{\partial h_d}{\partial y} & \frac{\partial h_\theta}{\partial y} & \frac{\partial h_s}{\partial y} & \frac{\partial h_\varphi}{\partial y} \\ \frac{\partial h_d}{\partial \dot{x}} & \frac{\partial h_\theta}{\partial \dot{x}} & \frac{\partial h_s}{\partial \dot{x}} & \frac{\partial h_\varphi}{\partial \dot{x}} \\ \frac{\partial h_d}{\partial \dot{y}} & \frac{\partial h_\theta}{\partial \dot{y}} & \frac{\partial h_s}{\partial \dot{y}} & \frac{\partial h_\varphi}{\partial \dot{y}} \end{bmatrix} = \begin{bmatrix} \frac{\partial h_d}{\partial x} & \frac{\partial h_\theta}{\partial x} & \multicolumn{2}{c}{\multirow{2}{*}{0}} \\ \frac{\partial h_d}{\partial y} & \frac{\partial h_\theta}{\partial y} & & \\ \multicolumn{2}{c}{\multirow{2}{*}{0}} & \frac{\partial h_s}{\partial \dot{x}} & \frac{\partial h_\varphi}{\partial \dot{x}} \\ & & \frac{\partial h_s}{\partial \dot{y}} & \frac{\partial h_\varphi}{\partial \dot{y}} \end{bmatrix}$$

with

$$\begin{aligned} \frac{\partial h_d}{\partial x} &= \frac{2x}{(x^2+y^2)^{\frac{1}{2}}} & \frac{\partial h_s}{\partial \dot{x}} &= \frac{2\dot{x}}{\left(\dot{x}^2+\dot{y}^2\right)^{\frac{1}{2}}} \\ \frac{\partial h_d}{\partial y} &= \frac{2y}{(x^2+y^2)^{\frac{1}{2}}} & \frac{\partial h_s}{\partial \dot{y}} &= \frac{2\dot{y}}{\left(\dot{x}^2+\dot{y}^2\right)^{\frac{1}{2}}} \\ \frac{\partial h_\theta}{\partial x} &\frac{-y}{(x^2+y^2)} & \frac{\partial h_\varphi}{\partial \dot{x}} &= \frac{-\dot{y}}{\left(\dot{x}^2+\dot{y}^2\right)} \\ \frac{\partial h_\theta}{\partial y} &= \frac{x}{(x^2+y^2)} & \frac{\partial h_\varphi}{\partial \dot{y}} &= \frac{\dot{x}}{\left(\dot{x}^2+\dot{y}^2\right)} \end{aligned} \tag{8}$$

Since a single timestep may not be reliable, a small window of track points is used to create an association score

$$reg_{assoc-score} = \sum_{k=0}^{win_{size}-1} \chi^2_{Reg}(n-k). \tag{9}$$

The implementation of the metric in the assignment is scaled so that the distances do not become zero in the exponent and cause computational zeros to be formed. This causes the order of the matrix to have an effect on the selection. The scale factor is in the exponent is shown in Eq. (10) where the subscript indicates a two-norm.

$$\|HP_{pltA}H^T + HP_{pltB}H^T\|_2 \tag{10}$$

3 Feature Analysis

Each of the elements of the feature vector in Eq. (4) provides a portion of the score. Each has an impact on the score which varies based on the rotation. To analyze this impact, each feature will be applied alone to the association problem for a variety of rotation angles. The measure to be used is the correct association.

A three-track scenario, as defined in [5], is used. As seen in Fig. 3, there are two sensor platforms, each with range-bearing sensors. Platform 1 is located at 300 m north of the origin point and 100 m east of the origin point. Its sensor has a range accuracy of 0.1 m and a bearing accuracy of 0.0975°. Platform 2 is located at 0m north of the origin and -100 m east of the origin. Its sensor's range accuracy is 0.05 m, and its bearing accuracy is 0.3°. Target 1 performs an S-curve maneuver over the scenario. The second target flies a straight-line trajectory with constant speed at the beginning of the scenario followed by a constant acceleration. Target 3 decelerates over a straight-line trajectory. The window size was selected to be 5.

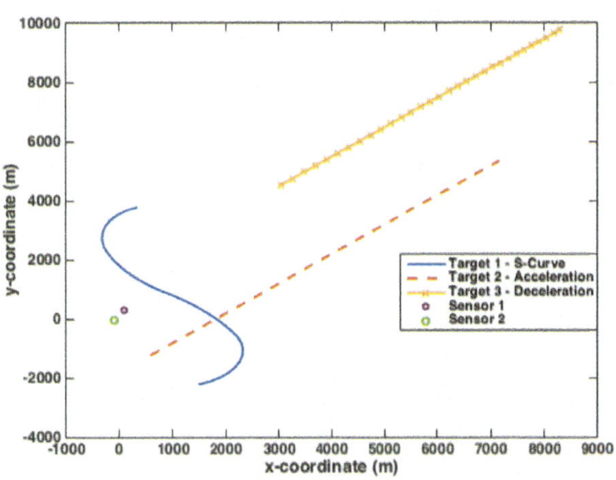

Fig. 3. A three-target and two-sensor scenario

The performance of each feature is measured for a variety of rotations. These rotations are 0°, 2°, 5°, 10°, 15°, 20°, 45°, 60°, and 90°.

In Tables 1, 2, 3 and 4, the performance was measured for each feature for the aforementioned rotation angles. The results were broken down into four different elements: Correct Association, Missed Associations – Straight Line Target to Straight Line Target, Missed Associations – Straight Line Target to S-curve Target, All Associations Wrong.

Table 1. Distance travel feature association performance

Sensor bias	Track associations: 30			
Angle (deg.)	Mis-assoc: None (All correct)	Mis-assoc: Straight-to-Straight	Mis-assoc: Straight-to- S-Curve	Mis-assoc: All
0	18	0	12	0
2	18	0	12	0
5	22	0	8	0
10	30	0	0	0
15	30	0	0	0
20	29	0	1	0
45	30	0	0	0
60	26	0	4	0
90	18	0	12	0

Table 2. Direction change feature association performance

Sensor bias	Track associations: 30			
Angle (deg.)	Mis-assoc: None (All correct)	Mis-assoc: Straight-to-Straight	Mis-assoc: Straight-to- S-Curve	Mis-assoc: All
0	2	9	8	11
2	1	9	9	11
5	2	8	15	5
10	12	2	13	3
15	14	2	11	3
20	19	0	8	3
45	26	0	2	2
60	29	0	0	1
90	2	10	7	11

Table 3. Speed change feature association performance

Sensor bias	Track associations: 30			
Angle (deg.)	Mis-assoc: None (All correct)	Mis-assoc: Straight-to-Straight	Mis-assoc: Straight-to- S-Curve	Mis-assoc: All
0	25	0	5	0
2	30	0	0	0
5	30	0	0	0
10	30	0	0	0
15	30	0	0	0
20	30	0	0	0
45	28	0	2	0
60	28	0	2	0
90	25	0	5	0

Table 4. Heading feature association performance

Sensor bias	Track associations: 30			
Angle (deg.)	Mis-assoc: None (All correct)	Mis-assoc: Straight-to-Straight	Mis-assoc: Straight-to- S-Curve	Mis-assoc: All
0	20	9	1	0
2	6	23	1	0
5	7	22	1	0
10	19	10	1	0
15	26	2	2	0
20	26	1	2	1
45	22	1	2	5
60	20	1	8	1
90	4	1	2	23

As can be seen, the most effective single features are the distance-traveled feature, d, and the speed-change feature, s. These features would provide 81.9% and 94.8% correct associations on average. The least effective as a solo feature is direction change, θ, at 39.6% average correct associations.

4 Feature Weighting Results

A variety of weighting values for the chi-squared vector features were compared. Based on the analysis of Sect. 3, Features 1 and 3 were considered exceedingly useful. They were each given equal weights for all of the comparisons. The direction and heading features, Features 2 and 4, were varied from values between 0 and 1. Table 5 lists the correct associations for each angle and an average percentage of correct associations.

As the results show, the weighting can have a small improvement in performance. If feature weighting were to be used in implementation, a weighting with Features 1 and 3 equally weighted and Feature 2 weighted less than Features 1 and 3 but more than Feature 4 would be best. All features should be used. The analysis shows that the technique is effective up to 20 degrees of rotation in most cases and the outrageous heading error of 90 degrees has deleterious effects. If the 90-degree heading error is not considered, the top four performing weighting schemes increase to between 95% to 98.75% accuracy.

Table 5. Correct associations based on feature weighting

Angle (deg.)	Feature weights (4)						
	1	1	1	1	1	1	1
	1	0	0	0.5	0.5	0.25	0.5
	1	1	1	1	1	1	1
	1	0	0.5	0.5	0	0.5	0.25
0	29	21	30	30	20	30	30
2	29	19	30	30	21	30	30
5	28	24	29	30	26	30	30
10	29	30	29	30	27	30	30
15	29	30	29	30	29	29	30
20	30	30	26	28	28	28	29
45	26	30	25	29	30	27	30
60	28	25	22	25	28	24	28
90	7	21	7	6	20	7	8
Avg.	87.0	85.2	84.1	88.2	84.8	87.0	90.7

5 Conclusion

A new target registration-association scheme was analyzed to determine the effectiveness of the features selected. The analysis showed that individual features of position change and speed acceleration were the most impactful. When combined in different weightings, the least impactful individual feature was quite useful. Weighting of the features used in the association metric can have an impact. This impact appears to be noticeable but relatively small, approximately 3% absolute improvement.

All features, the distance travelled, the direction change, the speed change, and the heading angle, should be used in the association score. Further research to improve the representation where the features both translation and rotation are invariant and improved weighting techniques are being considered.

References

1. Blackman, S., Popoli, R.: Design and Analysis of Modern Tracking System. Artech House, Boston (1999)
2. Bar-Shalom, Y., Blair, W.D. (eds.): Multitarget - Multisensor Tracking: Applications and Advances, vol. III. Artech House, Boston (2000)
3. Stubberud, S.C.: Track registration using image correlation. In: Proceedings of the 2015 Aerospace Conference, Big Sky, MT, pp. 1–7 (2015)
4. Stanek, C.J., Javidi, B., Yanni, P.: Performance assessment of frequency plane filters applied to track association and sensor registration. Proc. SPIE **5094**, 323–328 (2003)
5. Stanek, C.J.: Gridlocking and correlation methods and arrangements. U.S. Patent US680399B2 (2004)
6. Stubberud, S.C., Kramer, K.A.: A target registration association technique. In: Submitted to the IEEE Transactions of Aerospace Electronic Systems (2020)
7. Giernacki, W.: Robust control with optimization of robustness index. In: 2012 IEEE International Conference on Systems, Man, and Cybernetics (SMC), Seoul, South Korea, pp. 2389–2394 (2012)

Investigation of Wind Energy Impact on Power Systems Stability Using Lyapunov Exponents

Mojtaba Amiri[1], Maryam Dehghani[1], Alireza Khayatian[1], Mohsen Mohammadi[2], Navid Vafamand[1(✉)], and Jalil Boudjadar[3]

[1] School of Electrical and Computer Engineering, Shiraz University, Shiraz, Iran
n.vafamand@shirazu.ac.ir, navidvafamand@gmail.com
[2] School of Mechanical Engineering, Shiraz University, Shiraz, Iran
[3] Department of Engineering, Aarhus University, Aarhus, Denmark

Abstract. The world tendency is to replace all generators that use fossil fuels by renewable energy. Although this is a very important step toward keeping the environment clean, this paper shows that by replacing wind energy by the conventional generators, the relative stability becomes worse. This does not mean that the paper is against such replacement, but makes an alert to be aware of such important issue and to implement appropriate controllers for increasing the relative stability of the power systems equipped with renewable energy sources. In this work, a data-driven approach is used to assess the stability of power grids by just using voltage time-series data. Lyapunov exponent (LE) index is presented as an effective tool to analyze complex systems stability. If the maximum LE (MLE) is positive (negative), then the system is unstable (stable). In the present paper, the MLE is computed for a finite-time interval for the sample system of the New England 39-Bus system. In the end, the effect of renewable wind energy on the stability of the system has been investigated and by substituting the wind farm in different areas of the system, the impact of adding a wind farm to the system in different locations is found.

Keywords: Lyapunov exponent · Power system · Stability assessment · Wind energy

1 Introduction

Evaluating the complex power system stability is a critical issue which can be done through various approaches [1]. The model equations of power systems are highly nonlinear and these systems are high order systems. There are many approaches to assess the stability of systems. Many of these methods use the system model. On the other hand, the nonlinearity of the power systems, as well as the high order of the systems, is itself a strong reason for the high computational burden which requires expensive hardware and increases cost. Therefore, the methods that do not use model equations to analyze stability can be very effective. Because of the importance and complexity of these systems, the online and model-free voltage stability assessment is very important.

H. Selvaraj et al. (Eds.): ICSEng 2020, LNNS 182, pp. 12–22, 2021.
https://doi.org/10.1007/978-3-030-65796-3_2

In [2], Liu et al. used LE to evaluate the behavior of power systems for the first time. By using measurement data, they have measured the transient chaos of the power system with LE. That method is applied on a three-machine system. In [3], an applicable LE-based approach is suggested for the online stability evaluation of rotor angle in the presence of perturbations. The method presented in [3] is an extension of the results achieved by Lui et al. [2]. In [3], the relation of the MLE with the rotor angle stability in power systems is studied. The effectiveness of that method for the 200-Bus system has been investigated. In [4], a data-driven approach is developed to assess the voltage stability in power systems. The algorithm of [4] is an extension of the method proposed by Rosenstein et al. [5], in which, the LE is used as an indicator for determining stability. The authors of [6] proposed an algorithm to evaluate the rotor angle stability of a power system. That approach is not a model-based approach and can use phasor measurement unit (PMU) fuzzy measurement data. In that algorithm, the LE measures the degree of convergence or divergence of the rotor angle trajectories. In [7], a robust stability analysis is performed for a power system by the means of the LE to analyze the stability of the voltage and rotor angle. The author of [7] calls that method as "THRUST". Contrary to references [8, 9] that have been used only for small systems, the THRUST has been also utilized to evaluate the stability of a small system (39-Bus) and a very large system (13029-Bus). In [10], Wei et al. proposed a data-based approach using LE theory to analyze rotor angle stability. The MLE is obtained by a method based on the recursive least squares from the measured data of the rotor angles. The results of that study were applied to the Northeast Power Coordinating Council 140-Bus and the New England 39-Bus systems and have yielded good results. In [11], Ge et al. presented an improved method of voltage stability assessment. The effect of voltage fluctuations on the computation procedure of the LE is studied and a phase correction method is presented to compensate for the negative effects of voltage fluctuations. Simulations have been carried out on the Guangdong Province Power Network of China.

In this research, the LE-based method is used for stability assessment. LE is introduced by Lyapunov [12] and is a powerful and efficient method to analyze the stability of a power system [13]. LE is calculated by two methods: I) model-based, and II) model-free (data-driven) [14, 15]. This paper uses a model-free (data-driven) and online approach of computation of the MLE to analyze the stability of the power system using only voltage time-series data. If the MLE value is positive, the system is unstable and if the LE gets a negative value, the system is stable. In other words, this paper is an extension of [16–19] in assessing system stability using online data from system, such as PMU data. Moreover, the use of renewable energies power plants instead of fossil fuel power plants to reduce pollution and costs is increasing. Replacing a synchronous generator with a renewable energy system reduces the relative stability of the power system [20–23]. In the present paper, the effect of replacing a wind farm with a synchronous generator on the stability of a New England 39-Bus system is investigated. The effect of replacing a wind farm with all of the synchronous generators of the system is investigated by using an online model-free approach of MLE computation for several different cases. Power system simulations are done by DIgSILENT.

The paper is organized as follows. The low-inertia systems are represented in Sect. 2. Section 3 represents the Date-Driven-based stability assessment approach.

In Sect. 4, simulation results on the New England 39-Bus system are presented to assess the effect of substituting a wind farm with a specific synchronous generator of the system. Finally, the conclusions are in Sect. 5.

2 Low-Inertia Systems

The main purpose of synchronous generators is to generate active power in order to adjust the frequency and voltage. The inertia of the synchronous generators affects the dynamic behavior of the system shortly after fault or disturbance in the power balance. Research has shown that low-inertia generators such as wind turbines that can be used in a power system instead of synchronous generators, cause the system to be unable to maintain voltage and frequency of the system in the desired value [20]. Here, a system with low inertia is examined shortly after a fault occurs. The equation of a conventional system with both synchronous and non-synchronous generators is as follows [20]:

$$M\dot{\omega}(t) = P_s(t) + P_{ns}(t) - P_l(t) - P_j(t), \tag{1}$$

where M shows the total inertia of the system, $\omega(t)$ is average frequency, P_{ns} and P_s show the active power of non-synchronous and synchronous generators, respectively. P_l and P_j show the loads and losses, respectively. Bigger values of M increase the system's control ability to and robustness against load fluctuations and faults. By increasing the number of non-synchronous generators, the value of M is reduced, causing an increase in the frequency fluctuations and even system failure to maintain the voltage at a certain value. In order investigate this effect, a data-driven tool is used.

3 Data-Driven-Based Stability Assessment

Lyapunov introduced two methods to analyze the systems stability: I) LE-based approach, II) Lyapunov function-based approach [12]. The second method is very common, but it is a model-based approach and for stability analysis should have an accurate model of the system. Additionally, in the second method, finding a proper Lyapunov function is very challenging. On the other hand, the LE-based method can be used in both model-based and model-free approaches. In this paper, a model-free approach is used to assess the stability of the power system in an online procedure. LEs can convey information about the system relative stability. The largest LE is known as MLE [24]. The sign of MLE is used for stability investigation. Negative MLE refers to a stable system, while positive MLE shows instability.

The LE generally shows the rate of separation and closeness of two trajectories which are initially so close to each other. Two trajectories which are starting from very close initial conditions may exponentially diverge or may converge to a finite set, such as a fixed point, a limit cycle, a periodic or quasi-periodic motion, or a chaotic attractor. Briefly, if the value of λ is positive, it indicates that the two trajectories are separated and if the value of λ is negative, two trajectories will become close to each other. This is called chaotic if the initial conditions are perturbed and the system becomes unpredictable. Attractors of a system with one or more positive LEs are called strange or chaotic. The LE is highly dependent on the initial conditions of the system.

The pioneer numerical algorithm to compute the MLE of the system from time series was introduced in [5], which was unable to compute the MLE of the big data and also this algorithm is not robust to delay time changes. System information can be also obtained from system state-space equations. In the words, it is not enough to only use set of time-series data or simply a measurement variable. To solve this problem, the measurement variables should be increased [25] by embedding data in the m-dimensional delay coordinates. Suppose voltage time series data is available in the form $\{v_1, v_2, \ldots, v_k\}$, the delay coordinates are as follows:

$$\boldsymbol{V}_i = \left(v_i; v_{i+\tau}; \ldots; v_{i+(m-1)\tau}\right) \rightarrow \boldsymbol{V} = (\boldsymbol{V}_1, \boldsymbol{V}_2, \ldots, \boldsymbol{V}_M)^T, \tag{2}$$

in which $i = 1, \ldots, M$, $M = k - (m-1)\tau$, m is the embedding dimension and τ is the reconstruction delay time. If the number of data is infinite and the data are not noisy, there is no obstacle in choosing any value of τ, but the data are often noisy and finite [26]. The method of calculating m is well described in [27], and here the value of m is chosen equal to one.

The MLE computation algorithm of this paper is adapted from [4]. MLE is calculated from the time-series data of the grid voltages. In this paper the MLE is shown by λ. Data-driven MLE computation algorithm is provided in Fig. 1, where $\lambda(k\Delta t)$ computes the MLE at time $k\Delta t$. $V_t = \left(v_t^1, v_t^2, \ldots, v_t^n\right)$ in (2) displays the time-series data of the n bus voltages of the system. If the value of the MLE in (2) becomes positive (negative), the system is unstable (stable).

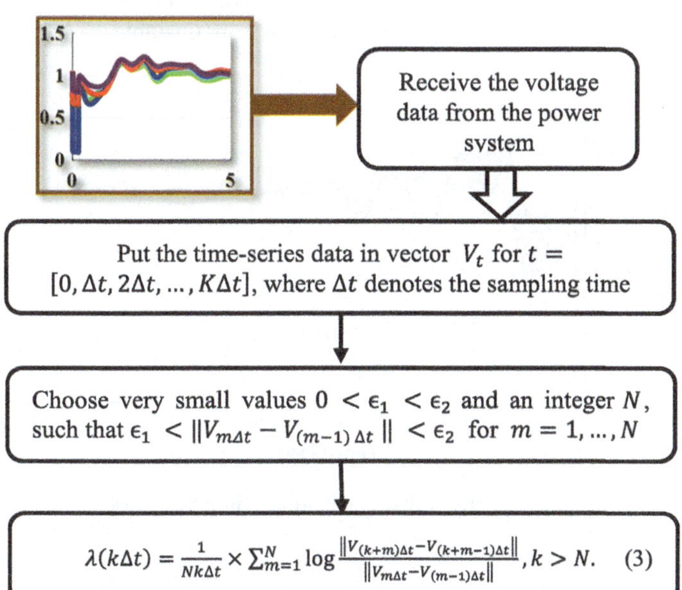

Fig. 1. Data-driven MLE computation algorithm

4 Simulation Results

The New England 39-Bus system is considered as the case study and simulated in DIgSILENT for different cases. First of all, stability analysis of this system is investigated for six cases. Then, by using MLE, the impact of wind energy on the system stability is investigated for three cases. The New England 39-Bus System has 10 generators, 12 transformers, 39 buses, and 34 lines. This system is a reduced order of the transmission system in the USA [28].

By using the data-driven approach presented in Fig. 1 to compute the MLE, the stability is analyzed in six cases which are mentioned below. The MLE is computed only using bus voltage time-series data.

Case 1) A short-circuit at 16th bus is simulated and it is cleared after 180 ms.
Case 2) A short-circuit at 16th bus is simulated and it is cleared after 200 ms.
Case 3) A short-circuit at 31th bus is simulated and it is cleared after 210 ms.
Case 4) A short-circuit at 16th bus is simulated and it is cleared after 230 ms.
Case 5) A short-circuit at the beginning of Line 2–3 is simulated and it is cleared after 230 ms.
Case 6) A short-circuit at the beginning of Line 2–3 is simulated and it is cleared after 250 ms.

The 39 bus voltage trajectories in case 1 are shown in Fig. 2 for 10 s. The MLE evolution of the system in case 1 is shown in Fig. 3. The value of the MLE in this case became negative and we can realize that the system is stable. It should be mentioned that the voltage magnitudes are shown in p. u.

The 39 bus voltage trajectories in case 2 are shown in Fig. 4 for 5 s. Figure 5 shows the MLE evolution of the system in case 2. The value of the MLE, in this case, became positive and one can realize that the system is unstable.

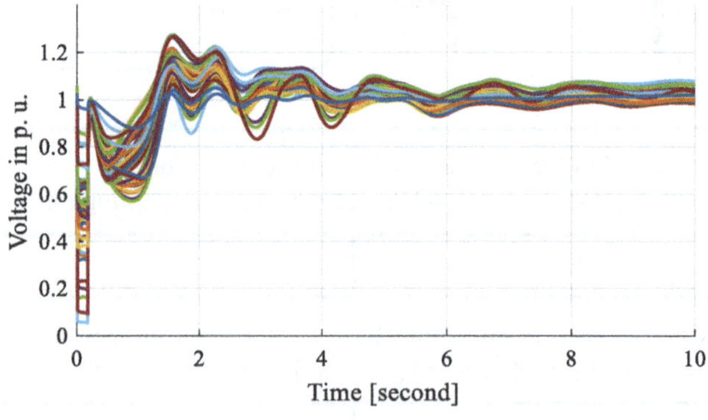

Fig. 2. 39 bus voltage magnitudes in p.u. in case 1

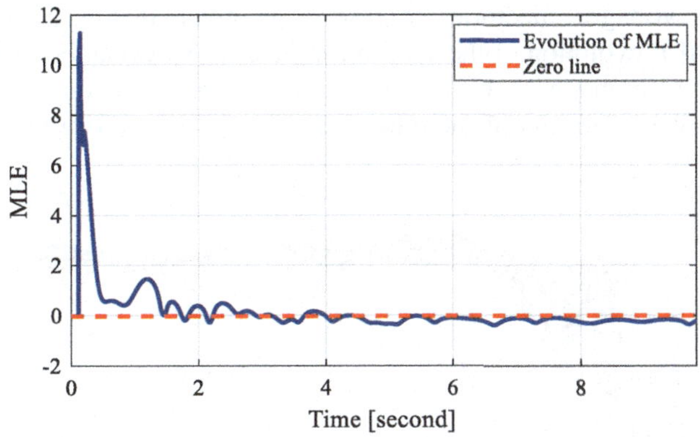

Fig. 3. The evolution of the MLE of the system in case 1

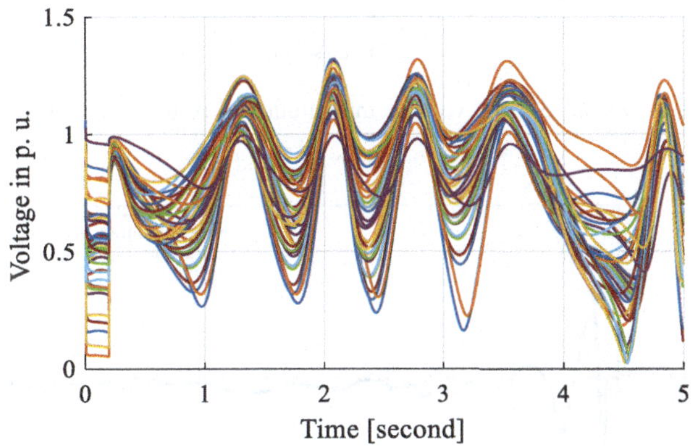

Fig. 4. 39 bus voltage magnitudes in p. u. in case 2

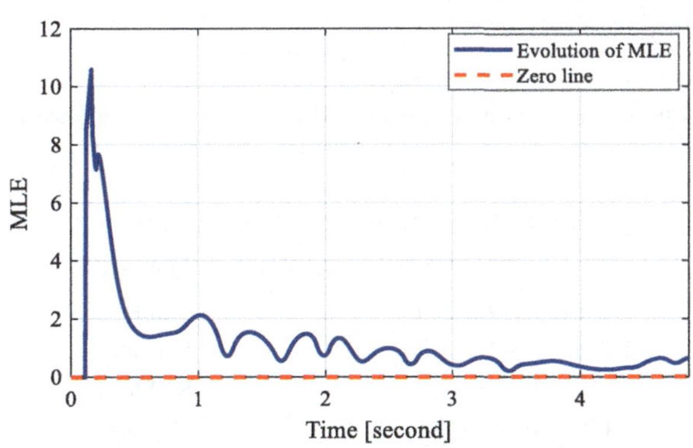

Fig. 5. The evolution of the MLE of the system in case 2

The 39 bus voltage trajectories in case 3 are given in Fig. 6 for 5 s. Figure 7 illustrates the MLE evolution of the system for the case 3. As can be seen in Fig. 7, the value of the MLE, in this case, became negative and one can realize that the system is stable. The simulation results of the other cases for 5 s are shown in Table 1.

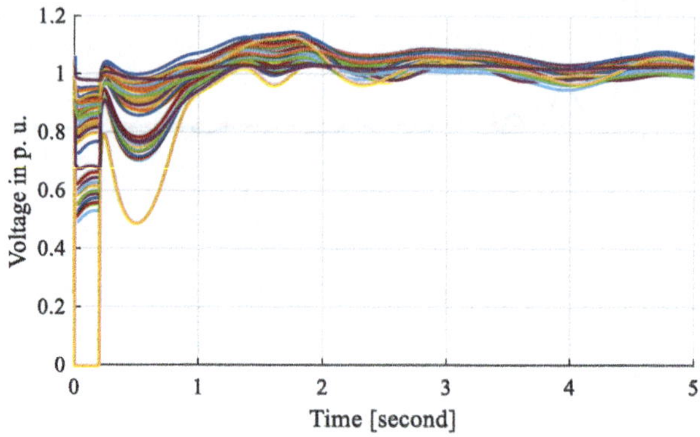

Fig. 6. 39 bus voltage magnitudes in p. u. in case 3

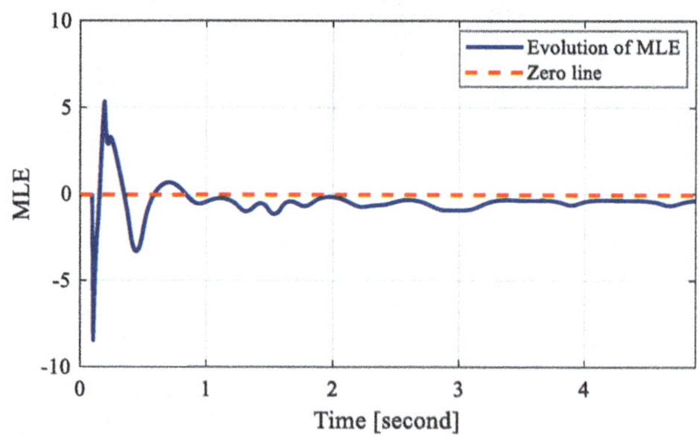

Fig. 7. The evolution of the MLE of the system in case 3

Table 1. Simulation results of six cases

Cases	The MLE values after 5 s	Stability
Case 1	−0.3144	Stable
Case 2	+0.6408	Unstable
Case 3	−0.3316	Stable
Case 4	+0.4984	Unstable
Case 5	−0.6119	Stable
Case 6	+0.5918	Unstable

For an investigation of wind energy impact on the power system stability, each of the synchronous generators of the 39-Bus system is replaced by a wind farm, respectively, and the impact of this substitution is investigated in three cases. It should be mentioned that the New England 39-Bus system is stable in these three cases when the substitution is not done. By calculating the MLE of the system using voltage time-series data, we can analyze the stability of the system in each case. The three cases are as follows:

Case 1) A short-circuit at 16th Bus is simulated and is cleared after 180 ms.
Case 2) A short-circuit at 31th Bus is simulated and is cleared after 210 ms.
Case 3) A short-circuit at the beginning of "Line 02–03" is simulated and by switching off the faulted line, this fault is cleared after 230 ms.

Wind farms are connected to the New England 39-Bus system in parallel with synchronous generators as given in Fig. 8. Each of the synchronous generators is replaced with a wind farm specified by the active power generation capacity same as the original generator and the impact of this substitution is investigated by computation of the MLE of the system using voltage time-series data in 5 s for 3 mentioned cases. The results of this substitution of the wind farm with synchronous generators are shown in Table 2. The table shows values of the MLE after 5 s.

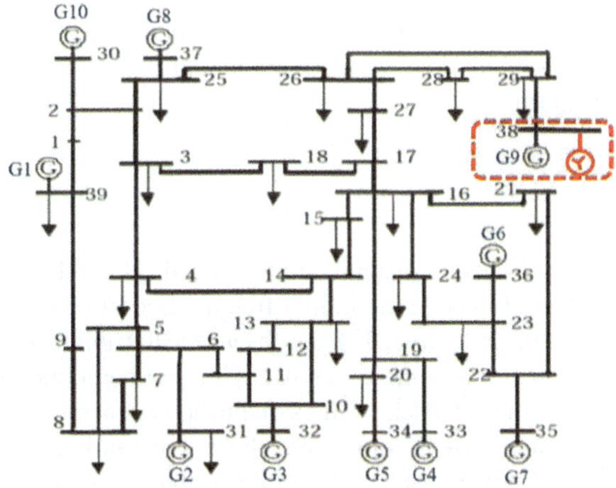

Fig. 8. Parallel connection of wind farm with a synchronous generator to the system

Table 2. The MLE values of the system in different cases after 5 s

	MLE		
	Case 1	Case 2	Case 3
Without DFIG	**−0.3144**	**−0.3316**	**−0.6119**
Replacing with G1	+0.6591	+0.6971	+0.7543
Replacing with G2	**−0.1594**	**−0.1654**	**−0.1691**
Replacing with G3	+0.5441	+0.3658	+0.3702
Replacing with G4	+0.6927	+0.4920	+0.4721
Replacing with G5	+0.6556	+0.4469	+0.3719
Replacing with G6	+0.7842	+0.4183	+0.3702
Replacing with G7	+0.1241	+0.5167	+0.5016
Replacing with G8	+0.6477	+0.5318	+0.5549
Replacing with G9	**−0.2114**	**−0.2773**	**−0.3160**
Replacing with G10	+0.4785	+0.6119	+1.0570

The MLE values in the table show that with the displacement of the wind farm in areas of the system, in most cases, the system becomes unstable and, in some cases, the relative stability is decreased.

As shown in Table 2, by substitution of 39 buses system synchronous generators in most cases, the MLE became positive and in two cases reduced relative system stability. According to this table, the Lyapunov exponent theory can be utilized as a tool for locating the wind farm.

5 Conclusion

In this paper, first, the low-inertia systems in introduced. Then, the Lyapunov exponent theorem is introduced and by using the model-free method of computation of the MLE, the stability of the system is analyzed. It is observed that this method can assess the stability of power systems in an online manner with low computational burden. Then, by replacing the wind farm with each of the synchronous generators of a power system, system stability is evaluated. It is found that these substitutions may reduce relative stability or make the system unstable. This study proves the importance of controller design strategies when the renewable energy sources are added to the power system.

References

1. Liu, M., Bizzarri, F., Brambilla, A.M., Milano, F.: On the impact of the dead-band of power system stabilizers and frequency regulation on power system stability. IEEE Trans. Power Syst. **34**, 3977–3979 (2019)
2. Liu, C.-W., Thorp, J.S., Lu, J., Thomas, R.J., Chiang, H.-D.: Detection of transiently chaotic swings in power systems using real-time phasor measurements. IEEE Trans. Power Syst. **9** (3), 1285–1292 (1994)

3. Yan, J., Liu, C.-C., Vaidya, U.: PMU-based monitoring of rotor angle dynamics. IEEE Trans. Power Syst. **26**(4), 2125–2133 (2011)
4. Dasgupta, S., Paramasivam, M., Vaidya, U., Ajjarapu, V.: Real-time monitoring of short-term voltage stability using PMU data. IEEE Trans. Power Syst. **28**(4), 3702–3711 (2013)
5. Rosenstein, M.T., Collins, J.J., De Luca, C.J.: A practical method for calculating largest Lyapunov exponents from small data sets. Physica D **65**(1–2), 117–134 (1993)
6. Dasgupta, S., Paramasivam, M., Vaidya, U., Ajjarapu, V.: PMU-based model-free approach for real-time rotor angle monitoring. IEEE Trans. Power Syst. **30**(5), 2818–2819 (2014)
7. Khaitan, S.K.: THRUST: a Lyapunov exponents based robust stability analysis method for power systems. In: 2017 North American Power Symposium (NAPS), pp. 1–6. IEEE (2017)
8. Verdejo, H., Vargas, L., Kliemann, W.: Stability of linear stochastic systems via Lyapunov exponents and applications to power systems. Appl. Math. Comput. **218**(22), 11021–11032 (2012)
9. Dasgupta, S., Paramasivam, M., Vaidya, U., Ajjarapu, V.: PMU-based model-free approach for short term voltage stability monitoring. In: 2012 IEEE Power and Energy Society General Meeting, pp. 1–8. IEEE (2012)
10. Wei, S., Yang, M., Qi, J., Wang, J., Ma, S., Han, X.: Model-free MLE estimation for online rotor angle stability assessment with PMU data. IEEE Trans. Power Syst. **33**(3), 2463–2476 (2017)
11. Ge, H., et al.: An improved real-time short-term voltage stability monitoring method based on phase rectification. IEEE Trans. Power Syst. **33**(1), 1068–1070 (2017)
12. Lyapunov, A.M.: The general problem of the stability of motion. Int. J. Control **55**(3), 531–534 (1992)
13. Amiri, M., Dehghani, M., Khayatian, A., Mohammadi, M.: Lyapunov exponent based stability assessment of power systems. In: 2019 6th International Conference on Control, Instrumentation and Automation (ICCIA), pp. 1–5. IEEE (2019)
14. Khodadadi, H., Khaki-Sedigh, A., Ataei, M., Jahed-Motlagh, M.R.: Applying a modified version of Lyapunov exponent for cancer diagnosis in biomedical images: the case of breast mammograms. Multidimensional Syst. Signal Process. **29**(1), 19–33 (2016)
15. Pikovsky, A., Politi, A.: Lyapunov exponents: a tool to explore complex dynamics. Cambridge University Press, New York (2016)
16. Dehghani, M., Shayanfard, B., Khayatian, A.R.: PMU ranking based on singular value decomposition of dynamic stability matrix. IEEE Trans. Power Syst. **28**(3), 2263–2270 (2013)
17. Shayanfard, B., Dehghani, M., Khayatian, A.: Optimal PMU placement for full observability and dynamic stability assessment. In: 2011 19th Iranian Conference on Electrical Engineering, pp. 1–6. IEEE (2011)
18. Mohammadi, H., Khademi, G., Dehghani, M., Simon, D.: Voltage stability assessment using multi-objective biogeography-based subset selection. Int. J. Electric. Power Energy Syst. **103**, 525–536 (2018)
19. Mohammadi, H., Khademi, G., Simon, D., Dehghani, M.: Multi-objective optimization of decision trees for power system voltage security assessment. In: 2016 Annual IEEE Systems Conference (SysCon), pp. 1–6. IEEE (2016)
20. Milano, F., Dörfler, F., Hug, G., Hill, D.J., Verbič, G.: Foundations and challenges of low-inertia systems. In: 2018 Power Systems Computation Conference (PSCC), pp. 1–25. IEEE (2018)
21. Toor, A., et al.: Energy and performance aware fog computing: a case of DVFS and green renewable energy. Future Gener. Comput. Syst. **101**, 1112–1121 (2019)
22. Khooban, M.H., Vafamand, N., Boudjadar, J.: Tracking control for hydrogen fuel cell systems in zero-emission ferry ships. Complexity **2019** (2019)

23. Vafamand, N., Khooban, M.H., Dragičević, T., Boudjadar, J., Asemani, M.H.: Time-delayed stabilizing secondary load frequency control of shipboard microgrids. IEEE Syst. J. **13**(3), 3233–3241 (2019)
24. Best, E.A.: Stability Assessment of Nonlinear Systems Using the Lyapunov Exponent. Ohio University, Athens (2003)
25. Takens, F.: Detecting strange attractors in turbulence. In: Dynamical Systems and Turbulence, Warwick 1980, pp. 366–381. Springer (1981)
26. Fraser, A.M., Swinney, H.L.: Independent coordinates for strange attractors from mutual information. Phy. Rev. A **33**(2), 1134 (1986)
27. Kennel, M.B., Brown, R., Abarbanel, H.D.: Determining embedding dimension for phase-space reconstruction using a geometrical construction. Phys. Rev. A **45**(6), 3403 (1992)
28. Pai, M.: Energy Function Analysis for Power System Stability. Springer, New York (2012)

Motor Current Signature Analysis Using Shapelet

Maryam Iraji[1], Maryam Dehghani[1], Mohsen Mohammadi[2], Navid Vafamand[1](✉), and Jalil Boudjadar[3]

[1] Electrical and Computer Engineering, Shiraz University, Shiraz, Iran
n.vafamand@shirazu.ac.ir
[2] School of Mechanical Engineering, Shiraz University, Shiraz, Iran
[3] Department of Engineering, Aarhus University, Aarhus, Denmark

Abstract. In the contemporary industrial world, induction motors have special place in the technological advancements. Therefore, detecting faults in these systems is gaining more and more attentions. This paper proposes a new approach for the fault detection and classification in induction motors. In the developed approach, shapelet concept is employed. The main distinguished advantage of shapelet is that they can extract features in dataset based on a local shape-similarity. To make the suggested shapelet algorithm more accurate and efficient, a new criterion for the minimum and maximum number of all classes members is introduced. Compared to the previous shapelet feature extraction techniques, the proposed technique has high accuracy and speed. For case study of a sample induction motor, which is performed in healthy and broken rotor bar situations, the shapelet approach is exploited to identify any possible faults. Simulation results show the satisfactory performance of the suggested shapelet technique for detecting broken rotor bar fault.

Keywords: Shapelet algorithm · Fault detection · Induction motor faults

1 Introduction

Fault detection methods become one of the most critical issues as in the most parts of industry, faults are likely to occur and it can cause devastating damage in equipments. In this situation using methods that can detect fault accurately and fast is very important. Shapelet is a machine learning method that can find local shape-based differences in signals accurately and fast. So Shapelet method can be employed as one of the most accurate way to detect faults. In 2009, Eamonn Keogh et al. introduced shapelet concept as a new classification method for the first time [1]. After that pioneer work, the shapelet attracts a lot of attention because of its advantages. For instance, the shapelet can classify datasets high accurately; it can gain some insight into the data; and it works in both time and frequency domains [1]. Some extensions of the approach of [1] are logical-shapelet concept [2], shapelet transform [3], unsupervised-shapelet (U-shapelet), which is a novel method for clustering data [4], and multivariate shapelet detection (MSD) for multivariate time series [5]. As Shapelet algorithm can recognize local shape differences, it has been successfully applied on various systems to find

H. Selvaraj et al. (Eds.): ICSEng 2020, LNNS 182, pp. 23–33, 2021.
https://doi.org/10.1007/978-3-030-65796-3_3

faulty behaviors [6–8]. However, the process of selecting a proper shapelet is very time-consuming and needs a high computational burden.

Until now, a wide variety of fault detection methods have been presented for the induction motors (IMs). The IMs are widely exploited in industry, as they are highly robust and have low cost maintenance, power efficiency and reliability. Though, continuous care should be taken for their maintenance as the environmental situation could cause serious faults, like broken rotor bar or shorted stator winding in the IM systems [9–11]. The fault detection methods can be classified in two ways; *invasive* and *non-invasive* methods [12]. Motor current signature analysis (MCSA) is one of the popular non-invasive methods. Cusido et al. developed a method that use the convolution of wavelet-based functions [13]. However, to the best knowledge of the authors, the application of shapelet method in the IMs is not investigated yet, which is the main motivation of this work.

In this paper, shapelet algorithm with new criteria for the maximum and minimum number of all class members, is employed for online fault detection and classification in the IMs. These criteria are defined based on the distance of timeseries from each other and the number of timeseries that are closer to shapelet candidate than the other ones. To show the merits of the suggested approach, a train dataset is made by simulating a normal IM. Then, a type of fault is exerted on the train dataset. The proposed shapelet algorithm with new criteria is used to discover proper shapelet in order to detect and classify occurring faults of the IM, accurately. Finally, the chosen shapelet is utilized to detect faults on a numerical IM system.

The rest of the paper is organized as follow; in Sects. 2 and 3, the study tools for this paper such as the theory of shapelet and the model of IMs are elaborated. In Sect. 4, new criteria for shapelet algorithm and the novel method for fault detection and classification are proposed. In Sect. 5, the case study is presented that consists an IM with healthy condition and broken rotor fault. Then, the shapelet are utilized in each of these situations and fault is detected. Section 6 consists of the conclusion about using shapelet for fault detection.

2 Theory of Shapelet

A time series is a temporal ordered list of real-valued data that is recorded in a fixed interval. Suppose there is a set of n time series $T = \{T_1 T_2 ... T_n\}$, where each T_i has m samples $T_i = \langle t_{i1} t_{i2} ... t_{im} \rangle$. For the sake of simplicity, it is assumed that all series in dataset have the same length. Though, inspired by [14], the results of this paper can be extended to the case that the series do not have the same length. Shapelet is a subset of one of the time series in the dataset and it is employed to extract a specific feature of each class, in order to find the best touchstone for achieving fast and accurate classification. Shapelet is used for finding local shape-based similarity, so shapelet must be normalized and non-variable to scale and offset [14]. Generally, the shapelet discovery has three main steps

2.1 Generating Candidates

There are time series with length m and the goal is to find a candidate with length l and l can change in the range of $[min, max]$, while $min \geq 3$ and $max \leq m$. All normalized subseries of T_i that have l samples are put in $W_{i,l}$ and $W_{i,l}$ for all time series of T are collected in W_l. Finally, collection of W_l for all possible amount of l for dataset T is W.

$$W_{i,l} = \{\langle t_{i,1}, \ldots, t_{i,l} \rangle, \langle t_{i,2}, \ldots, t_{i,l+1} \rangle, \ldots, \langle t_{i,m-l+1}, \ldots, t_{i,m} \rangle\} \tag{1}$$

$$W_l = \{W_{1,l} \cup W_{2,l} \cup \ldots \cup W_{n,l}\} \tag{2}$$

$$W = \{W_{min} \cup \ldots \cup W_{max}\} \tag{3}$$

2.2 Shapelet Distance Calculation

The distance between two subseries with the same length by square Euclidean distance is calculated as

$$dist(S, R) = \sum_{i=1}^{m} (s_i - r_i)^2 \tag{4}$$

where S and R are time series, s_i and r_i are their components and m is their length. For computing the distance between a time series by length m and a subseries by the length l, while $l \leq m$, minimum amount of squared Euclidean distance between normalized subseries S and all normalized subseries inside T_i with the same length as the S (each of then is named as R) must be found.

$$d_{Si} = \min_{R \in W_{i,l}} dist(S, R) \tag{5}$$

The list of distance between the shapelet candidate S and the time series T is:

$$D_S = \langle d_{S,1}, d_{S,2}, \ldots, d_{S,n} \rangle \tag{6}$$

2.3 Shapelet Assessment

The purpose of shapelet assessment is to determine how accurate this shapelet classifies the dataset based on D_S. For specifying the quality of shapelet, information gain (IG) is used as a basic method for the shapelet assessment [7]. In this manner, D_S is put in order and IG must be calculated for each split point sp, where a split point is the average between each two consecutive distances in D_s. For calculating the IG of each *sp*, D_S must be divided into A_S and B_S. The A_S consists of all distances of $D_S < sp$ and the B_S consists of the rest of distances. So, the IG of *sp* is

$$IG(D_S, sp) = H(D_S) - \left(\frac{|A_S|}{|B_S|}H(A_S) + \frac{|B_S|}{|D_S|}H(B_S)\right) \tag{7}$$

where $|D_S|$ is the number of all members in D_S and $H(D_S)$ is defined as

$$H(D_S) = -p(A_S)\log(p(A_S)) - p(B_S)\log(p(B_S)) \tag{8}$$

where $p(A_s)$ and $p(B_s)$ are the proportion of objects in classes A_s and B_s with respect to the overall dataset, respectively as

$$p(A_s) = |A_s|/|D_s|; p(B_s) = |B_s|/|D_s| \tag{9}$$

The IG for the candidate S, (i.e. IG_S) is

$$IG_S = \max_{sp \in D_s} IG(D_S s, p) \tag{10}$$

Optimal Split Point (OPS) is the *sp* for which the IG_S has the maximum value.

3 Modeling Faults in Induction Motors

In this section, dynamical representation of the IM in the healthy and broken rotor bars situations are presented. An IM is an extremely symmetrical electromagnetic system. Fault leads to a specified percent of asymmetry. So, broken rotor bars by inducting a specific degree of unbalancing are modelled in the rotor parameters [15].

3.1 Model of Healthy Induction Machine

In the *qd0* reference frame, the IM system is modelled by the following differential equations of the voltages:

$$V_s^{qd0} = r_s^{qd0} i_s^{qd0} + \omega \begin{bmatrix} 0 & 1 & 0 \\ -1 & 0 & 0 \\ 0 & 0 & 0 \end{bmatrix} \lambda_s^{qd0} + \frac{d\lambda_s^{qd0}}{dt} \tag{11}$$

$$V_r^{qd0} = r_r^{qd0} i_r^{qd0} + (\omega - \omega_r) \begin{bmatrix} 0 & 1 & 0 \\ -1 & 0 & 0 \\ 0 & 0 & 0 \end{bmatrix} \lambda_r^{qd0} + \frac{d\lambda_r^{qd0}}{dt} \tag{12}$$

where $\omega = d\theta/dt$, $\omega_r = d\theta_r/dt$, $r_{s,r}^{qd0} = r_{s,r}I_3$ and

$$\begin{bmatrix} \lambda_s^{qd0} \\ \lambda_r^{qd0} \end{bmatrix} = \begin{bmatrix} L_{11} & 0 & 0 & L_{14} & 0 & 0 \\ 0 & L_{22} & 0 & 0 & L_{25} & 0 \\ 0 & 0 & L_{33} & 0 & 0 & 0 \\ L_{41} & 0 & 0 & L_{44} & 0 & 0 \\ 0 & L_{52} & 0 & 0 & L_{55} & 0 \\ 0 & 0 & 0 & 0 & 0 & L_{66} \end{bmatrix} \begin{bmatrix} i_s^{qd0} \\ i_r^{qd0} \end{bmatrix} \tag{13}$$

The elements of L_{ij} are elaborated in [15].

3.2 Model of Induction Machine with Broken Rotor Bars

Broken rotor bar causes asymmetry in the IM rotor and it leads to non-symmetrical stator current distribution, so unbalanced current is the reason of breaking other rotor bars during long time [11]. In broken rotor bar fault in induction motor, rotor becomes asymmetric by changing in r_r

$$r_r^{new} = r_r + \begin{bmatrix} r_{r11} & r_{r12} & r_{r13} \\ r_{r21} & r_{r22} & r_{r23} \\ r_{r31} & r_{r31} & r_{r33} \end{bmatrix} \tag{14}$$

The elements of Δr_r^{qd0} are found in [15].

4 Proposed Approach

The proposed method has two main parts. The first part is the offline process that consists of modeling the IM and its faults in order to make a comprehensive train dataset. In addition, new criteria for specifying the number of members in all classes are proposed for the shapelet algorithm described in Sect. 2. In the second part, the obtained shapelet is employed to detect faults online.

4.1 Offline Step

This step consists of two main processes: *generating train dataset* and *shapelet discovery*. The first requirement for finding shapelet is to generate a comprehensive train dataset that includes all kinds of time series from all operating situations. For the application of the fault detection of IMs, the train dataset consists of two healthy IM and IM with broken rotor bar (BRB) fault working situations. A sample filtered and normalized Fourier transform of stator current of phase a for healthy IM and an IM with broken rotor bars is shown in Fig. 1. It is obvious from Fig. 1 that whenever each type of fault is applied to the IM, several peaks appear in frequencies around the supply frequency. The difference in their shapes is a good criterion for detection and classification of occurring faults. Therefore, shapelet can be utilized to classify the differences in the shapes of the time series.

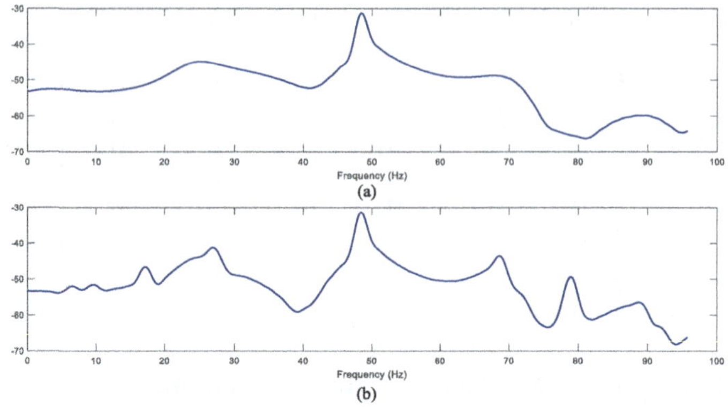

Fig. 1. Filtered and normalized Fourier transform of stator current: (a) healthy condition (b) IM with broken rotor bars.

In the conventional shapelet algorithm, which is given in Sect. 2, the number of members in the classes are not predefined. For a dataset that consists of several classes two of which have with negligible differences, if no limit is exerted on the maximum number of members in each class, the shapelet algorithm may not distinguish these two classes. Otherwise, if no limit is determined for the minimum number of members in each class, a sample class with various members and small difference among its own members may be considered as a different class. Thereby, it is necessary to choose the number of classes members systematically. To achieve this goal, the minimum number of each class member is considered as Min_n.it can be written as [10]:

$$Min_n = D/(k+1) \tag{16}$$

where k is the number of classes, n is the number of members for each class and D is the number of all members in dataset.

The maximum number of each class is considers as Max_n and it can be calculated by

$$M = \max((k-1)D/2k - 1, D/k) \tag{17}$$

$$Max_n = \max(M, Min_n + floor(M - Min_n/2)) \tag{18}$$

The algorithm of calculating Min_n and Max_n is shown in Fig. 2. As can be seen in Fig. 2, the Min_n are calculated from (16) and at the first step. After that, M must be calculated from (17) and the value of Max_n is achieved by it. The algorithm of IG calculation based on the Min_n and Max_n is provided in Fig. 3. In comparison with the conventional IG algorithm, this algorithm considers a range for numbers of objects in each class. In the algorithm that is shown in Fig. 3, if the number of remaining objects is less than Max_n, all of them must be put in D_A and D_B is provided with artificial objects that do not exist in the dataset. Also, the split point (sp) is considered as a summation of maximum distance and 2/3 of the last shapelet's split point.

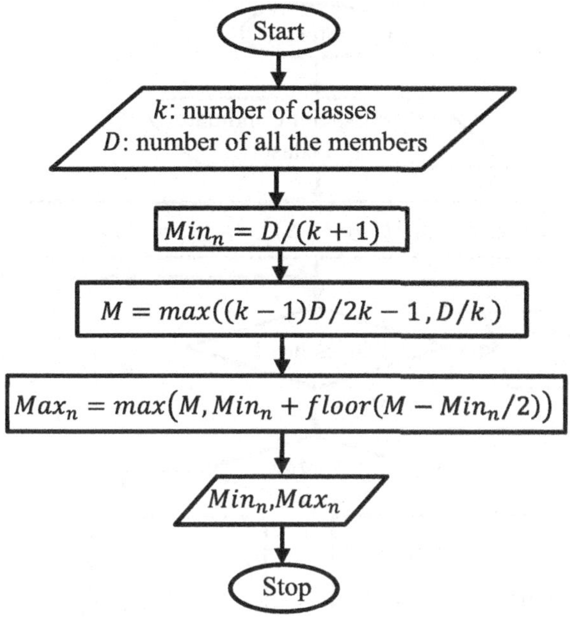

Fig. 2. The algorithm of calculating Min_n and Max_n

4.2 Online Step

In this step, first it must be determined whether fault is happened in IM or not. If yes, is it BRB or none of them? In order to do it, a detection and classification method is proposed. In this method, firstly, the distance vector between Fourier transform of stator current and shapelet are calculated. Secondly, each array of this vector is compared to the threshold of the corresponding shapelet. If the array is not less than threshold, it must be eliminated. Third, minimum of the remaining arrays is chosen and its corresponding shapelet is considered as this stator current time series feature. If this shapelet is a representative of healthy IM class, it means fault has not happened. If the chosen shapelet belongs to IM with BRB fault class, it means this fault has happened in IM. If all of the arrays are eliminated, it means IM is faulty but the type of fault is not broken rotor bar. The algorithm of the proposed method is shown in Fig. 4.

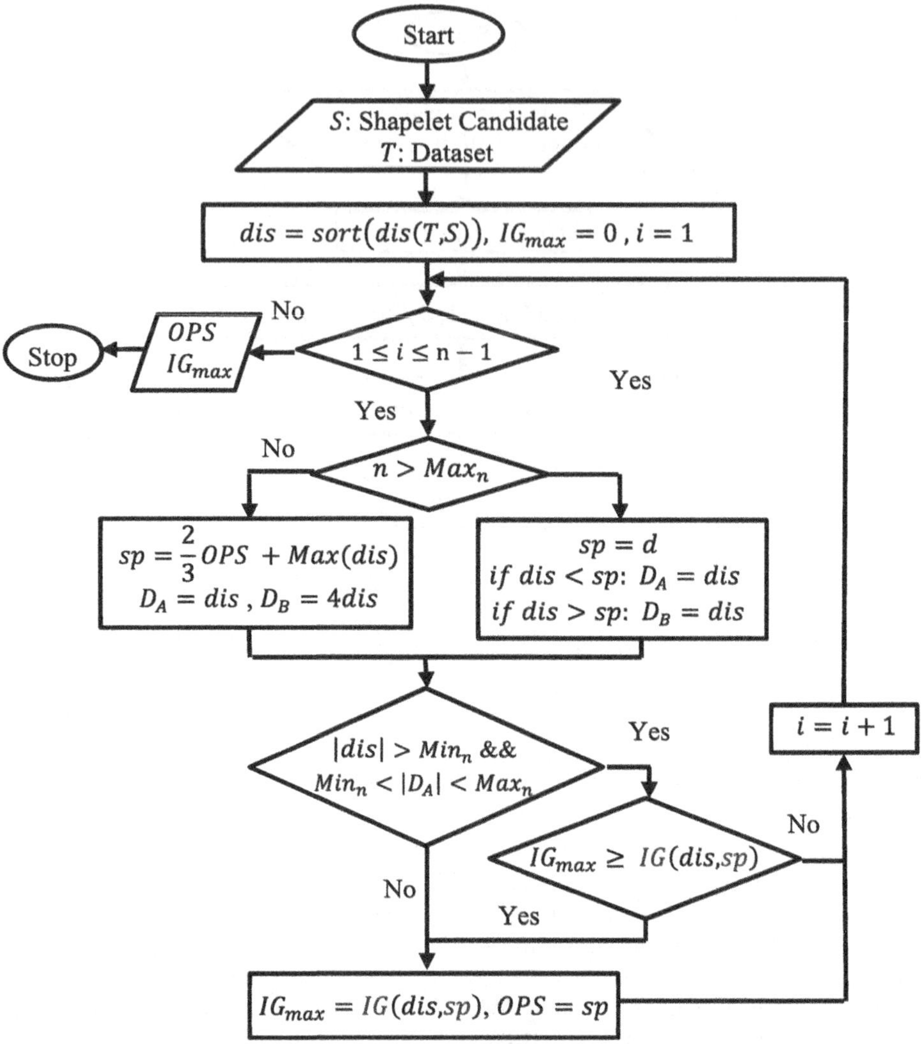

Fig. 3. The algorithm of calculating *IG* based on Min_n and Max_n

5 Case Study

At the first step, a 50 Hz, 6 poles, three phase induction motor is modelled in healthy situation based on [13]. A dataset of Fourier transform of stator current for phase a, for healthy condition and 6 broken rotor bar condition are made for train dataset. It consists of 50 series in frequency domain. This dataset is shown in Fig. 5.

Shapelet Algorithm with new criteria is employed in this step, in order to find best shapelet for accurate classification. For each condition of the induction motor, a specific shapelet is found. The shapelets of healthy motor and motor with BRB fault are presented in Fig. 6. It is obvious that the shapelet of IM with broken rotor bar fault consists peaks of signal that caused by fault. on the other hand, the shapelet of Healthy IM consists smooth signal without peaks. After finding Shapelets, a test dataset with 800 time series is made for performance evolution. This dataset consist of 400 timeseries in healthy situation and 400 timeseries for IM with 6 broken rotor bars. The test dataset is shown in Fig. 7. Finally, discovered Shapelets are used to classify test dataset, the algorithm of the proposed method for classification is shown in Fig. 4.

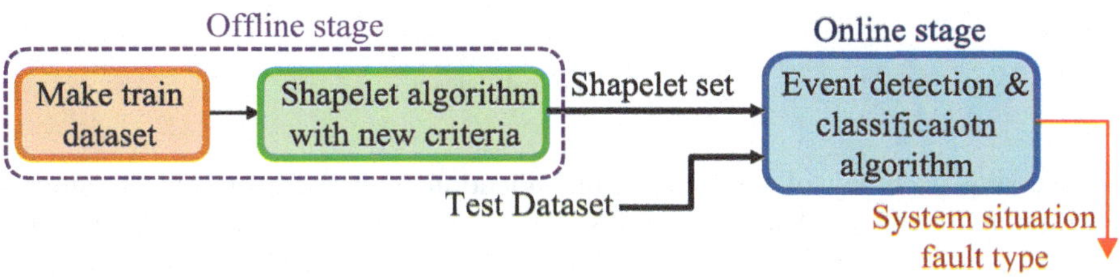

Fig. 4. Algorithm of the proposed method

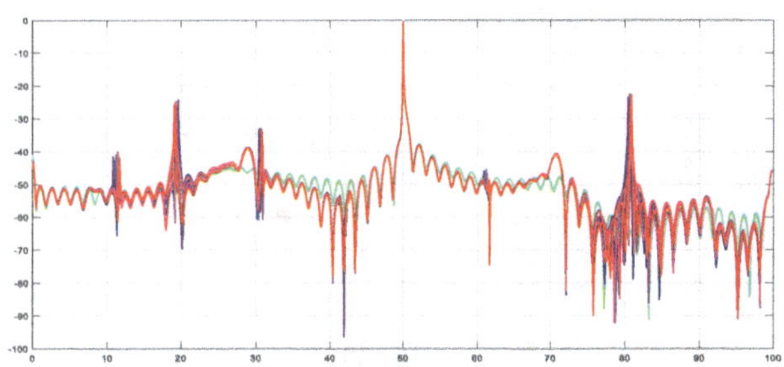

Fig. 5. The train dataset

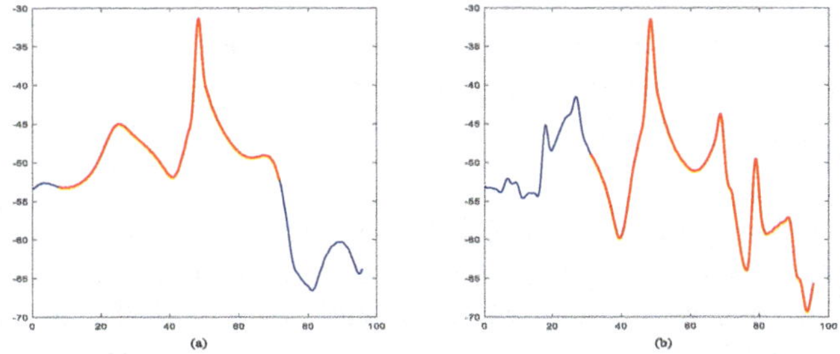

Fig. 6. (a) Shapelet of healthy IM (b) Shapelet of IM with 6 broken rotor bars

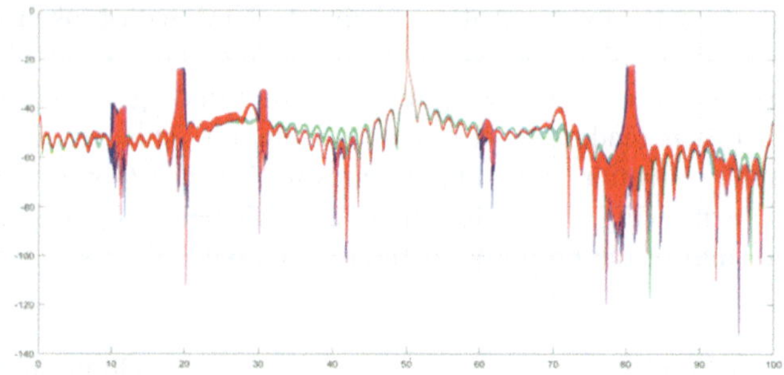

Fig. 7. The test dataset

Assessing accuracy of test dataset classification with the proposed classification algorithm and ordinary shaplet algorithm is shown in Fig. 8. In this diagram, red and blue columns are shown the classification accuracy of Shapelet algorithm with new criteria for the minimum and maximum number of all classes' members and ordinary Shapelet algorithm, respectively. So, it is obvious this method is totally successful in achieving its purposes. Therefore, this method can be applied for accurate and fast fault detection and it can classify faults easily.

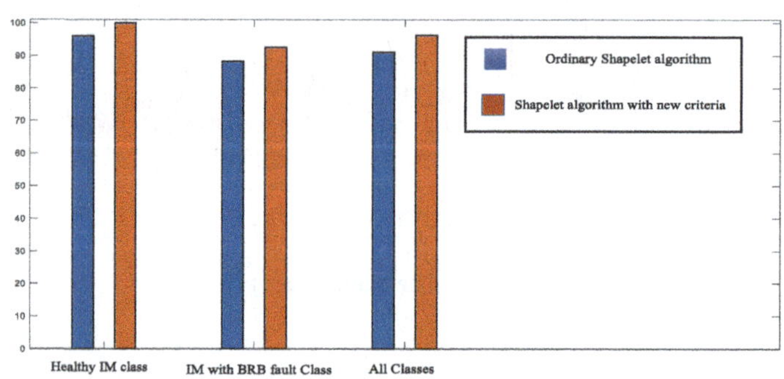

Fig. 8. The classification accuracy

6 Conclusion

In this paper, an induction motor in healthy and broken rotor situations is simulated. A train dataset consisting Fourier transform of stator current of phase a, is made and its features are extracted by Shapelet algorithm with new criteria for the minimum and maximum number of all classes' members. The results show that these criteria make shapelet algorithm more accurate and the classification is more efficient. In this paper, Shapelet is used as a new fault detection and classification algorithm. This algorithm can classify stator current of induction motor in healthy and faulty conditions. Besides, it can make a difference between faults, too. As a result, it is found out that shapelets can be used for feature extraction in order to detect the induction motor faults easily, accurately and efficiently. So it can be seemed as a novel prosperous technique for fault detection and classification.

References

1. Ye, L., Keogh, E.: Time series shapelets: a new primitive for data mining. In: Proceedings of the 15th ACM SIGKDD (2009)
2. Mueen, A., Keogh, E.J., Young, N.: Logical-shapelets: an expressive primitive for time series classification. In: Proceedings of the 17th ACM SIGKDD International Conference on Knowledge Discovery and Data Mining, pp. 1154–1162 (2011)
3. Lines, J., Davis, L., Hills, J., Bagnall, A.: A shapelet transform for time series classification. In: Proceedings of the 18th ACM SIGKDD (2012)
4. Zakaria, J., Mueen, A., Keogh, E.: Clustering time series using unsupervised-shapelets. In: IEEE International Conference on Data Mining (2012)
5. Ghalwash, M.F., Obradovic, Z.: Early classification of multivariate temporal observations by extraction of interpretable shapelets. BMC Bioinform. **13** (2012)
6. Jazizadeh, F., Afzalan, M., Wang, J.: Self-configuring event detection for electricity disaggregation. In: Proceedings of the 4th International Workshop on Non-Intrusive Load Monitoring, University of Texas, Austin, TX, USA (2018)
7. Brahma, S., Kavasseri, R., Cao, H., Chaudhuri, N.R., Alexopoulos, T., Cui, Y.: Real-time identification of dynamic events in power systems using PMU data, and potential applications—models, promises, and challenges. IEEE Trans. Power Delivery **32**(1), 294–301 (2016)
8. Xie, S., et al.: Online long-term voltage stability assessment based on time series shapelet extraction. In: 2018 IEEE Innovative Smart Grid Technologies-Asia (ISGT Asia). IEEE (2018)
9. Hassan, O.E., Amer, M., Abdelsalam, A.K., Williams, B.W.: Induction motor broken rotor bar fault detection techniques based on fault signature analysis–a review. IET Electric Power Appl. **12**(7), 895–907 (2018)
10. Bindu, S., Thomas, V.V.: Diagnoses of internal faults of three phase squirrel cage induction motor—a review. In: Proceedings of the International Conference on Advances in Energy Conversion Technologies, Manipal, India, pp. 48–54, January 2014
11. Dwivedi, S.K., Dannehl, J.: Modeling and simulation of stator and rotor faults of induction motor and their experimental comparison. In: 2017 IEEE 11th International Symposium on Diagnostics for Electrical Machines, Power Electronics and Drives (SDEMPED). IEEE (2017)
12. Rangel-Magdaleno, J., Ramirez-Cortes, J., Peregrina-Barreto, H.: Broken bars detection on induction motor using mcsa and mathematical morphology: an experimental study. In: 2013 IEEE International Instrumentation and Measurement Technology Conference (I2MTC), pp. 825– 829, May 2013
13. Cusido, J., Romeral, L., Garcia, A., Ortega, J., Riba, J.: On-line fault detection method for induction machines based on signal convolution. Eur. Trans. Electr. Power **21**(1), 475–488 (2011)
14. Hills, J., Lines, J., Baranauskas, E., Mapp, J., Bagnall, A.: Classification of time series by shapelet transformation. Data Mining Knowl. Disc. **28**(4), 851–881 (2013)
15. Chen, S., Zivanovic, R.: Modelling and simulation of stator and rotor fault conditions in induction machines for testing fault diagnostic techniques. Euro. Tran. Elect. Power **20**(1), 611–629 (2009)

A Prototype Telepresence Robot for Use in the Investigation of Ebola and Lassa Virus Threatened Villages in Nigeria

S. D. Monk[1(✉)], E. Jorgensen[1], A. McCulloch[1], R. Peacock[1], I. Sangprachatanaruk[1], J. Azeta[2], C. A. Bolu[1], and C. J. Taylor[1]

[1] Engineering Department, Lancaster University, Lancaster LA1 4YW, UK
s.monk@lancaster.ac.uk

[2] College of Engineering, Covenant University, 10 Idiroko Road, Canaan Land, Ota, Ogun State, Nigeria

Abstract. The article investigates the idea of low-cost, telepresence-based mobile robots for eventual use within villages and rural areas in Nigeria, where diseases such as the Ebola Virus Disease (EVD) and Lassa Haemorrhagic Fever (LHF) are common, yet human intervention is constrained due to the great risk of transmission through bodily fluids. To illustrate the concept and practical issues arising, a systems design approach is taken to identify some of the engineering requirements; and, in the focus of this article, a prototype has been developed at Lancaster University. The robotic device is semi-humanoid in that the upper half features two 7-DOF manipulators, designed in part to resemble human operation, while the lower half consists of a four-wheeled base, prioritising ease of operation and reliability over the flexibility offered by a leg-based system.

Keywords: Telepresence · First-Person Viewer · Kinect · Unsafe environment · Ebola

1 Introduction

In 1976, a new filovirus was identified in what is now the Democratic Republic of Congo, which was named the Ebola Virus Disease (EVD) after the Ebola River. A 90% mortality rate was recorded, although this had dropped to around 50% by the time of the 2014 Western African outbreak (which effected Liberia, Sierra Leone, Guinea, Senegal and Nigeria) with improved medical facility development [1–4]. In this breakout, a total of 5481 cases and 2946 deaths were reported, a mortality rate of 54% [5]. Like EVD, Lassa haemorrhagic fever (LHF) is highly transmittable and, therefore, caring for patients with the fever poses a significant risk of transmission. EVD is transmitted through direct contact with body fluids such as sweat, saliva, urine, vomitus, tears, breast milk, stool and respiratory secretions of an infected patient during the severe stage of the disease [6, 7]. Some preventive precautions against direct contact include the wearing of protective clothing and the use of standard infection control measures, such as the sterilization of equipment [8]. There is currently no

H. Selvaraj et al. (Eds.): ICSEng 2020, LNNS 182, pp. 34–48, 2021.
https://doi.org/10.1007/978-3-030-65796-3_4

standard treatment for EVD [9–11] outside of supportive care therapies, such as fluid preservation and balancing of electrolytes, oral and intravenous nutritional therapy, maintaining blood pressure and oxygen, and treating other complications that may arise such as secondary infections [4]. It is vital to avoid contact with infected patients, as healthcare related infection was noted as a major cause of transmission in the 2014 outbreak [7]. Techniques to prevent transmission of EVD to health care providers (HCPs) include simple procedures such as putting on and taking off Personal Protective Equipment (PPE) after treatment. This becomes uncomfortable after long periods of use, due to the high temperature and humid conditions, leading to potential human error and reduced effective working hours.

For these reasons, a robotic human stand-in able to perform at least some of the tasks of the people currently required to enter this environment would be greatly beneficial. Human entry time could be significantly reduced if a robotic platform could undertake non-clinical tasks such as the bringing in of general supplies (food and liquids), the removal of rubbish generated (discarded drink cans and food cartons), the cleaning and removal of soiled bedsheets and the removal of human wastes such as urine or faeces. Human-robot interactions have been of interest to the automation community since the 1980's where words like "telerobotics" were first widely used [12] and interactions between humans and robots became more developed [13]. Subsequently, numerous potential applications have been investigated, including rescue robots developed in response to the events of September 11th 2001 [14] as well as other disasters and incidents [15]. Despite obvious barriers such as ethical concerns, robotics used in healthcare has also seen a significant volume of work [16]. Prototype mobile robot telepresence systems have included the PRoP [17, 18] and the MeBot [19], with off-the-shelf versions now available to the general public on websites like Amazon. There are also relatively simple telepresence platforms used to stand in for vocations such as tour guides [20], aids for the blind [21] and remote meeting attendance [22]. More complex examples of telepresence robots include Toyota's 32-DOF T-HR3 robot that precisely replicates a seated user's movements using torque sensors [23, 24], and the humanoid robot developed by Whitney et al. [25]. Clinical platforms such as the da Vinci robot [26] must obviously be considered here, although the cost is in the region of $2,000,000 - well out of scope for a medium income country such as Nigeria. Other robots of interest to anyone working in the area of hospital orderly robots may be military robots such as the 'Dragon Runner' and the 'Robo Sally' [27, 28], the ChiCaRo (Child Care Robot) [29] used to interact with babies and toddlers and the co-operative robot work by Ilias [30]. Various tele-robotic systems have also been developed to operate in extreme environments that are inhospitable to humans, such as space exploration [31], bomb disposal [32] and nuclear decommissioning [33–35].

This article highlights the motivation and associated research challenge for the use of mobile telepresence robots during EVD and LHF outbreaks. The research aim is to investigate the engineering design requirements, and to build and test in the laboratory a prototype device capable of simple rubbish removal. Standard COTS platforms with two arms and a wheeled base are typically far too expensive to be affordable to the poorly funded and over-burdened Nigerian health service. Hence, an important design challenge is to provide a relatively low-cost system for future R&D purposes i.e. the design should be based on off–the–shelf components that have readily available equivalents in the countries effected. The specific objective is to develop a low-cost tele-operated robot that mimics an operator's movements and performs simple operations, such as traversing a flat floor and picking up small and lightweight objects up to a maximum mass of 100 grams. The first prototype will achieve mobility by straightforward wheeled operation.

2 Device Concept

Initially, the following requirements have been developed in order to enable the platform to carry out the tasks required:

- Performance optimisation of the proposed device from a technical standpoint must be balanced with a pragmatic realisation that it should be as cheap and as simple as possible, with parts widely available, so that the robot is not only easy to construct anywhere around the world but also straightforward to fix or debug by relatively unskilled labour.
- The device must operable by an unskilled and remotely located person using a Microsoft Kinect V2 based vision system or similar equivalents (available for < £100) to allow robotic mimicry of human movements.
- To optimise the ability of the platform to mimic human abilities, the robot will be of approximately small human height (1.5 m), and feature two arm sized manipulators.
- The platform should be capable of manoeuvring over flat ground at a speed of roughly 3 km/hr (slow to average human walking speed), not exceed a mass of 30 kg, feature two manipulator arms capable of mimicking human movements and sustaining 100 g payloads.
- A First-Person Viewer (FPV) system involving the use of goggles attached to the operator's head will allow remote vision of the environment, with the camera attached to the robot via two servo motors that are used to mimic the operator's movements.

An overarching block diagram of the robotic layout is shown in Fig. 1. Here, the operator and robot are linked via Radio Frequency wireless channels with three 14.8 V, 4 cell lithium polymer batteries utilised, one to power the wheels and one used for each manipulator (12 V is used as standard throughout the robot except where this option is unavailable, such as within the microcontroller where 5 V is utilised). Low cost and widely available RF XBee modules were selected for the prototype, since they can form a network to increase the range of the system and operate at 2.4 GHz, hence reducing interference with the 5.8 GHz headset used in the vision system. The XBee-S2S RF transceiver module was chosen due to the small form factor (24 × 27 × 0.5 mm). XBee protocol allows for point to multi-point configurations for networking purposes, with multiple channels to support multiple networks. They are also suitable for long distance applications, using a network of modules as beacons to reach the desired location. Another concern within the scope of this robot design is safety. There are no formally agreed and published guidelines regarding Nigerian robot safety standards, although there is a small volume of work published regarding robot safety generally e.g. [36, 37]. Most industrial safety standards [38] focus on isolating the robot from humans, something that will be implemented here implicitly. The use of PAN IDs and addresses within the Xbee RF communication protocol used here, allows the user to effectively lock the module to prevent data being received from other devices.

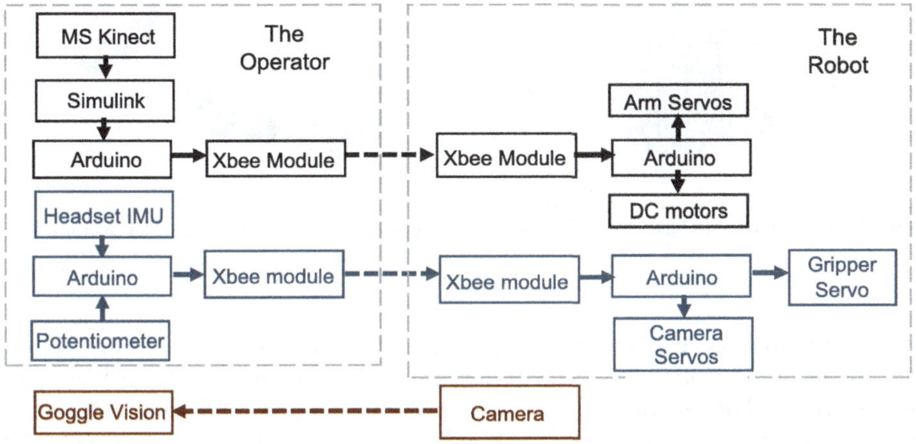

Fig. 1. Block diagram of the prototype showing the communication pathways.

3 Mechanical Design

Considering the need for reliability within a remote location, a four-wheel based design was employed for the prototype. The construction can be summarised into three parts: the wheelbase, the spine and the arms. The wheel base and spine are designed in a straightforward manner with four, 15 cm diameter wheels attached to 5 W brushed DC motors with a 50:1 gearbox attached, each one thus producing a stall torque of 4.12 N. m at 100 RPM, corresponding to a maximum speed of 2.83 km hr^{-1}. The base has a width of 50 cm, a depth of 35 cm and a height of 15 cm. The spine is constructed in a similar way with a height of 150 cm as illustrated in Fig. 2(a). Finite Element Analysis (FEA) was used to evaluate extreme load case performance, with the results indicating that either aluminium or steel would both be durable enough for this application since very high concentrations of stress are unlikely. In fact, 1050A grade aluminium with a yield strength of 90 MPa and ultimate tensile strength of 120 MPa was chosen for the prototype.

Fig. 2(a). The wheels base and robot body.

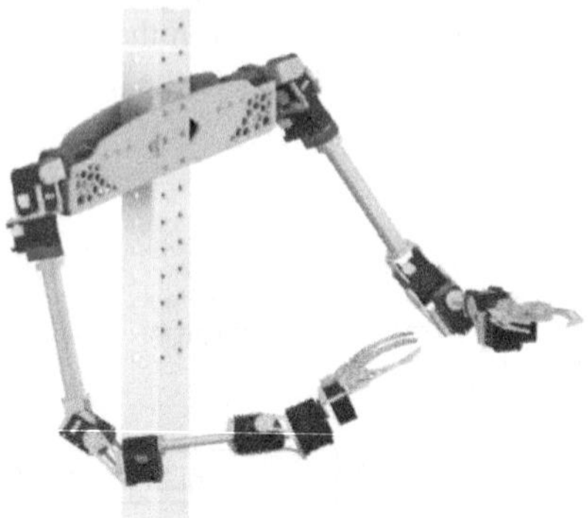

Fig. 2(b). Arms and shoulder system.

The manipulator arms comprise in-joint actuation and simple metal limbs to connect the joints, with an acrylic end effector designed to pick up objects, such as empty drinks cans or bottles. The arms are designed with brackets and 1050A aluminium metal, together with custom machined and additive manufactured components. The joints are actuated by Dynamixel servo motors to achieve an 8-Degree of Freedom (DOF) arm overall (3-DOF in the shoulder, 2-DOF in the elbow and 3-DOF in the wrist and hand) as shown in Fig. 2(b). Each arm uses one 7.3 N.m torque dynamixel MX-64 motor in the shoulder, with three 3.1 N.m MX-28 motors and three 1.5 N.m AX-12 motors used through the rest of the arm, except for the gripper itself which is actuated with a 1.1 N.m towerpro mg995 motor. Servos such as the Dynamixels offer high-resolution performance with daisy-chain connections, enabling easier logistical control. The prototype is thus comparatively easy to create and maintain although an alternative is to utilise lower-cost DC motors, albeit requiring a more complex arm design than the present prototype, featuring embellishments such as gearing. The main performance criteria of the constituent parts within the arm is that the stall torque can maintain any requested joint angle at maximum load conditions. In this regard, Fig. 3 illustrates the manipulator arms in the horizontal configuration, where the maximum load will be experienced by three of the motors A2, A4 and A6. MX represents the mass of the payload utilised, while lengths L1 to L5 are the links between the centre of masses M1 to MX, all shown in Table 1. Subsequent torque calculations are shown in Table 2. The motors all possess enough torque to successfully perform the operations required, although motor A2 is clearly the limiting node, as this may be run at 90% of capacity when a load of 100 g is used.

Table 1. Mass and length of each manipulator component.

Object	Mass (g)	Length (mm)
M1	188	–
L1	52	75
M2	188	–
L2	17	246
M3	156	–
L3	15	181
M4	132	–
L4	0	30
M5	159	–
L5	0	90
MX	100	–

Table 2. Maximum and available torque for each manipulator motor.

Motor	Maximum N m	Available N m	Capacity %
A1	2.78	7.3	38
A2	2.78	3.1	90
A3	1.08	3.1	35
A4	1.08	3.1	35
A5	0.38	1.5	25
A6	0.38	1.5	25

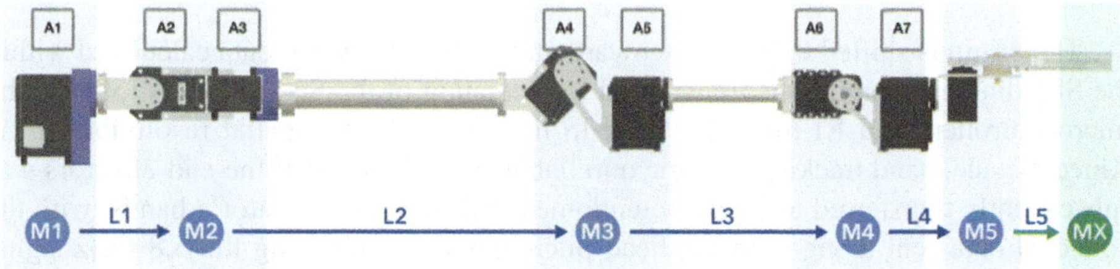

Fig. 3. The motors in each manipulator at maximum load configuration.

4 Motor Control

Entry level Arduino microcontroller boards such as the "Mega" and "Nano" (featuring ATmega2560 and ATmega328 microcontrollers respectively) are selected for the prototype because of their relatively low price and wide availability in Nigeria. Periodically, data sent from the Kinect are not received in the order anticipated, causing significant problems in the development of the robot control system. Hence, a data encryption technique was introduced to the code, in which a number ID was assigned to each segment of data sent. A Kinect V2 sensor (depth range 0.8 m to 4.0 m, a vertical angle limit of 43.5° and a horizontal angle limit of 57°) was selected for the prototype. The sensor allocates 25 joints per person, as illustrated by Fig. 4, with C# used to manipulate the output together with OpenNI APIs – an open source package for 3D sensing utilising middleware libraries. The Simulink toolbox ‘Simulink Real-Time’ was used with metadata obtained using a ‘From Video Device’ block. The metadata includes ‘JointPositions’; a 25 × 3 × 6 matrix indicating the location of the 25 joints of a tracked body in 3-D space. This contains firstly the joint number, secondly the X, Y and Z coordinates of the positions of the joints relative to the Kinect in meters, and thirdly; code to represent which of the six bodies is being considered.

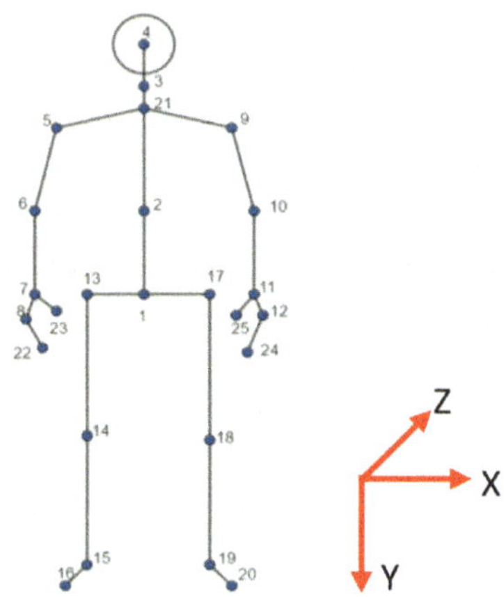

Fig. 4. The points of the human body as utilised by the Kinect.

The origin is shifted 0.75 m downward and to the left. Angle data calculated within the Simulink environment is transmitted for actuation of the Servo motors [39] via the microcontrollers and RF modules. Due to limitations in the spatial resolution of the Kinect device, hand tracking became unreliable and so control of the end-effectors was subsequently transferred to linear potentiometers held in the operator’s hands, with the collected data sent along with the head pitch and yaw data using the 5.8 GHz signal channel. The shoulder and elbow of the robot are illustrated in Figs. 5(a) and 5(b) respectively, with the gripper mechanism shown in Fig. 5(c). Human shoulders, elbows and wrists feature far greater flexibility than is currently mechanically realisable and, therefore, two servo motors are used to represent each of the shoulder and elbow joints.

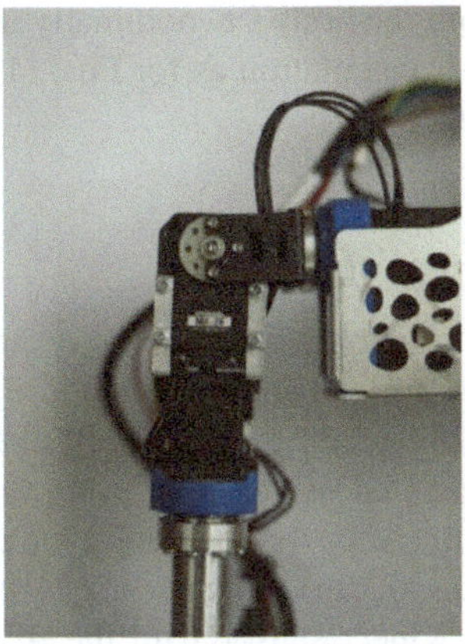

Fig. 5(a). Robot shoulder joint.

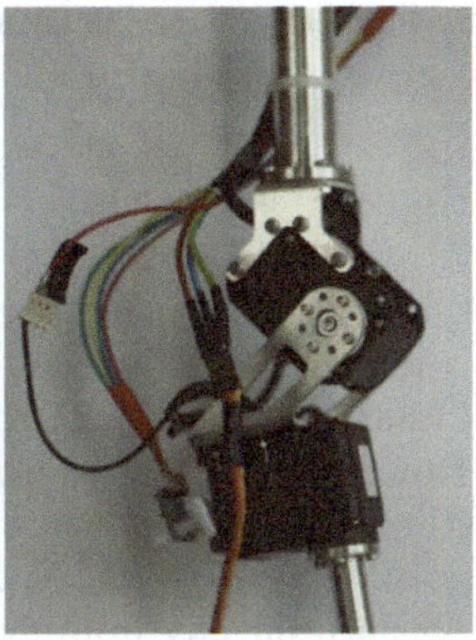

Fig. 5(b). Robot elbow joint.

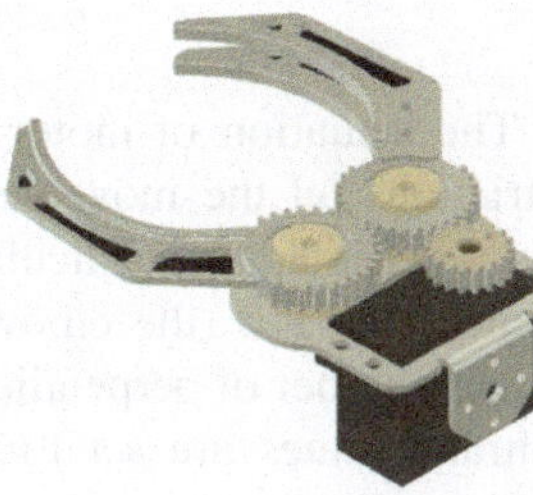

Fig. 5(c). The gripper.

Motor A1 is in the shoulder of the robot and controls the degree to which the manipulator arm swings forward and back i.e. the Y and Z orientation. The angle associated with motor A1 (θ_{A1}) is calculated in a two-dimensional perspective using straightforward trigonometry as illustrated in Eqs. (1) and (2).

$$V_{upper} = \begin{bmatrix} a \\ b \\ c \end{bmatrix} = \begin{bmatrix} J10_x - J9_x \\ J10_y - J9_y \\ J10_z - J9_z \end{bmatrix} \quad (1)$$

$$\theta_{A1} = tan^{-1}\left(\frac{c}{b}\right) \quad (2)$$

where V_{upper} represents the vector of the upper arm, with a, b and c defined as components of the vector between joints J9 (the shoulder) and J10 (the elbow) as illustrated in Eq. (1). Here, $J10_y$ for example represents the Y component of point J10 in Fig. 6(a).

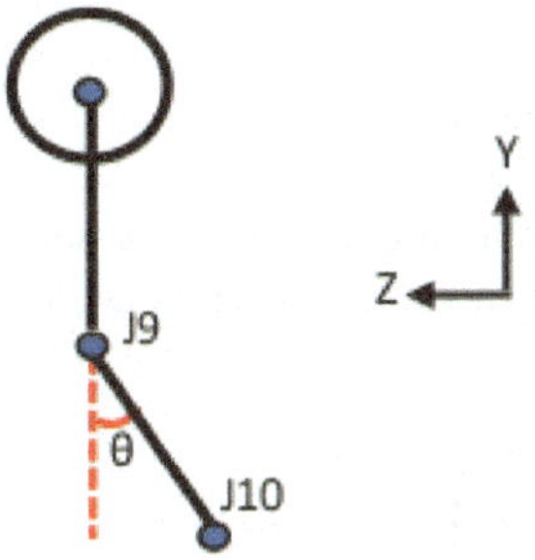

Fig. 6(a). Manipulator movement (Motor A1).

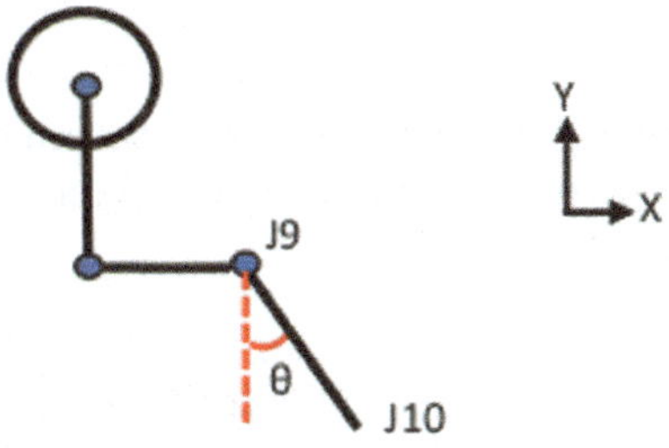

Fig. 6(b). Manipulator movement (Motor A2).

Motor A2 causes movement of the upper arm away from the body i.e. nominally in the X and Y planes, as shown by Fig. 6(b). Hence, a similar calculation as for Eqs. (1) and (2) can be used. The angle θ_{A2} is determined in Eq. (3).

$$\theta_{A2} = tan^{-1}\left(\frac{a}{b}\right) \tag{3}$$

The actuation of motor A3 represents the upper arm rotation. Since there are no restrictions on the movement, a reference axis is required to determine the angle of rotation. The angle is calculated from the reference axis to the vector from point J9 (the shoulder) to J10 (the elbow). However, in a three-dimensional system, there are an infinite number of perpendicular lines that can be found with such a vector, hence some arbitrary values are set to fix a solution. Here, the vector of the upper arm (1), is considered as an input vector obtained directly from the Kinect sensor with the reference vector which is orthogonal to the upper arm V_{perp} in Eq. (4). Similar to the upper arm vector in Eq. (1), V_{lower} is defined as the vector between the elbow (J10) and wrist (J11) as defined in Eq. (5).

$$V_{perp} = \begin{bmatrix} u \\ v \\ w \end{bmatrix} = \begin{bmatrix} 1 \\ 0 \\ \frac{-a}{c} \end{bmatrix} \tag{4}$$

$$V_{lower} = \begin{bmatrix} p \\ q \\ r \end{bmatrix} = \begin{bmatrix} J11_x - J10_x \\ J11_y - J10_y \\ J11_z - J10_z \end{bmatrix} \tag{5}$$

At this juncture, two assumptions are made, firstly that the elbow of the operator will always go forward of the body rather than backwards (i.e. c will always be positive), and secondly that a, b and c will all be non-zero. This realistic constraint simplifies the calculation of the perpendicular reference vector. Once the reference axis has been determined, the rotation angle between the lower arm and the reference axis is determined as in Eq. (6), while ignoring the Y-component, since the rotation will be relative to a vertical and reference axis,

$$\theta_{A3} = cos^{-1}\left(\frac{p * u + r * w}{\sqrt{p^2 + r^2} * \sqrt{u^2 + w^2}}\right) \tag{6}$$

The angle of motor A4 is determined in a similar manner, except that here the relevant vector is considered between the upper and lower arm components as defined in Eq. (7)

$$\theta_{A4} = cos^{-1}\left(\frac{p * a + r * c}{\sqrt{p^2 + r^2} * \sqrt{a^2 + c^2}}\right) \tag{7}$$

Motor A8 determines the degree of grip and is directly altered using a linear potentiometer in the hand of the operator, as noted above. Motor A7 determines the roll of the hand and uses feedback from the IMU on the operator's hand to ensure the manipulator hand is always parallel to ground. Similarly, motor A6 manages the pitch of the gripper and uses another axis from the gyroscope to calculate this requirement. Finally, motor A5 controls the rotation of the lower arm and achieves this by assuming the gripper/hand will always be required to face forwards i.e. parallel to the Z-axis. Hence, data from the IMU will allow consistent updating of this information to ensure the gripper stays in the required orientation. Due to noise in the Kinect data, the measured positions generally fluctuate by a small amount, leading to noticeable fluctuations in the calculated angle of between 1° to 4°. To address this issue, a filter based on comparing the current angle with the previously sampled angle was designed. Finally, the driving system of the robot uses a signal measured from movement of the operator's feet. The movement is registered by the Kinect and converted to a value that dictates how the robot moves. There are two signals for the driving system, namely a driving signal (forward/backward) and a rotating signal (left/right). The position of each foot was measured with the spine base as a reference point.

5 Operator Vision

A 2-DOF system was developed for this robotic prototype, with pitch and yaw being controlled independently and roll not implemented. This action simplifies the mounting design and allows for more stable code development. A Fatshark 600TVL CMOS camera was utilised with the corresponding Fatshark Predator V2 goggles providing a feeling of immersion in the environment to the user. In order to determine the operator's head orientation, an MPU6050 three-axis accelerometer and gyroscope is used. Two RF data streams are thus operational between the user goggles and the robot camera: transmission of the visual image from the camera to the goggles; and the orientation of the goggles which is transmitted towards the servo motors that subsequently determine the orientation of the camera. The 3D printed camera mount design consists of four parts: a base plate, two sections for the main body and one part to hold the camera and pitch servo, as shown in Fig. 7(a). The Fatshark goggles as shown in Fig. 7(b) are employed as the mounting point for the MPU6050, the microcontroller and XBee module used for RF transmission, with the MPU6050 Jarzebski library used to determine the IMU data. A threshold value was established via preliminary experimentation, and programmed into the software, to allow the user to specify the level of movement required to register as head movement as opposed to noise. A filter was developed to combine the accurate, but less stable, gyroscope data with the less accurate, but rather more stable accelerometer data. Another problem encountered was that the value returned by the gyroscope can drift from the actual value due to the gyroscope being integral. The position is thus only known relative to its previous position with the integral of the error causing the drift. This was partially rectified by the design of another filter, which reduces the drift to a rate of 1/min. Further improvements are realised by altering the threshold value, such that very small movements were ignored by the system completely.

Fig. 7(a). The CMOS camera and servo.

Fig. 7(b). Goggles, IMU, Nano, Xbee RF.

6 Testing

Initial testing of the robot was undertaken in order to determine whether the device traverses the floor when the operators foot is lifted from the floor and orientated either (i) straight ahead to indicate that both motors should drive forward or (ii) to one side to indicate that one of the sets of motors (left or right) should operate and thus the robot will orientate accordingly. These tests confirm that the Kinect can detect movement in the operator and the software algorithm implemented translates this into robotic actuation at a rate of around 2.4 km.hr^{-1}.

The second part of the test involved the lifting up of a drinks bottle containing 100 ml of liquid. The Kinect is successfully able to identify movement in the individual constituents of the operators arm (shoulder, elbow and wrist) and translate this into movement. The information transmitted from the MPU6050 allowed the hand of the robot to operate smoothly, providing operator movement is kept quite languid and with no sudden acceleration. The combination of these data streams from the Kinect and the IMU allowed the end effector to be positioned around the plastic bottle. The linear potentiometer held by the operator allowed gripping motion to occur and thus the object was grasped and subsequently lifted successfully.

Inaccuracies in the exact position of the end-effector are sometimes observed which are ascribed as due to two major factors. Firstly, deviations in dimensions between the arms of the operator and the arms of the robot, and secondly an angular offset in the servo motors. Furthermore, due to latency within the system, the end-effector occasionally displays a jolting motion, impacting on its ability to perform precise movements. Although this is not an issue with a durable object such as a drinks bottle, it may be a hindrance if more precise movements are required in a future iteration of the device. This latency is likely caused by several small contributions from various elements in the system, such as the Kinect, the signal processing, sending data and IDs, receiving data, assigning data and writing the data to the specific motors. Slight jolting motion is generated as the different angles are assigned causing the motors to move to the next point and stop repetitively. To diminish this, the rate of the motors is reduced as much as possible to create smoother motion. Finally, once the operator has placed the end effector/gripper around the object, visual feedback from the camera is successfully used to determine the grip required through a simple trial and error method.

7 Discussion

The prototype device built and tested in Lancaster has proven to be fit for basic tasks within a laboratory environment. In future work, the device will be recreated on site at Covenant University in Nigeria in order to mimic 'real life' scenarios. The present work has concentrated on determining a compromise between utilisation of parts that are most suitable for the task (from a performance and engineering design perspective) and parts which are readily available worldwide, as well as being cheap enough for easy replacement and simple enough for future use and alteration by non-experts. Whilst the robot reported in this article is useable within laboratory conditions, a device for use *in-situ* for realistic conditions in Nigeria will require several improvements.

Torque within the Dynamixel MX-28 motors used in the robot manipulator arms was a constant problem, with servo motors operating near to their limits, a situation that could lead to their required replacement over time. As noted above, the MX-28 motor used in the shoulder experiences the greatest torque demand to availability ratio, although any of these motors could become damaged and thus could require replacement at any time. Unfortunately, the Dynamixel motors are relatively expensive (~£200 each) and are not generally found locally in Nigeria. These servo motors were chosen for the prototype because of their relatively high torque and ease and precision of control. Use of simpler and more common DC motors will necessitate a redesign of the arms, a stage that is likely to require some gearing. Furthermore, the current wheel base is only suitable for relatively flat areas, whereas the addition of embellishments such as larger pneumatic wheels or an active suspension system might allow movement over less structured terrains.

There is also scope for implementation of a haptic feedback system in the robot gripper to enable suitable grip strength at all times. Currently, the only aspect of the robot that could be described as semi-autonomous is the gripper, since it features a 3D gyroscope on top of it ensuring that the end effector stays parallel to earth. However, a force feedback system would enable further implementation of semi-autonomous control. The robot will also need to be ruggedized, so that it can survive harsh conditions when used. The device is currently operational only in lab conditions and over relatively short periods of time. Repairs are likely to be difficult within field conditions and thus close to 100% reliability will be essential. Also essential is a system of wireless communication which can be operational wherever required. Basic RF communications modules were used in the present research for convenience, but this may not be suitable in field conditions and over significant distances.

Depth perception is not enabled here and is not something that would be an obvious necessity for implementation as 3D vision systems involving two cameras are widely available although this leads to greater complications within the coding requirement as well as possible increases in latency. The simplicity of the robot in its current state means that using simple proportional control is adequate. However, it is anticipated that future iterations of the device would further evolve in terms of the control system utilised. Another consideration to consider in future work may be in the societal and financial effect such a device may have, if ever fully incorporated into Nigerian contagious disease wards. Potential loss of employment for local people as well as simple mistrust of technology in some cultures, may provide a greater barrier than expected.

Finally, significantly tougher testing is required both in the laboratory and in real-life conditions. A possible testing methodology may be the NASA task load index (NASA TLX) assessment which requires users to subjectively rate a task on six different areas. Groups such as Marturi et al. [40], have used the human factors approach while investigating the effect of adding a level of autonomy to direct tele-operation systems. The index features questionnaires answered by the participants who subjectively rate the workload on 6 different sub-scales: mental demand, physical demand, temporal demand, effort, frustration level and overall performance.

8 Conclusions

A tele operational robot prototype for supporting health care providers during outbreaks of Ebola and Lassa virus disease has been designed and built. It can successfully mimic an operator's movements and perform simple operations such as traversing a flat floor and picking up small and light objects, up to a maximum mass of 100 grams. The robot designed at Lancaster University and reported in this article will be recreated with the partner university in Nigeria for further testing, initially in the context of an indoor environment. However, several changes and implementations are required in the design of the robot before it can be utilised *in situ* for the intended long-term application, namely investigations of disease threatened villages. The upgrades most needed are in the ruggedization of the device to facilitate its operation in real-life conditions at high levels of reliability. These issues need addressing whilst also considering cost and other local practical constraints, and these are the subject of on-going research by the authors.

Acknowledgements. This work is supported in the UK by the EPSRC project EP/R02572X/1.

References

1. World Health Organization Ebola Response Roadmap Situation Reports. http://www.who.int/csr/disease/ebola/situation-reports/archive/en. Accessed 26 May 2020
2. Sanchez, A., Geisbert, T.W., Feldmann H.: Filoviridae: Marburg and Ebola viruses. Fields Virology. 5th Edn. Lippincott Williams & Wilkins, Philadelphia, pp. 1409–1448 (2007)
3. Report of an International Commission: Ebola haemorrhagic fever in Zaire. Bull. World Health Organ. **56**(2), 271–293 (1976)
4. Tseng, C.P., Chan, Y.J.: Overview of ebola virus disease. J. Chin. Med. Assoc. **78**(1), 51–55 (2015)
5. Alexander, K.A., Sanderson, C.E., Marathe, M., Lewis, B.L., Rivers, C.M., Shaman, J., Drake, J.M., Lofgren, E., Dato, V.M., Eisenberg, M.C., Eubank, S.: What factors might have led to the emergence of ebola in West Africa?. PLOS Neglected Trop. Dis. **9**(6) (2015)
6. Fauci, A.S.: Ebola-underscoring the global disparities in health care resources. N. Engl. J. Med. **371**(12), 1084–1086 (2014)

7. Moghadam, S.R.J., Omidi, N., Bayrami, S., Moghadam, S.J., Seyed Alinaghi., S.: Ebola viral disease: a review literature. Asian Pac. J. Trop. Biomed. **5**(4), 260–267 (2015)
8. Ogbu, O., Ajuluchukwu, E., Uneke, C.J.: Lassa fever in West African sub-region: an overview. J. Vector Borne Dis. **44**(1), 1 (2007)
9. World Health Organization, Ebola Virus Disease factsheet. http://www.who.int/news-room/fact-sheets/detail/ebola-virus-disease. Accessed on 26 May 2020
10. Althaus, C.L., Low, N., Musa, E.O., Shuaib, F., Gsteiger, S.: Ebola virus disease outbreak in Nigeria: transmission dynamics and rapid control. Epidemics **11**, 80–84 (2015)
11. Weyer, J., Grobbelaar, A., Blumberg, L.: Ebola virus disease: history, epidemiology and outbreaks. Curr. Infect. Dis. Rep. **17**(5), 21 (2015)
12. Sheridan, T.: Human supervisory control of robot systems. In: Proceedings of the IEEE International Conference on Robotics and Automation, vol. 3, pp. 808–812. IEEE (1986)
13. Harless, M., Donath, M.: An intelligent safety system for unstructured human/robot interaction. In: Proceedings of the Robots Conference and Exposition, 117–120 (1985)
14. Murphy, R.R.: Rescue robotics for homeland security. Commun. ACM **47**(3), 66–68 (2004)
15. Murphy, R.R.: A decade of rescue robots. In: IEEE/RSJ International Conference on Intelligent Robots and Systems, pp. 5448–5449. IEEE (2012)
16. Kristoffersson, A., Eklundh, K.S., Loutfi, A.: Measuring the quality of interaction in mobile robotic telepresence: a pilot's perspective. Int. J. Soc. Robot. **5**(1), 89–101 (2013)
17. Paulos, E., Canny, J.: Designing personal tele-embodiment. In: Proceedings of the IEEE International Conference on Robotics and Automation, pp. 3173–3178 (1998)
18. Paulos, E., Canny, J.: Social tele-embodiment: understanding presence. Auton. Robots **11** (1), 87–95 (2001)
19. Adalgeirsson, S.O., Breazeal, C.: MeBot: a robotic platform for socially embodied presence. In: Proceedings of the 5th ACM/IEEE International Conference on Human-Robot Interaction (HRI), Osaka, Japan, pp. 15–22 (2010)
20. Burgard, W., Trahanias, P., Hähnel, D., Moors, M., Schulz, D., Baltzakis, H., Argyros, A.: Tele-presence in populated exhibitions through web-operated mobile robots. Auton. Robots **15**(3), 309–316 (2003)
21. Chaudary, B., Paajala, I., Keino, E., Pulli, P.: Tele-guidance based navigation system for the visually impaired and blind persons. In: eHealth vol. 360, pp. 9–16. Springer, Cham (2017)
22. Double robotics website. https://www.doublerobotics.com/. Accessed on 26 May 2020
23. IEEE Spectrum article. https://spectrum.ieee.org/automaton/robotics/humanoids/toyota-gets-back-into-humanoid-robots-with-new-thr3. Accessed 26 May 2020
24. Toyota website. https://newsroom.toyota.co.jp/en/detail/19666346. Accessed 26 May 2020
25. Whitney, J.P., Chen, T., Mars, J., Hodgins, J.K.: A hybrid hydrostatic transmission and human-safe haptic telepresence robot. In: IEEE International Conference on Robotics and Automation, Stockholm, Sweden, pp. 690–695 (2016)
26. Bodner, J., Wykypiel, H., Wetscher, G., Schmid, T.: First experiences with the da Vinci™ operating robot in thoracic surgery. Eur. J. Cardio-thoracic Surg. **25**(5), 844–851 (2004)
27. Quinetiq Dragon Runner. https://www.qinetiq-na.com/products/unmanned-systems/dragon-runner. Accessed 26 May 2020
28. Xsens website. https://www.xsens.com/cases/robo-sally-bomb-disposal-robot. Accessed on 26 May 2020
29. Abe, K., Shiomi, M., Pei, Y.C., Zhang, T.Y., Ikeda, N., Nagai, T.: ChiCaRo: tele-presence robot for interacting with babies and toddlers. Adv. Robot. **32**(4), 176–190 (2018)
30. Ilias, B., Shukor, S.A., Yaacob, S., Adom, A.H., Razali, M.M.: A nurse following robot with high speed kinect sensor. ARPN J. Eng. Appl. Sci. **9**(12), 2454–2459 (2014)

31. Ardanuy, P., Otto, C., Head, J., Powell, N., Grant, B., Howard, T.: Telepresence enabling human and robotic space exploration and discovery: antarctic lessons learned. In: Space, Long Beach California, USA, p. 6756 (2005)
32. Kron, A., Schmidt, G., Petzold, B., Zah, M., Hinterseer, P., Steinbach, E.: Disposal of explosive ordnances by use of a bimanual haptic telepresence system. In: IEEE International Conference on Robotics and Automation, New Orleans, LA, USA, vol. 2, pp. 1968–1973 (2004)
33. Talha, M., Ghalamzan, E.A.M., Takahashi, C., Kuo, J., Ingamells, W., Stolkin, R.: Towards robotic decommissioning of legacy nuclear plant: Results of human-factors experiments with tele-robotic manipulation, and a discussion of challenges and approaches for decommissioning. In: IEEE International Symposium on Safety, Security, and Rescue Robotics, Lausanne, Switzerland, pp. 166–173 (2016)
34. West, C., Monk, S., Montazeri, A., Taylor, C.J.: A vision-based positioning system with inverse dead-zone control for dual-hydraulic manipulators. In: UKACC 12th International Conference on Control, Sheffield, UK, IEEE (2018)
35. Tsitsimpelis, I., Taylor, C.J., Lennox, B., Joyce, M.J.: A review of ground-based robotic systems for the characterization of nuclear environments. Progress Nuclear Energy **111**, 109–124 (2019)
36. Colgate, E., Bicchi, A., Peshkin, M.A., Colgate, J.E.: Safety for physical human-robot interaction. In: Springer Handbook of Robotics pp. 1335–1348. Springer (2008)
37. Kulić, D., Croft, E.: Pre-collision safety strategies for human-robot interaction. Auton. Robots **22**(2), 149–164 (2007)
38. RIA/ANSI R15.06—1999 American National Standard for Industrial Robots and Robot Systems—Safety Requirements. American National Standards Institute. New York
39. Hussein, M.A., Ali, A.S., Elmisery, F., Mostafa, R.: Motion control of robot by using kinect sensor. Res. J. Appl. Sci. Eng. Tech. **8**(11), 1384–1388 (2014)
40. Marturi, N., Rastegarpanah, A., Takahashi, C., Adjigble, M., Stolkin, R., Zurek, S., Kopicki, M., Talha, M., Kuo, J.A., Bekiroglu, Y.: Advanced robotic manipulation for nuclear decommissioning: a pilot study on tele-operation and autonomy. In: Proceedings of the International Conference on Robotics and Automation for Humanitarian Applications, Kerala, India, pp. 1–8 (2016)

Hamiltonian-Based Libration Point Orbit Control on Manifold of Constant Energy

Keum W. Lee[1] and Sahjendra N. Singh[2(✉)]

[1] Department of Electronic Engineering, Catholic Kwandong University, Gangneung, Gangwon 25601, Republic of Korea
kwlee@cku.ac.kr

[2] Department of Electrical and Computer Engineering, University of Nevada Las Vegas, Las Vegas, NV 89154, USA
sahjendra.singh@unlv.edu

Abstract. The circular restricted three-body problem (CR3BP) is important from the theoretical as well as a practical viewpoint. In this paper, the objective is to control a spacecraft along stable or unstable libration point orbits. For this purpose, a manifold of constant energy is specified. Then, a Hamiltonian-based state variable feedback control law is designed for regulating the spacecraft to attain the specified level of energy. Through the Lyapunov analysis, asymptotic convergence of the system energy to prescribed level is established. For this multi-dimensional constant energy manifold, a variety of periodic and quasi-periodic libration point orbits can exist. Interestingly, the spacecraft can be controlled using a small control force applied to the spacecraft. Simulation results are presented, which show that in the closed-loop system, the spacecraft follows libration point orbits on the manifold of prescribed energy.

1 Introduction

The study of the motion of spacecraft in the gravitational field of multiple massive bodies has been of considerable interest to researchers in the field of astrodynamics. The solution of this general problem is difficult. For this reason, the simplified circular restricted three-body Problem (CR3BP), in which it is assumed that the spacecraft has no effect on the motion of two primaries, has received much attention. Furthermore, in the simplified CR3BP, it is assumed that the primaries are in a uniform circular motion about their barycenter. For the CR3BP, Euler discovered three collinear equilibrium points (L_1, L_2, L_3), and Lagrange found two equilateral equilibrium points (L_4, L_5). These are also termed libration or Lagrangian points ($L_1, ..., L_5$). Consider, for an example, the motion of a spacecraft of relatively small mass, in the gravitation field of the Earth-Moon system. In the CR3BP, the Earth and Moon rotate about the barycenter, and the libration points remain fixed in the rotating coordinate frame of the Earth and Moon. Considerable effort has been made in the past to determine natural periodic (Halo) orbits and quasi-periodic trajectories (Lissajous

H. Selvaraj et al. (Eds.): ICSEng 2020, LNNS 182, pp. 49–60, 2021.
https://doi.org/10.1007/978-3-030-65796-3_5

trajectories) in the vicinity of the libration points. Farquhar [1,2] proposed the use of libration points and orbits around libration points for lunar far-side communications. By placing a satellite at L_1 point and another in Halo orbit at L_2 point, a continuous communication coverage of the lunar surface would be possible. Further, effort was made to determine Halo and Lissajous trajectories [3–5]. (Readers may refer to recent survey papers on orbital dynamics and libration point orbit (LPO) control [6,7]).

For station-keeping, control system can be designed (i) by exploiting the geometric structure of the phase space around an orbit [8–11] or (ii) by applying modern control theory [12–14]. A Hamiltonian preserving control law has been designed for the CR3BP problem [11]. Although theoretically, station-keeping at an LPO, or at a quasi-periodic Lissajous trajectory, does not require any propulsion system, the controllers of [8–11] are sensitive to model uncertainties. Therefore, the authors of [11] used a combination of a Hamiltonian-based control signal with a robustifying control signal derived from a disturbance estimator (extended observer). There exist a variety of modern control techniques that can be used for uncertain systems. The state feedback controllers in Refs. [12–14] were designed based on H^∞ and disturbance rejection control techniques. (Additional references can be found in [6,7].)

The Hamiltonian-based controller of [11] requires explicit expression of the reference LPO, obtained by solving the nonlinear CR3BP. To avoid any knowledge of the solution of the CR3BP, one can pose a simpler problem of steering a satellite to a manifold of constant energy. Such a formulation can achieve swinging (oscillating) control of a spacecraft on a specified manifold. Fradkov and Pogromsky [15] considered the problem of energy control of Hamiltonian systems. In this paper, the objective is to design a feedback controller for steering a spacecraft to a reference manifold in the gravitational field of two massive bodies.

The contribution of this paper is two-fold. First, a control law is designed for controlling the trajectory of a spacecraft from its initial state to a manifold of constant energy. Here, the Hamiltonian of the CR3BP is selected to define the constant energy surface. Through the Lyapunov analysis, asymptotic convergence of the system energy to the prescribed level of energy is established. Of course, on a multi-dimensional manifold of constant energy, there exist a variety of oscillating trajectories, which are confined on the manifold. Interestingly, regulation of the trajectories can be accomplished using propulsive force of small magnitudes. Once the trajectory evolves on the manifold, control force becomes zero. Second, simulation results for the Earth-Moon-spacecraft system are obtained. These results show the existence a variety of orbits on the manifold of constant energy. It is found that these orbits on the constant energy surface depend on the initial state of the spacecraft and the feedback parameters of the controller.

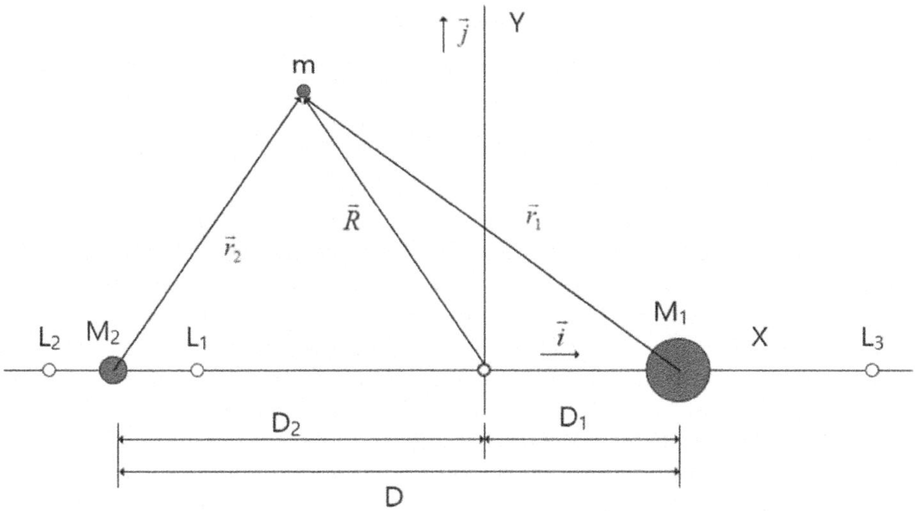

Fig. 1. Circular restricted three-body problem.

2 Dynamics of CR3BP

Figure 1 shows a spacecraft of negligible mass and two primary bodies (such as the Earth and Moon, or the Sun-Earth and Moon) of masses M_1 and M_2, respectively. A rotating coordinate frame with its basis vectors $(\hat{i}, \hat{j}, \hat{k})$ centered at the barycenter (which is at distance D_1 from M_1) is also shown in the figure. It is assumed that the primary bodies are rotating in circular orbits about the barycenter with uniform angular velocity:

$$n = \sqrt{\frac{G(M_1 + M_2)}{D^3}} \tag{1}$$

where D is the constant distance between the two primaries, and G is the universal gravitational constant. Define $M = M_1 + M_2$ and $\rho = M_2/M$. It will be convenient to use time in units of $(1/n)$, and distances in units of D. Then, the nondimensional position coordinates (x, y, z) of the spacecraft (m), with respect to the barycenter in the rotating frame, can be shown to satisfy the following Eqs. [14]:

$$\ddot{x} = 2\dot{y} + x - \frac{(1-\rho)(x-\rho)}{r_1^3} - \frac{\rho(x+1-\rho)}{r_2^3} + u_1$$

$$\ddot{y} = -2\dot{x} + y - \frac{(1-\rho)y}{r_1^3} - \frac{\rho y}{r_2^3} + u_2$$

$$\ddot{z} = -\frac{(1-\rho)z}{r_1^3} - \frac{\rho z}{r_2^3} + u_3 \tag{2}$$

where r_i is the distance of the spacecraft from M_i, $i = 1, 2$:

$$r_1 = \sqrt{(x-\rho)^2 + y^2 + z^2}$$

$$r_2 = \sqrt{(x+1-\rho)^2 + y^2 + z^2}$$

For the Earth-Moon-spacecraft system, the collinear points are L_1 (cislunar), L_2 (translunar), and L_3. Besides these, there are two equilateral points (L_4, L_5) (not shown in the figure). By the linear analysis, one can show that the collinear Lagrangian points are unstable, but the equilateral libration points are stable. The Lagrange points L_1 and L_2 are important for future space missions. Satellites placed at L_1 and L_2 can serve as communication platforms.

We are interested in deriving a control law for regulating the spacecraft, originating in the vicinity of the translunar libration point to a constant energy manifold in the vicinity of L_2. The manifolds of interest will be parameterized using selected constant values of Hamiltonian of the system. Of course, this derivation of the control law will be applicable to other stable or unstable libration point orbits (LPOs). It should be noted that for this multi-dimensional manifold - defined by a constant Hamiltonian - one can expect oscillating orbits of a variety of patterns.

3 Libration Point Orbit Control

The derivation of control law for the LPO control is based on the Hamiltonian of the system. For controlling the spacecraft on a manifold of constant energy, the expression of the Hamiltonian is necessary.

3.1 Hamiltonian for the CR3BP

Define $x_s = (x, y, z)^T \in R^3$ as the position coordinate vector of the spacecraft. The kinetic energy of a spacecraft in the gravitational field of the two primaries is:

$$T(x_s, \dot{x}_s) = \frac{1}{2}\dot{x}_s^T \dot{x}_s + b^T \dot{x}_s + \frac{1}{2}(x^2 + y^2) \tag{3}$$

where $b(x_s) = (-y, x, 0)^T$, and the potential energy P is:

$$P(x_s) = -\frac{1-\rho}{r_1} - \frac{\rho}{r_2} \tag{4}$$

The Lagrangian has the expression:

$$L(x_s, \dot{x}_s) = T - P = \frac{1}{2}\dot{x}_s^T \dot{x}_s + b^T \dot{x}_s + \frac{1}{2}(x^2 + y^2) - P \tag{5}$$

The generalized momenta $p = (p_1, p_2, p_3)^T \in R^3$ is defined by:

$$p = \frac{\partial L}{\partial \dot{x}_s} = \dot{x}_s + b \tag{6}$$

Next, solving for $\dot{x}_s$, Eq. (6) gives

$$\dot{x}_s = p - b = p + [y, -x, 0]^T \tag{7}$$

Now, eliminating $\dot{x}_s$ in (5) by using Eq. (7), and noting that $||b||^2 = (x^2 + y^2)$, one obtains Lagrangian of the form:

$$L(x_s, p) = \frac{1}{2}(p - b)^T(p - b) + b^T(p - b) + \frac{1}{2}(x^2 + y^2) - P$$

$$= \frac{1}{2}||p||^2 - \frac{1}{2}||b||^2 + \frac{1}{2}(x^2 + y^2) - P = \frac{1}{2}||p||^2 - P \tag{8}$$

The Hamiltonian of the system is defined as:

$$H = \left[\frac{\partial L}{\partial \dot{x}_s}\right]^T \dot{x}_s - L$$

$$= p^T \dot{x}_s - L \tag{9}$$

Then, H as a function of p and x_s takes the form:

$$H(p, x_s) = p^T(p - b) - L = p^T(p - b) - \frac{1}{2}||p||^2 + P$$

$$= \frac{1}{2}||p||^2 - p^T b + P \tag{10}$$

Alternatively, the generalized momenta p can be eliminated using Eq. (6) to obtain H as :

$$H = \frac{1}{2}||\dot{x}_s + b||^2 - (\dot{x}_s + b)^T b + P$$

$$= \frac{1}{2}(\dot{x}_s + b)^T(\dot{x}_s - b) + P$$

$$= \frac{1}{2}(\dot{x}^2 + \dot{y}^2 + \dot{z}^2) - \frac{1}{2}(x^2 + y^2) - \frac{(1 - \rho)}{r_1} - \frac{\rho}{r_2} \tag{11}$$

By taking the partial derivative of H in Eq. (10), with respect to x_s and p, it can be shown that:

$$\dot{x}_s = \frac{\partial H}{\partial p} = p + [y, -x, 0]^T \tag{12}$$

$$\dot{p} = -\frac{\partial H}{\partial x_s} = [p_2, -p_1, 0]^T - \frac{\partial P}{\partial x_s} \tag{13}$$

By differentiating Eq. (12), substituting $\dot{p}$ from Eq. (13), and eliminating p using Eq. (6), one can obtain the second-order equation for $\ddot{x}_s$ as given in Eq. (2). Thus, Eqs. (12) and (13) provide a state variable representation of Eq. (2). For a given constant H^*, a constant, $H(x_s, p) = H^*$ defines a manifold of constant energy in the state space with coordinates (x_s, p).

Next, a control law will be derived such that $H(p(t), x_s(t))$ asymptotically converges to H^*. For a given H^*, define a manifold as:

$$\mathcal{M} = \{(x_s^T, p^T)^T \in R^6 : H(x_s, p) = H^*\} \tag{14}$$

Of course, in general, convergence of $(H - H^*)$ to zero, does not necessarily imply that the trajectory $(x_s(t), p(t))$ will converge to $\mathcal{M}$, unless for the chosen H^*, $\mathcal{M}$ is a compact set [15]. It should be noted that $H(x_s, p)$ is not a positive definite function. However, if the trajectory happens to be bounded, then it will tend to the set $\mathcal{M}$.

3.2 A Control Law for Swinging Control on $\mathcal{M}$

The derivation of control law is based of the Lyapunov method. For the purpose of control, a modified Hamiltonian:

$$H_m(x_s, p, u) = H - H_1 = H - x_s^T u \tag{15}$$

is considered, where $u = (u_1, u_2, u_3)^T \in R^3$ is the control force vector and H_1 is called an interaction Hamiltonian [15]. Then, one obtains:

$$\dot{x}_s = \frac{\partial H_m}{\partial p} = \frac{\partial H}{\partial p}$$

$$\dot{p} = -\frac{\partial H_m}{\partial x_s} = -\frac{\partial H}{\partial x_s} + u \tag{16}$$

Consider a Lyapunov function:

$$V(x_s, p) = \frac{1}{2}(H(p, x_s) - H^*)^2 \tag{17}$$

Differentiating V along the solution of the system (16) gives:

$$\begin{aligned}\dot{V} &= (H - H^*)\left[\left(\frac{\partial H}{\partial p}\right)^T \dot{p} + \left(\frac{\partial H}{\partial x_s}\right)^T \dot{x}_s\right] \\ &= (H - H^*)\left[\left(\frac{\partial H}{\partial p}\right)^T \left(-\frac{\partial H}{\partial x_s} + u\right) + \left(\frac{\partial H}{\partial x_s}\right)^T \frac{\partial H}{\partial p}\right] \\ &= (H - H^*)\dot{x}_s^T u\end{aligned} \tag{18}$$

For regulating V to zero, one selects u_i as:

$$u_i = -\gamma_i \dot{x}_{si}^{\nu_1}(H - H^*)^{\nu_2}, i = 1, 2, 3 \tag{19}$$

where γ_i is positive, and ν_1 and ν_2 are positive odd integers. Then, Eq. (18) gives:

$$\dot{V} = -\sum_{i=1}^{3}[\gamma_i \dot{x}_{si}^{\nu_1+1}(H - H^*)^{\nu_2+1} \leq 0 \tag{20}$$

because $\nu_j + 1$ are even integers.

Here $V(x_s, p) \geq 0$, and $\dot{V} \leq 0$. This implies that V approaches a finite value $V(\infty) \leq V(x_s(0), p(0))$. Suppose the trajectory $(x_s(t), p(t))^T$ remains bounded. Then, its positive limit set Γ^+ will be bounded, and V will tend to a constant value on Γ^+ [16]. Therefore, $\dot{V}$ will be zero on Γ^+. Now, in view of Eq.(20), this will imply that on Γ^+, either $\dot{x}_s = 0$ or $H - H^* = 0$. If $\dot{x}_s = 0$, then $\ddot{x}_s = 0$, which will imply the converge of the trajectory to an equilibrium point. However, if $H - H^*$ is zero, its derivative will be zero. This will imply the convergence of trajectory on $\mathcal{M}$.

3.3 A Saturating Control Law

One can alternatively choose a smooth saturating control law as:

$$u_i = -\gamma_i \tanh(\delta_i (H - H^*) \dot{x}_{si}), i = 1, 2, 3 \tag{21}$$

where $\gamma_i > 0$, $\delta_i > 0$, and $\dot{x}_s = (\dot{x}_{si}, \dot{y}_{si}, \dot{z}_{si})^T$. Then, substituting this control input in Eq. (18) gives:

$$\dot{V} = -\sum_{i=1}^{3} [(H - H^*) \dot{x}_{si} \gamma_i \tanh(\delta_i (H - H^*) \dot{x}_{si})] \leq 0 \tag{22}$$

Because $\xi \tanh(\delta_i \xi) \geq 0$ for all $\xi \in R$. Equation (21) shows that control magnitude does not exceed the value γ_i. This completes the control law design.

Remark 1: A sufficient condition under different assumptions has been presented in Ref. [15] for controlling the energy of Hamiltonian systems. Here, it was assumed that the trajectory remained bounded. It will be seen in next section that the controller does accomplish regulation of the energy, but requires proper selection of H^* for station-keeping.

4 Simulation Results

This section presents simulated responses of the closed-loop system (Eq. (2) with unconstrained control law Eq. (19)), as well as the system (Eq. (2) with saturating control law (21)). For the Earth (M_1)-Moon (M_2) system, the parameters given in [14] are: $\rho = 0.01215$, $D_1 = 4674$ [km], $D_2 = 380073$ [km], $D = R_{12} = D_1 + D_2$ [km], and $n = 2.661699 \times 10^6$ [rad/s]. The nondimensional coordinate vector of L_2 point, with respect to the barycenter, is $(x_u = -1.15568, y_u = 0, z_u = 0)$. The design parameters in the control law are $\gamma_i = 10^{-6}/(D * n^2) = 0.3669$, $\nu_i = 1$ $(i = 1, 2)$, $\delta_1 = 0.01, \delta_2 = 0.1$, and $\delta_3 = 1$. A Halo orbit given in [14] based on linearization has coordinates (with respect to the L_2 point):

$$x_H^T(t) = [-A_x \sin \omega_{xy} t, -A_y \cos \omega_{xy} t, A_z \sin \omega_z t] \tag{23}$$

where $\omega_{xy} = 1.86265, \omega_z = 1.78618, k_{uu} = 2.91261$, $A_x = 0.00312, A_y = k_{uu} A_x$, and $A_z = A_y$. The initial conditions of spacecraft close to the Halo orbit (expressed relative to the barycenter) were assumed to be of the form $x_s(0) \approx ((x_u, 0, 0)^T + x_H^T(0))$ and $\dot{x}_s(0) \approx \dot{x}_H(0)$. The selected two initial conditions were:

$$X_{ica} = (x_s(0)^T, \dot{x}_s^T(0)) = [-1.1557, -0.0091, 0, -0.0058, 0, 0.0162] \quad (24)$$

$$X_{icb} = (x_s(0)^T, \dot{x}_s^T(0)) = X_{ica} + [-0.003, 0.002, 0.004, 0, 0, 0] \quad (25)$$

The value of the Hamiltonian at the L_2 point, denoted as H_{L2}^*, can be evaluating by setting $x = x_u$, $y = z = 0$, and $\dot{x}_s = 0$ in Eq. (11). It should be noted that the coordinates $(x - x_u, y, z)$ plotted in all of the figures are shown relative to the libration point L_2.

***Case A*. Swinging control:** $H^* = H_{L2}^*$, $(x_s^T(0), \dot{x}_s^T(0)) = X_{ica}$
The responses of the closed-loop system (Eq. (2) and (19)) for $H^* = H_{L2}^*$ and $(x_s^T(0), \dot{x}_s^T(0)) = X_{ica}$ are shown in Figs. 2 and 3. One observes quasi-periodic orbit. After initial transients, H converges to H^*. The maximum control acceleration is within 5×10^{-5} [km/s^2].

***Case B*. Swinging control:** $H^* = H_{L2}^* - 0.0005$, $(x_s^T(0), \dot{x}_s^T(0)) = X_{ica}$
Simulation was done to steer the spacecraft to a lower energy manifold by setting $H^* = H_{L_2}^* - 0.0005$. It is seen that control law (19) is effective in regulating the energy (see Figs. 4 and 5), but the orbits are of different pattern.

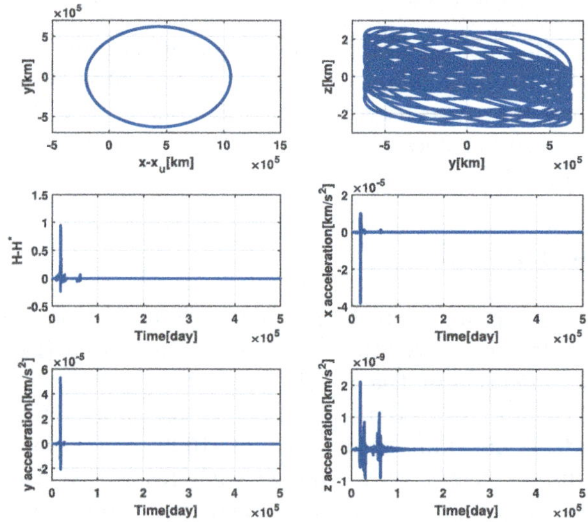

Fig. 2. Swinging control: $H^* = H_{L2}^*$, $(x_s^T(0), \dot{x}_s^T(0)) = X_{ica}$

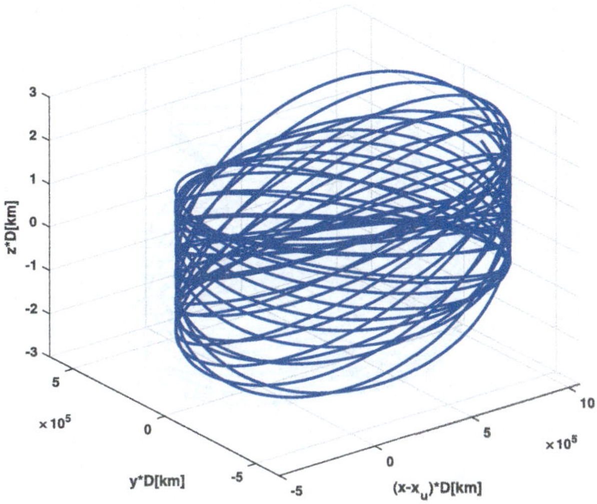

Fig. 3. Orbit of Case A in 3-D space

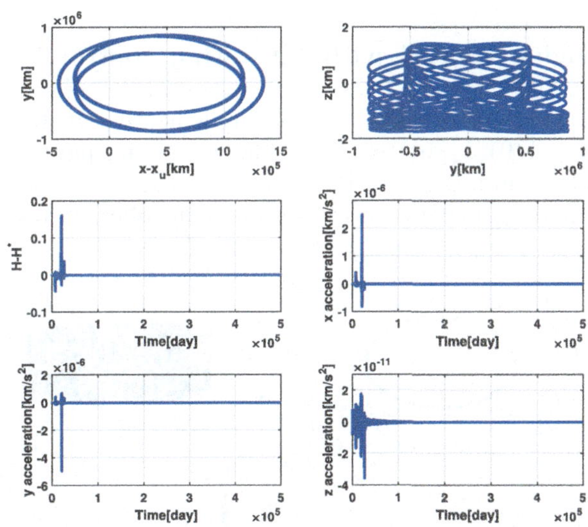

Fig. 4. Swinging control: $H^* = H^*_{L2} - 0.0005$, $(x_s^T(0), \dot{x}_s^T(0)) = X_{ica}$

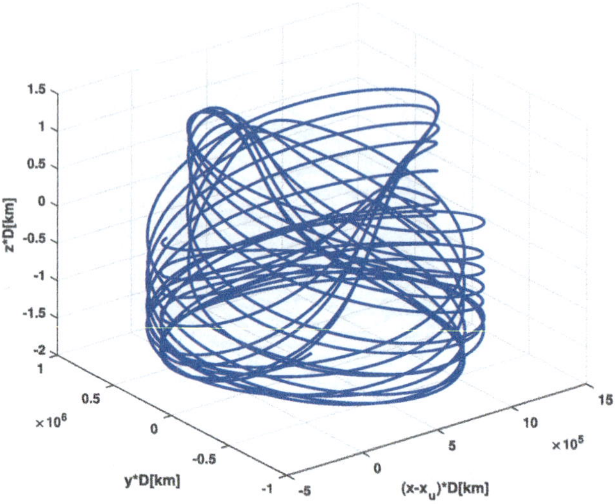

Fig. 5. Orbit of Case B in 3-D space

***Case C*. Swinging control with saturating control law:** $H^* = H^*_{L2}$, $(x_s^T(0), \dot{x}_s^T(0)) = X_{ica}$

Next, the saturating control law 21 was implemented. It was assumed that $H^* = H^*_{L2}$ and $(x_s^T(0), \dot{x}_s^T(0)) = X_{ica}$. The remaining parameters of Case A were retained. One can observe in Fig. 6 that compared to case A, regulation is achieved with smaller control acceleration and controller saturates briefly.

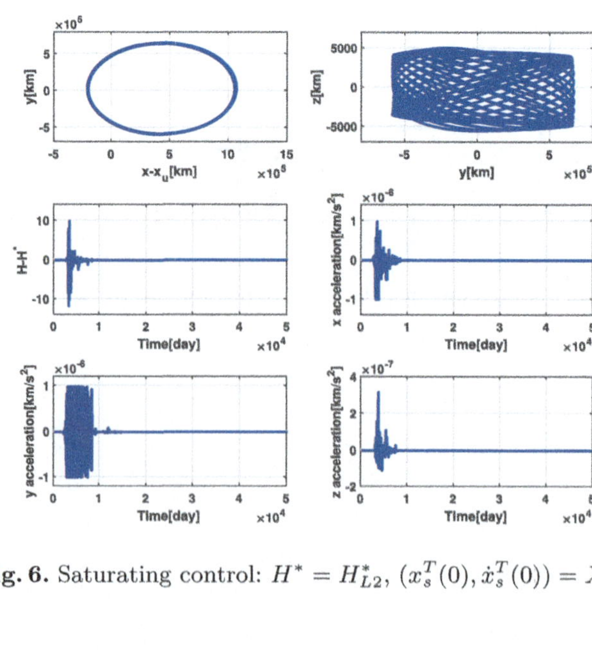

Fig. 6. Saturating control: $H^* = H^*_{L2}$, $(x_s^T(0), \dot{x}_s^T(0)) = X_{ica}$

***Case D.* Swinging control:** $H^* = H^*_{L2}$, $(x_s^T(0), \dot{x}_s^T(0)) = X_{icb}$

Finally, simulation was done similar to Case A (with control law Eq. (19)), but the initial state was assumed to be X_{icb}. It is interesting to see that now, for the choice of a different initial condition, one obtains an entirely different orbit (see Figs. 7 and 8).

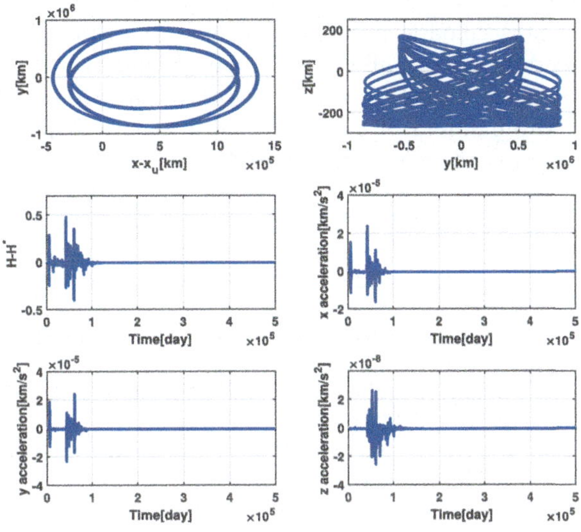

Fig. 7. Swinging control: $H^* = H^*_{L2}$, $(x_s^T(0), \dot{x}_s^T(0)) = X_{icb}$

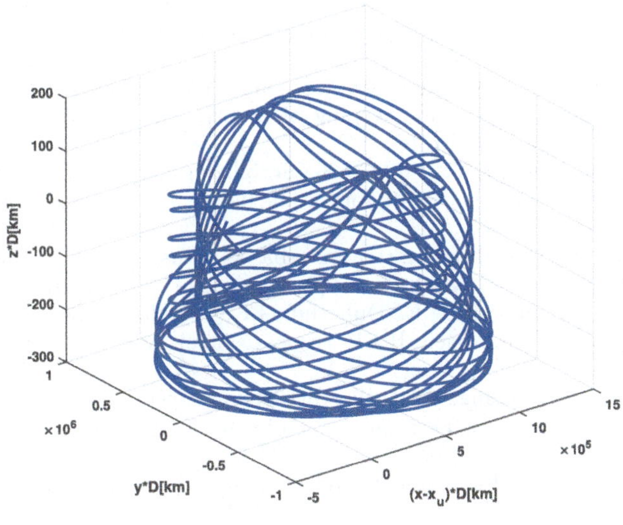

Fig. 8. Orbit of Case D in 3-D space

5 Conclusion

In this paper, the control of a spacecraft of the restricted circular three-body problem along stable or unstable libration point orbits confined to a manifold was considered. The Hamiltonian of the system was chosen to define a manifold of constant energy. Then, a Hamiltonian-based state variable feedback control law was designed for regulating the spacecraft to attain the specified level of energy. Through the Lyapunov stability analysis, asymptotic convergence of the system energy to prescribed level was established. Interestingly, it is possible to control the spacecraft using small control forces. Simulation results showed that in the closed-loop system, the spacecraft asymptotically followed quasi-periodic orbits on the chosen manifold. It was seen that a variety of quasi-periodic orbits (Lissajous trajectories) exist on the manifold.

References

1. Farquhar, R.M.: The Control and Use of Libration-Point Satellite. NASA Technical report TR R-346 (1970)
2. Farquhar, R.M.: The Utilization of Halo Orbits in Advanced Lunar Operations. NASA TN D-6345 (1971)
3. Farquhar, E.W., Kamel, A.A.: Quasi-periodic orbits about the translunar libration point. Celest. Mech. **7**, 458–473 (1973)
4. Richardson, D.L.: Halo orbit determination for the ISEE-3 mission. J. Guid. Contr. Dyn. **3**, 543–548 (1980)
5. Howell, K.C., Pernicka, H.J.: Numerical determination of Lissajous trajectories in the restricted three-body problem. Celest. Mech. **41**, 107–124 (1988)
6. Xu, M., Liang, Y., Ren, K.: Survey on advances in orbital dynamics and control for libration points. Prog. Aerosp. Sci. **82**, 24–35 (2016)
7. Shirobokov, M., Trofimov, S., Ovchinnikov, M.: Survey of station- keeping techniques for libration point orbits. J. Guidance Control Dyn. **40**(5), 1085–1105 (2017)
8. Simo, C., Gomez, G., Llibre, J., Martinez, R., Rodriguez, J.: On the optimal station keeping control of halo orbits. Acta Astronaut. **15**(6–7), 391–397 (1987)
9. Farres, A., Jorba, A.: Station-keeping of a solar sail around a halo orbit. Acta Astronaut. **94**, 527–539 (2014)
10. Scheeres, D.J., Hsiao, F.Y., Vinh, N.X.: Stabilizing motion relative to an unstable orbit: applications to spacecraft formation flight. J. Guidance Control Dyn. **26**, 62–73 (2003)
11. Jung, S., Kim, Y.: Hamiltonian structure-based robust station-keeping for unstable libration point orbits. J Guid. Contr. Dyn. **42**(12), 1–18 (2019)
12. Kulkarni, J.E., Campbell, M.E., Dullerud, G.E.: Stabilization of spacecraft flight in halo orbits: an H^∞ approach. IEEE Trans. Control Syst. Technol. **14**(3), 572–578 (2006)
13. Lou, Z., Zhang, K., Wang, Y.: Active disturbance rejection station-keeping control for solar-sail libration-point orbits. J. Guidance Control Dyn. **39**(8), 1917–1921 (2016)
14. Wie, B.: Space Vehicle Dynamics and Control. AIAA., Alexander Bell Dr., VA. (2008)
15. Fradkov, A.L., Pogromsky, A.Y.: Introduction to Control of Oscillations and Chaos. World Scientific, Singapore (1998)
16. Khalil, H.K.: Nonlinear Systems. Prentice Hall, Upper Saddle River (2002)

State and Control Input Limitations in Sliding Mode Control with System Uncertainties

Marek Jaskuła(✉) and Piotr Leśniewski

Institute of Automatic Control, Łodź University of Technology, 18/22 Stefanowskiego Street, 90-924 Łodź, Poland
Marek.Jaskula@zut.edu.pl

Abstract. This paper presents the approach that limits the state variables and the control input in the sliding mode technique. The plant dynamics is perturbed by both external disturbances and the system's uncertainties. Nevertheless, the convergence rate is designed in such a way to ensure the swiftest, monotonic convergence of the sliding variable to zero, in the same time, keeping the required constraints. The variable structure controller is designed using the reaching law technique and switching hyperplane definition.

Keywords: Continuous time systems · Control signal limitation · State variables limitation · Sliding mode control · Reaching law technique

1 Introduction

Recently, among variety of the variable structure control strategies, the sliding mode approach has settled to be a popular method, especially to deal with the plants perturbed by the external disturbances and system's uncertainties [3,11,13,14]. It was a consequence of the major advantages of this designing technique, which are: effectiveness, robustness and computational efficiency. The basis of this strategy can be followed back till 1950's, when Emelyanov examined the features of the switching controllers. The results were published in [10], inspiring next researchers to become interested in this way of design [23]. At first the sliding mode approach was used to control only continuous time systems [8,9]. Notwithstanding, in time, it has been successfully developed also onto the discrete time domain [2,20,24].

The major idea of this control strategy is to drive the system's representative point to the predefined switching hyperplane and then make it glide along the mentioned surface, up to the moment when the demand state is achieved. Therefore, the researchers used to divide the regulation process, specifying two separated phases: the reaching phase and the sliding phase. Moreover, designing the sliding mode controller can be realized by one of two different methods.

H. Selvaraj et al. (Eds.): ICSEng 2020, LNNS 182, pp. 61–70, 2021.
https://doi.org/10.1007/978-3-030-65796-3_6

We can propose the control signal and further prove the stability of the sliding motion or use the so-called reaching law technique. The second approach focuses on predetermining the dynamics of the sliding variable and was first implemented by Gao and Hung to continuous time systems in [12]. Since then, this strategy was extended to the discrete time systems and found popularity, as a consequence of augmented control over the sliding variable during the reaching phase [1,4–7,17,19,21,22,25].

Among the variety of the different reaching laws there are few that consider the strict state variables and control signal limitations. Whats more, they usually have altering presumptions on the perturbations. Thus, in current paper, we investigate this problem, extending our previous researches [15,16]. We analyze the continuous time plant perturbed by both the external disturbances and system's uncertainties and require the swiftest, monotonic convergence of the sliding variable to zero, simultaneously keeping the demanded constrains. Moreover, we have weakened the sufficient condition, guaranteeing mentioned properties, by taking into account the initial conditions.

Usually, the parameters of the reaching law can be optimized to get certain magnitude of the control input. However, in most cases, there is no direct equation that specifies mentioned relationship and it is done by computer tuning. Notwithstanding, the control strategy presented in this work guarantees the fastest convergence of the representative point to the predefined sliding hyperplane without additional optimization.

2 Plant Model

In this paper we consider the continuous time plant subjected to unknown, bounded external disturbances and system's uncertainties:

$$\dot{\boldsymbol{x}}(t) = \left(\boldsymbol{A} + \widetilde{\boldsymbol{A}}\right)\boldsymbol{x}(t) + \boldsymbol{b}u(t) + \boldsymbol{d}(t). \tag{1}$$

Matrix $\boldsymbol{A}$ is a known state matrix, matrix $\widetilde{\boldsymbol{A}}$ represents system's uncertainties, vector $\boldsymbol{x}(t)$ is the state vector, vector $\boldsymbol{b} = [b_1, \ldots, b_n]^T$ is the input vector ($b_i \neq 0, i \in \{1, \cdots, n\}$), $\boldsymbol{d}(t)$ is the disturbance vector, and lastly scalar $u(t)$ is the control input. We assume that the external disturbances and system's uncertainties are bounded in the following way: $|d_i(t)| < D_i$, $|a'_{ij}| < u_{ij}$ for $i, j \in \{1, \cdots, n\}$, where a'_{ij} are elements of matrix $\widetilde{\boldsymbol{A}}$. Let us point out that matrix $\widetilde{\boldsymbol{A}}$ is not only unknown, but also can be time varying. For the sake of clarity we introduce $\boldsymbol{U} = [u_{ij}]$. Now, we define the sliding hyperplane as

$$s(t) = \boldsymbol{c}^T\boldsymbol{x}(t) = 0. \tag{2}$$

Vector $\boldsymbol{c}^T = [c_1, \ldots, c_{n-1}, 1]$ must be selected in such a way to guarantee $\boldsymbol{c}^T\boldsymbol{b} \neq 0$, otherwise the control signal cannot be derived. Further, for the purpose of predetermining the evolution of the sliding variable s, we introduce reaching law

$$\dot{s}(t) = -K\operatorname{sgn}[s(t)] + \boldsymbol{c}^T\boldsymbol{d}(t) + \boldsymbol{c}^T\widetilde{\boldsymbol{A}}\boldsymbol{x}(t). \tag{3}$$

In order to achieve the monotonic, finite time sliding variable convergence to zero the convergence rate K must be designed to overcome the impact of both external disturbances and system's uncertainties. Taking into consideration (1), (2) and (3) we derive the control input as

$$u(t) = -\left(\boldsymbol{c}^T \boldsymbol{b}\right)^{-1} \left\{\boldsymbol{c}^T \boldsymbol{A} \boldsymbol{x}(t) + K \operatorname{sgn}\left[\boldsymbol{c}^T \boldsymbol{x}(t)\right]\right\}. \tag{4}$$

Let us point out the fact that despite the presence of unknown terms in the reaching law formula, the controller is still feasible. Further, to ensure finite time reaching phase, we must guarantee that

$$-\operatorname{sgn}[s(t)]\,\dot{s}(t) \geq \lambda \tag{5}$$

where $\lambda > 0$, is true up to the time instant when the sliding motion begins. Applying the reaching law formula (3) we rewrite (5) as

$$K - \boldsymbol{c}^T \boldsymbol{d}(t) \operatorname{sgn}[s(t)] - \boldsymbol{c}^T \widetilde{\boldsymbol{A}} \boldsymbol{x}(t) \geq \lambda > 0. \tag{6}$$

Now, we denote the maximal possible influence of the external perturbations on the sliding variable as

$$D_{\max} = |c_1|\,D_1 + \ldots |c_{n-1}|\,D_{n-1} + D_n. \tag{7}$$

By using the above denotation (6) is satisfied if

$$K - D_{\max} - \boldsymbol{c}^T \widetilde{\boldsymbol{A}} \boldsymbol{x}(t) \geq \lambda. \tag{8}$$

In case when (8) is true, the representative point will begin the sliding motion after the finite time. However, because of perturbations and system's uncertainties, the precise moment of reaching the sliding hyperplane is unknown.

3 State and Control Input Limitations

In this section we will design the convergence rates K_u and K_i, $i \in \{1, \ldots, n\}$, that are the biggest possible and result in maintaining the control input and i-th state variable limitations: $|u(t)| \leq r_u$, $|x_i(t)| \leq r_i$, respectively, where $r_u > 0, r_i > 0, t \geq 0$.

Firstly, let us notice that if the sliding variable diminishes monotonically to zero and the limitations are maintained, the state will belong to the set

$$\mathbf{X} = \{\boldsymbol{x} \in \Re^n : \forall_{i \in \{1, \cdots, n\}} |x_i| \leq r_i, |\boldsymbol{c}^T \boldsymbol{x}| \leq \boldsymbol{c}^T \boldsymbol{x}_0\}, \tag{9}$$

where $\boldsymbol{x}_0 = \boldsymbol{x}(0)$ satisfies given restrictions. Considering that the convergence rate is selected large enough we obtain the maximal possible impact of system's uncertainties on the sliding variable

$$U_{\max} = \max_{\boldsymbol{x} \in X} \{\boldsymbol{c}^T \boldsymbol{U} \boldsymbol{x}(t)\}. \tag{10}$$

Now, we move on to the issue of the control signal constraint. Using the formula for the control input (4) we obtain

$$-r_u \leq -\left(\boldsymbol{c}^T\boldsymbol{b}\right)^{-1}\left\{\boldsymbol{c}^T\boldsymbol{A}\boldsymbol{x}(t) + K\mathrm{sgn}\left[\boldsymbol{c}^T\boldsymbol{x}(t)\right]\right\} \leq r_u. \tag{11}$$

By rewriting the above inequality one can get

$$r_u - \left(\boldsymbol{c}^T\boldsymbol{b}\right)^{-1}\boldsymbol{c}^T\boldsymbol{A}\boldsymbol{x}(t) \geq \left(\boldsymbol{c}^T\boldsymbol{b}\right)^{-1}K\mathrm{sgn}\left[\boldsymbol{c}^T\boldsymbol{x}(t)\right] \geq -r_u - \left(\boldsymbol{c}^T\boldsymbol{b}\right)^{-1}\boldsymbol{c}^T\boldsymbol{A}\boldsymbol{x}(t). \tag{12}$$

Let us bring into attention the fact that up to the moment when the sliding phase starts, we have $\mathrm{sgn}\left[s(t)\right] = const.$ That is why, we can calculate the convergence rate K_u, ensuring the swiftest convergence of sliding variable to zero in the same time limiting the control input as

$$K_u = \left|\boldsymbol{c}^T\boldsymbol{b}\right| r_u - \mathrm{sgn}\left[\boldsymbol{c}^T\boldsymbol{x}(t)\right]\boldsymbol{c}^T\boldsymbol{A}\boldsymbol{x}(t). \tag{13}$$

Subsequently, we move to the issue of keeping $x_i(t)$ from violating the state variable restriction. We start from multiplying both sides of (1) by $\boldsymbol{e}_i$ (i-th versor) and taking into account the control input formula (4) obtaining

$$\dot{x}_i(t) = \boldsymbol{e}_i\left(\boldsymbol{A} + \widetilde{\boldsymbol{A}}\right)\boldsymbol{x}(t) - b_i\left(\boldsymbol{c}^T\boldsymbol{b}\right)^{-1}\left\{\boldsymbol{c}^T\boldsymbol{A}\boldsymbol{x}(t) - K\mathrm{sgn}\left[\boldsymbol{c}^T\boldsymbol{x}(t)\right]\right\} + d_i(t). \tag{14}$$

When the representative point reaches the limitation $x_i(t) = r_i$, then we have to ensure condition $\dot{x}_i(t) \leq 0$. Otherwise, if $x_i(t) = -r_i$, we require $\dot{x}_i(t) \geq 0$. Summarizing, we demand that $\mathrm{sgn}\left[x_i(t)\right]\dot{x}_i(t) \leq 0$. Now, we derive the biggest possible impact of system's uncertainties of the i-th state variable getting

$$U_i = \max_{\boldsymbol{x}\in X}\{\boldsymbol{e}_i\boldsymbol{U}\boldsymbol{x}(t)\}. \tag{15}$$

Considering the perturbations influencing the i-th state variable, we design K_i, that results in maintaining the given state limitation, in the following way

$$\begin{aligned} K_i = &\,\mathrm{sgn}\left[\boldsymbol{c}^T\boldsymbol{x}(t)\right] b_i^{-1}\left(\boldsymbol{c}^T\boldsymbol{b}\right)\left\{\boldsymbol{e}_i\boldsymbol{A}\boldsymbol{x}(t) + \mathrm{sgn}\left[x_i(t)\right](D_i + U_i)\right\} \\ &- \mathrm{sgn}\left[\boldsymbol{c}^T\boldsymbol{x}(t)\right]\boldsymbol{c}^T\boldsymbol{A}\boldsymbol{x}(t). \end{aligned} \tag{16}$$

It is important to notice that depending on the sign of the term multiplying K in (14) and $\mathrm{sgn}\left[x_i(t)\right]$ we can increase or decrease the convergence rate and still guarantee the fulfillment of the $i-th$ state variable limitation. Hence, in order to select the biggest possible K_i, we need to consider two cases. If

$$\mathrm{sgn}\left[b_i\boldsymbol{c}^T\boldsymbol{b}\boldsymbol{c}^T\boldsymbol{x}(t)\,x_i(t)\right] = 1 \tag{17}$$

is satisfied, then increasing K above level (16) will not result in exceeding the state variable limitation. However, decreasing K below (16) must effect in violating the x_i restriction. We can also observe that all signs of considered term cannot equal zero when the i-th state variable is on the boundary and the reaching phase has not finished. Moreover, for the sake of further analysis, we will use the denotation $K_i = K_i^+$ in case when (17) is satisfied and $K_i = K_i^-$ elseways. However, we want to point out that we still consider a single convergence rate, according to the i-th state variable restriction.

4 Regulation Method and Its Properties

The above section presented the design of the convergence rates connected with control input and state variables limitations. Now, we will discuss the control methodology and formulate sufficient conditions that guarantee the swiftest, monotonic, finite time convergence of the sliding variable to zero, simultaneously maintaining mentioned constraints. First of all, we must ensure that the control input limitation is large enough to let the representative point approach to the sliding hyperplane i.e.

$$K_u - D_{\max} - U_{\max} \geq \lambda > 0 \tag{18}$$

is true in the whole reaching phase.

Theorem 1. *In order to ensure that (18) holds it is sufficient that*

$$\left|\boldsymbol{c}^T \boldsymbol{b}\right| r_u \geq sgn\left[\boldsymbol{c}^T \boldsymbol{x}(t)\right] \boldsymbol{c}^T \boldsymbol{A}\boldsymbol{x}(t) + D_{\max} + U_{\max} + \lambda \tag{19}$$

is fulfilled when $x(t)$ belongs to the set X.

Proof. Substituting the formula for K_u into (18) we get (19). Taking into account the fact that the initial point belongs to the admissible set and that we approach the switching hyperplane it is sufficient to consider only set X.

Moreover, in the case when the representative point is on the i-th state variable restriction and (17) is true the convergence rate connected with control input limitation must exceed K_i^+. Otherwise, at least one of considered constraints will be violated. Bottom line, we require that

$$K_u \geq K_i^+ \tag{20}$$

is satisfied for all $i \in \{1, \cdots, n\}$ in the whole reaching phase.

Theorem 2. *In order to ensure that (20) holds it is sufficient that*

$$|b_i| r_u \geq sgn\left[x_i(t)\right]\left\{\boldsymbol{e}_i \boldsymbol{A}\boldsymbol{x}(t) + sgn\left[x_i(t)\right](D_i + U_i)\right\} \tag{21}$$

is fulfilled for all $i \in \{1, \ldots, n\}$, when $K_i = K_i^+$, $|x_i(t)| = r_i$ and $x(t)$ belongs to the set X.

Proof. We begin the proof with multiplying the left and the right side of (21) by the expression $\left|b_i^{-1}\right|\left|\boldsymbol{c}^T \boldsymbol{b}\right|$ getting

$$\left|\boldsymbol{c}^T \boldsymbol{b}\right| r_u \geq \operatorname{sgn}\left[x_i(t)\right]\left|b_i^{-1}\right|\left|\boldsymbol{c}^T \boldsymbol{b}\right|\left\{\boldsymbol{e}_i \boldsymbol{A}\boldsymbol{x}(t) + \operatorname{sgn}\left[x_i(t)\right](D_i + U_i)\right\}. \tag{22}$$

Now, acknowledging the fact that (17) is true we obtain $\operatorname{sgn}\left[b_i\left(\boldsymbol{c}^T \boldsymbol{b}\right) \boldsymbol{c}^T \boldsymbol{x}(t)\right] = \operatorname{sgn}(x_i)$. Hence, one can get

$$\left|\boldsymbol{c}^T \boldsymbol{b}\right| r_u \geq b_i^{-1} \boldsymbol{c}^T \boldsymbol{b} \operatorname{sgn}\left[\boldsymbol{c}^T \boldsymbol{x}(t)\right]\left\{\boldsymbol{e}_i \boldsymbol{A}\boldsymbol{x}(t) + \operatorname{sgn}\left[x_i(t)\right](D_i + U_i)\right\}. \tag{23}$$

Considering the formulas for K_u and K_i we diminish both sides by the term $\operatorname{sgn}\left[\boldsymbol{c}^T \boldsymbol{x}(t)\right] \boldsymbol{c}^T \boldsymbol{A}\boldsymbol{x}(t)$ obtaining (20). Analogously to the previous proof, we only need to consider the set X.

Further step is to guarantee that the situation $K_i^- < K_j^+$ will not happen at the intersection of the i-th and j-th limits. If so, at least one of the state variable constraints would be exceeded. To deal with this issue we formulate the next theorem.

Theorem 3. *In order to ensure that during the reaching phase $K_i^- \geq K_j^+$ holds on the intersection of the i-th and j-th limits it is sufficient that*

$$\begin{aligned} &\left\{\left|b_i^{-1}\right| sgn\left[x_i\left(t\right)\right] \boldsymbol{e}_i + \left|b_j^{-1}\right| sgn\left[x_j\left(t\right)\right] \boldsymbol{e}_j\right\} \boldsymbol{A}\boldsymbol{x}(t) \\ &\qquad \leq -\left[\left|b_i^{-1}\right|\left(D_i + U_i\right) + \left|b_j^{-1}\right|\left(D_j + U_j\right)\right] \end{aligned} \tag{24}$$

is fulfilled for all $i, j \in \{1, \ldots, n\}$, $i \neq j$, when $|x_i| = r_i, |x_j| = r_j$ and $x(t)$ belongs to the set X.

Proof. First of all, we multiply the left and the right side of inequality (24) by the expression $-\left|\boldsymbol{c}^T\boldsymbol{b}\right|$ getting

$$\begin{aligned} &-\left|\boldsymbol{c}^T\boldsymbol{b}\right|\left\{\left|b_i^{-1}\right| \operatorname{sgn}\left[x_i\left(t\right)\right] \boldsymbol{e}_i + \left|b_j^{-1}\right| \operatorname{sgn}\left[x_j\left(t\right)\right] \boldsymbol{e}_j\right\} \boldsymbol{A}\boldsymbol{x}\left(t\right) \\ &\geq \left|\boldsymbol{c}^T\boldsymbol{b}\right|\left[\left|b_i^{-1}\right|\left(D_i + U_i\right) + \left|b_j^{-1}\right|\left(D_j + U_j\right)\right]. \end{aligned} \tag{25}$$

Acknowledging the fact that if $K_i = K_i^-$ and $K_j = K_j^+$, then from (17) we conclude that $\operatorname{sgn}\left[b_i^{-1}\boldsymbol{c}^T\boldsymbol{b}\boldsymbol{c}^T\boldsymbol{x}\left(t\right)\right] = -\operatorname{sgn}\left[x_i\left(t\right)\right]$, $\operatorname{sgn}\left[b_j^{-1}\boldsymbol{c}^T\boldsymbol{b}\boldsymbol{c}^T\boldsymbol{x}\left(t\right)\right] = \operatorname{sgn}\left[x_j\left(t\right)\right]$. Hence, rewriting the previous condition we obtain

$$\begin{aligned} &\left\{\boldsymbol{c}^T\boldsymbol{b}\operatorname{sgn}\left[\boldsymbol{c}^T\boldsymbol{x}\left(t\right)\right] b_i^{-1}\boldsymbol{e}_i - \boldsymbol{c}^T\boldsymbol{b}\operatorname{sgn}\left[\boldsymbol{c}^T\boldsymbol{x}\left(t\right)\right] b_j^{-1}\boldsymbol{e}_j\right\} \boldsymbol{A}\boldsymbol{x}\left(t\right) \\ &\geq -\operatorname{sgn}\left[x_i\left(t\right)\right] \operatorname{sgn}\left[\boldsymbol{c}^T\boldsymbol{x}\left(t\right)\right] b_i^{-1}\boldsymbol{c}^T\boldsymbol{b}\left(D_i + U_i\right) \\ &+ \operatorname{sgn}\left[x_j\left(t\right)\right] \operatorname{sgn}\left[\boldsymbol{c}^T\boldsymbol{x}\left(t\right)\right] b_j^{-1}\boldsymbol{c}^T\boldsymbol{b}\left(D_j + U_j\right). \end{aligned} \tag{26}$$

Further, we distinguish elements associated with either i-th and j-th convergence rate, placing them on different sides of the inequality, getting

$$\begin{aligned} &\operatorname{sgn}\left[\boldsymbol{c}^T\boldsymbol{x}\left(t\right)\right] b_i^{-1}\boldsymbol{c}^T\boldsymbol{b}\boldsymbol{e}_i\boldsymbol{A}\boldsymbol{x}\left(t\right) + \operatorname{sgn}\left[x_i\left(t\right)\right] \operatorname{sgn}\left[\boldsymbol{c}^T\boldsymbol{x}\left(t\right)\right] b_i^{-1}\boldsymbol{c}^T\boldsymbol{b}\left(D_i + U_i\right) \\ &\geq \operatorname{sgn}\left[\boldsymbol{c}^T\boldsymbol{x}\left(t\right)\right] b_j^{-1}\boldsymbol{c}^T\boldsymbol{b}\boldsymbol{e}_j\boldsymbol{A}\boldsymbol{x}\left(t\right) + \operatorname{sgn}\left[x_j\left(t\right)\right] \operatorname{sgn}\left[\boldsymbol{c}^T\boldsymbol{x}\left(t\right)\right] b_j^{-1}\boldsymbol{c}^T\boldsymbol{b}\left(D_j + U_j\right). \end{aligned} \tag{27}$$

Finally, we subtract from both sides of the above inequality the expression $\operatorname{sgn}\left[\boldsymbol{c}^T\boldsymbol{x}\left(t\right)\right] \boldsymbol{c}^T\boldsymbol{A}\boldsymbol{x}\left(t\right)$ and apply the formulas for the convergence rates obtaining $K_i^- \geq K_j^+$. Same as in the previous proofs, we only need to take into consideration the states belonging to the set X.

In the end, similarly to requirement (18), we need to guarantee that for every state constraint the convergence rate K_i is large enough to overcome the impact of external disturbances and system's uncertainties. However, due to the possibility of rising the convergence rate in case when $K_i = K_i^+$, we only demand that

$$K_i^- - D_{\max} - U_{\max} \geq \lambda > 0 \tag{28}$$

is satisfied for all $i \in \{1, \cdots, n\}$ in the whole reaching phase.

Theorem 4. *In order to ensure that (28) holds it is sufficient that*

$$sgn\left[x_i\left(t\right)\right]\left[\left|b_i^{-1}\right|\left|\boldsymbol{c}^T\boldsymbol{b}\right|\boldsymbol{e}_i - sgn\left(b_i\boldsymbol{c}^T\boldsymbol{b}\right)\boldsymbol{c}^T\right]\boldsymbol{A}\boldsymbol{x}(t) \\ + \left|b_i^{-1}\right|\left|\boldsymbol{c}^T\boldsymbol{b}\right| D_i \leq -\lambda - D_{\max} - U_{\max} \tag{29}$$

is fulfilled for every $i \in \{1, \ldots, n\}$, *when* $K_i = K_i^-$, $|x_i| = r_i$ *and* $x(t)$ *belongs to the set* X .

Proof. Taking into consideration that (17) is not true we get that $\text{sgn}\left[\boldsymbol{c}^T\boldsymbol{x}\left(t\right)\right] = -\text{sgn}\left[b_i\boldsymbol{c}^T\boldsymbol{b}x_i\left(t\right)\right]$ and $\text{sgn}(x_i) = -\text{sgn}\left[b_i\boldsymbol{c}^T\boldsymbol{b}\boldsymbol{c}^T\boldsymbol{x}\left(t\right)\right]$. Therefore, we can rewrite (29) in the following form

$$\left\{-\text{sgn}\left[\boldsymbol{c}^T\boldsymbol{x}\left(t\right)\right]b_i^{-1}\boldsymbol{c}^T\boldsymbol{b}\boldsymbol{e}_i + \text{sgn}\left[\boldsymbol{c}^T\boldsymbol{x}\left(t\right)\right]\boldsymbol{c}^T\right\}\boldsymbol{A}\boldsymbol{x}\left(t\right) \\ - \text{sgn}\left[\boldsymbol{c}^T\boldsymbol{x}\left(t\right)x_i\left(t\right)\right]b_i^{-1}\boldsymbol{c}^T\boldsymbol{b}D_i \leq -\lambda - D_{\max} - U_{\max}. \tag{30}$$

By transforming the above condition we get (28). Similar to preceding proofs, we just have to analyze the case for with the state is in the set X.

Acknowledging the fact that every theorem presented in the section contains a linear inequality, one can use linear programming to examine whether the conditions are true or not.

Now, we will move to the details of the regulation strategy that results in the swiftest, monotonic and finite time convergence of the sliding variable to zero in the same time guaranteeing the control input and state limitations.

We select $K = K_u$ when the state is in the interior of the admissible set or when the state is on the single/multiple boundaries for all of which (17) is fulfilled. Otherwise, we choose $K = \min\{K_u, K_{i_1}^-, K_{i_2}^-, \ldots, K_{i_l}^-\}$. After the reaching phase is done and the sliding motion begins, the convergence rate should be set at least at the same level as the sum of the maximal influence of the external disturbances and the system's uncertainties on the sliding variable i.e $K \geq D_{\max} + U_{\max}$.

Whats more, our regulation strategy has an asset over the common methodology to make a simple emergency reaction when the boundary is reached. By checking the theorems the designer will know in advance if all the constraints hold.

5 Simulation Example

For the purpose of verifying the theoretical background presented in this paper, we consider the simulation example. To begin with, we determine the continuous time plant parameters in the following manner:

$$\boldsymbol{A} = \begin{bmatrix} 2 & 1 \\ 1 & -1 \end{bmatrix}, \quad \boldsymbol{b} = \begin{bmatrix} 1 \\ 0.5 \end{bmatrix}, \quad \boldsymbol{c}^T = \begin{bmatrix} 0.25 & 1 \end{bmatrix},$$

$$\widetilde{\boldsymbol{A}} = \begin{bmatrix} 0.02\sin(100t) & 0 \\ 0 & 0.01\sin(200t) \end{bmatrix}, \quad \boldsymbol{d}(t) = \begin{bmatrix} 5(-1)^{\lfloor 20t \rfloor} & 10(-1)^{\lfloor 10t \rfloor} \end{bmatrix}.$$

Further, we calculate the biggest possible impact of the external disturbances and system's uncertainties on the sliding variable getting: $D_{\max} = 11.25, U_{\max} = 0.425$. The switching line is designed for the purpose of satisfying $\boldsymbol{c}^T\boldsymbol{b} \neq 0$ and obtaining the stability in sliding phase. Moreover, we require fulfilling the following limitations: $|x_1(t)| \leq 80$, $|x_2(t)| \leq 40$ and $|u(t)| \leq 200$ for $t \geq 0$. Our goal is to drive the representative point monotonically and in finite time to the switching line simultaneously keeping the control input and state from violating given restrictions. We select the initial point at the boundary of the admissible set to show the advantages of the proposed regulation strategy: $x_1(0) = 70, x_2(0) = -40$. At first, let us notice that for such parameters of the system every theorem presented in the previous section holds. However, if the conditions were considered at the whole admissible set, then they would not be true (for example (21) at point $[-80; -40]$). Therefore, taking into account the impact of the initial point (i.e. considering set (9)) directly weaken the sufficient condition.

Figure 1 depicts the state trajectory. Starting from the initial state, the representative point drives to the first state variable constraint $x_1 = 80$ and glides along it. Subsequently, the representative point reaches the switching line and sustains the sliding motion up to the moment when the demand state is attained. Further, the control input is shown at the Fig. 2. At the very beginning of the regulation process the control signal is set at its maximal magnitude which is a consequence of implementing the convergence rate K_u. After that, the control input varies for the purpose of maintaining the state at the given limitation. Finally, when the sliding phase is achieved the convergence rate is selected at the level $K = D_{\max} + U_{\max}$ in order to maintain the sliding motion and reduce the chattering. Let us notice that we do not establish any prior presumptions on the external disturbances and system's uncertainties. Next, Fig. 3 illustrates the

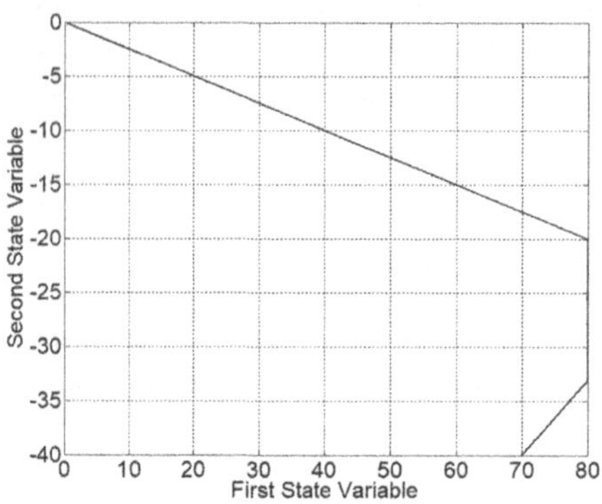

Fig. 1. State trajectory

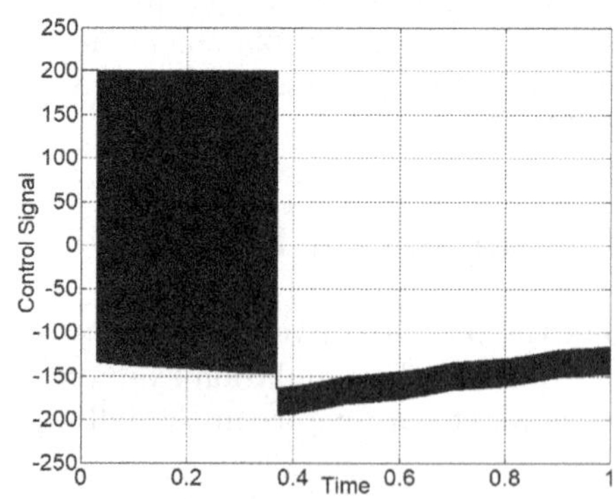

Fig. 2. Control signal

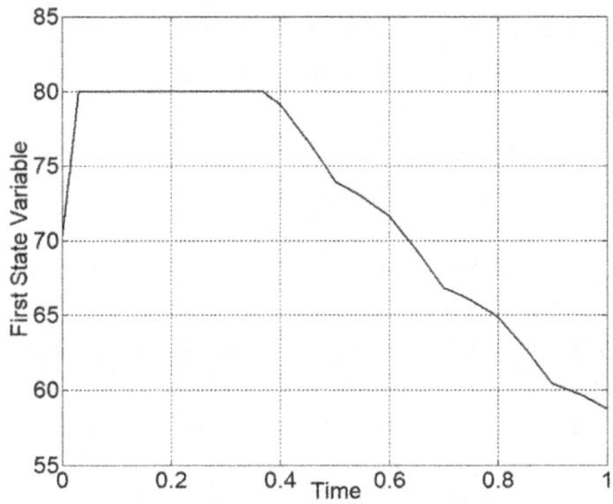

Fig. 3. First state variable

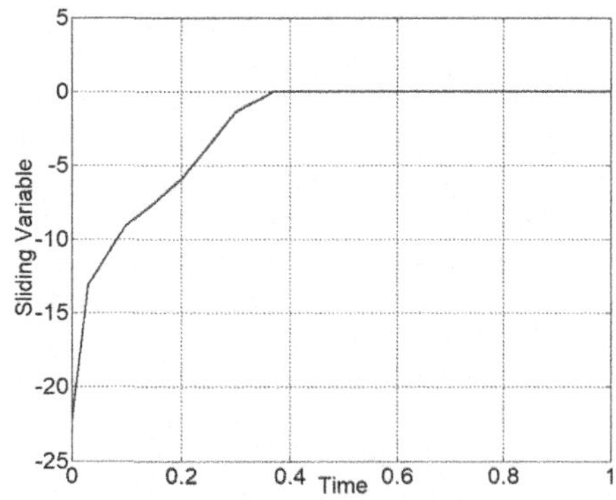

Fig. 4. Sliding variable

evolutions of the first state variables. It can be seen that demanded state limitation holds for the whole regulation process. Lastly, Fig. 4 reveals that the finite time, monotonic convergence of the sliding variable to zero has been successfully achieved.

6 Conclusions

In this paper the issue of keeping the control input and the state variables from violating the imposed restrictions in the sliding mode control methodology was studied. The continuous time plant, influenced by the bounded external disturbances and unknown system's uncertainties, was taken into consideration. The reaching law method was used to derive the control signal formula and the sliding variable convergence rate was designed to obtain the swiftest, monotonic and finite time convergence of the representative point to the predefined sliding hyperplane. Whats more, the sufficient condition guaranteeing mentioned properties has been weakened by taking into account the initial state. Finally, the theoretical results were verified by the simulation example. In future, we aim to extrapolate our research also to some classes of nonlinear systems.

References

1. Bartoszewicz, A.: A new reaching law for sliding mode control of continuous time systems with constraints. Trans. Inst. Meas. Control **37**, 515–521 (2015)
2. Bartoszewicz, A.: Discrete-time quasi-sliding-mode control strategies. IEEE Trans. Ind. Electron. **45**, 633–637 (1998)
3. Bartoszewicz, A., Kaynak, O., Utkin, V.I., Sliding mode control in industrial applications: Sliding mode control in industrial applications. IEEE Trans. Ind. Electron **55**, 3805–4103 (2008)
4. Bartoszewicz, A., Latosiński, P.: Reaching law for DSMC systems with relative degree 2 switching variable. Int. J. Control **90**, 1626–1638 (2017)

5. Bartoszewicz, A., Leśniewski, P.: Reaching law approach to the sliding mode control of periodic review inventory systems. IEEE Trans. Autom. Sci. Eng. **11**, 810–817 (2014)
6. Bartoszewicz, A., Leśniewski, P.: Reaching law-based sliding mode congestion control for communication networks. IET Proc. Control Theory Appl. **8**, 1914–1920 (2014)
7. Chakrabarty, S., Bandyopadhyay, B.: A generalized reaching law with different convergence rates. Automatica **63**, 34–37 (2016)
8. DeCarlo, R., Żak, S., Mathews, G.: Variable structure control of nonlinear multivariable systems: a tutorial. Proc. IEEE **76**, 212–232 (1988)
9. Draženović, B.: The invariance conditions in variable structure systems. Automatica **5**, 287–295 (1969)
10. Emelyanov, S.V.: Variable Structure Control Systems. Nauka, Moscow (1967)
11. Fallaha, C.J., Saad, M., Kanaan, H.Y., Al-Haddad, K.: Sliding-mode robot control with exponential reaching law. IEEE Trans. Ind. Electron. **58**, 600–610 (2011)
12. Gao, W., Hung, J.C.: Variable structure control of nonlinear systems: a new approach. IEEE Trans. Ind. Electron. **40**, 45–55 (1993)
13. Gao, P., Zhang, G., Ouyang, H., Mei, L.: An adaptive super twisting nonlinear fractional order PID sliding mode control of permanent magnet synchronous motor speed regulation system based on extended state observer. IEEE Access **8**, 53498–53510 (2020)
14. Hou, H., Yu, X., Xu, L., Cao, Z.: Finite-time continuous terminal sliding mode control of servo motor systems. IEEE Trans. Ind. Electron. **67**, 5647–5656 (2020)
15. Jaskuła, M., Leśniewski, P.: Constraining state variables and control signal via sliding mode control approach. IEEE Access **8**, 111475–111481 (2020)
16. Jaskuła, M., Pietrala, M., Leśniewski, P.: The problem of state constraints in designing the discrete time sliding mode controller. Pomiary Automatyka Robotyka **4**, 15–22 (2017)
17. Junejo, A.K., Xu, W., Ismail, M., M., Liu, Y.: Adaptive speed control of PMSM drive system based a new sliding-mode reaching law. IEEE Trans. Power Electron. (2020, early access)
18. Ma, H., Li, Y.: A novel dead zone reaching law of discrete-time sliding mode control with disturbance compensation. IEEE Trans. Ind. Electron. **67**, 4815–4825 (2020)
19. Mija, S., Susy, T.: Reaching law based sliding mode control for discrete MIMO systems. In: IEEE International Conference on Control, Automation, Robotics and Vision, pp. 1291–1296 (2010)
20. Milosavljević, C.: General conditions for the existence of a quasi sliding mode on the switching hyperplane in discrete variable structure systems. Autom. Remote Control **46**, 307–314 (1985)
21. Niu, Y., Ho, D.W.C., Wang, Z.: Improved sliding mode control for discrete-time systems via reaching law. IET Control Theory Appl. **4**, 2245–2251 (2010)
22. Song, L., Gong, S., Tian, Y.: Study of reaching law approach for discrete-time variable structure control system. In: IEEE International Conference on Modelling, Identification and Control, Wuhan, China, pp. 206–210 (2012)
23. Utkin, V.I.: Variable structure systems with sliding modes. IEEE Trans. Autom. Control **22**, 212–222 (1977)
24. Utkin, V.I., Drakunow, S.V.: On discrete-time sliding mode control. In: IFAC Conference on Nonlinear Control, pp. 484–489 (1989)
25. Wang, Y., Feng, Y., Zhang, X., Liang, J.: A new reaching law for anti disturbance sliding-mode control of PMSM speed regulation system. IEEE Trans. Power Electron. **35**, 4117–4126 (2020)

Pointing Orientations for Manipulator End-Effectors

Douglas R. Isenberg(✉)

Embry-Riddle Aeronautical University, Prescott, AZ 86301, USA
isenberd@erau.edu
http://robotics.pr.erau.edu

Abstract. This paper discusses the problem of pointing of robotic end-effectors. A formal definition of pointing orientations is first described in terms of both direction cosine matrices and unit-quaternions. The paper then describes a numerical inverse kinematics technique that can be used to find solutions that point an end-effector in a desired direction. This technique is extended to the related problem of pointing an end-effector towards a desired location. Both techniques are demonstrated by examples applied to a Denso VP-6242 six degree of freedom manipulator.

Keywords: Pointing attitudes · Unit-quaternions · Inverse kinematics

1 Introduction

It is conventional to define robotic manipulation tasks in terms of a task-space position and an end-effector orientation. However, there are many manipulation tasks that do not require a three-dimensional orientation to be fully defined. For example, a welding task generally does not require a full specification of end-effector orientation. This task can be sufficiently defined by simply specifying an end-effector pointing direction.

Several benefits are obtained from defining a task in terms of a task-space position and an end-effector pointing direction. First, the task is inherently simpler to define. The general orientation problem requires the specification of the direction of three basis vectors. In contrast, pointing requires the specification of the direction of one vector. This can be beneficial to technicians and other practitioners who are not well versed in the mathematical quantification of three-dimensional motion. This is especially true considering that the orientation interpolation and path generation problem is still an open area of study [1,2]. Second, it is possible to perform useful tasks in the region between the manipulator's dexterous workspace and its reachable workspace. Since full dexterity is not required for pointing, a manipulator's actions are no longer bound by the size of its dexterous workspace. Third, the common six degree-of-freedom manipulator becomes redundant with respect to its task. Pointing only requires two degrees-of-freedom whereas obtaining a desired three-dimensional orientation requires three. This increases the number of joint-space solutions and offers

H. Selvaraj et al. (Eds.): ICSEng 2020, LNNS 182, pp. 71–80, 2021.
https://doi.org/10.1007/978-3-030-65796-3_7

the possibility of strategically selecting a solution that optimizes a performance measure.

Surprisingly, there is not an extensive amount of literature available on the topic of robotic end-effector pointing. Wie and Chung realized the importance of the topic to space applications and performed a few studies with respect to time-optimal control in the early 1990s, [3–5]. However, they primarily considered a planar two-link manipulator and were focused on the optimal control problem as opposed to the forward and inverse kinematic analysis in a 3-dimensional task space.

This paper does address the kinematic analysis and is organized as follows. First, pointing orientations are mathematically defined in terms of both direction cosine matrices and unit-quaternions. After this, a numerical technique for solving the inverse kinematics problem is described for problems that involve a desired pointing vector in lieu of an orientation. The technique is then extended to include cases for which the end-effector must remain pointing towards a desired location regardless of its position. These techniques are then applied to the inverse kinematics problem of a Denso VP-6242 six degree-of-freedom manipulator.

2 Pointing Orientations

Consider Fig. 1 below. Here, there are two coordinate frames, A and B. The 3×3 direction cosine matrix, ${}^{A}\boldsymbol{T}_{B}$, describes the orientation of frame B relative to A. Specifically, the columns of ${}^{A}\boldsymbol{T}_{B}$ describe the unit-basis vectors of frame B measured with respect to frame A. Assume that unit-vector ${}^{A}\boldsymbol{d}$ is fixed in frame A and is measured with respect to frame A. Additionally, assume that unit-vector ${}^{B}\boldsymbol{p}$ is fixed in frame B and measured with respect to frame B. In this paper, pointing orientations are defined as all of the orientations of frame B that align vector ${}^{B}\boldsymbol{p}$ with ${}^{A}\boldsymbol{d}$.

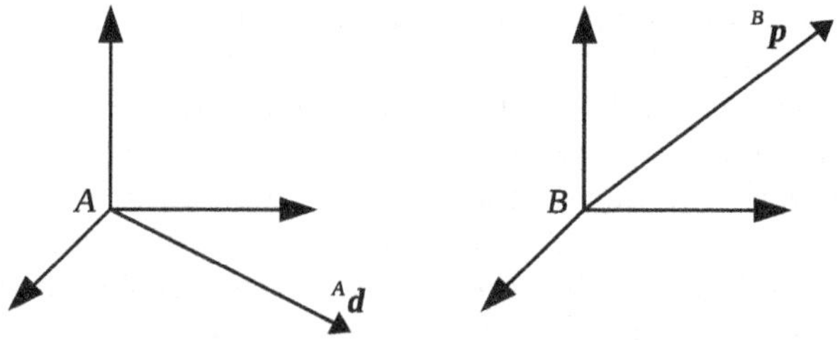

Fig. 1. Problem statement: define all orientations of B that align ${}^{A}\boldsymbol{d}$ with ${}^{B}\boldsymbol{p}$.

An orientation of B that aligns ${}^{B}\boldsymbol{p}$ with ${}^{A}\boldsymbol{d}$ can be defined by an axis-angle rotation. The angle of rotation is:

$$\alpha = acos(({}^{A}\boldsymbol{T}_{B}\,{}^{B}\boldsymbol{p})^{T}\,{}^{A}\boldsymbol{d}). \tag{1}$$

The unit-axis of rotation is:

$$\boldsymbol{u} = \begin{cases} \dfrac{{}^A\boldsymbol{T}_B\, {}^B\boldsymbol{p} \times {}^A\boldsymbol{d}}{\| {}^A\boldsymbol{T}_B\, {}^B\boldsymbol{p} \times {}^A\boldsymbol{d}\|} & for\ ({}^A\boldsymbol{T}_B\, {}^B\boldsymbol{p})^T\, {}^A\boldsymbol{d} \neq \pm 1 \\ {}^A\boldsymbol{d} & for\ ({}^A\boldsymbol{T}_B\, {}^B\boldsymbol{p})^T\, {}^A\boldsymbol{d} = 1 \\ \dfrac{{}^A\boldsymbol{T}_B\, {}^B\boldsymbol{p} \times \boldsymbol{r}}{\| {}^A\boldsymbol{T}_B\, {}^B\boldsymbol{p} \times \boldsymbol{r}\|} & for\ ({}^A\boldsymbol{T}_B\, {}^B\boldsymbol{p})^T\, {}^A\boldsymbol{d} = -1 \end{cases}, \tag{2}$$

where $\boldsymbol{r}$ can be any unit-vector that is not parallel to ${}^B\boldsymbol{p}$. Rather than selecting an $\boldsymbol{r}$ at random, a vector that is not parallel to ${}^B\boldsymbol{p}$ can be obtained by reordering the elements of ${}^B\boldsymbol{p}$ in the following manner:

$$\boldsymbol{r} = \begin{cases} \left[-p_y\ -p_z\ -p_x\right]^T & for\ {}^B\boldsymbol{p} \neq \pm \left[\frac{1}{\sqrt{3}}\ \frac{1}{\sqrt{3}}\ \frac{1}{\sqrt{3}}\right]^T \\ \left[1\,0\,0\right]^T & otherwise \end{cases} \tag{3}$$

A direction cosine matrix that describes an orientation of B relative to A that aligns ${}^B\boldsymbol{p}$ with ${}^A\boldsymbol{d}$ is then obtained from the axis-angle rotation matrix:

$$rot(\boldsymbol{u}, \alpha) = \begin{bmatrix} u_x^2(1-c_\alpha)+c_\alpha & u_xu_y(1-c_\alpha)-u_zs_\alpha & u_xu_z(1-c_\alpha)+u_ys_\alpha \\ u_xu_y(1-c_\alpha)+u_zs_\alpha & u_y^2(1-c_\alpha)+c_\alpha & u_yu_z(1-c_\alpha)-u_xs_\alpha \\ u_zu_x(1-c_\alpha)-u_ys_\alpha & u_yu_z(1-c_\alpha)+u_xs_\alpha & u_z^2(1-c_\alpha)+c_\alpha \end{bmatrix}. \tag{4}$$

It should be stressed that this orientation is just one of an infinite number of orientations that accomplish this goal. Once ${}^B\boldsymbol{p}$ and ${}^A\boldsymbol{d}$ are aligned, then any subsequent rotation about the newly oriented ${}^B\boldsymbol{p}$ vector keeps ${}^B\boldsymbol{p}$ and ${}^A\boldsymbol{d}$ aligned. Similarly, any rotation about ${}^A\boldsymbol{d}$ also keeps ${}^B\boldsymbol{p}$ and ${}^A\boldsymbol{d}$ aligned. This is depicted in Fig. 2.

Let ${}^A\hat{\boldsymbol{T}}_B$ denote all of the direction cosine matrices that align the two vectors. It can then be stated that:

$${}^A\hat{\boldsymbol{T}}_B = rot(\boldsymbol{u}, \alpha) rot({}^B\boldsymbol{p}, \theta) \tag{5}$$

or

$${}^A\hat{\boldsymbol{T}}_B = rot({}^A\boldsymbol{d}, \phi) rot(\boldsymbol{u}, \alpha) \tag{6}$$

where θ and ϕ are arbitrary angles of rotation.

These pointing orientations can also be expressed in terms of a unit-quaternion. Let the unit-quaternion that describes the orientations of B relative to A that align ${}^B\boldsymbol{p}$ with ${}^A\boldsymbol{d}$ be denoted by ${}^A\hat{\boldsymbol{q}}_B$. Knowing the axis of rotation, $\boldsymbol{u}$, from above, and the corresponding angle of rotation, α, this quaternion is constructed from the Euler-parameters as either

$${}^A\hat{\boldsymbol{q}}_B = \begin{bmatrix} cos(\alpha/2) \\ \boldsymbol{u} sin(\alpha/2) \end{bmatrix} \circ \begin{bmatrix} cos(\theta/2) \\ {}^B\boldsymbol{p} sin(\theta/2) \end{bmatrix} \tag{7}$$

or

$${}^A\hat{\boldsymbol{q}}_B = \begin{bmatrix} cos(\phi/2) \\ {}^A\boldsymbol{d} sin(\phi/2) \end{bmatrix} \circ \begin{bmatrix} cos(\alpha/2) \\ \boldsymbol{u} sin(\alpha/2) \end{bmatrix}. \tag{8}$$

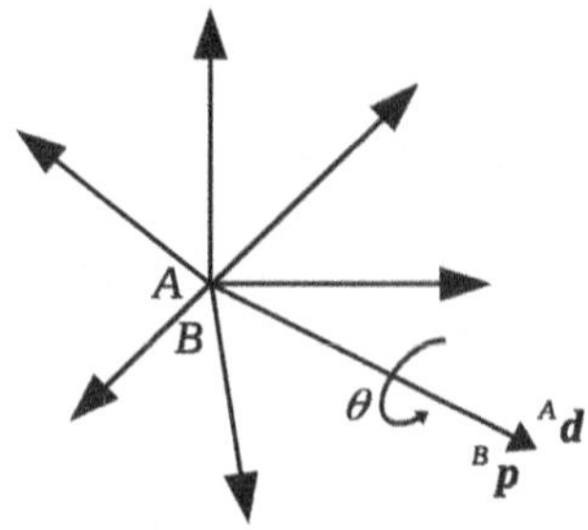

Fig. 2. Once $^B\boldsymbol{p}$ and $^A\boldsymbol{d}$ are aligned, any rotation of B about $^B\boldsymbol{p}$ or $^A\boldsymbol{d}$ keeps these vectors aligned.

Here, the $\circ$ operator is utilized to denote the quaternion product, which is defined for any two quaternions, $\boldsymbol{a}$ and $\boldsymbol{b}$ as

$$\boldsymbol{a} \circ \boldsymbol{b} = \begin{bmatrix} a_0 & -a_1 & -a_2 & -a_3 \\ a_1 & a_0 & -a_3 & a_2 \\ a_2 & a_3 & a_0 & -a_1 \\ a_3 & -a_2 & a_1 & a_0 \end{bmatrix} \boldsymbol{b}. \tag{9}$$

3 Inverse Kinematic Solution for Pointing Orientations

The expressions in the previous section clearly show that there is an unconstrained parameter, θ or ϕ, that exist when expressing a pointing orientation. While it is possible to formulate an over-determined inverse kinematic problem for a manipulator with six or more degrees of freedom using one of these expressions by placing, θ, in the solution space, a much simpler and lower order problem formulation is possible.

Let frame I denote the base frame or inertial frame of a robotic manipulator. Furthermore, let frame E denote a frame that is fixed with respect to the manipulator's end-effector. Assume that unit-vector $^E\boldsymbol{p}$ defines the pointing direction of a tool or object measured in the end-effector frame. Additionally, assume $^I\boldsymbol{d}$ defines the desired pointing direction of the tool or object in the inertial frame. The objective is to find the configuration of the robot, $\boldsymbol{\gamma}$, that makes $^E\boldsymbol{p}$ parallel to $^I\boldsymbol{d}$ while simultaneously positioning the end-effector location, $^I_I\boldsymbol{r}_E$, to a desired location, $^I_I\boldsymbol{r}_D$. This is depicted below in Fig. 3.

This problem is formulated by defining error metrics for the end-effector position and for the direction of the pointing vector. The positional error is defined as the arithmetic difference between the end-effector's actual location and its desired location, $^I_I\boldsymbol{r}_E(\boldsymbol{\gamma}) - ^I_I\boldsymbol{r}_D$. Here, the explicit dependence of $^I_I\boldsymbol{r}_E$ on the configuration has been shown. The pointing vector error is succinctly defined in terms of the inner product. Since both $^E\boldsymbol{p}$ and $^I\boldsymbol{d}$ are unit-vectors, one needs only to project one vector onto the other and measure any deviation of the resulting value from 1. Before doing so, however, the vector measurements must be made

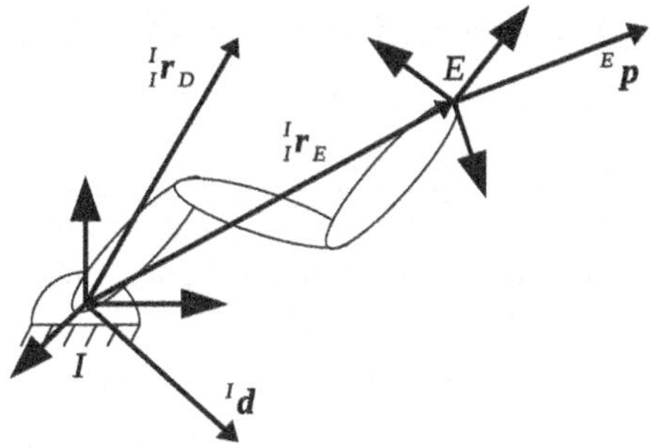

Fig. 3. The inverse kinematic problem involves finding the configuration, γ, that makes $^E\boldsymbol{p}$ parallel to $^I\boldsymbol{d}$ and moves $^I_I\boldsymbol{r}_E$ to $^I_I\boldsymbol{r}_D$.

in the same frame. This requires knowledge of the end-effector's orientation relative to the inertial frame, $^I\boldsymbol{T}_E(\gamma)$, which is also an explicit function of the manipulator's configuration, in order to transform the vector's measurement from frame E to frame I. Collecting these two measures into a single vector function produces:

$$\boldsymbol{F}(\gamma) = \begin{bmatrix} ^I_I\boldsymbol{r}_E(\gamma) - ^I_I\boldsymbol{r}_D \\ 1 - {}^I\boldsymbol{d}^T\, {}^I\boldsymbol{T}_E(\gamma)\, {}^E\boldsymbol{p} \end{bmatrix}. \tag{10}$$

Solutions to (10) can be found using the closed-loop inverse kinematics technique in which the the problem is formulated such that the inverse kinematic solution is an equilibrium point of a stable differential equation [6,7]. Alternatively, the inverse kinematic problem can be solved using classical numerical methods [8,9]. This is the approach taken in the current paper.

If an inverse kinematic solution exist, then it is a root of (10). Given a properly designed six degree-of-freedom manipulator, this problem is underdetermined. The problem is also highly nonlinear, though it can usually be solved with a Newton-Raphson root finding method that employs the minimum-norm pseudo-inverse of the following Jacobian:

$$\boldsymbol{J}(\gamma) = \frac{\partial \boldsymbol{F}(\gamma)}{\partial \gamma}. \tag{11}$$

The root-solving iteration is then:

$$\gamma^{(k+1)} = \gamma^{(k)} - \boldsymbol{J}\left(\gamma^{(k)}\right)^+ \boldsymbol{F}\left(\gamma^{(k)}\right). \tag{12}$$

Here, the + superscript is used to denote the pseudo-inverse and k is used as the iteration index variable. It should be noted that the pseudo-inverse will result in the smallest $\|\gamma^{(k+1)} - \gamma^{(k)}\|$ on each iteration.

4 Inverse Kinematic Solution for Pointing Towards Locations

Suppose that the objective is to point a tool or an object attached to a robot's end-effector towards a desired location. Such a situation can occur in the applications of robotic painting and visual inspection, among others. A graphical depiction of this situation is shown in Fig. 4.

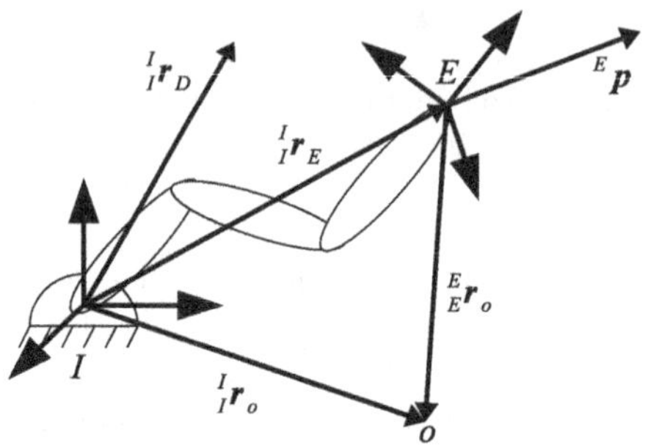

Fig. 4. The inverse kinematic problem for pointing towards a location involves finding the configuration, γ, that aligns ${}^{E}\boldsymbol{p}$ with ${}^{E}_{E}\boldsymbol{r}_O$ and moves ${}^{I}_{I}\boldsymbol{r}_E$ to ${}^{I}_{I}\boldsymbol{r}_D$.

As before, one goal is to move the end-effector to ${}^{I}_{I}\boldsymbol{r}_D$. This part of the problem is identical to what was previously formulated. The other goal is to align ${}^{E}\boldsymbol{p}$ with ${}^{E}_{E}\boldsymbol{r}_O$. It should be emphasized that ${}^{E}_{E}\boldsymbol{r}_O$ is not fixed in frame E. Notice from the forward kinematics that:

$$ {}^{I}_{I}\boldsymbol{r}_O = {}^{I}_{I}\boldsymbol{r}_E + {}^{I}\boldsymbol{T}_E \, {}^{E}_{E}\boldsymbol{r}_O. \tag{13} $$

So,

$$ {}^{E}_{E}\boldsymbol{r}_O(\gamma) = {}^{I}\boldsymbol{T}_E(\gamma)^T \left({}^{I}_{I}\boldsymbol{r}_O - {}^{I}_{I}\boldsymbol{r}_E(\gamma) \right). \tag{14} $$

An error metric for quantifying any deviation from the alignment goal is then formulated by the difference between 1 and the projection of ${}^{E}\boldsymbol{p}$ onto a unit-vector in the direction of ${}^{E}_{E}\boldsymbol{r}_O(\gamma)$. Thus, a single function whose roots are a solution to this problem is:

$$ \boldsymbol{F}(\gamma) = \begin{bmatrix} {}^{I}_{I}\boldsymbol{r}_E(\gamma) - {}^{I}_{I}\boldsymbol{r}_D \\ 1 - {}^{E}\boldsymbol{p}^T \left(\frac{{}^{E}_{E}\boldsymbol{r}_O(\gamma)}{\| {}^{E}_{E}\boldsymbol{r}_O(\gamma) \|} \right) \end{bmatrix} \tag{15} $$

5 End-Effector Pointing with a Six Degree-of-Freedom Manipulator

In order to illustrate the effectiveness of the two end-effector pointing techniques described above, the techniques are applied to a Denso VP-6262 manipulator with a SMC MHS3-25D pneumatic gripper. This robot has six joint angles, $\theta_1, \theta_2, \ldots \theta_6$. Thus,

$$\gamma = \left[\theta_1\ \theta_2\ \theta_3\ \theta_4\ \theta_5\ \theta_6\right]^T. \tag{16}$$

The forward kinematics for the manipulator are succinctly captured by the six relative position vectors and six relative rotation matrices shown in Table 1.

Table 1. Relative position vectors and rotations for denso VP-6242.

Vector	x (mm)	y (mm)	z (mm)	DCM	Rotation
${}^I_I\boldsymbol{r}_1$	0	0	155	${}^I\boldsymbol{T}_1$	$rotz(\theta_1)$
${}^1_1\boldsymbol{r}_2$	0	0	123.5	${}^1\boldsymbol{T}_2$	$roty(\theta_2)$
${}^2_2\boldsymbol{r}_3$	0	0	210	${}^2\boldsymbol{T}_3$	$roty(\theta_3)$
${}^3_3\boldsymbol{r}_4$	−75	0	86	${}^3\boldsymbol{T}_4$	$roty(-\pi/2)rotx(\theta_4)$
${}^4_4\boldsymbol{r}_5$	125	0	0	${}^4\boldsymbol{T}_5$	$roty(\theta_5)$
${}^5_5\boldsymbol{r}_E$	132.5	0	0	${}^5\boldsymbol{T}_E$	$rotx(\theta_6)$

Here, the rotation matrices are defined for any angle ϕ as:

$$rotx(\phi) = \begin{bmatrix} 1 & 0 & 0 \\ 0 & cos(\phi) & -sin(\phi) \\ 0 & sin(\phi) & cos(\phi) \end{bmatrix},$$

$$roty(\phi) = \begin{bmatrix} cos(\phi) & 0 & sin(\phi) \\ 0 & 1 & 0 \\ -sin(\phi) & 0 & cos(\phi) \end{bmatrix},$$

and

$$rotz(\phi) = \begin{bmatrix} cos(\phi) & -sin(\phi) & 0 \\ sin(\phi) & cos(\phi) & 0 \\ 0 & 0 & 1 \end{bmatrix}.$$

In general, the orientation of the Nth link, expressed as a direction cosine matrix, is obtained as:

$${}^I\boldsymbol{T}_N = {}^I\boldsymbol{T}_1 \prod_{i=2}^{N} {}^{i-1}\boldsymbol{T}_i \tag{17}$$

So, the end-effector's orientation is obtained as a function of the joint angles θ_n for $n = 1, 2, \ldots 6$ as:

$${}^I\boldsymbol{T}_E = {}^I\boldsymbol{T}_1\, {}^1\boldsymbol{T}_2\, {}^2\boldsymbol{T}_3\, {}^3\boldsymbol{T}_4\, {}^4\boldsymbol{T}_5\, {}^5\boldsymbol{T}_E. \tag{18}$$

Similarly, the position of the end-effector, expressed as a function of the joint angles, is

$${}^I_I\boldsymbol{r}_E = {}^I_I\boldsymbol{r}_1 + {}^I\boldsymbol{T}_1\, {}^1_1\boldsymbol{r}_2 + {}^I\boldsymbol{T}_2\, {}^2_2\boldsymbol{r}_3 + {}^I\boldsymbol{T}_3\, {}^3_3\boldsymbol{r}_4 + {}^I\boldsymbol{T}_4\, {}^4_4\boldsymbol{r}_5 + {}^I\boldsymbol{T}_5\, {}^5_5\boldsymbol{r}_E \tag{19}$$

In the following examples, it is assumed that ${}^E\boldsymbol{p} = \left[0\ 1\ 0\right]^T$. It should be noted that this vector is orthogonal to the axis of rotation, relative to link 5, of the end-effector.

5.1 Pointing Orientations

A list of end-effector positions and pointing directions are shown in Table 2 along with the joint-space solutions obtained by solving (10) with the Newton-Raphson method. In addition to this data, Fig. 5 depicts the manipulator with its end-effector located at these desired positions and pointing in the desired direction. In the figure, a red dot is utilized to denote the desired end-effector location, a red-line is utilized to indicate the direction of ${}^{E}\boldsymbol{p}$, and a black line is utilized to indicate the direction of ${}^{I}\boldsymbol{d}$. As shown by the tabulated data and the graphical illustration, the goal of simultaneously moving the end-effector to a desired location and pointing it in a desired direction is accomplished.

Table 2. Desired locations, pointing directions, and inverse kinematic solution

point	x (mm)	y (mm)	z (mm)	${}^{I}d_x$	${}^{I}d_y$	${}^{I}d_z$	θ_1 (rad)	θ_2 (rad)	θ_3 (rad)	θ_4 (rad)	θ_5 (rad)	θ_6 (rad)
1	200	150	80	1	0	0	0.614	0.842	2.239	1.262	0.058	−5.356
2	200	0	40	0.707	0	0.707	0.620	1.350	2.298	1.349	−1.117	−4.780
3	200	−10	80	−0.707	0.707	0	−1.053	1.291	2.121	1.396	1.260	2.844
4	100	−200	80	−0.707	0.707	0	−1.487	1.147	1.950	0.858	0.975	3.165
5	100	−300	100	0	0	1	−1.681	1.418	1.714	1.590	1.569	4.722
6	0	−300	200	0	0.707	0.707	−1.997	1.109	1.748	1.247	1.736	4.203

5.2 Pointing Towards a Location

To demonstrate the end-effector reaching a desired position while also pointing towards a desired location, the points in Table 3 are utilized. Also included in Table 3 is the inverse kinematic solution. Figure 6 graphically depicts the data in Table 3. Here, the desired end-effector location is depicted with a red dot, the direction of the end-effector, ${}^{E}\boldsymbol{p}$, is depicted with a red line with exaggerated length, and the pointing locations are depicted with a red "x". Again, the inverse kinematic problem formulation and numerical solution technique achieve the desired goal.

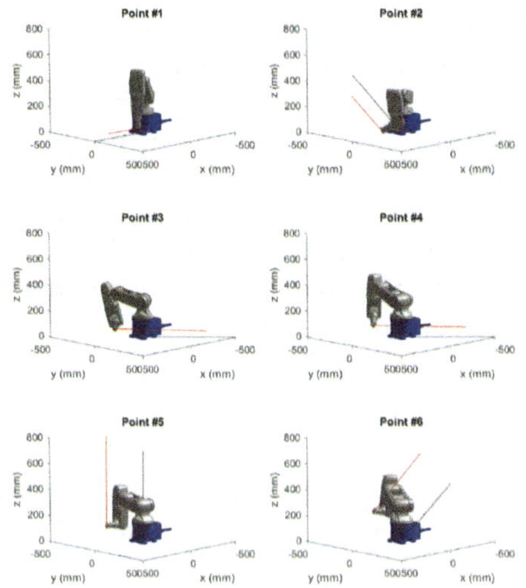

Fig. 5. Inverse kinematic solutions of the Denso VP-6242 end-effector given desired location and pointing orientation.

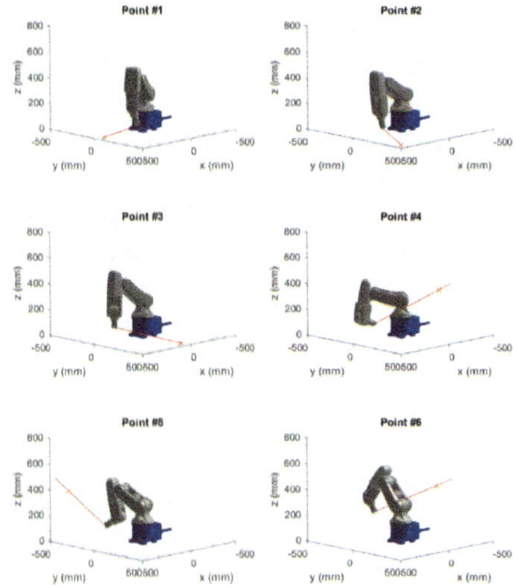

Fig. 6. Inverse kinematic solutions of the Denso VP-6242 end-effector given desired location and pointing location.

Table 3. Desired locations, pointing locations, and inverse kinematic solution

point	x (mm)	y (mm)	z (mm)	x_o (mm)	y_o (mm)	z_o (mm)	θ_1 (rad)	θ_2 (rad)	θ_3 (rad)	θ_4 (rad)	θ_5 (rad)	θ_6 (rad)
1	200	150	80	400	0	0	0.479	0.892	2.174	1.712	0.342	−1.370
2	200	0	40	400	400	0	0	1.017	2.254	0.001	0.066	0.623
3	200	−10	80	0	400	0	−0.282	0.875	2.239	−1.011	0.369	1.916
4	100	−200	80	0	400	400	−0.738	1.337	1.689	−0.683	1.303	2.470
5	100	−300	100	400	−400	400	−1.273	0.981	2.360	−0.061	−1.180	−0.809
6	0	−300	200	−400	0	400	−1.370	0.623	1.916	2.470	−0.809	1.309

6 Conclusion

Many robot manipulation tasks can be specified without defining the full pose for a robotic end-effector. For tasks that involve manipulation of symmetric objects or pointing towards an object in a desired direction, an end-effector pointing direction can be specified rather than an orientation. This can potentially simplify a robotic technician's task as a pointing vector can be defined without requiring knowledge of the various orientation parameterizations. This paper has shown how to define pointing attitudes with both direction cosine matrices and unit-quaternions. Additionally, the paper has shown how to formulate and numerically solve inverse kinematics problems when the task space coordinates are specified as end-effector positions and both pointing directions and pointing locations.

Typically, analytical inverse kinematic solutions for six degree-of-freedom manipulators are difficult to find or non-existent. One avenue for future work may involve examining whether or not adopting pointing orientations would lead to analytical inverse kinematic solutions. Other areas that should be examined are both path and trajectory generation for pointing orientations. Of particularly usefulness would be a polynomial-based method for generating pointing orientation splines in a manner that analogous to creating splines for Cartesian positions with cubic polynomials. Additionally, the ideas of pointing orientations should be applied to the task-space feedback control problem.

References

1. Neubauer, M., Muller, A.: Smooth orientation path planning with quaternions using b-splines. In: Proceedings of International Conference on Intelligent Robots and Systems, pp. 2087–2092 (2015)
2. Grassman, R.M., Burgner-Kahrs, J.: Quaternion-based smooth trajectory generator for via poses in SE(3) considering kinematic limits in cartesian space. IEEE Robot. Autom. Lett. **4**, 4192–4199 (2019)
3. Wie, B., Chuang, C., Sunkel, J.: Minimum-time pointing control of a two-Link manipulator. J. Guidance **13**(5), 867–873 (1990)
4. Chuang, C.: Optimal pointing control of robotic manipulators with state inequality constraints. J. Guidance **16**(5), 966–969 (1990)
5. Chuang, C.: Time-optimal pointing control of robotic manipulators with state-variable inequality-constraints. In: Proceedings of 29th Aerspace Sciences Meeting (1991)
6. Sciavicco, L., Siciliano, B.: A solution algorithm to the inverse kinematic problem for redundant manipulators. IEEE J. Robot. Automat. **4**(4), 403–410 (1988)
7. Siciliano, B.: A closed-loop inverse kinematic scheme for on-line joint-based robot control. Robotica **8**, 231–243 (1990)
8. Goldenberg, A., Benhabib, B., Fenton, R.: A Complete generalized solution to the inverse kinematics of robots. IEEE J. Robot. Autom. **1**(1), 14–20 (1985)
9. Buss, S.: Introduction to Inverse Kinematics with Jacobian Transpose, Pseudoinverse and Damped Least Squares Methods (2004)

Non-switching Reaching Law Based DSMC Strategies in the Context of Robustness Comparison

Paweł Latosiński(✉) and Monika Herkt

Institute of Automatic Control, Łódź University of Technology,
18/22 Stefanowskiego St., 90-924 Łódź, Poland
pawel.latosinski@p.lodz.pl

Abstract. Discrete-time sliding mode control (DSMC) is well known for ensuring robustness of systems subject to external disturbances and model uncertainties. The main idea behind DSMC is to first specify a switching hyperplane according to which the control signal is changed. Then, the control signal with variable structure drives the system representative point to a specified vicinity of the switching surface. That vicinity is usually referred to as quasi-sliding mode band (QSMB) and its width is one of the most common indicators of system robustness in discrete-time SMC. Indeed, a narrower QSMB directly implies better robustness of the system with respect to uncertainties. However, analysis of sliding mode performance of the system only provides information about values of the sliding variable, which is a constructed system output. A more useful performance index would be based on state error, but explicit limits of individual state variables are difficult to obtain in SMC. The purpose of this paper is to highlight the proportional relation between these two quality criteria (QSMB and state errors) found in recently published literature. In this paper we focus specifically on strategies using non-switching type reaching laws, which does not require switching across the sliding hyperplane in every subsequent step. Aforementioned relation is considered for a certain class of discrete-time systems both in theory and in simulation example.

1 Introduction

Sliding mode control is well known for ensuring robustness of various nonlinear systems with respect to external disturbance and parameter uncertainties. Moreover, it offers valuable properties such as reduction of the problem dimension and computational efficiency. This methodology has been introduced in the 1960s [1–3] for continuous time systems and then further developed by numerous authors [4–6]. The main purpose of sliding mode control is to combine the advantages of different control structures in the same control process depending on which structure is more efficient for the current system state. To that end, sliding hyperplane is created to determine the switching point between the two possible structures of the control signal.

H. Selvaraj et al. (Eds.): ICSEng 2020, LNNS 182, pp. 81–92, 2021.
https://doi.org/10.1007/978-3-030-65796-3_8

In practical terms, most control processes are applied digitally so their structure changes based on a set frequency. Therefore, for practical applications continuous version of sliding mode control appeared to be sub-optimal. For that reason discrete-time sliding mode control (DSMC) has been developed [7–10] as it proved to be more adequate for real-life applications. However, DSMC does not guarantee complete insensitivity with respect to disturbance, one can only expect the system representative point to stay in a certain vicinity of its target position during quasi-sliding phase. As proved in [11], by making additional assumptions concerning continuity of the disturbance function and its derivatives one can reduce the width of this vicinity for digitized systems.

Although in continuous sliding mode control, definition of sliding motion is clearly determined, in DSMC it is not as intuitive. Because of a finite frequency of switching between different control structures, definition used in continuous-time SMC is not viable for discrete-time case. Hence, another characterizations have been researched. In literature, two different definitions of this so-called quasi-sliding motion have been adopted. First Gao et al. [12] introduced the concept of switching type quasi-sliding modes. According to this definition, sliding variable has to change its sign between every two sampling moments starting from some finite time instant in which it crosses switching hyperplane for the first time. Among many authors using this explanation of quasi-sliding motion, one can mention [13–23]. Another definition of quasi-sliding mode has been used, for instance, in [14,24–26]. These non-switching type strategies require the sliding variable to stay in a certain vicinity of zero starting from some finite time instant, but not to cross the switching hyperplane in every subsequent sampling period. This relaxation imposed on the definition of quasi-sliding mode lets one achieve narrower quasi-sliding mode bands (QSMB) and can reduce the control effort as a result. On the other hand, it does not guarantee that the system representative point ever reaches its desired position. Thus, both types of quasi-sliding motion have their strengths and weaknesses. In order to obtain the best results in practical applications, the choice of switching or non-switching strategy should be based on concrete application requirements. In this paper we concentrate on strategies using non-switching type approach.

In order to design a sliding mode control strategy, the crucial step is to define a proper sliding hyperplane. Then, we obtain the control signal which provides stability of controlled system and assures that system representative point converges to the hyperplane. Several authors [9,27] adopt the traditional approach, in which they first postulate a specific formula for the control signal and then prove that it fulfills the aforementioned requirements. However, plenty of authors [13–26,29,30] apply an alternative method called reaching law approach, first proposed for continuous sliding mode by Gao et al. [31] and then extended on DSMC in [12]. In this technique, we first specify the desired course of the sliding variable using the so-called reaching law which is recursive equation that guarantees fulfillment of reaching conditions. Afterwards, using this reaching law and information about concrete system dynamics, we derive the formula describing the control signal which ensures the pre-defined course of the variable. Strategies designed using the reaching law approach will be the main focus of this paper.

In this paper we will research the relation between QSMB and maximum error of individual state variables, particularly in reference to non-switching type reaching law based sliding mode control strategies. Both of these values can be interpreted as measurements of controlled system robustness. State error is more direct in this matter, however QSMB width is less complex to obtain. We will consider five strategies based on non-switching reaching laws in the direction of showing aforementioned proportional relation.

2 Preliminaries

Let us take into account the class of discrete systems with following dynamics

$$\boldsymbol{x}(k+1) = \boldsymbol{A}\boldsymbol{x}(k) + \boldsymbol{b}u(k) + \boldsymbol{h}f(k), \tag{1}$$

where vector $\boldsymbol{x}$ describes the system state, $\boldsymbol{A}$ is the state matrix, u is the control signal, f is external disturbances attached to that system and $\boldsymbol{b}, \boldsymbol{h}$ are vectors representing the input channels for control signal and disturbance, respectively.

Let us assume that the unknown disturbances are bounded by some a priori known values, which is a common assumption in robust control. In particular, for any $k \in \mathbb{N}$ the lower bound of $f(k)$ is f_{L} and its upper bound is f_{U}.

In order to design a sliding mode controller, we first need to define an appropriate switching variable s. Let us consider a linear function

$$s(k) = \boldsymbol{c}^{\mathrm{T}}\boldsymbol{e}(k) = \boldsymbol{c}^{\mathrm{T}}(\boldsymbol{x}_{\mathrm{d}}(k) - \boldsymbol{x}(k)), \tag{2}$$

where $\boldsymbol{e}(k)$ is a difference between the target state $\boldsymbol{x}_{\mathrm{d}}(k)$ and the actual state variables $\boldsymbol{x}(k)$. Further in this paper, we will select the target state x_{d} that remains constant during the whole control process.

In order to get a stable response of the system, vector $\boldsymbol{c}$ needs to be selected in a way which ensures that all eigenvalues of the closed-loop system state matrix

$$\boldsymbol{H} = \boldsymbol{A} - \boldsymbol{b}(\boldsymbol{c}^{\mathrm{T}}\boldsymbol{b})^{-1}\boldsymbol{c}^{\mathrm{T}}\boldsymbol{A} \tag{3}$$

are placed inside the unit circle. There are many techniques of selecting such a vector [12,32,33]. In this paper we will assume that its choice guarantees a dead-beat performance of the closed-loop system, which occurs when the poles of $\boldsymbol{H}$ are placed in zero.

The objective of any sliding mode control is to drive the system representative point onto the selected switching hyperplane $s = 0$. To that end, one can use the conventional approach [9,27] or the more recent reaching law approach [13–30]. According to conventional technique, after defining sliding hyperplane as (2), we would have to postulate control signal which guarantees stability of the plant and convergence of the system representative point to this switching surface. Then, we would examine if these conditions are fulfilled. In this paper however, we will consider strategies originating from the reaching law approach exclusively. This approach involves predefining sliding variable dynamics using a recursive equation called reaching law, and then applying it to design an appropriate control signal. We will now present the reaching law based controller design process in more detail.

Let us first introduce the following notation

$$F(k) = \boldsymbol{c}^{\mathrm{T}}\boldsymbol{h}f(k), \quad F_{\mathrm{L}} = \boldsymbol{c}^{\mathrm{T}}\boldsymbol{h}f_{\mathrm{L}}, \quad F_{\mathrm{U}} = \boldsymbol{c}^{\mathrm{T}}\boldsymbol{h}f_{\mathrm{U}}.$$

Hence the average disturbance and its maximum deviation from this average can be expressed as

$$F_{\mathrm{A}} = \frac{F_{\mathrm{U}} + F_{\mathrm{L}}}{2} \quad \text{and} \quad F_{\mathrm{D}} = \frac{F_{\mathrm{U}} - F_{\mathrm{L}}}{2}.$$

As mentioned before, in this paper we will research sliding mode control strategies that have been developed using reaching law approach. This approach involves designing specific dynamics of the sliding variable in subsequent time instant and then applying these dynamics to design an appropriate control signal. In general, we can express the reaching law using the following recursive formula

$$s(k+1) = \Gamma(\boldsymbol{x}(k)) - F(k), \tag{4}$$

where Γ is a function defined to provide evolution of the sliding variable requested by designer. Function Γ is recursive in this case, as it uses the value of sliding variable from the previous step. This reaching law can now be used to synthesize the control signal. Considering system dynamics (1) we obtain

$$s(k+1) = \boldsymbol{c}^{\mathrm{T}}\boldsymbol{x}_{\mathrm{d}} - \boldsymbol{c}^{\mathrm{T}}\boldsymbol{A}\boldsymbol{x}(k) - \boldsymbol{c}^{\mathrm{T}}\boldsymbol{b}u(k) - F(k). \tag{5}$$

Then comparing (4) and (5) we get

$$\boldsymbol{c}^{\mathrm{T}}\boldsymbol{x}_{\mathrm{d}} - \boldsymbol{c}^{\mathrm{T}}\boldsymbol{A}\boldsymbol{x}(k) - \boldsymbol{c}^{\mathrm{T}}\boldsymbol{b}u(k) - F(k) = \Gamma(\boldsymbol{x}(k)) - F(k).$$

Finally, solving this equation for u we obtain

$$u(k) = -(\boldsymbol{c}^{\mathrm{T}}\boldsymbol{b})^{-1}\left(\boldsymbol{c}^{\mathrm{T}}\boldsymbol{A}\boldsymbol{x}(k) - \boldsymbol{c}^{\mathrm{T}}\boldsymbol{x}_{\mathrm{d}} + \Gamma(\boldsymbol{x}(k))\right). \tag{6}$$

It is easy to notice that derived formula for control signal does not contain unknown disturbance term, hence in this form it is applicable to the system.

Next section will be devoted to presentation of several selected strategies which use the reaching law approach.

3 Selected Strategies

In this section we will present sliding mode control strategies ensuring non-switching type of quasi-sliding motion. As opposed to more conventional switching type discrete-time strategies [12], the ones presented in this chapter aim to eliminate undesirable oscillations (chattering) in the sliding phase while improving robustness of the system with respect to disturbance.

3.1 A. Bartoszewicz, 1998

First of the considered methods is the one proposed in [24]. This strategy aims to drive the system representative point into sliding surface gradually and in certain number of steps specified by chosen parameter. The reaching law determining the strategy is of the form

$$s(k+1) = s_{\mathrm{d}}(k+1) + F_{\mathrm{A}} - F(k), \tag{7}$$

where s_{d} is a function defined by the designer, which needs to satisfy the following conditions in case of $s(0) > 2F_{\mathrm{D}}$:

1. $s_{\mathrm{d}}(0) = s(0)$,
2. $s_{\mathrm{d}}(k) \cdot s_{\mathrm{d}}(0) \geq 0$ for any $k \geq 0$,
3. $s_{\mathrm{d}}(k) = 0$ for any $k \geq k^*$,
4. $|s_{\mathrm{d}}(k+1)| < |s_{\mathrm{d}}(k)| - 2F_{\mathrm{D}}$ for any $k < k^*$

and be equal to zero for any $k < k^*$ in case of $s(0) \leq 2F_{\mathrm{D}}$. The idea behind this control method is to use desired sliding surface represented by function s_{d}. This way, we can gradually move the sliding variable from its initial value $s(0)$ to 0. One of the possible functions satisfying above conditions was proposed in the paper and can be described as follows

$$s_{\mathrm{d}}(k) = \frac{k^* - k}{k^*} \cdot s(0)$$

for $k \in \{0, 1, ..., k^*\}$ if $s(0) > 2F_{\mathrm{D}}$ and $s_{\mathrm{d}}(k) = 0$ for any k if $s(0) \leq 2F_{\mathrm{D}}$. In order to make the control strategy applicable, k^* has to be chosen according to inequality $k^* < s(0)/2F_{\mathrm{D}}$.

Using the reaching law approach, we can determine control signal which guarantees the specified sliding variable behaviour. It is expressed as

$$u(k) = -(\boldsymbol{c}^{\mathrm{T}}\boldsymbol{b})^{-1}\left(\boldsymbol{c}^{\mathrm{T}}\boldsymbol{A}\boldsymbol{x}(k) - \boldsymbol{c}^{\mathrm{T}}\boldsymbol{x}_{\mathrm{d}} + s_{\mathrm{d}}(k+1) + F_{\mathrm{A}}\right).$$

It has been shown in [24], that applying such a strategy to the system (1), sliding variable will be bounded to a specified vicinity of 0 for any $k \geq k^*$. This QSMB is described as $\left\{\boldsymbol{x} : \left|s(\boldsymbol{x}(k))\right| \leq F_{\mathrm{D}} = \Delta_{\mathrm{I}}\right\}$. It is significantly less value than one determining QSMB obtained by switching type reaching law based strategies such as [12, 19–23], hence robustness of control is improved.

3.2 A. Bartoszewicz, 1998 (Modified Version)

In the same paper, author has also introduced another sliding mode control strategy, which is essentially a modification of (7). Compared to the previous strategy, the proposed modification additionally provides further decrease of the QSMB width in case of systems with slowly changing disturbance. The proposed reaching law can be described as

$$s(k+1) = s_{\mathrm{d}}(k+1) - \sum\nolimits_{i=0}^{k}\left(s(i) - s_{\mathrm{d}}(i)\right) + F_{\mathrm{A}} - F(k), \tag{8}$$

where s_d is the same function as the one chosen for previous strategy. We can notice that sliding variable is modified in every step by the sum of deviations of its actual value from the desired one in all preceding time instants. Inclusion of this sum lets us further reduce the disturbance impact on system sliding motion.

Adopting reaching law approach for (8) we obtain control signal

$$u(k) = -\left(\mathbf{c}^\mathrm{T}\boldsymbol{b}\right)^{-1}\left(\mathbf{c}^\mathrm{T}\boldsymbol{A}\boldsymbol{x}(k) - \mathbf{c}^\mathrm{T}\boldsymbol{x}_\mathrm{d} + s_\mathrm{d}(k+1) - \sum\nolimits_{i=0}^{k}\left(s(i) - s_\mathrm{d}(i)\right) + F_\mathrm{A}\right).$$

This control method appears especially useful for systems with limited disturbance rate, which means that there exists d_F such as $|F(k+1) - F(k)| \leq d_\mathrm{F}$ for any $k \geq 0$ (which results from the bounded disturbance assumption). QSMB for this strategy has been shown not to cross that disturbance rate limit, so it can be presented as $\left\{\boldsymbol{x} : \left|s\left(\boldsymbol{x}(k)\right)\right| \leq d_\mathrm{F} = \Delta_\mathrm{II}\right\}$, what can increase robustness of control in comparison to unmodified version in the case of $d_\mathrm{F} < F_\mathrm{D}$. The last inequality describes slowly changing disturbance attached to the plant.

3.3 S. Kurode, B. Bandyopadhyay and P. Gandhi, 2011

The third strategy considered in this paper is a linear adaptive reaching law proposed in [25]. The purpose of this approach is to make the movement of the system representative point towards the sliding hyperplane gradual and prevent overshoot in initial time instants of the control process. It can be expressed by the following reaching law

$$s(k+1) = \beta(s(k)) \cdot s(k) + F_\mathrm{A} - F(k), \tag{9}$$

where $\beta\left(s(k)\right) = \beta_1 \exp\left(-1/|s(k)|\right) + \beta_2$. Parameters β_1 and β_2 need to satisfy the following conditions in order to ensure a stable sliding motion of the system

$$\beta_1, \beta_2 > 0, \qquad \beta_1 + \beta_2 < 1, \qquad \beta_2 < \beta_1.$$

Values of function $\beta(s(k))$ get smaller as the distance of the sliding variable from the sliding hyperplane decreases. Hence, when sliding variable is far from $s = 0$, value of s in next step is calculated using respectively greater fraction of its previous value than when s is in closer vicinity of switching hyperplane. This property ensures decrease of overshoot and faster convergence of s to the vicinity of zero.

Adopting reaching law approach for this strategy we obtain the control signal of the form

$$u(k) = -\left(\mathbf{c}^\mathrm{T}\boldsymbol{b}\right)^{-1}\left(\mathbf{c}^\mathrm{T}\boldsymbol{A}\boldsymbol{x}(k) - \mathbf{c}^\mathrm{T}\boldsymbol{x}_\mathrm{d} + \beta\left(s(k)\right)s(k) + F_\mathrm{A}\right).$$

The value of QSMB limitation has not been explicitly provided in the paper, however basing on (9) we can conclude that quasi-sliding mode will stay in following vicinity of zero $\left\{\boldsymbol{x} : \left|s\left(\boldsymbol{x}(k)\right)\right| \leq F_\mathrm{D} = \Delta_\mathrm{III}\right\}$.

3.4 Y. Niu, D. W. C. Ho and Z. Wang, 2010

The next reaching law we consider in this paper is the one proposed in [26]. This strategy aims to lead the system representative point to vicinity of sliding surface bounded by a priori known number calculated basing on assumed limitations of disturbance. It aims to achieve this in finite time with no need to adjust any additional parameters. Reaching law describing this method is of the form

$$s(k+1) = s(k) + F_{\mathrm{A}} - F_{\mathrm{D}} \cdot \operatorname{sgn}\big(s(k)\big) - F(k). \tag{10}$$

It does not require any choice of parameters as it contains only ingredients corresponding to previous value of sliding variable and disturbance constraints. Hence the purpose of this method is just to modify natural dynamics of sliding variable against disturbance effecting the plant.

By applying the reaching law approach, we achieve control signal

$$u(k) = -\left(\boldsymbol{c}^{\mathrm{T}}\boldsymbol{b}\right)^{-1}\Big(\boldsymbol{c}^{\mathrm{T}}\boldsymbol{A}\boldsymbol{x}(k) - \boldsymbol{c}^{\mathrm{T}}\boldsymbol{x}_{\mathrm{d}} + s(k) + F_{\mathrm{A}} - F_{\mathrm{D}} \cdot \operatorname{sgn}\big(s(k)\big)\Big).$$

Guaranteed impassable band for quasi-sliding mode can be specified as

$$\left\{\boldsymbol{x} : \big|s\big(\boldsymbol{x}(k)\big)\big| \leq 2F_{\mathrm{D}} = \Delta_{\mathrm{IV}}\right\}.$$

3.5 A. Bartoszewicz and P. Leśniewski, 2015

The last of the examined strategies originated from [14] aims to reduce overshoot and control effort. The strategy is represented by the following reaching law

$$s(k+1) = \big(1 - q(s(k))\big) \cdot s(k) + F_{\mathrm{A}} - F(k), \tag{11}$$

where $q\big(s(k)\big) = s_0/\left(|s(k)| + s_0\right)$ and $s_0 > F_{\mathrm{D}}$. The factor $1 - q\big(s(k)\big)$ acts similarly to $\beta(s(k))$ in strategy described by (9). Function q is designed to ensure gradual driving of the system representative point into sliding surface. Because $1 - q(s(k))$ decreases as $s(k)$ distance to switching hyperplane becomes smaller, the value of sliding variable in next step is specified accordingly, so far position of s as compared to zero causes greater dependence on it, but smaller $|s(k)|$ leads to lesser significance of this ingredient. It supports reducing overshoot and obtaining faster convergence of the sliding variable to the vicinity of zero.

According to the reaching law approach, we derive the following control signal

$$u(k) = -\left(\boldsymbol{c}^{\mathrm{T}}\boldsymbol{b}\right)^{-1}\Big(\boldsymbol{c}^{\mathrm{T}}\boldsymbol{A}\boldsymbol{x}(k) - \boldsymbol{c}^{\mathrm{T}}\boldsymbol{x}_{\mathrm{d}} + \big(1 - q(s(k))\big)s(k) + F_{\mathrm{A}}\Big).$$

QSMB guaranteed by using this control is expressed as

$$\left\{\boldsymbol{x} : \big|s\big(\boldsymbol{x}(k)\big)\big| \leq s_0 F_{\mathrm{D}}/\left(s_0 - F_{\mathrm{D}}\right) = \Delta_{\mathrm{V}}\right\}.$$

The next section is devoted to establishing a relation between the QSMB width and the maximum errors of all state variables.

4 Relation Between QSMB and Maximum State Variables Errors

In this section we will focus on the relation between QSMB and maximum absolute values of individual state errors. This topic has been recently researched in [18,21–23]. We will use notation for closed-loop nominal system matrix introduced before in (3). Vector $\boldsymbol{c}$ is selected to ensure a dead-beat response of the system, which means that all eigenvalues of $\boldsymbol{H}$ are placed in zero. Thereafter we will use definition of reaching law proposed in (4) and the QSMB width obtained for strategy in that form will be denoted by Δ. We further denote vector $\boldsymbol{t}_i = [0\ldots0\,1\,0\ldots0]$, which has length n and 1 is placed on its i-th index.

In order to find the maximum state error, we first derive the formula for state errors vector in subsequent time instant, using (1), (3), (6) and definition $\boldsymbol{e}(k)$

$$\begin{aligned}\boldsymbol{e}(k+1) =& \boldsymbol{x}_{\mathrm{d}} - \boldsymbol{b}(\boldsymbol{c}^{\mathrm{T}}\boldsymbol{b})^{-1}\boldsymbol{c}^{\mathrm{T}}\boldsymbol{x}_{\mathrm{d}} - \boldsymbol{H}\boldsymbol{x}(k) + \boldsymbol{b}(\boldsymbol{c}^{\mathrm{T}}\boldsymbol{b})^{-1}s(k+1)\\ &- (1 - \boldsymbol{b}(\boldsymbol{c}^{\mathrm{T}}\boldsymbol{b})^{-1}\boldsymbol{c}^{\mathrm{T}})\boldsymbol{h}f(k).\end{aligned}$$

We can then calculate values of adequate error vectors for subsequent time instants by applying repeated substitution. Ultimately we achieve generic expression demonstrating the state error in certain ρ-th step as

$$\begin{aligned}\boldsymbol{e}(k+\rho) =& \boldsymbol{x}_{\mathrm{d}} - \boldsymbol{H}^0\boldsymbol{b}(\boldsymbol{c}^{\mathrm{T}}\boldsymbol{b})^{-1}\boldsymbol{c}^{\mathrm{T}}\boldsymbol{x}_{\mathrm{d}} - \boldsymbol{H}^1\boldsymbol{b}(\boldsymbol{c}^{\mathrm{T}}\boldsymbol{b})^{-1}\boldsymbol{c}^{\mathrm{T}}\boldsymbol{x}_{\mathrm{d}} - \ldots\\ &- \boldsymbol{H}^{\rho-1}\boldsymbol{b}(\boldsymbol{c}^{\mathrm{T}}\boldsymbol{b})^{-1}\boldsymbol{c}^{\mathrm{T}}\boldsymbol{x}_{\mathrm{d}} - \boldsymbol{H}^{\rho}\boldsymbol{x}(k) + \boldsymbol{H}^0\boldsymbol{b}(\boldsymbol{c}^{\mathrm{T}}\boldsymbol{b})^{-1}s(k+\rho)\\ &+ \boldsymbol{H}^1\boldsymbol{b}(\boldsymbol{c}^{\mathrm{T}}\boldsymbol{b})^{-1}s(k+\rho-1) + \ldots + \boldsymbol{H}^{\rho-1}\boldsymbol{b}(\boldsymbol{c}^{\mathrm{T}}\boldsymbol{b})^{-1}s(k+1)\\ &- \boldsymbol{H}^0(1 - \boldsymbol{b}(\boldsymbol{c}^{\mathrm{T}}\boldsymbol{b})^{-1}\boldsymbol{c}^{\mathrm{T}})\boldsymbol{h}f(k+\rho-1)\\ &- \boldsymbol{H}^1(1 - \boldsymbol{b}(\boldsymbol{c}^{\mathrm{T}}\boldsymbol{b})^{-1}\boldsymbol{c}^{\mathrm{T}})\boldsymbol{h}f(k+\rho-2)\\ &- \ldots - \boldsymbol{H}^{\rho-1}(1 - \boldsymbol{b}(\boldsymbol{c}^{\mathrm{T}}\boldsymbol{b})^{-1}\boldsymbol{c}^{\mathrm{T}})\boldsymbol{h}f(k).\end{aligned}$$

Vector $\boldsymbol{c}$ is selected so that matrix $\boldsymbol{H}$ is nilpotent, which means that $\boldsymbol{H}^{\rho} = \boldsymbol{0}$ for any $\rho \geq n$. Thus, we can present formula of error in n-th and next steps as

$$\begin{aligned}\boldsymbol{e}(k+\rho) =& \boldsymbol{x}_{\mathrm{d}} - \sum_{j=0}^{n-1} \boldsymbol{H}^j\boldsymbol{b}(\boldsymbol{c}^{\mathrm{T}}\boldsymbol{b})^{-1}\boldsymbol{c}^{\mathrm{T}}\boldsymbol{x}_{\mathrm{d}} + \sum_{j=0}^{n-1} \boldsymbol{H}^j\boldsymbol{b}(\boldsymbol{c}^{\mathrm{T}}\boldsymbol{b})^{-1}s(k+\rho-j)\\ &- \sum_{j=0}^{n-1} \boldsymbol{H}^j(1 - \boldsymbol{b}(\boldsymbol{c}^{\mathrm{T}}\boldsymbol{b})^{-1}\boldsymbol{c}^{\mathrm{T}})\boldsymbol{h}f(k+\rho-1-j).\end{aligned} \tag{12}$$

Our intention is to access every individual element of the state errors vector instead of what we have now. To that end, we multiply vector (12) by $\boldsymbol{t}_i$ to obtain for every $i \in \{1, ..., n\}$ the following expression

$$e_i(k+\rho) = x_{\mathrm{d},i} - w_i + \sum_{j=0}^{n-1} z_{i,j}s(k+\rho-j) - \sum_{j=0}^{n-1} v_{i,j}f(k+\rho-1-j),$$

where $x_{\mathrm{d},i} = \boldsymbol{t}_i \cdot \boldsymbol{x}_{\mathrm{d}}$,

$$\begin{aligned}&w_i = \sum_{j=0}^{n-1} \boldsymbol{t}_i\boldsymbol{H}^j\boldsymbol{b}(\boldsymbol{c}^{\mathrm{T}}\boldsymbol{b})^{-1}\boldsymbol{c}^{\mathrm{T}}\boldsymbol{x}_{\mathrm{d}}, \quad z_{i,j} = \boldsymbol{t}_i\boldsymbol{H}^j\boldsymbol{b}(\boldsymbol{c}^{\mathrm{T}}\boldsymbol{b})^{-1},\\ &\text{and} \quad v_{i,j} = t_i \cdot \boldsymbol{H}^j(1 - \boldsymbol{b}(\boldsymbol{c}^{\mathrm{T}}\boldsymbol{b})^{-1}\boldsymbol{c}^{\mathrm{T}})\boldsymbol{h}.\end{aligned} \tag{13}$$

Ultimately, we can derive the limitations of absolute values of state errors obtained above as

$$e_{i,\max} = |x_{\mathrm{d},i} - w_i| + \Delta \cdot \sum_{j=0}^{n-1} |z_{i,j}| + \max\{|f_{\mathrm{L}}|, |f_{\mathrm{U}}|\} \sum_{j=0}^{n-1} |v_{i,j}|. \tag{14}$$

It is easy to notice, that constants $\sum_{j=0}^{n-1} |z_{i,j}|$ and $\sum_{j=0}^{n-1} |v_{i,j}|$ are positive and hence we can deduce that for every $i = 1, \ldots, n$ maximum state error $e_{i,\max}$ and QSMB limitation Δ are proportional, so increasing one of them results in another one increase. This conclusion provides the way to compare different strategies obtaining practically useful information about state errors deviations with no need to count them separately. Instead, it is enough to calculate QSMBs of the compared strategies.

It is worth to observe that if system satisfies matching condition, which means $\boldsymbol{b} = \boldsymbol{h}$ and is commonly used assumption in literature, then $(1-\boldsymbol{b}(\boldsymbol{c}^{\mathrm{T}}\boldsymbol{b})^{-1}\boldsymbol{c}^{\mathrm{T}})\boldsymbol{h} = \mathbf{0}$ and hence $v_{i,j} = 0$ for any pair of i, j. It reduces (14) to

$$e_{i,\max} = |x_{\mathrm{d},i} - w_i| + \Delta \cdot \sum_{j=0}^{n-1} |z_{i,j}|. \tag{15}$$

Simulation example confirming this property will be examined in next section.

5 Simulation Example

In this section, we will provide simulation example for the purpose of comparison of considered strategies in context of proportional relation between QSMB and state errors. Let us consider system described by following dynamics

$$\dot{\boldsymbol{x}}(t) = \begin{bmatrix} 0 & 1 & 0 \\ 0 & 0 & 1 \\ 0 & -0.5 & 0 \end{bmatrix} \boldsymbol{x}(t) + \begin{bmatrix} 0 \\ 0 \\ 1 \end{bmatrix} u(kT) + \begin{bmatrix} 0 \\ 0 \\ 1 \end{bmatrix} f(t). \tag{16}$$

The disturbance applied to the plant will be characterised as

$$f(k) = \begin{cases} 5 \cdot \lfloor k/5 \rfloor, & k \in [0, 15) \\ -5 \cdot \lfloor (k-20)/5 \rfloor, & k \in [15, 35) \end{cases}.$$

The objective of control in this simulation is to effort system state from its initial location $\boldsymbol{x}(0) = \begin{bmatrix} 100 & -10 & 50 \end{bmatrix}^{\mathrm{T}}$ into the target position $\boldsymbol{x}_{\mathrm{d}} = \begin{bmatrix} 0 & 0 & 0 \end{bmatrix}^{\mathrm{T}}$.

Proceeding discreatization on undisturbed version of the plant (16) we obtain discrete system of the form

$$\boldsymbol{x}(k+1) = \begin{bmatrix} 1 & 0.9187 & 0.4795 \\ 0 & 0.7602 & 0.9187 \\ 0 & -0.4594 & 0.7602 \end{bmatrix} \boldsymbol{x}(k) + \begin{bmatrix} 0.1625 \\ 0.4795 \\ 0.9187 \end{bmatrix} u(k).$$

In accordance to assumption made, let us select vector defining sliding variable as $\boldsymbol{c}^{\mathrm{T}} = \begin{bmatrix} 2.8982 & 2.8981 & 1 \end{bmatrix}$ to satisfy mentioned conditions which guarantee dead-beat performance of the system. The parameters for individual strategies are chosen as follows. For strategy of Sect. 3.1 and its modified version described in Sect. 3.2 we select $k^* = 5$. For control method described in Sect. 3.3 we set parameters $\beta_1 = 0.2$ and $\beta_2 = 0.1$. Reaching law introduced in Sect. 3.4 does not require any parameters setting. Finally, strategy presented in Sect. 3.5 got parameter $s_0 = 1000$. Adopting respective formulas of QSMB provided for each of the strategies in Sect. 3 we calculate their values as follows

$$\Delta_{\mathrm{I}} = 27.7930,\ \Delta_{\mathrm{II}} = 13.8965,\ \Delta_{\mathrm{III}} = 27.7930,\ \Delta_{\mathrm{IV}} = 55.5859,\ \Delta_{\mathrm{V}} = 28.5875.$$

Further, applying determined numbers into (14) we obtain maximum errors of first, second and third state variables respectively guaranteed by individual strategies. Thus, for strategy from Sect. 3.1 these numbers are: 66.5281, 44.1103 and 13.6130. For its modified version placed in Sect. 3.2 they are: 33.2641, 22.0551 and 6.8065. Next strategy, presented in Sect. 3.3 gave maximum errors: 66.5281, 44.1103 and 13.6130. Strategy of Sect. 3.4 returned values: 133.0560, 88.2204 and 27.2260. The last method, which Sect. 3.5 was devoted to, appeared with the following numbers: 68.4299, 45.3712 and 14.0022. Brief analysis of provided data let us notice that following relation $\Delta_{\mathrm{II}} < \Delta_{\mathrm{I}} = \Delta_{\mathrm{III}} < \Delta_{\mathrm{V}} < \Delta_{\mathrm{IV}}$ is indeed adequate to relations between different strategies maximum errors of respective state variables.

The results presented above are also illustrated in Fig. 1 and 2. First plot shows the comparison of control signals in all of these five strategies and Fig. 2 displays courses of first state variable error.

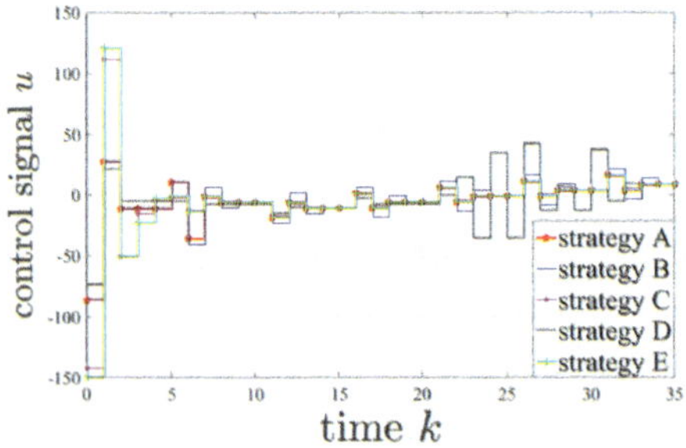

Fig. 1. Control signal

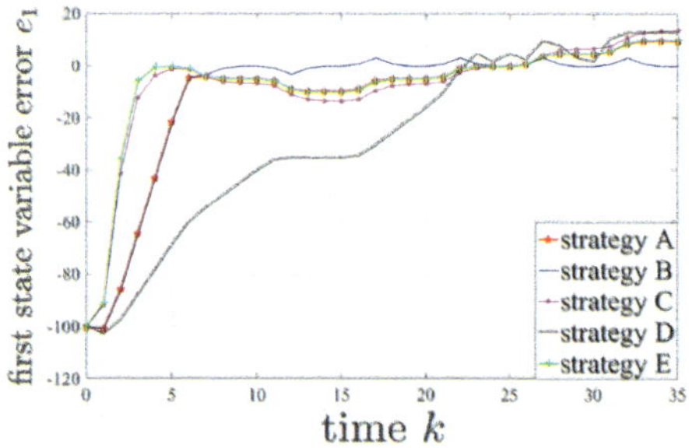

Fig. 2. Control error of first state variable

6 Conclusions

In this paper we examined the relation between quasi-sliding mode band width and maximum errors of individual state variables. It has been shown that the relation is proportional which means that narrower QSMB results in smaller maximum deviations of all state variables from their desired values. Firstly, we described this comparison method in theoretical manner. Afterwards, we provided simulation example presenting discussed theory in practical application. For that end, we considered five sliding mode strategies based on non-switching reaching laws, applied to digitized plant with external disturbance.

References

1. Emelyanov, S.V.: Variable structure Control Systems. Nauka, Moscow (1967)
2. Draženović, B.: The invariance conditions in variable structure systems. Automatica **5**(3), 287–295 (1969)
3. Utkin, V.: Variable structure systems with sliding modes. IEEE Trans. Autom. Control **22**(2), 212–222 (1977)
4. Hung, J.Y., Gao, W., Hung, J.C.: Variable structure control: a survey. IEEE Trans. Ind. Electron. **40**(1), 2–22 (1993)
5. Spurgeon, S.K.: Sliding mode observers: a survey. Int. J. Syst. Sci. **39**(8), 751–764 (2008)
6. Latosiński, P: Sliding mode control based on the reaching law approach - a brief survey. In: 22nd International Conference on Methods and Models in Automation and Robotics (MMAR), pp. 519–524 (2017)
7. Milosavljević, D.: General conditions for the existence of a quasi-sliding mode on the switching hyperplane in discrete variable structure systems. Automat. Remote Contr. **46**(3), 307–314 (1985)
8. Utkin, V.I., Drakunov, S.V.: On discrete-time sliding modes. IFAC Proc. **22**(3), 273–278 (1989)
9. Furuta, K.: Sliding mode control of a discrete system. Syst. Contr. Lett. **14**(2), 145–152 (1990)
10. Spurgeon, S.K.: Hyperplane design techniques for discrete-time variable structure control systems. Int. J. Control **55**(2), 445–456 (1992)
11. Su, W.-C., Drakunov, S.V., Özguner, Ü.: An $O(T^2)$ boundary layer in sliding mode for sampled-data systems. IEEE Trans. Autom. Control **45**(3), 482–485 (2000)
12. Gao, W., Wang, Y., Homaifa, A.: Discrete-time variable structure control systems. IEEE Trans. Ind. Electron. **42**(2), 117–122 (1995)
13. Yuan, L., Shen, J., Xiao, F., Wang, H.: A novel reaching law approach of quasi-sliding-mode control for uncertain discrete-time systems. J. Cent. South Univ. **19**, 2514–2519 (2012). https://doi.org/10.1007/s11771-012-1304-y
14. Bartoszewicz, A., Leśniewski, P.: New switching and nonswitching type reaching laws for SMC of discrete time systems. IEEE Trans. Control Syst. Technol. **24**, 670–677 (2015)
15. Chakrabarty, S., Bartoszewicz, A.: Improved robustness and performance of discrete time sliding mode control systems. ISA Trans. **65**, 143–149 (2016)
16. Ma, H., Wu, J., Xiong, Z.: Discrete-time sliding-mode control with improved quasi-sliding-mode domain. IEEE Trans. Ind. Electron. **63**, 6292–6304 (2016)
17. Latosiński, P., Bartoszewicz, A.: Discrete time sliding mode controllers with relative degree one and two switching variables. J. Franklin Inst. **355**(15), 6889–6903 (2018)
18. Bartoszewicz, A., Latosiński, P.: Generalization of Gao's reaching law for higher relative degree sliding variables. IEEE Trans. Autom. Control **63**(9), 3173–3179 (2018)
19. Bartoszewicz, A., Latosiński, P.: Discrete time sliding mode control with reduced switching - a new reaching law approach. Int. J. Robust Nonlinear Control **26**, 47–68 (2014)

20. Leśniewski, P., Bartoszewicz, A.: Inverse tangent based switching type reaching law for discrete time sliding mode control systems. In: European Control Conference, pp. 2390–2395 (2015)
21. Bartoszewicz, A., Adamiak, K.: Model reference discrete-time variable structure control. Int. J. Adapt. Control Signal Process. **32**, 1440–1452 (2018)
22. Bartoszewicz, A., Adamiak, K.: A reference trajectory based discrete time sliding mode control strategy. Int. J. Appl. Math. Comput. Sci. **29**(3), 517–525 (2019)
23. Bartoszewicz, A., Adamiak, K.: Discrete time sliding mode control with a desired switching variable generator. IEEE Trans. Autom. Control **65**(4), 1807–1814 (2020)
24. Bartoszewicz, A.: Discrete time quasi-sliding mode control strategies. IEEE Trans. Ind. Electron. **45**(4), 633–637 (1998)
25. Kurode, S., Bandyopadhyay, B. Gandhi, P.: Discrete sliding mode control for a class of underactuated systems. In: Annual Conference of the IEEE Industrial Electronics Society, pp. 3936–3941 (2011)
26. Niu, Y., Ho, D.W.C., Wang, Z.: Improved sliding mode control for discrete-time systems via reaching law. IET Control Theory Appl. **4**(11), 2245–2251 (2010)
27. Sarpturk, S.Z., Istefanopulos, Y., Kaynak, O.: On the stability of discrete-time sliding mode control systems. IEEE Trans. Autom. Control **AC–32**(10), 930–932 (1987)
28. Furuta, K., Kim, S.B.: Pole assignment in a specifies disk. IEEE Trans. Autom. Control **AC–32**(5), 423–427 (1987)
29. Veselić, B., Peruničić-Draženović, B., Milosavljević, Č.: Improved discrete-time sliding-mode position control using Euler velocity estimation. IEEE Trans. Ind. Electron. **57**(11), 3840–3847 (2010)
30. Bartoszewicz, A.: A new reaching law for sliding mode control of continuous time systems with constraints. Trans. Inst. Measur. Control **37**(4), 515–521 (2015)
31. Gao, W., Hung, J.C.: Variable structure control of nonlinear systems: a new approach. IEEE Trans. Ind. Electron. **40**(1), 45–55 (1993)
32. Janardhanan, S., Kariwala, V.: Multirate output feedback based LQ optimal discrete-time sliding mode control. IEEE Trans. Autom. Control **53**(1), 367–373 (2008)
33. Bartoszewicz, A., Żuk, J.: Discrete time sliding mode flow controller for multi-source single-bottleneck connection-oriented communication networks. J. Vibrat. Control **15**(11), 1745–1760 (2009)

Intelligent Systems

Different Approaches to Threshold Estimation for Variation Mode Decomposition-Based Signal Denoising

Krzysztof Brzostowski(✉) and Jerzy Świątek

Faculty of Computer Science and Management, Wrocław University of Science and Technology, Wyb. Wyspiańskiego 27, 50-370 Wrocław, Poland
Krzysztof.Brzostowski@pwr.edu.pl, Jerzy.Swiatek@pwr.edu.pl

Abstract. In the paper, the problem of threshold estimation to the needs of the thresholding operators-based denoising algorithm is considered. The typical approach to design the Variational Mode Decomposition-based signal denoising algorithm uses thresholding operators. The fundamental problem of applying thresholding operators is the problem of threshold estimation. In the paper we investigate the performance of three threshold value estimators, i.e, universal threshold estimator, SURE Shrink estimator, and Bayes Shrink estimator.

The performance of the methods combining Variational Mode Decomposition algorithm with soft and hard thresholding in which the above-mentioned estimators are used was tested on four synthetic signals, i.e. Bumps, Blocks, Doppler, and Piece-Regular. The obtained results show that for almost all cases applying Bayes Shrink estimator allow designing algorithm predominate over methods.

Keywords: Non-linear signals processing · Mode decomposition · Thresholding

1 Introduction

Due to the physical limitations of sensing devices, observed signals become prone to noise during its acquisition. Noise is a kind of distortion that hinders the process of signals observation. Suppression of noise in measurements forms an elementary basis in areas of signal analysis and processing.

The typical approach to suppress unwanted components in measurements signal denoising based on filters. Unfortunately, filter-based methods have some limitations. One of them is the inability to separate signal and noise in the same frequency band. Moreover, the typical filter methods have poor performance in the case of non-stationary and short-term transient signals.

An alternative to filter methods are approach based on wavelet transform. Generally, the wavelet-based denoising algorithm is better comparing to the filter-based approach. Unfortunately, the method is sensitive to periodic noise

H. Selvaraj et al. (Eds.): ICSEng 2020, LNNS 182, pp. 95–104, 2021.
https://doi.org/10.1007/978-3-030-65796-3_9

and it is difficult to select the exact value of threshold [1]. Moreover, the wavelet-based denoising algorithm relying on the predefined basis functions.

In [11] the authors proposed a method to signal decomposition named Empirical Mode Decomposition (EMD). EMD is an approach to signal decomposition, in which a signal is decomposed into a set of simple and intrinsic signals (IMF) without the need for a priori system information (unlike the wavelet-based approach). The filter bank properties of Empirical Mode Decomposition allow us to design an effective denoising algorithm.

EMD method is widely used for analyzing non-stationary signals. However, there are many problems to use EMD. One of them is the problem of mode mixing, i.e. the existence of different time scales in a single IMF. Another is that Empirical Mode Decomposition still has not to sound mathematical theory.

Variational Mode Decomposition (VMD) is an adaptive and non-recursive signal analysis method [7]. Compared to EMD, the VMD approach has a strong mathematical background. This helps to improve the accuracy of signal separation and noise robustness [10]. Moreover, Variational Mode Decomposition effectively eliminates the effect of mode mixing during the decomposition process [18]. The signal determined by applying VMD is named mode.

Recently, in the paper [13] the authors propose an approach to signal denoising combining VMD with thresholding operators. The general idea of the proposed approach is to decompose the signal by applying the VMD method. Then, the thresholding operator is applied to each obtained mode. The denoised signal is estimated by summing up the thresholded modes.

One of the fundamental problems of thresholding techniques is the problem of estimating the optimal value of the threshold. In this paper, we investigate three different approaches to determine the value of the threshold, i.e., a universal threshold, a SURE Shrink threshold, and a Bayes Shrink threshold.

2 Basic Ingredients

2.1 Variational Mode Decomposition

VMD is a multi-resolution analysis method to extract from signal a number of modes and their central frequencies. In general, the extraction procedure bases on iterative scheme to find the optimal solution of the variational model of each finite bandwidth in which the components are adaptively separated according to their respective centre frequencies [7]. The objective function and the constrained variational problem has the form:

$$\begin{aligned} \{u_k^*\}, \{\omega_k^*\} &= \min_{\{u_k\},\{\omega_k\}} \left\{ \sum_{k=1}^{K} \| \frac{\partial}{\partial t} \left[\left(\delta(t) + \frac{j}{\pi t} \right) * u_k(t) \right] \exp^{-j\omega_k t} \|_2^2 \right\}, \\ & s.t. \sum_{k=1}^{K} u_k(t) = y(t) \end{aligned} \tag{1}$$

where t is the time index, $y(t)$ stands for the signal to be decomposed, $\delta(t)$ is the Dirac distribution, u_k is shorthand notations for k-th IMF from the set of all IMFs, i.e. $u_K = \{u_1, u_2, \ldots, u_K\}$ and ω_k stands for k-th centre frequencies of k-th IMF, i.e. $\omega_K = \{\omega_1, \omega_2, \ldots, \omega_K\}$. The symbol $*$ denotes convolution.

By introducing the Lagrangian multiplier λ and the second penalty factor α (data-fidelity constraint parameter), the constrained variational problem (1) can be transformed into an unconstrained variational problem:

$$\begin{aligned} L(\{u_k\},\{\omega_k\},\lambda) =& \alpha \sum_{k=1}^{K}\left\{\sum_{k=1}^{K} \| \frac{\partial}{\partial t}\left[\left(\delta(t)+\frac{j}{\pi t}\right) * u_k(t)\right] \exp^{-j\omega_k t} \|_2^2 \right\} \\ &+ \|y(t) - \sum_{k=1}^{K} u_k(t)\|_2^2 + \left\langle \lambda(t), y(t) - \sum_{k=1}^{K} u_k(t) \right\rangle \end{aligned} \tag{2}$$

In order to solve the unconstrained optimization problem, Eq. (2) can be split into two different sub-problems, i.e, minimization of $u_k(t)$ and $\omega_k(t)$. Division of the problem (2) leads us to:

$$u_k^{(n+1)} \leftarrow \arg\min_{u_k} L(u_{i<k}^{(n+1)}, u_{i\geqslant k}^{(n)}, \omega_i^{(n)}, \lambda^{(n)}) \tag{3a}$$

$$\omega_k^{(n+1)} \leftarrow \arg\min_{\omega_k} L(u_i^{(n+1)}, \omega_{i<k}^{(n+1)}, \omega_{i\geqslant k}^{(n)}, \lambda^{(n)}) \tag{3b}$$

where $k = 1, 2, \ldots, K$.

2.2 Sparsity-Based Signal Denoising

In our investigation we assume the following discrete model of measured signal $y(n)$:

$$y(n) = s(n) + \xi(n), \quad n = 1, 2, \ldots, N \tag{4}$$

where $s(n)$ is the noise-free signal and $\xi(n)$ denotes the additive white Gaussian noise.

For the sake of simplicity, the components of model (4) in the rest of the paper will be represented as vectors in $\mathbb{R}^N$, i.e., $\mathbf{y} \in \mathbb{R}^N$, $\mathbf{s} \in \mathbb{R}^N$, and $\boldsymbol{\xi} \in \mathbb{R}^N$.

Based on (4), we formulate the following unconstrained optimization problem:

$$\tilde{\mathbf{s}} = \arg\min_{\mathbf{s}}\{F(\mathbf{s}) = \frac{1}{2}||\mathbf{y} - \mathbf{s}||_2^2 + \eta\phi(\mathbf{s})\}, \tag{5}$$

where $\phi(\mathbf{s}) : \mathbb{R}^N \rightarrow \mathbb{R}$ is the regularization term, and $\eta > 0$ is the regularization parameter.

The introduced optimization problem (5) can be reformulated as convex or non-convex. The convex approaches are based on sparsity-promoting convex penalty functions (e.g., ℓ_2 norm), while non-convex approaches are based on non-convex regularizers (e.g., ℓ_p pseudo norm with $p < 1$). Typical formulation of optimization problem (5) with ℓ_2 norm leads to Tikhonov regularization [2,16].

Instead of the norm ℓ_2, in the optimization task (5), we can use ℓ_0 [20]. For this norm, the problem (5) is transformed into ℓ_0-norm optimization task

$$\tilde{\mathbf{s}} = \arg\min_{\mathbf{s}}\{F(\mathbf{s}) = \frac{1}{2}||\mathbf{y} - \mathbf{s}||_2^2 + \eta||\mathbf{s}||_0\}. \tag{6}$$

To find the approximation of solution for the introduced optimization problem (6), the following operator can be applied [14]

$$\tilde{\mathbf{s}} = \text{hard}(\mathbf{s}, \eta) = \begin{cases} \mathbf{s} & |\mathbf{s}| > \eta \\ 0 & |\mathbf{s}| \leq \eta. \end{cases} \tag{7}$$

Equation (7) is called the *hard thresholding operator.*

Moreover, instead of ℓ_0 we can use ℓ_1 norm. This implies the following penalty function $\phi(\mathbf{s}) = ||\mathbf{s}||_1$. Based on new penalty term the optimization problem (5) has the form

$$\tilde{\mathbf{s}} = \arg\min_{\mathbf{s}}\{F(\mathbf{s}) = \frac{1}{2}||\mathbf{y} - \mathbf{s}||_2^2 + \eta||\mathbf{s}||_1\}. \tag{8}$$

The solution of the problem (8) can be determined by [14]

$$\tilde{\mathbf{s}} = \text{soft}(\mathbf{s}, \eta) = \begin{cases} \text{sgn}\,(\mathbf{s})\,(|\mathbf{s}| - \eta) & |\mathbf{s}| > \eta \\ 0 & |\mathbf{s}| \leq \eta, \end{cases} \tag{9}$$

where $||\cdot||_1$ represents ℓ_0 norm of $\mathbf{s}$.

The function $\text{soft}(\mathbf{s}, \eta)$ is called the *soft thresholding operator.*

2.3 VMD-Based Signal Denoising with Thresholding Operator

As it was shown in [13], the hard and soft thresholding technique can be used to solve the problem of signal denoising in combination with Variational Mode Decomposition [10]. In this work, authors proposed an approach applying hard and soft threshold operators to modified VMD-related modes:

$$\tilde{u}_k = \text{hard}(\hat{u}_k, \eta_k) = \begin{cases} \hat{u}_k & |\hat{u}_k| > \eta_k \\ 0 & |\hat{u}_k| \leq \eta_k. \end{cases} \tag{10}$$

and

$$\tilde{u}_k = \text{soft}(\hat{u}_k, \eta_k) = \begin{cases} \text{sgn}\,(\hat{u}_k)\,(|\hat{u}_k| - \eta_k) & |\hat{u}_k| > \eta_k \\ 0 & |\hat{u}_k| \leq \eta_k, \end{cases} \tag{11}$$

where $k = 1, 2, \ldots, K$ is the k-th IMF. η_k denotes the threshold of the k-th IMF.

In hard thresholding, each value of k-th mode is compared against the threshold value and lower value is replaced by zero. It is worth mentioning that the hard threshold can lead to abrupt changes in processed signals. Moreover, it generates artifacts in the denoised signal. In turn, soft thresholding-based signal denoising tends to smooth the signal.

Both, for hard and soft thresholding one of the fundamental problems is the problem of determining the optimal value of the threshold η (see Eqs. 10, 11).

2.4 Threshold Estimation

One of the most difficult problems in applying thresholding operators to the Variational Mode Decomposition-based denoising algorithm is to find out the optimal value of the threshold η_k for each k-th determined mode. On the one hand, a small value of threshold can keep the data related to the noise and it results in a signal which is still noisy. On the other hand, when the threshold has a large value it may lead to shrinking the maximum portion of data. It can cause the losing of important components of the signal. In the literature, some methods to estimate the value of thresholds for thresholding operators are proposed. In this paper we take into consideration three popular approaches, i.e., a universal threshold [6,12,19] a SURE Shrink threshold [4,5], and a Bayes Shrink threshold [3,9] and applied to the VMD-based denoising algorithm.

The estimation of universal threshold based on the following formula

$$\eta_k = \sigma_k \sqrt{2 \ln N} \tag{12}$$

where N is the number of data points (see Eq. 4), while σ_k is standard deviation of noise in k-th mode.

In practical settings, the formula (12) can be approximated with [19]:

$$\eta_k = E_k \sqrt{2 \ln N} \tag{13}$$

where E_k is determined from

$$E_1 = \frac{\text{median}(|\hat{u}_1(n)|)}{0.6745} \tag{14a}$$

$$E_k = \frac{E_1^2}{\gamma} \rho^{-k}, \quad k = 2, 3, \ldots, K \tag{14b}$$

where n represents n-th data point and ρ and γ are parameters which values have to be estimated. In [8] the authors proposed $\gamma = 0.719$ and $\rho = 2.01$.

Originally, SURE Shrink was proposed for the wavelet-based method. It is an adaptive approach to estimate the threshold where its value is estimated based on Stein's Unbiased Risk Estimator (SURE). The proper formula is given as the following optimization task

$$\eta_k = \arg\min_{\mu \geq 0} SURE(\mu_k, \hat{u}_k) \tag{15}$$

where μ_k is mean value of each k-th mode. The $SURE$ function has the form

$$SURE(\mu, \hat{u}_k) = K - 2\{k : |\hat{u}_k| \leq \mu_k\} + \sum_{k=1}^{K} \min(|\hat{u}_k|, \mu_k)^2 \tag{16}$$

$\hat{u}_k$ is the k-th mode and K is the number of modes. As pointed out in [5], when the modes are not sparse, this thresholding method is applied. Otherwise the universal threshold can be used.

Bayes Shrink is another approach to adaptive data driven method for threshold estimation. This method assumes that the noise is spreaded as a generalized Gaussian distribution in each mode. It also finds a threshold which minimizes the Bayesian risk. This empirical approach to estimate threshold is determined applying following formula

$$\eta_k = \frac{\sigma_n^2}{\sigma_y} = \frac{\sigma_n^2}{\sqrt{\max(\frac{1}{K}\sum_{k=1}^{K} u_k^2 - \sigma_n^2, 0)}} \tag{17}$$

where σ_n is the standard deviation of noise, σ_y represents the standard deviation of signal, and K is the number of modes. This method adapts signal to noise ratio in each mode and it uses a robust estimator of noise variance as median absolute value of the modes. The noise variance is estimated similarly to 14a.

3 Experiments

To quantitatively evaluate the denoising performance of the proposed method, some typical signals are generated. To simulate real-world signals, we added to the synthetic data noise, then we applied the proposed approach and some reference methods for comparison purposes. The performance of the proposed method, in the presented studies, is evaluated based on the signal-to-noise-ratio

$$SNR = 10\log_{10}\left(\frac{\sum_{n=1}^{N} s(n)}{\sum_{n=1}^{N}(s(n) - \tilde{y}(n))^2}\right) \tag{18}$$

where N is the number of data points, $s(n)$ is the noise-free (original) signal and $\tilde{y}(n)$ stands for denoised signal.

3.1 Synthetic Data Analysis

In this study we test the performance of the proposed method based on simulated signals (i.e. noise-corrupted synthetic signals). We consider Bumps (see Fig. 1), Blocks (Fig. 2), Doppler (Fig. 3), and Piece-Regular (Fig. 4) signals of length 4096 samples. The signals are generated by the *Wavelab* function *MakeSignal* [15].

The generated synthetic signals have different SNR (0, 5, 10, and 15 [dB]) and contaminated by white Gaussian noise.

The proposed VMD-based denoising methods with the different algorithms to threshold estimation were compared with the Non-Local Means algorithm (NLM) and two wavelet-based denoising methods. The first wavelet-based method is based on a hard threshold operator (DWT-HT), while the second one based on soft thresholding (DWT-ST).

Variational Mode Decomposition-based signal denoising with the use of hard thresholding and universal threshold estimator we named VMD-HUT. The algorithm based on hard thresholding and SURE estimator is named VMD-HST. In

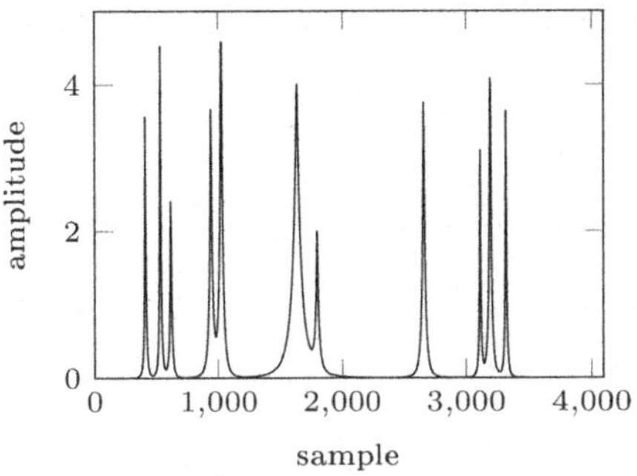

Fig. 1. The clear Bumps signal

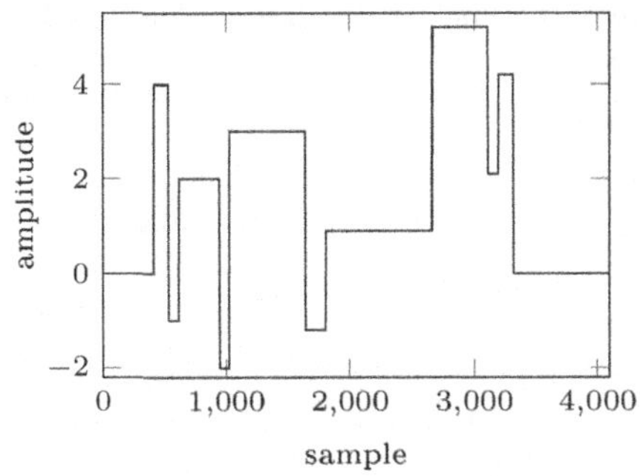

Fig. 2. The clear Blocks signal

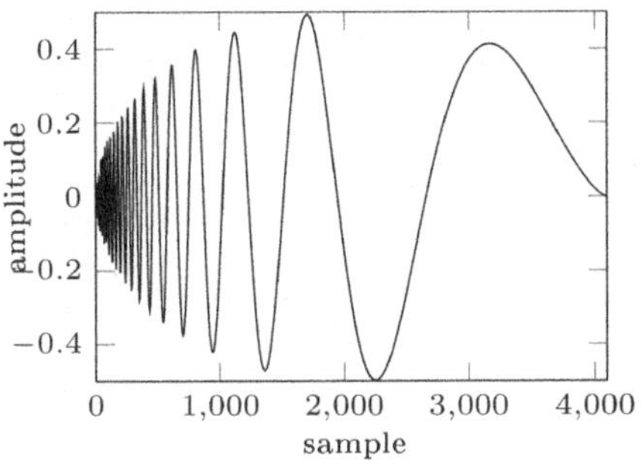

Fig. 3. The clear Doppler signal

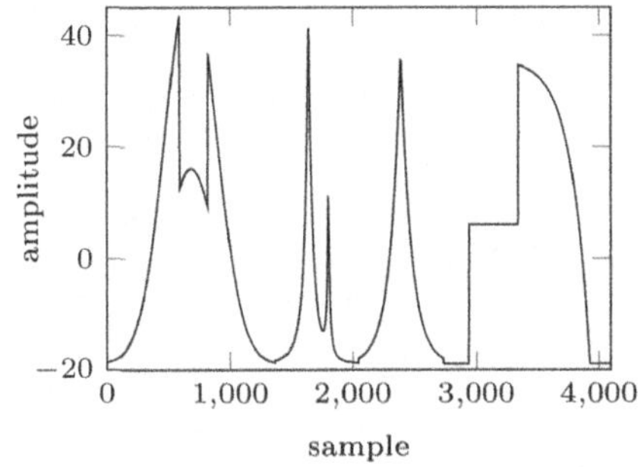

Fig. 4. The clear Piece-Regular signal

turn, an approach based on hard threshold and Bayes Shrink we called VMD-HBT. Similarly, the nomenclature is proposed for VMD-algorithm combined with soft threshold and different threshold estimators: VMD-SUT (universal threshold estimator), VMD-SST (based on SURE Shrink) and VMD-SBT (applying Bayes Shrink).

In our studies, wavelet-based approaches (DWT-HT and DWT-ST) utilized Matlab's function *wden*. While Non-Local Means algorithm implementation based on [17]. NLM algorithm's parameters, i.e., the *bandwidth*, the *patch half-width*, and the *neighbourhood half-width* are determined empirically [17].

To verify the effectiveness of the proposed method, we carried out 30 trials for noisy synthetic signals.

3.2 Simulation Experiments

The results of numerical experiments are presented in Table 1, 2, 3, 4. We tested Bumps, Blocks, Doppler, and Piece Regular signals (see Fig. 1, 2, 3, 4) with SNR varying from 0 to 15 [dB].

Table 1. The results of Doppler signal denoising

		SNR			
		0	5	10	15
NLM	mean	8,799	13,186	18,992	23,653
	std	0,143	0,009	0,275	0,278
DWT-HT	mean	8,973	13,514	18,752	23,485
	std	0,346	0,207	0,263	0,254
DWT-ST	mean	8,973	13,514	18,763	23,321
	std	0,346	0,207	0,265	0,238
VMD-HUT	mean	10,711	13,363	**20,516**	**25,899**
	std	0,445	0,053	0,378	0,333
VMD-HST	mean	8,852	12,964	18,789	23,989
	std	0,391	0,038	0,286	0,276
VMD-HBT	mean	12,141	**14,496**	20,443	24,944
	std	0,469	0,044	0,444	0,272
VMD-SUT	mean	10,092	13,753	19,461	23,587
	std	0,302	0,135	0,282	0,278
VMD-SST	mean	8,524	12,972	18,491	23,613
	std	0,301	0,079	0,246	0,251
VMD-SBT	mean	**12,354**	14,404	20,239	24,572
	std	0,435	0,060	0,395	0,334

Table 2. The results of Bumps signal denoising

		SNR			
		0	5	10	15
NLM	mean	11,185	15,056	18,120	23,202
	std	0,513	0,301	0,239	0,221
DWT-HT	mean	8,657	13,219	17,730	22,320
	std	0,321	0,250	0,256	0,249
DWT-ST	mean	8,659	13,105	17,158	21,005
	std	0,322	0,228	0,187	0,168
VMD-HUT	mean	10,553	**15,353**	**20,454**	23,251
	std	0,210	0,246	0,420	0,639
VMD-HST	mean	**13,099**	15,065	20,389	22,978
	std	0,251	0,250	0,328	0,657
VMD-HBT	mean	10,474	13,987	19,143	**23,986**
	std	0,222	0,220	0,562	0,317
VMD-SUT	mean	10,652	14,921	19,900	22,758
	std	0,198	0,231	0,428	0,553
VMD-SST	mean	12,458	15,238	19,492	22,557
	std	0,240	0,256	0,340	0,605
VMD-SBT	mean	11,222	14,360	19,650	23,576
	std	0,093	0,227	0,327	0,429

Table 3. The results of Blocks signal denoising

		SNR			
		0	5	10	15
NLM	mean	4,094	15,099	19,290	24,398
	std	0,031	0,275	0,256	0,507
DWT-HT	mean	8,811	13,446	17,647	22,227
	std	0,333	0,249	0,228	0,251
DWT-ST	mean	8,811	13,461	17,508	21,182
	std	0,333	0,247	0,201	0,162
VMD-HUT	mean	10,147	15,599	18,865	19,760
	std	0,524	0,556	0,300	0,306
VMD-HST	mean	10,303	15,597	**19,417**	20,756
	std	0,470	0,451	0,230	0,151
VMD-HBT	mean	12,451	**16,473**	19,382	**20,953**
	std	0,960	0,368	0,300	0,112
VMD-SUT	mean	10,886	15,345	18,868	20,651
	std	0,410	0,395	0,187	0,144
VMD-SST	mean	10,824	15,684	19,197	20,701
	std	0,453	0,329	0,188	0,105
VMD-SBT	mean	**12,694**	16,425	19,366	20,887
	std	0,770	0,543	0,245	0,135

Table 4. The results of Piece-Regular signal denoising

		SNR			
		0	5	10	15
NLM	mean	0,008	5,001	10,002	15,004
	std	0,023	0,002	0,002	0,002
DWT-HT	mean	8,906	13,793	18,403	23,084
	std	0,274	0,271	0,250	0,304
DWT-ST	mean	8,916	13,808	18,374	22,620
	std	0,272	0,266	0,246	0,256
VMD-HUT	mean	9,284	14,830	19,278	23,083
	std	0,356	0,348	0,300	0,208
VMD-HST	mean	8,863	14,429	19,236	23,040
	std	0,359	0,268	0,293	0,196
VMD-HBT	mean	11,148	18,334	20,767	23,608
	std	0,789	0,424	0,352	0,277
VMD-SUT	mean	9,874	15,063	19,483	23,034
	std	0,371	0,278	0,247	0,189
VMD-SST	mean	8,877	14,477	19,386	23,227
	std	0,361	0,272	0,296	0,193
VMD-SBT	mean	**11,643**	**18,339**	**21,414**	**23,711**
	std	0,594	0,378	0,321	0,280

As can be seen, for the Piece-Regular signal the best results were obtained from the algorithm in which Bayes Shrink method to estimate threshold value is combined with a soft threshold operator, i.e., VMD-SBT. For the Block signal, the outcomes are not straightforward. However, we can see that the methods combining Bayes Shrink with soft and hard thresholding predominate over other methods. In this case, in almost all cases VMD-HBT and VMD-SBT give the best results of signal denoising.

In the case of the Doppler signal, for high SNR the best results are obtained from the method in which a hard threshold is combined with a universal threshold estimator. For the signal with low SNR again the methods combining Bayes Shrink with soft and hard threshold give the best results.

In the case of the high value of input SNR (SNR = 15 [dB]) the best results we obtain from VMD-HBT. However, for the SNR = 5 [dB] and 10 [dB] the best results give the method combining universal threshold estimator with hard thresholding operator. It is worth mentioning, that for the lowest SNR the best results are obtained from VMD-HST (algorithm in which SURE estimator is combined with hard thresholding operator).

4 Conclusions

In the paper, we design and investigate new algorithms to signal denoising which are based on Variational Mode Decomposition. The typical approach to VMD-based signal denoising applies a hard and soft thresholding operator. One of the fundamental problems in applying these operators is the problem of threshold estimation. In the work we investigate three popular methods to estimate the value of the threshold: universal threshold estimator, SURE Shrink, and Bayes Shrink.

The performance of the methods was tested on four synthetic signals disturbed by white Gaussian noise. To measure the quality of signal denoising we use SNR.

The results of our experiments show that Bayes Shrink estimator improves significantly the performance of the algorithm based on hard and soft thresholding. It is clear for Piece-Regular, Blocks, and Doppler signals. For Bumps and Doppler signal meaningful outcomes are noticeable as well. In turn, the worst results in almost all cases are obtained for methods in which hard or soft operators are combined with the SURE Shrink estimator.

The obtained results clearly show that the selection of threshold estimators is crucial elements of the designing process of denoising algorithms based on VMD and thresholding operators. The further direction of work is the investigation on signal disturbed by noise differs from white noise. It is important since in the most real-world data the noise has no characteristic of white noise.

References

1. Alfaouri, M., Daqrouq, K.: ECG signal denoising by wavelet transform thresholding. Am. J. Appl. Sci. **5**(3), 276–281 (2008)

2. Boyd, S., Boyd, S.P., Vandenberghe, L.: Convex Optimization. Cambridge University Press, Cambridge (2004)
3. Chipman, H.A., Kolaczyk, E.D., McCulloch, R.E.: Adaptive bayesian wavelet shrinkage. J. Am. Stat. Assoc. **92**(440), 1413–1421 (1997)
4. Dengwen, Z., Wengang, C.: Image denoising with an optimal threshold and neighbouring window. Pattern Recogn. Lett. **29**(11), 1694–1697 (2008)
5. Donoho, D.L., Johnstone, I.M.: Adapting to unknown smoothness via wavelet shrinkage. J. Am. Stat. Assoc. **90**(432), 1200–1224 (1995)
6. Donoho, D.L., Johnstone, J.M.: Ideal spatial adaptation by wavelet shrinkage. Biometrika **81**(3), 425–455 (1994)
7. Dragomiretskiy, K., Zosso, D.: Variational mode decomposition. IEEE Trans. Signal Process. **62**(3), 531–544 (2013)
8. Flandrin, P., Gonçalves, P., Rilling, G.: EMD equivalent filter banks, from interpretation to applications. In: Hilbert-Huang Transform and its Applications, pp. 57–74. World Scientific (2005)
9. Hashemi, M., Beheshti, S.: Adaptive noise variance estimation in Bayesshrink. IEEE Signal Process. Lett. **17**(1), 12–15 (2009)
10. Hu, H., Zhang, L., Yan, H., Bai, Y., Wang, P.: Denoising and baseline drift removal method of MEMS hydrophone signal based on VMD and wavelet threshold processing. IEEE Access **7**, 59913–59922 (2019)
11. Huang, N.E., Shen, Z., Long, S.R., Wu, M.C., Shih, H.H., Zheng, Q., Yen, N.C., Tung, C.C., Liu, H.H.: The empirical mode decomposition and the Hilbert spectrum for nonlinear and non-stationary time series analysis. Proc. Roy. Soc. Lond. Ser. A Math. Phys. Eng. Sci. **454**(1971), 903–995 (1998)
12. Kopsinis, Y., McLaughlin, S.: Development of EMD-based denoising methods inspired by wavelet thresholding. IEEE Trans. Signal Process. **57**(4), 1351–1362 (2009)
13. Lahmiri, S.: Comparative study of ECG signal denoising by wavelet thresholding in empirical and variational mode decomposition domains. Healthc. Technol. Lett. **1**(3), 104–109 (2014)
14. Rish, I., Grabarnik, G.: Sparse Modeling: Theory, Algorithms, and Applications. CRC Press, Boca Raton (2014)
15. Wavelab 850. http://statweb.stanford.edu/~wavelab/. Accessed 1 May 2020
16. Tikhonov, A.N., Arsenin, V.I.: Solutions of Ill-posed Problems, vol. 14. Winston, Washington, DC (1977)
17. Tracey, B.H., Miller, E.L.: Nonlocal means denoising of ECG signals. IEEE Trans. Biomed. Eng. **59**(9), 2383–2386 (2012)
18. Xiao, Q., Li, J., Sun, J., Feng, H., Jin, S.: Natural-gas pipeline leak location using variational mode decomposition analysis and cross-time-frequency spectrum. Measurement **124**, 163–172 (2018)
19. Yang, G., Liu, Y., Wang, Y., Zhu, Z.: EMD interval thresholding denoising based on similarity measure to select relevant modes. Sig. Process. **109**, 95–109 (2015)
20. Yang, Z., Ling, B.W.K., Bingham, C.: Joint empirical mode decomposition and sparse binary programming for underlying trend extraction. IEEE Trans. Instrum. Meas. **62**(10), 2673–2682 (2013)

Multi-criteria Optimization of Freight Parking Planning Along a Single Corridor

Grzegorz Filcek(✉) and Maciej Hojda

Faculty of Computer Science and Management, Wrocław University of Science and Technology, Wyb. Wyspianskiego 27, 50-370 Wrocław, Poland
{grzegorz.filcek,maciej.hojda}@pwr.edu.pl

Abstract. Freight transportation across the European Union suffers from severe losses caused by increased cargo thefts, especially at parking stops. Inclusion of safety among the criteria considered in transportation planning processes aims at limiting the scope of those losses. Further inclusion of driver-specific criteria such as parking comfort can increase the quality of drivers' rest, increase the work efficiency, and, in the long term, reduce the turnover rate. In this paper, a multicriteria optimization problem is formulated in order to reconcile the multiple opposing facets of freight parking planning. The problems is to select, from a predetermined set, parking spots along a route while ensuring that working hours regulations are satisfied. The problem is solved with the use of a Branch and Bound approach which is tested empirically for various parking options.

Keywords: Multi-criteria decision-making · Freight transportation · Freight route planning · Parking area safety

1 Introduction

Majority of goods transportation is done on the surface, where freight transportation plays a critical role [1,13]. In recent years it became apparent that it is undesirable to limit freight optimization to readily apparent criteria such as cost of transportation, distance of travel or travel duration. Omitting basic quality-of-life requirements, including driver comfort and driver safety, results in increased turnover rate and cargo loss [10]. This problem remains a substantial issue even at the end of the decade, especially in the European Union (EU) [13].

A recent study [13] reports that in the EU alone, direct losses from cargo thefts in 2019 were estimated to exceed 8 billion Euro. As robberies, hijacks and human trafficking become more prominent, especially in parking situations, it becomes necessary to provide increased safety measures to protect both, the transported cargo and the lives of the truck drivers. While work on increasing the safety of parking spots across the EU is underway, access to SSTPA's (Safe and Secure Truck Parking Areas) remains limited. Raising the requirement for safety of parking options reduces the availability of eligible parking spots and

H. Selvaraj et al. (Eds.): ICSEng 2020, LNNS 182, pp. 105–114, 2021.
https://doi.org/10.1007/978-3-030-65796-3_10

results in undesirable conflicts, such as double parking, which reduce the overall quality of drivers' working conditions. Double parking is especially common in non-adapted parking areas such as those in city centres [14].

Truck driver working hours are subject to strict regulations introduced with the goal of reducing driver fatigue [10] which helps prevent on-road accidents. Exact legislation depends on the region of the world, such as the USA [4] or the EU [5]. In the EU, the main act regulating the working hours is the (EC) 561/2006 [2] (with subsequent updates) and local implementations of the 2002/15/EC directive [3]. Taken jointly, the working-hour requirements and the limited availability of parking spots along a typical freight route results in a difficult decision-making problem, especially when traditional criteria such as transportation time and cost are optimized alongside driver safety and driver comfort.

Multi-criteria optimization in freight transportation has been a recent topic of interest. In [16] authors consider a problem of fleet sizing with different stakeholders such as shareholders, customers and employees. Each stakeholder is represented by a different criterion. Authors of [9] consider a truck driver scheduling problem with different types of idling options. Options depend on whether an EPS or APU is available. While the resulting optimization problem includes (USA-specific) working hours constraints, the quality criterion is only cost-oriented. In the paper [12] authors solve a vehicle-routing problem with cost, noise, pollution and fuel consumption as the criteria. This is followed up in [11] where a so-called green version of the joint travelling salesman and the transportation problem was considered. Vehicle routing within a terminal is considered in [6] where the authors try to minimize transportation cost, transportation time and total emissions.

More recently, in [7] authors consider a joint problem of vehicle selection, transportation job selection and vehicle routing. Considered is an aggregated quality criterion where safety, comfort, profit of the driver, gain of the business owner, fitness of the vehicle for the job and lateness are considered. Finally, worth mentioning is a work [15] where authors schedule long-haul truck transportation considering both, weekly hour limits and parking availability. However, only total trip duration is considered.

None of the research presented here considers a problem of parking area selection for freight routes that also focuses on driver security and comfort. This work hopes to fill in that gap. In the paper we formulate the problem of multi-criteria optimization of parking spot selections for individual truck drivers in a single corridor. Optimization is done with regards to driver-specific quality-of-life criteria such as driver parking safety and driver parking comfort. Those criteria are optimized jointly, together with the more common total travel time and total travel cost. In the paper it is shown how to use a Branch-and-Bound approach to quickly generate a set of non-dominated solutions for the user to pick from. The work focuses on Europe and EU-specific legislation is considered.

The model in this paper can be easily applied to the universal standards such as those outlined in [13] where four different security levels of parking spaces are established (bronze, silver, gold and platinum). The requirements for each level depends on perimeter security (i.e. barriers, CCTV coverage), parking area parameters (i.e. lighting, lane marking), entry and exit facilities (i.e. turnstiles) and staff availability and training. Regarding service, recognized are facilities such as toilets, showers, electric plugs, food dispensers and others. Based on the security level and service facilities, one can assign different numerical values representing the relative safety and relative comfort of each parking space.

1.1 EU Working Hours Regulations

In the European Union the driver has to take 45 min break after each 4.5 h of driving. This time can be divided in two breaks lasting at least 15 and 30 min. First break must be used during the 4.5 h drive, whereas the second one must start no later than after the total of 4.5 h of driving extended by the time of the first break. Moreover, a driver is allowed to have a resting break after 6 h of work which last 30 or 45 min depending on working hours conditions. If working hours are less than 9, then 30 min are allowed, otherwise 45 min are required. This break can be divided into periods, all lasting not less than 15 min, which must be used within 6 h of work time. The total driving time cannot exceed 10 h a day at most twice a week. Otherwise, the driving time should last no more than 9 h. After work time for the day is finished, the daily resting break should last no less than 9 h. This can happen at most three times in a period of two weeks. Otherwise, the daily resting time last should last at least 11 h, which can be divided in two periods lasting 3 h during the working time and 9 after working time.

To reflect those regulations, typical resting breaks may last 15, 30, 45, 180, 540, or 660 min. We consider only the working time and in consequence, we do not plan daily breaks (9 and 11 h breaks),

2 Mathematical Model

Let us consider a list of P possible parking stops $\mathbf{PL} = (0, 1, 2, ..., P)$ along the route of the truck driver, where he may stop for a break. Each parking stop i is described by safety rate SR_i, comfort rate CR_i, maximal parking time allowed MPT_i and the probability $A_i(t)$ that a free parking space is available at time of a day t. Break time duration index is selected from set $\mathbf{BT} = \{0, 1, 2, ..., MBT\}$, where for each option $k \in \mathbf{BT}$ a parking cost $PC_{i,k}$ is specified. Direct travel time $TT_{i,j}$ and direct travel cost $TC_{i,j}$ between parking stops i and j $(i < j)$. Let $bt_k \in \{0, 15, 30, 45, 180\}$, where $k \in \mathbf{BT}$ denotes the time of kth break, which is implied by driver's working hours regulations during one working day. It is assumed that the driver starts his journey at hour t_0 after wt_0 hours of work since the last daily break, dh minutes of driving this day, dt_0 minutes of driving since the last resting break, st_0 minutes since last 15 min break. Let some of higher level decisions made and concerning driver's working hours be given by the following indicators. If the driver may drive for 10 h a day $ewh = 1(0)$

(otherwise). If the driver should take 3 h of partial daily resting break during his journey $ddb = 1(0)$ (otherwise). If the driver is allowed to take short 9 h daily resting break $nhb = 1(0)$ (otherwise), and if $ddb = 1$ then $nhb = 1$. For details, see Table 1.

2.1 Decision Variables

Let us introduce an integer decision vector $\mathbf{x} = [x_i]_{i \in \mathbf{PL}}$, where $x_i = k(0)$ if the break time bt_k is planned at parking stop i, (if the parking stop i will not be used).

2.2 Auxiliary Variables

Let us introduce the following auxiliary variables. The binary vector $\mathbf{y} = [y_i]_{i \in \mathbf{PL}}$ represents the decision stating which parking stops are planned to be used for making the break, where $y_i = 1(0)$ if the parking stop i is planned to be used (otherwise). The binary vector $\mathbf{z} = [z_i]_{i \in \mathbf{PL} \setminus \{P\}}$ represents driver's route, where $z_i \in \mathbf{PL}$ is the number of the next planned parking stop after parking stop i. The vector $\mathbf{t} = [t_i]_{i \in \mathbf{PL}}$ represents the time when the driver arrives at parking stop i. The vector $\mathbf{dt} = [dt_i]_{i \in \mathbf{PL}}$ represents the driving time since the last driver's resting break when arriving at parking stop i. The vector $\mathbf{wt} = [wt_i]_{i \in \mathbf{PL}}$ represents the working time since the last driver's daily resting break when arriving at parking stop i.

Table 1. Notation

Variable	Description
P	-a number of possible parking stops
$\mathbf{PL}$	-a list of parking stops
SR_i	-a safety rate of parking stop i
CR_i	-a comfort rate of parking stop i
MPT_i	-the maximal parking time allowed (in minutes) at parking stop i
$A_i(t)$	-a probability of free parking space at time t at parking stop i
$PC_{i,k}$	-a parking cost at parking stop i for each break time k
MBT	-a number of possible time breaks
$\mathbf{BT}$	-a set of break time indices
$TT_{i,j}$	-a travel time between parking stops i and j $(i < j)$
$TC_{i,j}$	-a travel cost between parking stops i and j $(i < j)$
bt_k	-the kth break time (in minutes)
t_0	-a number of minutes of a day when the journey begins
wt_0	-a number of minutes of work since the last daily resting break
dh	-a number of driving minutes since last daily resting break
dt_0	-a number of driving minutes since the last resting break
st_0	-a number of minutes since last 15 min break
ewh	-an indicator of allowance to drive up to 10 h

(*continued*)

Table 1. (*continued*)

Variable	Description
ddb	-an indicator of the need of taking 3 h break during working hours
nhb	-an indicator of possibility of taking 9 h daily break
$\mathbf{x}$	-a decision vector
x_i	-a decision variable indicating which break time bt_k is planned at parking stop i
$\mathbf{y}$	-an auxiliary decision vector indicating which parking stops are planned to be used
y_i	-an auxiliary decision variable indicating if the parking stop i is planned to be used
$\mathbf{z}$	-an auxiliary integer decision vector of consecutive parking stops, representing the driver's route
z_i	-an auxiliary decision variable indicating the number of the next planned parking stop after parking stop i
$\mathbf{t}$	-an auxiliary integer decision vector of times at which the driver arrives at parking stops
t_i	-an auxiliary decision variable indicating the time (in minutes) at which the driver arrives at parking stop i
$\mathbf{dt}$	-an auxiliary integer decision vector of driving times, since the last resting break, at which the driver arrives at parking stops
dt_i	-an auxiliary decision variable indicating the driving time, since the last resting break, at which the driver arrives at parking stop i
$\mathbf{wt}$	-an auxiliary integer decision vector of working times, since the last daily break, at which the driver arrives at parking stops
wt_i	-an auxiliary decision variable indicating the working time, since the last daily break, at which the driver arrives at parking stop i
mts	-minimal safety of the parking stop
mtc	-minimal comfort of the parking stop

2.3 Evaluation Criteria

The decisions of which parking stops should be used for taking which break can be evaluated with the use of the following criteria.
1. Total travel time between origin and destination

$$TTT(\mathbf{x}) = t_P - t_0, \tag{1}$$

2. Total travel cost

$$TTC(\mathbf{x}) = \sum_{i \in \mathbf{PL} \setminus \{P\}} y_i \cdot (PC_{i,x_i} + TC_{i,z_i}), \tag{2}$$

3. Average parking safety

$$APS(\mathbf{x}) = \frac{\sum_{i \in \mathbf{PL}: x_i > 0} R_i \cdot bt_{x_i}}{\sum_{i \in \mathbf{PL}: x_i > 0} y_i \cdot bt_{x_i} + 1 - max_{i \in \mathbf{PL}} \mathbb{1}(x_i)}, \tag{3}$$

4. Worst parking safety

$$WPS(\mathbf{x}) = \min_{i \in \mathbf{PL}: x_i > 0} \{max_j SR_j, SR_i\}, \tag{4}$$

5. Average parking comfort

$$APC(\mathbf{x}) = \frac{\sum_{i \in \mathbf{PL}: x_i > 0} CR_i \cdot bt_{x_i}}{\sum_{i \in \mathbf{PL}: x_i > 0} bt_{x_i} + 1 - max_{i \in \mathbf{PL}} \mathbb{1}(x_i)}, \tag{5}$$

6. Worst parking comfort

$$WPC(\mathbf{x}) = \min_{i \in \mathbf{PL}: x_i > 0} \{max_j CR_j, CR_i\}, \tag{6}$$

7. Probability of not violating the working and driving time regulations due to lack of parking places at the chosen parking stops.

$$PVR(\mathbf{x}) = \min\{1, \prod_{i \in \mathbf{PL}: x_i > 0} A(t_i)\}, \tag{7}$$

2.4 Constraints

The following constraints ensure the correct values of auxiliary variables based on decision variables.

$$\bigwedge_{i \in \mathbf{PL}} y_i = \begin{cases} 0 & \text{iff } x_i = 0 \wedge i \in \mathbf{P} \setminus \{0, P\} \\ 1 & \text{iff } x_i > 0 \vee i \in \{0, P\} \end{cases} \tag{8}$$

$$\bigwedge_{i \in \mathbf{PL} \setminus \{P\}} z_i = \min\{j \in \mathbf{PL} : j > i \wedge y_j > 0\}, \tag{9}$$

$$\bigwedge_{i \in \mathbf{PL} \setminus \{P\}} t_{i+1} = y_i \cdot (TT_{i,z_i} + bt_{x_i}) + t_i \tag{10}$$

$$\bigwedge_{i \in \mathbf{PL} \setminus \{P\}} dt_{i+1} = y_i \cdot (TT_{i,z_i}) + \begin{cases} 0 \text{ if} & ((st_i < dt_i \wedge bt_{x_i} \geq 30) \\ & \vee (st_i \geq dt_i \wedge bt_{x_i} \geq 45)) \\ dt_i \text{ if} & ((st_i < dt_i \wedge bt_{x_i} < 30) \\ & \vee (st_0 \geq dt_0 \wedge bt_{x_i} < 45)) \end{cases} \tag{11}$$

$$\bigwedge_{i \in \mathbf{PL} \setminus \{P\}} wt_{i+1} = wt_i + y_i \cdot (TT_{i,z_i}) + \begin{cases} 0 & \text{if } bt_{x_i} \geq 180 \\ bt_{x_i} & \text{otherwise,} \end{cases} \tag{12}$$

$$\bigwedge_{i \in \mathbf{PL} \setminus \{P\}} st_{i+1} = y_i \cdot (TT_{i,z_i} + bt_{x_i}) + \begin{cases} 0 & \text{if } bt_{x_i} = 15 \\ st_i & \text{otherwise,} \end{cases} \tag{13}$$

The parking safety must be not less then the given value mts, i.e.,

$$\bigwedge_{i \in \mathbf{PL}} SR_i \geq y_i \cdot mts, \tag{14}$$

The travel comfort must be not less then the given value mtc, i.e.,

$$\bigwedge_{i \in \mathbf{PL}} CR_i \geq y_i \cdot mtc, \tag{15}$$

The break on a given parking stop cannot exceed the allowed maximal parking time at this stop.

$$\bigwedge_{i \in \mathbf{PL}} b_{x_i} \leq MPT_i, \tag{16}$$

The driving time longer than 4,5 h without 45 min break cannot be continued.

$$\bigwedge_{i \in \mathbf{PL} \setminus \{0\}} dt_i \leq 270, \tag{17}$$

The total driving time cannot be greater than 9 or 10 h

$$dh + t_P - t_0 - \sum_{i \in \mathbf{PL} \setminus \{P\}} bt_{x_i} \leq 540 + 60 \cdot ewh, \tag{18}$$

After 6 h of work or earlier, the total break time must be at least 45 min (there can be 3 × 15 min break). In the situation under consideration, the working time other than driving may occur only before the travel begins. Therefore, it should be checked if during the first 6 h of work, the sum of breaks is at least 45 min.

$$\sum_{i \in \mathbf{PL} \setminus \{0\} : wt_i < 360 - wt_0} bt_{x_i} \geq 45, \tag{19}$$

The 3 h break must be planned if ddb = 1

$$\sum_{i \in \mathbf{PL} : bt_{x_i} = 180} y_i \geq ddb, \tag{20}$$

The driver must arrive at the destination (parking stop P) not later than at the time when he starts his daily (9 or 11 h) resting break.

$$wt_0 + t_P - t_0 - x_P \leq 900 - (1 - nhb) * 180, \tag{21}$$

2.5 Problem Formulation

For the given P, MBT, SR_i, CR_i, MPT_i, MP_i, AFP_i, $PC_{i,k}$, $TT_{i,j}$, $TC_{i,j}$, bt_k, $i = 0, 1, 2, ..., P$, $k = 0, 1, 2, ..., MBT$, t_0, wt_0, dh, dt_0, st_0, ewh, ddb, nhb, mts, and mtc, find the parking stops for making breaks $\mathbf{x}$ minimizing the vector of criteria

$$\begin{aligned} \mathbf{Q}(\mathbf{x}) = [&TTT(\mathbf{x}), TTC(\mathbf{x}), -APS(\mathbf{x}), -WPS(\mathbf{x}), \\ &-APC(\mathbf{x}), -WPC(\mathbf{x}), -PVR(\mathbf{x})] \end{aligned} \tag{22}$$

subject to constraints (8)–(21)

3 Solution Algorithm

Given a limited time period for finishing the whole route (here, one working day), the maximum number of solutions to be evaluated in order to determine a Pareto-optimal set is (roughly) bounded by $O(P^k)$ where k is the maximum number of breaks that fit in the period. Even though this value does not grow exponentially with regards to the number of parking stops, it is still large enough to make a complete enumeration of all solutions infeasible. We use a Branch and Bound approach where the enumeration is subject to pruning of infeasible or dominated branches. The solution tree has a no parking breaks at the root and branches correspond to parking options at the next parking spot. The estimations of lower (or upper, in case of maximization) bounds for a given branch, for each criterion, are easily derived under the assumption that the reminder of the route is optimized with regards to a single criterion. As such, the lower bound on cost is the current cost, in the node of the tree added to the lowest possible cost to the end of the route. The lower bound on time is the current time added to the shortest possible duration of transportation to the end of the route. The same manner of estimation follows for all of the criteria. If the estimated best possible solution at a node of the decision tree is dominated by any of the solutions currently in the Pareto-set (candidates), then the whole branch is pruned.

4 Empirical Evaluation

We perform all of the experiments on a route between Warsaw and Kudowa, two cities in Poland. The route has a total travel time of 518 min, distance of 487.4 km and cost of 642 euros (parking not included). We consider 15 possible parking stops. Comforts and safety are rated from 1 to 10 (arbitrary values were chosen randomly) while availability (probability) was generated between 0.5 and 1 for each parking stop. We assume that all the parking stops permit long parking of at least 15 h and parking costs are 5 euro for breaks under one hour and 20 euro for longer breaks.

The experiments are performed for a number P of available parking spots selected randomly from the available total. The results, for two separate sets of P spots are presented in Table 2. The cases for which $P = 12$ are further expanded upon in Table 3 which summarizes the results with regards to individual quality criteria. All the experiments were preformed on a an isolated Windows virtual machine running Intel i7@4 GHz with a dedicated 30 GB of RAM using Python 3.6.

It is clear that the execution time of the algorithm grows significantly with even a slight increase of P. This does not prohibit the use of an average-class personal computer to generate a set of solutions on a day-to-day basis. However, high computational requirements even for relatively short routes makes it impossible to use the method in a real-time manner without employing more powerful workstations. The advantage of the proposed approach is that the computations can be easily parallelised by considering different branches of the decision tree separately.

Table 2. Results for varying P in two series of experiments

Series	P	8	9	10	11	12
First	Duration [s]	160	340	980	2916	6180
	Pareto set cardinality	467	442	454	646	784
Second	Duration [s]	56	196	568	1599	4865
	Pareto set cardinality	202	283	385	335	549

Generated Pareto fronts are numerous and varied which makes it difficult to examine them quickly. On the other hand, it provides ample choice for the decision maker although the manner of presentation is likely to significantly affect the usefulness of the results. To simplify the selection procedure, the front can be further pruned with the use of interactive selection methods, such as the Light Beam Search [8].

Table 3. Quality inervals for $P = 12$ for two experiments

Series		TTT	TTC	APS	WPS	APC	WPC	PV
First	Min	614	660	4.04	2	3.29	1	0.05
	Max	894	735	8.8	8	7.91	3	0.56
Second	Min	614	660	4.41	2	3.29	1	0.05
	Max	884	755	9	9	7.91	3	0.56

5 Conclusions

We have formulated a problem of selecting parking spots along a route for a freight transport. This multicriteria optimization problem can be solved in reasonable time with the use of Branch and Bound techniques. Obtained set of Pareto-optimal solutions can then be further perused to select a compromising solution for which a satisfactory guarantee of driver safety and driver comfort can be ensured. This is likely to improve working conditions of freight drivers and increase the security of the cargo.

Despite the expected benefits of the proposed approach, it is necessary to consider methods that decrease the time needed to generate the set of Pareto-optimal solutions. To ensure that the problem can be solved quickly is to ensure that the method can be applied to various real-life situations. Due to the inherent complexity of the problem, use of heuristic algorithms might be necessary.

References

1. Alam, A., Besselink, B., Turri, V., Martensson, J., Johansson, K.H.: Heavy-duty vehicle platooning for sustainable freight transportation: a cooperative method to enhance safety and efficiency. IEEE Control Syst. **35**(6), 34–56 (2015)
2. Parliament, European, Council, European: Regulation (EC) No 561/2006 of the European Parliament and of the Council of 15 March 2006 on the harmonisation of certain social legislation relating to road transport and amending Council Regulations (EEC) No 3821/85 and (EC) No 2135/98 and repealing Council Regulation (EEC) No 3820/85 (Text with EEA relevance) - Declaration. Official Journal of the European Union, L 102, 1–14 (2016)
3. Parliament, European, Council, European: Directive 2002/15/EC of the European Parliament and of the Council of 11 March 2002 on the organisation of the working time of persons performing mobile road transport activities. Official Journal of the European Union, L 80, 35–39 (2002)
4. Goel, A., Kok, L.: Truck driver scheduling in the United States. Transp. Sci. **46**(3), 297–438 (2012)
5. Goel, A.: Legal aspects in road transport optimization in Europe. Transp. Res. Part E **114**, 144–162 (2018)
6. Heilig, L., Lalla-Ruiz, E., Voß, S.: Multi-objective inter-terminal truck routing. Transp. Res. Part E **106**, 178–202 (2017)
7. Hojda, M., Zak, J., Filcek, G.: Multiple criteria optimization of the joint vehicle and transportation jobs selection and vehicle routing problems for a small road freight transportation fleet. Transp. Res. Procedia **30**, 178–187 (2018)
8. Jaszkiewicz, A., Słowiński, R.: The "light beam search" approach - an overview of methodology and applications. Eur. J. Oper. Res. **113**(2), 300–314 (1999)
9. Koç, C., Bektaş, T., Jabali, O., Laporte, G.: A comparison of three idling options in long-haul truck scheduling. Transp. Res. Part B **93**, 631–647 (2016)
10. Rancourt, M.È., Paquette, J.: Multicriteria optimization of a long-haul routing and scheduling problem. Multi-Criteria Decis. Anal. Optim. Learn. Decis. Support **21**(5–6), 239–255 (2012)
11. Sawik, B., Faulin, J., Pérez-Bernabeu, E.: Multi-objective traveling salesman and transportation problems with environmental aspects. Appl. Manage. Sci. **18**, 21–55 (2017)
12. Sawik, B., Faulin, J., Perez-Bernebeu, E.: A multicriteria analysis for the green VRP: a case discussion for the distribution problem of a Spanish retailer. Transp. Res. Procedia **22**, 305–313 (2017)
13. de Leeuw van Weenen, R., Newton, S., Menist, M., Mass, F., Penasse, D., Nielsen, M.: Study on Safe and Secure Parking Places for Trucks. Final Report. European Commission B-1049 Brussels (2019). https://doi.org/10.2832/067535
14. Vidal, T., Laporte, G., Matl, P.: A concise guide to existing and emerging vehicle routing problem variants. Eur. J. Oper. Res. **286**(2), 401–416 (2020)
15. Vital, F., Ioannou, P.: Long-haul truck scheduling with driving hours and parking availability constraints. In: IEEE Intelligent Vehicles Symposium, Paris, France, 9–12 June, pp. 620-625 (2019)
16. Żak, J., Redmer, A., Sawicki, P.: Multiple objective optimization of the fleet sizing problem for road freight transportation. J. Adv. Transp. **42**, 379–427 (2011)

Synthesis of Index Generation Function Using Linear and Functional Decomposition

Grzegorz Borowik[1](✉), Tadeusz Łuba[2], Dariusz Wąsicki[2], and Grzegorz Chmaj[3]

[1] SWPS University of Social Sciences and Humanities, Warsaw, Poland
gborowik@swps.edu.pl
[2] Warsaw School of Computer Science, Warsaw, Poland
tadluba@gmail.com
[3] Electrical and Computer Engineering Department, University of Nevada, Las Vegas, NV, USA
grzegorz.chmaj@unlv.edu

Abstract. Researchers have thoroughly studied decomposition in many different contexts, such as switching theory or data mining. In recent years, a renewed interest in this problem was caused by memory-based pattern matching circuits, particularly in the synthesis of Index Generation Functions. In this case, function is a composition of a linear function L and a general function G. The function L is implemented using ExOR gates, while G is usually realized using embedded memories.

In this paper, we show that the linear function reduces the number of variables to represent an index generation function, thus reducing memory size. However, as another efficient technique for logic synthesis using memories, a functional decomposition can be applied. The decomposition is a methodology of expressing a function of n variables as a bunch (collection) of functions of fewer variables. This paper presents an exact method of searching for a functional decomposition with a minimum number of variables by using the theory of r-admissibility. Therefore, the method proposed can be used for ROM-based synthesis, particularly for pattern matching and communication circuits.

Keywords: Index generation function · Functional decomposition · Linear decomposition · Logic cell

1 Introduction

Index generator units are digital hardware systems which development is related to the need for classifying and identifying data represented by long binary vectors (e.g., 48 bits). Their use includes distribution IP addresses, virus scanning, unwanted data detection, code conversions, etc.

The increasing number of scientific work published in articles and conference papers [1–4] is dedicated to such kind of synthesis method. One of the reasons

H. Selvaraj et al. (Eds.): ICSEng 2020, LNNS 182, pp. 115–127, 2021.
https://doi.org/10.1007/978-3-030-65796-3_11

for this situation is the specificity of the activity such systems boils down to the synthesis of strongly undefined Boolean functions. Their characteristic feature is a vast number of variables, and in consequence, the need to use non-volatile memory, such as ROM and their smaller equivalent in the form of LUT (Look-Up Table) logic cells [5]. The significant number of variables of Boolean functions describing how index generator works require the use of new methods of logic synthesis. New methods should be more focused on reducing arguments and decomposition. Only these procedures can meet the challenges putting by the need to precisely extract the right data from a vast mass of unnecessary data.

The synthesis of such functions is a big challenge for researchers and even the best teams in the world that perform compression and attribute reduction algorithms cannot solve this problem. For example, the Boolean function cited in [6] (10 vectors, 40 arguments) calculated by Sasao [3] is reduced to 5 variables. The Lightning program, developed by a graduate of the Distance Education Center at the Warsaw University of Technology, calculates 2261 4-argument reducts for this function [6]. It is worth noting that the said program can generate a truth table of this function for each reduct.

The research discussed in this paper concerns the methods of decomposing Boolean functions. In connection with the diversity of these methods, with their increasingly common and more important application, the article's primary purpose is to review these methods enabling the reader to make the right choice. For these reasons, both methods and algorithms of linear decomposition, as well as functional decomposition are discussed, providing opportunities for the development of universal methods for the synthesis of index generators.

2 Index Generation Functions

Increasing the efficiency of processes such as IP address distribution, virus scanning, undesirable data detection, etc. are currently the subject of intense research.

Due to technical limitations related to the limited number of memory inputs used to implement these processes, scientific research is directed towards the logic synthesis of digital circuits. The great significance of logic synthesis concerns the synthesis of the index generation function (IGF). The methods of creating the index generation function are characterized by diversified efficiency. The purpose of these methods is to select the right data from a vast amount of unnecessary data. Extracting the necessary data is difficult since the input data is usually binary vectors described by strongly unspecified Boolean functions with a significant number of arguments.

Functional dependence models in which the reduction of arguments and functional decomposition are mainly developed are used to implement logic synthesis methods. This approach is referred to as the standard synthesis methods. Heuristic methods use methods of deconvolution of finite element theory. They are characterized by developed computational complexity.

The Index Generation Function with weights k is a strongly unspecified Boolean function of the form:

$$f : D^n \to \{1, 2, \ldots, k\}, \text{ where } D^n \subseteq \{0,1\}^n \text{ and } |D^n| = k$$

Vectors belonging to the set D^n are called registered vectors. The function defined in this way can be implemented using the n-width ROM address bus, which stores q-bit data, where $q = \lceil \log_2 k \rceil$. The total memory size needed to perform this function is $2^n q$ bites. A characteristic feature of the Index Generation Function is a large number of input variables (n) and relatively not very numerous ($k < 2^n$) registered vectors set. Thanks to this property, it is possible to effectively reduce the number of inputs to the system ($p < n$).

The development of an effective method of minimizing the number of variables is crucial, determining the possibility of effective implementation of functions, e.g., in FPGA systems. A large number of input variables with a relatively small number of registered vectors creates the need for an effective logic synthesis of input variables reduction. It decides about the possibility of effective implementation of the function realized physically using the Index Generator Unit (IGU) in the structure proposed by Sasao Fig. 1 [3]. In the index generator built in this way, two memories are distinguished: main memory and auxiliary memory, moreover comparator and AND gate.

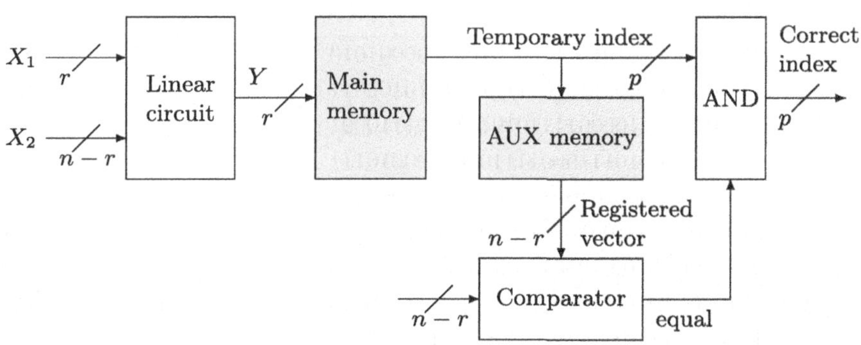

Fig. 1. Index Generation Unit – IGU

The variable selection block selects from among all variables the X_1 subset, representing the minimum number of variables (the so-called reduct) of the Index Generation Function (IGF) implemented. The quantity of the reduct is r. It is also the number of entries to the main memory. The input vector $X = X_1 + X_2$ appears on the main memory outputs. It is represented by a binary vector with the number of bits.

This index is a valid index of the input vector if it is a registered vector. And only then will the index appear on the outputs of the generator. In the case when the input vector is not a registered vector, on the outputs of the generator, a '0' will appear. Checking that the main memory outputs represent the correct index is implemented in auxiliary memory and comparator. The auxiliary memory contains the vectors represented by variables belonging to the X_2 subset. These vectors are fed to the comparator input and compared to the actual X_2 vector fed to the comparator's second input. The vector taken from the auxiliary memory will be different from the real vector X_2 if it is not part of the registered vector. Then the comparator with the output 0 will block the appearance of the vector created in the main memory on the AND gate outputs.

The operation of the index generator constructed in this way can be presented on the example of the function analyzed in the Sasao paper [3] – Table 1a. For this function, a 5-argument reduct was determined (Table 1b).

It is not difficult to notice that the reduct calculation method's efficiency determines the effectiveness of the index generator constructed in this way. For these reasons, in the summary of the paper [3], the development of a fast argument reduction algorithm was considered the essential task of scientific work in the field of pattern detection. It is worth emphasizing that such an algorithm has already been developed and published in [7]. The entire series of 4-argument reducts can be calculated using this algorithm. One of these solutions is given in Table 1c.

Table 1. a. An example of the Index Generation Function [3]; b. Indexes calculated for function [3]; c. Indexes calculated using the top-up method

a)		b)	c)
0110000100010000101001100001000100001010	1	01100	0000
0101111101101010001101011111011010100011	2	01011	1011
1111010101110111000011110101011101110001	3	11110	1111
0001111000010001011100011110000100010111	4	00011	1110
0011110000000100010100111100000001000101	5	00111	1010
0111001001000100100101110010010001001001	6	01110	0010
0010001110001111001000100011100011110010	7	00100	0100
1111111111010001111000100011100011110010	8	11111	1100
1110111000110001011011101110001100010110	9	11101	1101
1010000110100100001110100001101001000011	10	10100	0001

3 Linear Decomposition

In existing solutions, linear decomposition is limited to the synthesis of strongly undefined Boolean functions with specific properties: $|D^n| = k$. To simplify the entries, vectors $a, b \in \{0,1\}^n : F(a) \neq F(b)$, are represented by numbers $p, q \in K = \{1, ..., |D^n|\}$. In general, the problem can be reduced to indicate the conditions that must be met for function F to have decomposition with two-argument functions G:

$$F = H(G_1(X_1), G_2(X_2), \ldots, x_n), \text{ where } X_1 = \{x_{i_1}, x_{i_2}\}, X_2 = \{x_{i_3}, x_{i_4}\}$$

The G function can be called as g-argument, according to the general theory of decomposing Boolean functions. With such an approach, it is convenient to use an orderly and compact way to represent a function, the partition algebra discussed below.

Example 1. Consider the logic function ExDM1 defined in the Table 2, in which the symbols vectors $T: 1, 2, \ldots, 9$ represent $x_1, \ldots, x_7$ input variables, and F – symbolic output.

Table 2. ExDM1 function

T	x_1	x_2	x_3	x_4	x_5	x_6	x_7	F
1	1	0	0	0	1	1	0	1
2	0	1	0	0	1	0	1	2
3	1	0	1	0	0	0	0	3
4	1	0	1	0	1	1	0	4
5	1	1	1	0	1	0	1	5
6	1	0	0	0	1	0	1	6
7	1	0	1	1	1	1	0	7
8	1	1	0	1	1	1	0	8
9	1	1	1	0	1	1	1	9

For the function from Table 2 input partitions are as follows:

$$P(x_1) = \{\overline{2}, \overline{1,3,4,5,6,7,8,9}\}$$

$$P(x_2) = \{\overline{1,3,4,6,7}, \overline{2,5,8,9}\}$$

$$P(x_3) = \{\overline{1,2,6,8}, \overline{3,4,5,7,9}\}$$

$$P(x_4) = \{\overline{1,2,3,4,5,6,9}, \overline{7,8}\}$$

$$P(x_5) = \{\overline{3}, \overline{1,2,4,5,6,7,8,9}\}$$

$$P(x_6) = \{\overline{2,3,5,6}, \overline{1,4,7,8,9}\}$$

$$P(x_7) = \{\overline{1,3,4,7,8}, \overline{2,5,6,9}\}$$

For this example, the different values assumed by the input variable x_4 define the subsets of the vectors $\{1, 2, 3, 4, 5, 6, 9\}$ and $\{7, 8\}$. For vectors $1, 2, 3, 4, 5, 6$ and 9 – $x_4 = 0$, for $7, 8$ – $x_4 = 1$. Therefore, knowing that $x_4 = 0$, we know that the vector $1, 2, 3, 4, 5, 6$ or 9 was chosen, but we do not know which one. Similarly, knowing that $x_8 = 1$, we know that the symbol $1, 4, 7, 8$ or 9 is chosen, but we cannot indicate which of them specifically. In this way, the information is modeled by partitions [8].

To simplify the entries, vectors

$$a, b \in \{0,1\}^n : F(a) \neq F(b)$$

are represented by numbers $p, q \in K = \{1, \ldots, |D^n|\}$.

Let F be a multi-output Boolean function where the set of variables is $X = (x_1, \ldots, x_n)$, the set of vectors (minterms) is $M = (m_1, \ldots, m_t)$. By the C_{pq}, where p, q are vectors of M, for which $F(p) \neq F(q)$ we denote discernibility set, defined as follows:

$$C_{pq} = \{x \in X : x(p) \neq x(q)\} \text{ for } p, q = 1, \ldots, t, \text{ and } p < q.$$

The family of sets $C = (C_1, \ldots, C_n)$, we refer to RC_{pq} for all C_{pq}. We use the C_{pq} family set in calculating the pairs of variables that form linear decomposition.

The problem of determining appropriate conditions is based on the notion of a separating function. We say about the function g that it is a distributing function of p, q vector pairs if and only if p and q belong to different blocks of partition $P(g)$. In simple terms, it can be said that in this case, the function g allows the separation of vectors p and q.

The adapted definition shows that each unary function $g(x_j) = x_j$, where $x_j \in C_{pq}$ is the distribution of the p, q. Besides, if g separates the pair p, q, this pair also separates the function $\overline{g}$.

Using the above findings, a more convenient form of the decomposition condition is $F = H(g_1, g_2, \ldots, g_t)$ if and only if for each pair of the p, q such that (p, q) is not contained in the same characteristic block of P_F, there exists g_j separating the pair p, q. The advantages of this condition are used to formulate a methodology for the synthesis of systems in which the arguments with $X = (x_1, \ldots, x_n)$ are paired and entered into the H system through functors $g(x_i, x_j)$, i.e., $F = H(g_1, g_2, \ldots, g_t, x_a, x_b, \ldots)$.

Now, it can be proposed a method of carrying out logic synthesis as decomposition directly into the gates, in contrast to functional decomposition, in which the components are functions described by truth tables. It is possible to calculate the $P(g)$ partition for any function with a finite number of input variables and in the next step to delete all C_{pq} files for which g is not a distribution. In this way, it is calculated the F function decomposition receiving the form $F = H(G(x_{i_1}, \ldots, x_{i_t}), \ldots, x_{i_n})$. In this case, the number of H function arguments is at most $n - 1$, therefore the process reduces the function complexity.

Theorem 1. *The function $g = x_i \oplus x_j$ is a function that separates (distributes) a pair (p, q) if and only if $\{x_i, x_j\} \notin C_{pq}$ and $|\{x_i, x_j\} \cap C_{pq}| = 1$.*

The above theorem leads to a simple method of decomposing functions and determining $g = g(x_i, x_j)$ directly from C_{pq} sets and $P(x_i)$ divisions, e.g., divisions generated by the unary function $g(x_i) = x_i$.

Test 1. $G = x_i \oplus x_j$ is a decomposition of the F function if and only if $\{x_i, x_j\} \notin RC_{pq}$, where RC_{pq} is a family of C_{pq} files for which p and q belong to different output split blocks P_F.

Linear decomposition connects variables into pairs and enters them through ExOR gates to the primary system performing detection and classification (i.e., indexing) of the data. It is important because it leads to a further reduction of the number of entries in index generators. And probably for these reasons, linear decomposition is intensively used in the synthesis of Index Generation Functions [3,5,9].

In existing solutions, linear decomposition is limited to the synthesis of strongly indeterminate Boolean functions with a specific property: $|D^n| = k$. Limiting this problem to functions in which all output vectors are different is unnecessary because it can be reduced to a more general task of synthesizing Boolean systems of any outputs. In other words, it is better to assume that the number of the elements of output vectors Y set of the function F is $\leq |D^n|$.

In general, the problem can be reduced to the indication of the conditions that must be met for the F function to be decomposed with two-argument G functions

$$F = H(G_1(X_1), G_2(X_2), \ldots, x_n),$$

where: $X_1 = \{x_{i_1}, x_{i_2}\}$, $X_2 = \{x_{i_3}, x_{i_4}\}$.

According to the general theory of decomposition of Boolean functions, the G function is called a g-argument. We also assume that in linear decomposition, we use functions with a reduced number of arguments, i.e., minimal argument representations of the F function.

An important generalization of Theorem 1 concerns inseparable decomposition. In this case, an additional condition for the existence of decomposition with a pair of functions $\{x_i \oplus x_j, x_j \oplus x_k\}$ is to check whether the set of variables $\{x_i, x_j, x_k\}$ is contained in the family RC_{pq}.

Example 2. Consider the logic function ExDM1 defined in Table 2, where the symbols of vectors $T : 1, 2, \ldots, 9$ represent $x_1, \ldots, x_7$ input variables, and F – symbolic output. For the functions from Table 2, the COM family (RC_{pq}) limited to two element sets is as follows (COM denotes Complement):

$$(x_1; x_4); (x_1; x_5); (x_1; x_6); (x_1; x_7); (x_2; x_5);$$
$$(x_2; x_6); (x_3; x_5); (x_3; x_6); (x_3; x_7); (x_4; x_5).$$

The COM(RC_{pq}) contains a pair: x_1, x_4, so there exists a decomposition:

$$F = H(x_1 \oplus x_4, x_2, x_3, x_5, x_6, x_7).$$

The decomposition with $x_i \oplus x_j$ exists for all other COM(RC_{pq}) family pairs. However, it does not exist for every pair outside this set.

Concerning the non-disjoint decomposition of the logic function ExDM1 defined in Table 2, the three variable COM(RC_{pq}) subsets are:

$$(x_1; x_2; x_3); (x_1; x_2; x_4); (x_1; x_2; x_5); (x_1; x_2; x_6); (x_1; x_2; x_7); (x_1; x_3; x_4);$$
$$(x_1; x_3; x_5); (x_1; x_3; x_7); (x_1; x_4; x_5); (x_1; x_4; x_6); (x_1; x_4; x_7); (x_1; x_5; x_6);$$
$$(x_1; x_5; x_7); (x_1; x_6; x_7); (x_2; x_3; x_5); (x_2; x_4; x_5); (x_2; x_4; x_6); (x_2; x_5; x_6);$$
$$(x_3; x_4; x_5); (x_3; x_4; x_6); (x_4; x_5; x_7); (x_4; x_6; x_7); (x_5; x_6; x_7)$$

Thus, there exists a decomposition:

$$F = H(x_1 \oplus x_2, x_2 \oplus x_3, x_4, x_5, x_6, x_7).$$

Any other solution outside of the three-three variable COM (RC_{pq}) files, leads to contradictions.

4 Functional Decomposition

Functional decomposition is a method of logic synthesis adapted to implementation on memories. Therefore, its use in the design of digital circuits is becoming more and more significant. In [10,11], decomposition is defined as one of the most urgent tasks of logic synthesis. In particular, it is proposed to further develop the decomposition method involving using the so-called Decomposition Chart as it was discussed in [5]. In our opinion, the method proposed in [10] is much more developmental and then developed in [1]. The method's essence is the use of partition calculations, which makes it possible to formulate a convenient theorem in which a sufficient condition for decomposition consists in determining partitions on a set of numbered function table vectors.

Theorem 2. *Function* $F : D^n \to \{0,1\}^m$, $D^n \subseteq \{0,1\}^n$ *has the decomposition* $F = H(U, G(V, W))$ *only and only if there is a partition* $\Pi_G \geq P_V$ *such that* $P_U \cdot \Pi_G \leq P_F$.

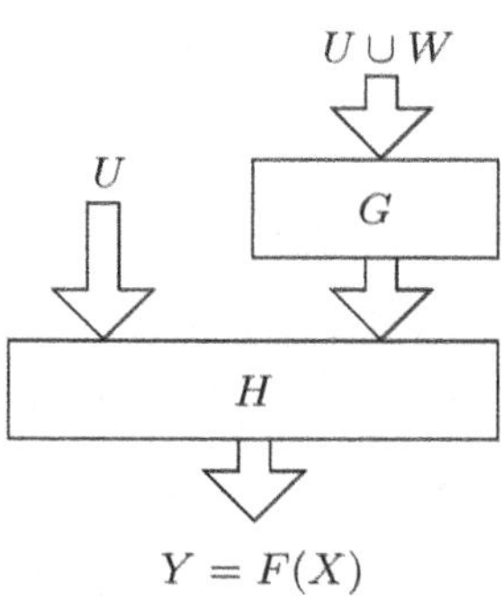

Fig. 2. Functional decomposition

The calculation algorithm can be reduced to the algorithm for calculating Maximal Compatibility Classes or to the graph coloring algorithm [8].

Two blocks B_i and B_j of partition $P_V = (B_1, \ldots, B_i, \ldots, B_j, \ldots, B_n)$ are compatible if partition $\gamma_{ij} = (B_1, \ldots, B_i, B_j, \ldots, B_n)$ obtained from P_V by merging B_i and B_j into a single block and meets $P_U \cdot \Pi_G \leq P_F$. Otherwise, these blocks are incompatible.

The advantage of this theorem is the possibility of using a graph coloring algorithm for calculating the Π_G partition, whose vertices are blocks B_i of partition P_V, and the edges are not matching pairs B_i, B_j, of these blocks.

Let's denote the number of elements in the largest quotient block τ and δ by $\gamma(\tau|\delta)$. Let $\Gamma(\tau|\delta) = \lceil \log_2 \gamma(\tau|\delta) \rceil$.

The set $\{P_1, P_2, \ldots, P_k\}$ is r-admissible for the partition P, where $r = k + \Gamma(P_1 \cdot \cdots \cdot P_k | P \cdot P_1 \cdot \cdots \cdot P_k)$. So you can calculate r-admissibility of two-block partitions P_i relative to P partition directly from P_i and P, simply by grouping the elements in the P_i partition block into subsets with the maximum number of elements belonging to the determined partition block P. Therefore, the quotient partition $P_i|P \cdot P_i$ is often simply marked by P_i writing the elements of this quotient in brackets.

If the set $\{P_1, P_2, \ldots, P_k\}$ is r-admissible ($k < r$) regarding P, there is a set of partitions $\{P_{k+1}, P_{k+2}, \ldots, P_r\}$, for which $P_1 \cdot \cdots \cdot P_k \cdot P_{k+1} \cdot \cdots \cdot P_r \leq P$. However, there is no set $\{P'_{k+1}, P'_{k+2}, \ldots, P'_{r-1}\}$ such that $P_1 \cdot \cdots \cdot P_k \cdot P'_{k+1} \cdot \cdots \cdot P'_{r-1} \leq P$. In other words, a necessary and sufficient condition for $P_1 \cdot \cdots \cdot P_k \cdot P_{k+1} \cdot \cdots \cdot P_m \leq P$ is m-admissibility of the set $\{P_1, P_2, \ldots, P_k\}$ regarding P.

It can also be shown that if $\{P_1, P_2, \ldots, P_k\}$ is m-admissible regarding P, then every subset of P is m'-admissible, where $m' \leq m$. It means that the necessary condition for $\{P_1, P_2, \ldots, P_k\}$ is m-admissible relative to P, is r-admissible of subsets of P such that $\forall P' \subseteq P : r(P') \leq m$. This simple statement allows to group easily 2-block partitions into sets of fixed admissibility.

Example 3. The way of selecting partitions with the same variables of the set U according to Theorem 2, we will discuss using the example of the ExDM2 function given in Table 3.

Table 3. ExDM2 Boolean function

	x_1	x_2	x_3	x_4	x_5	y
1	1	0	0	0	0	y_1
2	1	0	0	0	1	y_2
3	1	0	0	1	1	y_3
4	1	0	1	1	0	y_4
5	1	0	1	0	0	y_5
6	0	1	0	1	1	y_6
7	0	1	1	1	0	y_7
8	0	1	1	0	0	y_8
9	0	0	0	0	0	y_9
10	0	0	0	1	1	y_{10}
11	0	0	1	1	0	y_{11}
12	0	0	1	0	0	y_{12}

For this function, the partitions representing variables $x_1, \ldots, x_5$ are as follow:

$$P_1 = (\overline{1,2,3,4,5}, \overline{6,7,8,9,10,11,12}) \quad r = 1 + 3 = 4$$
$$P_2 = (\overline{1,2,3,4,5,9,10,11,12}, \overline{6,7,8}) \quad r = 1 + 4 = 5$$
$$P_3 = (\overline{1,2,3,6,9,10}, \overline{4,5,7,8,11,12}) \quad r = 1 + 3 = 4$$
$$P_4 = (\overline{1,2,5,8,9,12}, \overline{3,4,6,7,10,11}) \quad r = 1 + 3 = 4$$
$$P_5 = (\overline{1,4,5,7,8,9,11,12}, \overline{2,3,6,10}) \quad r = 1 + 3 = 4$$

So 4-admissible sets should be sought for P_1, P_3, P_4, P_5. Hence, further selection of the U set arguments consists in calculating r-admissibility for the pairs P_i, P_j from P_1, P_3, P_4, P_5.

$$P_1 \cdot P_3 = (\overline{1,2,3}; \overline{4,5}; \overline{6,9,10}; \overline{7,8,11,12}) \quad r = 4$$
$$P_1 \cdot P_4 = (\overline{1,2,5}; \overline{3,4}; \overline{8,9,12}; \overline{6,7,10,11}) \quad r = 4$$
$$P_1 \cdot P_5 = (\overline{1,4,5}; \overline{2,3}; \overline{7,8,9,11,12}; \overline{6,10}) \quad r = 5$$
$$P_3 \cdot P_4 = (\overline{1,2,9}; \overline{3,6,10}; \overline{5,8,12}; \overline{4,7,11}) \quad r = 4$$
$$P_3 \cdot P_5 = (\overline{1,9}; \overline{2,3,6,10}; \overline{4,5,7,8,11,12}) \quad r = 5$$
$$P_4 \cdot P_5 = (\overline{1,5,8,9,12}; \overline{2}; \overline{4,7,11}; \overline{3,6,10}) \quad r = 5$$

As a result of these calculations, we get information that only the $\{P_1, P_3, P_4\}$ set can be the 4-admissible set. So there is only one solution as $U = \{x_1, x_3, x_4\}$, where:

$$P_U = P_1 \cdot P_3 \cdot P_4 = (\overline{1,2}; \overline{3}; \overline{4}; \overline{5}; \overline{6,10}; \overline{9}; \overline{8,12}; \overline{7,11})$$
$$P_V = P_2 \cdot P_5 = (\overline{1,4,5,9,11,12}; \overline{2,3,10}; \overline{6}; \overline{7,8})$$

We create the partition $\Pi_{G'}$ in such a way that its first block is in form $\overline{1,4,5,9,11,12}$ and $\overline{6}$, while the other: $\overline{2,3,10}$ and $\overline{7,8}$. Therefore, we receive $\Pi_{G'} = (\overline{1,4,5,6,9,11,12}; \overline{2,3,7,8,10})$.

The table of G function (according to Fig. 3) is presented in Table 4. From this table, we calculate that the function modifying the address has the following form: $g = x_2\overline{x}_5 + \overline{x}_2 x_5 = x_2 \oplus x_5$. The function table H (presented in Table 5) is calculated from the product:

$$\begin{aligned} P(U) \cdot \Pi_{G'} &= (\overline{1,2}; \overline{3}; \overline{4}; \overline{5}; \overline{6,10}; \overline{9}; \overline{8,12}; \overline{7,11}) \cdot (\overline{1,4,5,6,9,11,12}; \overline{2,3,7,8,10}) \\ &= (\overline{1}; \overline{2}; \overline{3}; \overline{4}; \overline{5}; \overline{6}; \overline{10}; \overline{9}; \overline{12}; \overline{8}; \overline{11}; \overline{7}) \end{aligned}$$

Table 4. Function G

	x_2, x_5	g
$\overline{1,4,5,9,11,12}$	00	0
$\overline{6}$	11	0
$\overline{2,3,10}$	01	1
$\overline{7,8}$	10	1

Table 5. Function H

	x_1, x_3, x_4, g	Y
1	1000	y_1
2	1001	y_2
3	1011	y_3
4	1110	y_4
5	1100	y_5
6	0010	y_6
10	0011	y_{10}
9	0000	y_9
12	0100	y_{12}
8	0101	y_8
11	0110	y_{11}
7	0111	y_7

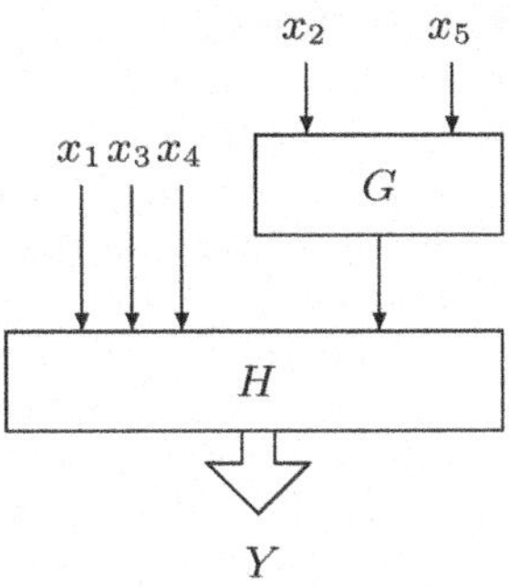

Fig. 3. Decomposition of ExDM2 Boolean function

5 Summary

The article deals with the issue of logic synthesis algorithms in hardware data analysis. Interest in hardware data analysis results from the specific development of programmable digital circuit technologies. The direct cause of this development was determined by Sasao in the fundamental work "Memory based logic synthesis". The title of this work clearly emphasizes the total change in the direction of the development of logic synthesis – so far focused mainly on realizations in gates structures.

There is no doubt that the current base of hardware implementation are memory structures, and the use of typical logic gates concerns the implementation of address modification systems for ROM. For these reasons, decomposition methods are increasingly important. In particular, linear decomposition and functional decomposition.

The system implementing linear decomposition includes a linear layer in which the variables are combined in pairs and a general layer into which the pairs of variables are introduced via ExOR gates – reducing the number of ROM inputs.

Functional decomposition is also an important method of logic synthesis used for implementation on ROMs. In functional decomposition, the input variables of the Boolean function are divided into two joint or disjoint subsets U and V for blocks H and G. The input variables of block H are variables directly from the set U and output variables of the block G (reduced variables of the set V). The choice of variables supports the concept of r-admissibility. R-admissibility allows determining variables of the decomposition with the smallest number of G block outputs.

References

1. Borowik, G.: Optimization on the complementation procedure towards efficient implementation of the index generation function. Appl. Math. Comput. Sci. **28**(4), 803–815 (2018). https://doi.org/10.2478/amcs-2018-0061
2. Borowik, G., Łuba, T., Klempous, R.: Comparison of algorithms for dimensionality reduction and their application to index generation functions. In: 2020 IEEE 15th International Conference of System of Systems Engineering (SoSE), pp. 283–288 (2020). https://doi.org/10.1109/SoSE50414.2020.9130484
3. Sasao, T.: Index generation functions. In: Logic Synthesis for Pattern Matching, EPFL Workshop on Logic Synthesis & Verification (2015)
4. Sasao, T., Matsuura, K., Iguchi, Y.: An algorithm to find optimum support-reducing decompositions for index generation functions. In: Design, Automation & Test in Europe Conference & Exhibition (DATE), 2017, pp. 812–817 (2017). https://doi.org/10.23919/DATE.2017.7927100
5. Sasao, T.: Memory-Based Logic Synthesis, 1st edn. Springer-Verlag, New York (2011). https://doi.org/10.1007/978-1-4419-8104-2
6. Białas, M.: Implementation of software supporting the reduction and classification of data. Engineering Thesis, Warsaw University of Technology, Warsaw (2018)
7. Borowik, G., Luba, T.: Fast algorithm of attribute reduction based on the complementation of Boolean function. In: Klempous, R., Nikodem, J., Jacak, W., Chaczko, Z. (eds.) Advanced Methods and Applications in Computational Intelligence. Topics in Intelligent Engineering and Informatics, vol. 6, pp. 25–41. Springer, Heidelberg (2014). https://doi.org/10.1007/978-3-319-01436-4_2
8. Brzozowski, J.A., Łuba, T.: Decomposition of Boolean functions specified by cubes. J. Multi-Valued Log. Soft Comput. **9**, 377–417 (2003)
9. Sasao, T.: Index generation functions: minimization methods. In: 2017 IEEE 47th International Symposium on Multiple-Valued Logic (ISMVL), pp. 197–206, May 2017. https://doi.org/10.1109/ISMVL.2017.22
10. Łuba, T., Selvaraj, H.: A general approach to Boolean function decomposition and its applications in FPGA-based synthesis. VLSI Des. **3**(3–4), 289–300 (1995)
11. Rokach, L., Maimon, O.: Data mining using decomposition methods. In: Maimon, O., Rokach, L. (eds.) Data Mining and Knowledge Discovery Handbook, pp. 981–998. Springer, Boston (2010). https://doi.org/10.1007/978-0-387-09823-4_51
12. Borowik, G., Łuba, T., Poźniak, K.: New trends in logic synthesis for both digital designing and data processing. In: Romaniuk, R.S. (ed.) Photonics Applications in Astronomy, Communications, Industry, and High-Energy Physics Experiments 2016, vol. 10031, pp. 1371–1381. International Society for Optics and Photonics, SPIE (2016). https://doi.org/10.1117/12.2249240
13. Łuba, T.: Decomposition of multiple-valued functions. In: Proceedings 25th International Symposium on Multiple-Valued Logic, pp. 256–261 (1995). https://doi.org/10.1109/ISMVL.1995.513540
14. Mazurkiewicz, T., Łuba, T.: Linear and non-linear decomposition of index generation functions. In: 2019 MIXDES – 26th International Conference "Mixed Design of Integrated Circuits and Systems", pp. 246–251 (2019). https://doi.org/10.23919/MIXDES.2019.8787031

15. Sasao, T.: A reduction method for the number of variables to represent index generation functions: s-min method. In: 2015 IEEE International Symposium on Multiple-Valued Logic, pp. 164–169, May 2015. https://doi.org/10.1109/ISMVL.2015.40
16. Wąsicki, D.: Gate based synthesis of Index Generation Functions. Master Thesis, Warsaw School of Computer Science, Warsaw (2019)

Differential Evolution with Taguchi Crossover Using Polar Coordinates

Peter Stubberud(✉)

University of Nevada Las Vegas, Las Vegas, NV 89154-4026, USA
Peter.Stubberud@UNLV.edu

Abstract. This paper compares the performance of two differential evolution algorithms. The algorithms are identical except that one of the algorithms is implemented in rectangular coordinates and the other algorithm is implemented using both rectangular and a cascade of terms in polar coordinates. Both algorithms use an elitist, ranking, random selection method and both two and three level Taguchi crossover. To compare the performance of the algorithms, both algorithms are applied to 13 commonly used global numerical optimization test functions, including a spherical, three hyper-ellipsoid, the sum of different powers, Rastrigin's, Schwefel's, Griewank's, Rosenbrock's valley, Styblinski-Tang, Ackley's Path, Price-Rosenbrock, and Eggholder's functions. The test results show that the algorithm that is implemented using both rectangular and polar coordinates performed better than the algorithm implemented using only rectangular coordinates.

Keywords: Differential evolution · Genetic algorithm · Taguchi crossover · Averaging crossover · Elitist selection · Ranking selection · Random selection · Polar coordinates · Rectangular coordinates · Compare

1 Introduction

The need for global optimization techniques are common in engineering and other scientific and mathematical fields. In engineering, global optimization techniques typically determine an optimal set of parameters that minimize an objective, or cost, function. Ideally, an objective function mathematically describes a problem's desired performance over a constrained solution space. In engineering, objective functions are often a norm of the difference between a system's performance and its ideal performance subject to a constraint such as the optimal parameters result in a stable system. As computing capabilities have increased, many researchers have developed heuristic optimization methods for solving constrained, multi-modal, optimization problems. These methods include simulated annealing [1], particle swarm optimization [2], colony optimization [3,4], genetic algorithms [5], and differential evolution [6,7]. Although no algorithm can solve

H. Selvaraj et al. (Eds.): ICSEng 2020, LNNS 182, pp. 128–138, 2021.
https://doi.org/10.1007/978-3-030-65796-3_12

all types of optimization problems [7,8], differential evolution has become popular in engineering optimization applications because of the method's efficiency and flexibility.

In many heuristic optimization methods, including differential evolution, objective functions are typically defined so that candidate parameters, or candidate solutions, are represented by vectors [6]. In many engineering applications such as applications involving system transfer functions, the elements of the solution vectors represent polynomial coefficients. However, small perturbations in polynomial coefficients typically result in significant perturbations in the roots of the polynomial [9]. Therefore, in engineering optimization problems that involve polynomial coefficients, using a cascade structure of polar coordinate terms that represent polynomial root locations could lead to optimal solutions that are less sensitive to the parameters that minimize an objective function, and as a result, this could improve the performance of an optimization algorithm.

Several papers have proposed using polar or spherical coordinates to improve the performance of differential evolution algorithms [10,11]. In this paper, the candidate solutions in the differential evolution optimization algorithm in [12] are redefined to represent a cascade of polar coordinate terms that represent the equivalent of polynomial root locations in polar coordinates. Section 2 describes both algorithms in detail. Both algorithms are applied to 13 commonly used global numerical optimization test functions that are listed in Table 1. These results are presented in Sect. 3.

2 Differential Evolution with Taguchi Crossover

The differential evolution algorithm in [12] and in this paper use an elitist, ranking, random selection method. Elitist selection methods guarantees that the fittest individual, or the candidate solution with the lowest cost, survives the selection process and also guarantee that the fittest individual is selected in all crossover and mutation operations. The candidate solutions that survive the selection process are randomly selected for mutation using a differential evolution operator to improve convergence, a differential evolution operator to improve diversity and a crossover operator that improves convergence and diversity. After generating mutations, candidate solutions from the entire population are randomly selected for two or three level Taguchi crossover. The algorithm can be summarized as follows:

- Initialize population ($K(0)$ candidate solutions)
- $n = 1$
- REPEAT
  - Ranking and stochastic selection (select $M(n)$ solutions)
  - Differential evolution mutation for convergence
  - Differential evolution mutation for diversity
  - Crossover for convergence and diversity
  - Ensure new solutions are in the solution space

Table 1. Optimization test functions and their solution spaces

Function	Solution space $x_k \in [\bullet, \bullet]$
$J_1(\mathbf{x}) = \sum_{k=1}^{N} x_k^2$	$[-5.12, 5.12]$
$J_2(\mathbf{x}) = \sum_{k=1}^{N} k x_k^2$	$[-5.12, 5.12]$
$J_3(\mathbf{x}) = \sum_{k=1}^{N} [k(x_k - 5k)]^2$	$[-500, 500]$
$J_4(\mathbf{x}) = \sum_{m=1}^{N} \sum_{k=1}^{m} x_k^2$	$[-65.536, 65.536]$
$J_5(\mathbf{x}) = \sum_{k=1}^{N} \lvert x_k \rvert^{k+1}$	$[-1, 1]$
$J_6(\mathbf{x}) = 10N + \sum_{k=1}^{N} \left[x_k^2 - 10\cos(2\pi x_k)\right]$	$[-5.12, 5.12]$
$J_7(\mathbf{x}) = 418.9828872724338N - \sum_{k=1}^{N} x_k \sin\left(\sqrt{\lvert x_k \rvert}\right)$	$[-500, 500]$
$J_8(\mathbf{x}) = 1 + \sum_{k=1}^{N} \frac{x_k^2}{4000} - \prod_{k=1}^{N} \cos\left[\frac{x_k}{\sqrt{k}}\right]$	$[-600, 600]$
$J_9(\mathbf{x}) = \sum_{k=1}^{N-1} \left[100(x_k^2 - x_{k+1})^2 + (x_k - 1)^2\right]$	$[-5, 10]$
$J_{10}(\mathbf{x}) = 39.16616570377142N + \sum_{k=1}^{N} \left(x_k^4 - 16x_k^2 + 5x_k\right)$	$[-5, 5]$
$J_{11}(\mathbf{x}) = 20 + e - 20e^{-0.2\sqrt{\frac{1}{N}\sum_{k=1}^{n} x_k^2}} - e^{\sum_{k=1}^{N} \frac{\cos(2\pi x_k)}{N}}$	$[32.768, 32.768]$
$J_{12}(\mathbf{x}) = 100(x_2 - x_1^2)^2 + \left[6.4(x_2 - 0.5)^2 - x_1 - 0.6\right]^2$	$[-5, 5]$
$J_{13}(\mathbf{x}) = 959.640662720851 - x_1 \sin\left(\sqrt{\lvert x_1 - x_2 - 47 \rvert}\right) - (x_2 + 47)\sin\left(\sqrt{\lvert \frac{x_1}{2} + x_2 + 47 \rvert}\right)$	$[-512, 512]$

  - Taguchi crossover ($K(n)$ solutions)
  - $n = n + 1$
- UNTIL (repeat until convergence)

where $K(0)$ is the number of randomly selected candidate solutions for the initial population, n is the algorithm's iteration number, $M(n)$ is the number of candidate solutions that survive the nth selection process, and $K(n)$ is the size of population after all mutation and crossover operations during the algorithm's nth iteration.

The main difference between the algorithm in [12] and this paper's algorithm is that the algorithm in [12] represents the candidate solutions in rectangular coordinates for all operations while this paper's algorithm represents the candidate solutions in rectangular coordinates for some operations and a cascade of terms in polar coordinates for other operations. For example, consider an optimization cost function, $J(\mathbf{x})$, where $\mathbf{x}$ is a 5 dimensional vector. For the algorithm in [12], the vector, $\mathbf{x}$, would be represented as $\left[x_1\; x_2\; x_3\; x_4\; x_5\right]^T$ for all operations while the algorithm in this paper represents the vector, $\mathbf{x}$, as $\left[r_1\; \theta_1\; r_3\; \theta_3\; x_5\right]^T$ where

$$r_k = \sqrt{x_k^2 + x_{k+1}^2} \qquad \text{and} \qquad \theta_k = \arctan\left(\frac{x_{k+1}}{x_k}\right)$$

for some of the algorithm's operations.

2.1 Population Initialization

In both algorithms, the initial population of candidate solutions is selected randomly so that the candidate solutions are uniformly distributed over the solution space. For a hyper-rectangular solution space, the initial population of candidate solutions can be chosen in rectangular coordinates and then converted to polar coordinates. In Matlab, an initial population of K candidate solutions can be generated over a hyper-rectangular solution space using

$$G = kron(\mathbf{X}_c, ones(1, K)) + diag(\mathbf{X}_s) * (rand(N, K) - 0.5);$$

where G is the matrix containing the $K(0)$ initial candidate solutions, N is the dimension of the solution space, $\mathbf{X}_c$ is a vector containing the rectangular coordinates of the center of the solution space, and $\mathbf{X}_s$ is a vector containing the size of each dimension of the solution space in rectangular coordinates.

For a hyper-ellipsoid solution space, the initial population of candidate solutions can be chosen in polar coordinates. In Matlab, an initial population of K candidate solutions can be generated over a hyper-ellipsoid solution space using

$$G(1:2:N,:) = kron(\mathbf{X}_r, ones(1, K)).*rand(ceil(N/2), K);$$
$$G(2:2:N,:) = 2 * pi * rand(floor(N/2), K);$$

where G is the matrix containing the $K(0)$ initial candidate solutions, N is the dimension of the solution space, $\mathbf{X}_r$ is a vector containing the radii of the hyper-eliptical solution space. If the solution space dimension, N is odd, then the radii in the Nth dimension are not associated with a phase and are all positive. Their signs can be randomly altered using

$$G(N,:) = G(N,:). * (-1).\hat{}\,randi(2, 1, K).$$

If the hyper-ellipsoid solution space is not centered at the origin, then initial candidate solutions can be moved in either rectangular or polar coordinates. In rectangular coordinates, the initial candidate solutions would be converted to rectangular coordinates, moved and then converted into polar coordinates. In polar coordinates, the candidate solutions of the form, $\left[r\; \theta\right]^T$ and centered at

$[0\ 0]^T$ could be moved to $[R\ \Theta]^T$ and centered at $[r_0\ \gamma]^T$ using the transformations,

$$R = \sqrt{r^2 + r_0^2 + 2rr_0\cos(\theta - \gamma)}$$
$$\Theta = \arctan\left(\frac{r\sin(\theta) + r_0\sin(\gamma)}{r\cos(\theta) + r_0\cos(\gamma)}\right).$$

2.2 Ranking and Stochastic Selection

At the beginning of both algorithms' nth iteration, each of the $K(n)$ candidate solutions, $\mathbf{x}_k(n)$ for $k = 1, 2, \cdots, K(n)$, is evaluated by the optimization criterion, $J(\mathbf{x}_k(n))$, and ranked according to its cost. After ranking the candidate solutions, $\mathbf{x}_k(n)$, they are sorted from highest cost to lowest cost and are assigned consecutive integers from 1 to $K(n)$; that is, the least fit candidate solution is assigned one and the most fit candidate solution is assigned $K(n)$. The selection probability for each candidate solution is assigned according to each solution's rank so that the selection probability, $P\big(\mathbf{x}_k(n)\big)$, of the kth ranked candidate solution, $\mathbf{x}_k(n)$, is

$$P\big(\mathbf{x}_k(n)\big) = \sum_{m=1}^{k} \Delta p_m \tag{1}$$

where

$$\Delta p_m = \frac{1}{K(n)}\left[\eta^- + (\eta^+ - \eta^-)\frac{m-1}{K(n)-1}\right] \tag{2}$$

and η^- is selected so $P\big(\mathbf{x}_1(n)\big) = \eta^-/K(n)$ which is the selection probability of the worst candidate solution [13].

Because this algorithm uses an elitist selection method, the best candidate solution is guaranteed to survive the selection process which implies that

$$P\big(\mathbf{x}_{K(n)}(n)\big) = 1. \tag{3}$$

Substituting (2) into (1) and the resulting equation into (3), implies that

$$\sum_{m=1}^{K(n)} \frac{1}{K(n)}\left[\eta^- + \left(\eta^+ - \eta^-\right)\frac{m-1}{K(n)-1}\right] = 1. \tag{4}$$

Solving (4), $\eta^+ = 2 - \eta^-$ where $0 < \eta^- < \eta^+$. The candidate solutions that survive the selection process are often referred to as the mating pool.

2.3 Differential Evolution Mutation for Convergence

For the differential evolution mutation for convergence operator, both algorithms randomly select solutions, except for the best candidate solutions, with a probability of $P_{DE1}(n)$. Because these are elitist algorithms, the best candidate solutions are always selected for this mutation operator. The differential evolution mutation for convergence operator produces a mutant candidate solution, $\mathbf{v}$, using

$$\mathbf{v} = \mathbf{x}_k(n) + R\left[\mathbf{x}_m(n) - \mathbf{x}_j(n)\right]$$

where $\mathbf{x}_k(n)$ is the candidate solution randomly selected for mutation, $\mathbf{x}_m(n)$ and $\mathbf{x}_j(n)$ are two randomly selected candidate solutions from the mating pool, and R is a uniformly distributed random number from the interval [0,1]. The difference between the algorithms is that the algorithm in [12] uses rectangular coordinates and the algorithm in this paper uses polar coordinates.

2.4 Differential Evolution Mutation for Diversity

For the differential evolution mutation for diversity operator, both algorithms randomly select solutions, except for the best candidate solutions, with a probability of $P_{DE2}(n)$. Because these are elitist algorithms, the best candidate solutions are always selected for this mutation operator. The differential evolution mutation for diversity operator uses rectangular coordinates to generate a mutant candidate solution, $\mathbf{v}$, so that

$$\mathbf{v} = \mathbf{x}_k(n) + \frac{1}{4} diag(\mathbf{R}) diag(\mathbf{X}) \left[\mathbf{x}_m(n) - \mathbf{x}_j(n)\right]$$

where $\mathbf{x}_k(n)$ is the candidate solution randomly selected for mutation, $\mathbf{x}_m(n)$ and $\mathbf{x}_j(n)$ are two randomly selected candidate solutions from the mating pool, $\mathbf{R}$ is a vector whose elements are uniformly distributed random number from the interval [0,1], and $\mathbf{X}$ is a vector containing half the size of each dimension of a rectangular solution space or the radii of an elliptic solution space.

2.5 Crossover for Convergence and Diversity

Crossover operators combine the characteristics of two or more other candidate solutions to create a new candidate solution. For this crossover operation, both algorithms randomly select solutions, except for the best candidate solutions, with a crossover probability of $P_{cr}(n)$. Because these are elitist algorithms, the best candidate solutions are always selected for this crossover operator. Each candidate solution selected for crossover generates two new candidate solutions.

To improve convergence, a new candidate solution, $\mathbf{v}$, is produced by averaging the selected candidate solution, $\mathbf{x}_k(n)$, with another randomly selected candidate solution, $\mathbf{x}_m(n)$, from the mating pool so that

$$\mathbf{v} = \left[\mathbf{x}_k(n) + \mathbf{x}_m(n)\right]/2.$$

The difference between the algorithms is that the algorithm in [12] uses rectangular coordinates and the algorithm in this paper randomly uses rectangular or polar coordinates to perform this operation.

In [12], the other candidate solution is generated to improve diversity by randomly changing the signs of the elements of the newly formed candidate solution, **v**, and then circularly shifting the elements by a uniformly distributed integer. In Matlab, this new vector, **w**, can be created by

$$w = sign(randi(2, N, 1) - 1.5). * circshift(v, randi(N))$$

where N is the dimension of the solution space. For the algorithm in this paper, the other candidate solution, **w**, is generated to improve diversity by randomly rotating the newly formed candidate solution, **v**, by a uniformly distributed random angle over the interval $[-\pi, \pi]$. In Matlab, this new vector, **w**, can be created by

$$\begin{aligned} &w = v; \\ &Rotate = 2 * pi * rand(N, 1) - pi; \\ &w(2:2:N) = w(2:2:N) + Rotate(2:2:N); \end{aligned}$$

2.6 Solution Space

In both algorithms, two methods are used to bring a mutant candidate solution, **v**, that falls outside the solution space back into the solution space. For a rectangular solution space, a mutant vector, **v**, created by the differential evolution mutation for diversity or the crossover for diversity operations is converted to rectangular coordinates, if necessary, and then brought back into the solution space using

$$\mathbf{v} = mod\left[\mathbf{v} - (\mathbf{X}_c - 0.5\mathbf{X}_s), \mathbf{X}_s)\right] + (\mathbf{X}_c - 0.5\mathbf{X}_s)$$

where $\mathbf{X}_c$ is a vector containing the rectangular coordinates of the center of the solution space, and $\mathbf{X}_s$ is a vector containing the size of each dimension of the solution space in rectangular coordinates. For an elliptical solution space centered at the origin, a mutant vector, **v**, created by the differential evolution mutation for diversity or the crossover for diversity operations is converted to polar coordinates, if necessary, and then brought back into the solution space using

$$\begin{aligned} r_k &= rem\left(r_k, r_{k\,\max}\right) \\ \theta_k &= \theta_k + \pi \end{aligned}$$

where rem is the remainder function, r_k is a radius that places the candidate solution outside of the solution space, $r_{k\,\max}$ is the maximum value of r_k that keeps the candidate solution inside the solution space and θ_k is the angle associated with the radius, r_k.

In both algorithms, a mutant vector that was created by the differential evolution for convergence or the crossover for convergence operations and lies outside the solution space is moved into the solution space by changing the mutant vector's elements that lie outside solution space to the nearest edge of the solution space. After the mutant solutions are brought back into the solution space, they are added to the mating pool.

2.7 Taguchi Crossover

Taguchi crossover is effective at generating a new candidate solution by selecting near optimal elements from two or more solution vectors. Taguchi crossover can greatly increase convergence rates [14,15]. Because this algorithm is an elitist algorithm, the best candidate solution is always selected for Taguchi crossover. The other solutions are selected randomly from the mating pool with a probability of P_{Tc}. For two level Taguchi crossover [16], another candidate solution is randomly selected from the mating pool. For three level Taguchi crossover, two other candidate solutions are randomly selected from the mating pool.

2.8 Managing Population Size

In both algorithms, the selection, mutation and crossover operations generate a random number of candidate solutions with each iteration of the algorithm. It can be shown [12] that the mean number, $E[K(n)]$, of the candidate solutions is

$$E[K(n)] = [(M(n) - 1)(1 + P_3(n)) + 4](1 + P_{Tc}) + 2 \tag{5}$$

where $E[K(n)]$ is the expected size of population after Taguchi crossover during the algorithm's nth iteration, $M(n)$ is the number candidate solutions that survive the nth selection process, and $P_3(n) = P_{DE1}(n) + P_{DE2}(n) + 2P_{cr}(n)$. To maintain the population's size, $K(n)$, at the population's initial size, $K(0)$, the probabilities of mutation and crossover must vary so that

$$E[K(n)] = K(0). \tag{6}$$

Substituting (5) into (6) and solving for $P_3(n)$,

$$P_3(n) = \frac{K(0) - 6 - 4P_{Tc}}{(1 + P_{Tc})(M(n) - 1)} - 1 \tag{7}$$

where P_{Tc} is assumed to be fixed and $P_3(n)$ varies. In both algorithms, the requirement in (7) is met by fixing $P_{DE1}(n)$ and $P_{DE2}(n)$, and letting

$$P_{cr}(n) = [P_3(n) - P_{DE1}(n) - P_{DE2}(n)]/2. \tag{8}$$

3 Optimization Test Function Results

To compare these algorithms, both algorithms were applied to 13 commonly used global numerical optimization test functions. The 13 test functions, $J_1(\mathbf{x})$ through $J_{13}(\mathbf{x})$, and their solution spaces are listed in Table 1. These functions include a spherical, three hyper-ellipsoid, the sum of different powers, Rastrigin's, Schwefel's, Griewank's, Rosenbrock's valley, Styblinski-Tang, Ackley's Path, Price-Rosenbrock, and Eggholder's functions. The first 11 functions, $J_1(\mathbf{x})$ through $J_{11}(\mathbf{x})$, are multidimensional functions and are tested for two dimensions ($N = 2$) and 34 dimensions ($N = 34$). Functions $J_{12}(\mathbf{x})$ and $J_{13}(\mathbf{x})$ are two dimensional functions and were only tested for $N = 2$. For all 13 functions, $J_{min} = 0$ where J_{min} is the global minimum value of the cost function. The algorithm was applied 1,000 times to each function and was assumed to converge when a solution, $\mathbf{x}_{opt}$, was determined so that $J(\mathbf{x}_{opt}) \leq J_{tol}$ where J_{tol} for each function is listed in Table 2. For all functions and dimensions, the initial population, $K(0)$, was set to 50, $\eta^- = 0.9$, $P_{DE1}(n) = 0.16$, $P_{DE2}(n) = 0.2$, $P_{Tc} = 0.24$ and $P_{cr}(n)$ was set each iteration according to (8). Both two level ($L = 2$) and three level ($L = 3$) Taguchi crossover were used. The results are listed in Table 2.

Table 2. Results for the optimization of the 2-D and 34-D test functions where $K(0) = 50$, $\eta^- = 0.9$ $P_{Tc} = 0.24$, $P_{DE1} = 0.16$, and $P_{DE2} = 0.2$ for two level ($L = 2$) and three level ($L = 3$) Taguchi crossover. Results are averages over 1,000 runs.

Test function	J_{tol}	Rectangular $N = 2$ Mean Iter. St Dev Iter.		Polar $N = 2$ Mean Iter. St Dev Iter.		Rectangular $N = 34$ Mean Iter. St Dev Iter.		Polar $N = 34$ Mean Iter. St Dev Iter.	
		$L = 2$	$L = 3$	$L = 2$	$L = 3$	$L = 2$	$L = 3$	$L = 2$	$L = 3$
$J_1(\mathbf{x})$	10^{-300}	48	34	42	32	122	96	106	83
		21	17	6	5	3	3	4	3
$J_2(\mathbf{x})$	10^{-300}	48	34	42	33	130	102	107	84
		20	17	6	5	4	3	4	3
$J_3(\mathbf{x})$	10^{-8}	40	33	41	33	348	179	327	182
		7	5	7	6	47	20	32	32
$J_4(\mathbf{x})$	10^{-300}	68	68	63	63	10675	5160	504	452
		34	33	8	8	1563	1310	446	71
$J_5(\mathbf{x})$	10^{-300}	43	33	43	35	1230	597	316	240
		18	16	6	5	140	92	19	14
$J_6(\mathbf{x})$	10^{-300}	32	24	25	20	80	60	74	52
		11	9	4	3	4	2	11	4
$J_7(\mathbf{x})$	10^{-10}	42	33	55	45	146	97	183	180
		5	4	9	8	12	7	34	18
$J_8(\mathbf{x})$	10^{-300}	39	35	59	60	90	76	76	62
		20	18	92	104	4	4	56	4
$J_9(\mathbf{x})$	10^{-2}	516	633	130	124	33270	17660	25140	18940
		969	1221	44	51	16850	19700	800	590
$J_{10}(\mathbf{x})$	10^{-11}	30	25	39	40	84	57	376	217
		3	3	19	45	4	2	990	790
$J_{11}(\mathbf{x})$	10^{-15}	43	28	43	34	108	68	90	66
		23	16	6	5	16	17	5	4
$J_{12}(\mathbf{x})$	10^{-2}	32	27	12	12	NA	NA	NA	NA
		90	87	15	13	NA	NA	NA	NA
$J_{13}(\mathbf{x})$	10^{-11}	1231	1382	495	524	NA	NA	NA	NA
		1610	1916	533	618	NA	NA	NA	NA

4 Summary and Conclusions

This paper compares two differential evolution algorithms that are identical except that one of the algorithms is implemented in rectangular coordinates and the other algorithm is implemented using both rectangular and polar coordinates. The algorithms are applied to 13 commonly used global numerical optimization test functions.

The results in Table 2 show that for the 2-D test functions, both algorithms performed similarly. The exceptions are test functions 7 and 8 where the rectangular coordinate algorithm performed a little better than the polar coordinate algorithm; however, the polar coordinate algorithm performed significantly better than the rectangular coordinate algorithm for test functions 9 and 13.

For the 34-D test functions, the results in Table 2 show that both algorithms performed similarly. The exceptions are test function 7 and 10 where the rectangular coordinate algorithm performed a little better than the polar coordinate algorithm; however, the polar coordinate algorithm performed significantly better than the rectangular coordinate algorithm for test functions 4, 5 and 9.

References

1. Chen, S., Istepanian, R., Luk, B.: Digital IIR design using adaptive simulated annealing. Digit. Signal Process. **11**, 241–251 (2001)
2. Bonyadi, M., Michalewicz, Z.: Particle swarm optimization for single objective continuous space problems: a review. Evol. Comput. **25**, 1–54 (2017)
3. Karaboğa, D.: An Idea Based on Honey Bee Swarm For Numerical Optimization. Erciyes University, Computer Engineering Department. Technical Report TR06 (2005)
4. Niu, B.: Bacterial colony optimization. Discret. Dyn. Nat. Soc. **698057**, 28 (2012)
5. Goldberg, D.: Genetic Algorithms in Search, Optimization and Machine Learning. Addison Wesley, Boston (1989)
6. Storn, R., Price, K.: Differential evolution - a simple and efficient heuristic for global optimization over continuous spaces. J. Global Optim. **11**, 341–359 (1997)
7. He, Y., Yuen, S.Y., Lou, Y., Zhang, X.: A sequential algorithm portfolio approach for black box optimization. Swarm vol. Comput. **44**, 559–570 (2019)
8. Wolpert, D.H., Macready, W.G.: No free lunch theorems for optimization. IEEE Trans. Evol. Comput. **1**(1), 67–82 (1997)
9. Proakis, J., Manolakis, D.: Digital Signal Processing Principles. Algorithms and Applications, 4th edn. Pearson Prentice Hall, New Jersey (2007)
10. Yang, J., Wei, J., Liu, J.: Adaptive differential evolution algorithm based on gradient and polar coordinates search strategies. In: 2015 11th International Conference on Computational Intelligence and Security, pp. 274–277 (2015)
11. Kuang, D., Zheng, J.: Stategies based on polar coordinates to keep diversity in multi-objective genetic algorithms. In: 2005 IEEE Congress on Evolutionary Computation, vol. 2, pp. 1276–1281 (2005)
12. Stubberud, P.: The design of digital IIR filters using a differential evolution optimization algorithm with Taguchi crossover. In: 2019 IEEE 10th Annual Ubiquitous Computing, Electronics & Mobile Communication Conference (UEMCON), pp. 462–468 (2019)

13. Whitley, D.: The GENITOR algorithm and selective pressure: why rank-based allocation of reproductive trials is best. In: Proceedings of the Third International Conference on Genetic Algorithms, pp. 116–123 (1989)
14. Leung, Y.: An orthogonal genetic algorithm with quantization for global numerical optimization. IEEE Trans. Evol. Comput. **52**, 41–53 (2001)
15. Tsai, J., Liu, T., Chou, J.: Hybrid Taguchi-genetic algorithm for global numerical optimization. IEEE Trans. Evol. Comput. **8**, 365–377 (2004)
16. Stubberud, P., Jackson, M.: A hybrid orthogonal genetic algorithm for global numerical optimization. In: International Conference on Systems Engineering, pp. 1–6 (2008)

Using Machine Learning Techniques to Estimate the Remaining Useful Life of a System with Different Types of Datasets

Carlos Lemus and Shahram Latifi(✉)

University of Nevada Las Vegas, Las Vegas, NV 89154, USA
lemus@unlv.nevada.edu, Shahram.latifi@unlv.edu

Abstract. Intelligent Prognostics and Health Management (PHM) technology aims to estimate the Remaining Useful Life (RUL) of a subsystem or a component using data collected by sensors. The rise in the complexity of systems requires new models to capture the relationship between the sensors and the RUL. Novel deep Convolutional Neural Networks (CNN) have been proposed as an approach for estimating the RUL. However, large amount of data is needed to use Machine Learning (ML) techniques. We explore current ML methods being used with different types of datasets and provide a conclusion on deciding what learning method works best with unique datasets. We find that, for most systems, the ML method used highly depends on the dataset and can greatly decrease the cost and increase the reliability.

Keywords: Deep learning · Degradation data · Fault detection · Convolutional neural networks · Prognostics and health management

1 Introduction

As the complexity of systems increases, periodic preventive and reactive maintenance becomes less feasible as a form of prognosis and health management (PHM). This becomes more apparent in engineering maintenance fields where the safety and reliability of the system is crucial for the preservation of human life or property. These include fields such as aerospace [1], automotive [2], industrial systems [3] and many others. As a result, there is a great need for better intelligent PHM in engineering maintenance to detect anomalies, diagnose faults, and predict the Remaining Useful Lifetime (RUL) of a system.

A PHM system is described as being able to first diagnosing and locating faulty components and secondly anticipating when faulty components require maintenance actions [4]. This can be achieved by proper planning when designing a system for PHM and should be done early in the project planning steps as it is often forgotten and is not a significant budget in operating budget [5]. It can be easily inferred that using PHM in a system would result in an improvement in reducing cost and repairs. Figure 1 demonstrates the results of a Tactical Stirling cryocoolers [6]. The cost of replacing a system after failure is $10M, after reaching a fixed age is $6.6M and after employing the PHM is about $3.6M. It is clear to see that being able to accurately.

H. Selvaraj et al. (Eds.): ICSEng 2020, LNNS 182, pp. 139–147, 2021.
https://doi.org/10.1007/978-3-030-65796-3_13

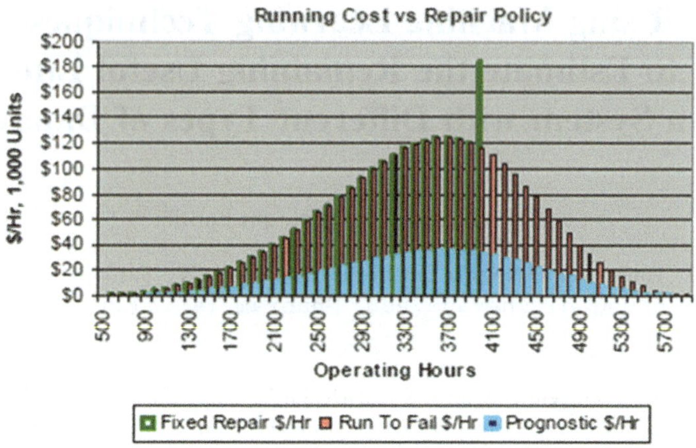

Fig. 1. Running benefits of using prognostics [6]

Machine learning (ML), a data-driven approach (one of three traditional PHM methods), has become a prevalent solution to predicting the RUL of a system. Predictions based on acquired data of the system's health can be made about the future health of a system [7]. Since sufficient prior knowledge of the degradation of the system components is needed to predict the RUL, a large amount of data from sensors monitoring the system is typically required. While most data is recorded over time, Texas A&M has developed DFA (Distribution Fault Anticipation) Technology to collect data in real-time to detect anomalies and identify the failed circuits [8]. As a result, DFA Technology that is tapped in real-time to electric-grid can prevent power outages or wildfires caused by predicting and preventing faults in the power lines. Being able to collect system data is the first step in system diagnostics using ML.

Data can come from many different systems with varying data that requires devising a unique ML model. For example, monitored data from sensors on the large commercial turbofan (C-MAPSS) by NASA [9] has been popular amongst researchers exploring Deep Convolutional Neural Networks (DCNN) in the prediction and classification of RUL. It contains more than 21 labeled sensor data of multiple extractable features. Most importantly, older traditional non-ML PHM algorithms have been used on this data that can show the improvements made with ML such as analytical behavior models [10], and multi-layer Perceptron [11].

However, different data sets require different methods with unique challenges of extracting useful representative features from the raw data. Furthermore, not every system will have the large amount of data required to train a ML model. In this case, it is still important to fall back on common PHM techniques still available. Deterioration of systems can still be modeled through Physics-Based Models (PbM) or other data driven models such as Stochastic and Statistical algorithms. For example, crack growth can be accurately modeled using stress intensity factor range to crack growth under fatigue stress regiment [12]. The tradeoff is that previous knowledge of the system is needed, but is useful for systems were sensor data may not be possible.

Sensor data has been collected from multiple systems including temperature, rolling bearing speeds, fuel-cells power charge, and even images from many different fields.

This paper studies the datasets using ML to predict RUL and identifies what is necessary when considering implementing ML in PHM to predict the RUL and degradation. Section 2 covers the popular NASA C-MAPSS used by many papers to implement DCNN for RUL. Section 3 investigates the use of time-frequency domain in the PRONOSTIA dataset to develop a Neural Network. Section 4 demonstrates a non-time series data that uses different ML methods. Section 5 introduces the complexity of implementing PHM in certain systems using different features for RUL. Section 6 discusses additional indirect approaches for predicting the degradation of a system. Section 7 concludes the paper by discussing the datasets and steps to be taken using neural networks for RUL in PHM.

2 C-MAPSS

As mentioned, NASA has provided a large amount of sensor data regarding the RUL of their large commercial turbofan. Zhang et al. first proposed discussing the application of an unsupervised artificial learning network to diagnose the failure degradation of the Turbofan Engine Degradation Simulation Dataset (C-MAPSS) presented by NASA [13]. They introduced the use of Deep Belief Networks (DBN) described as a multi-hidden layer neural network that uses multiple Restricted Boltzmann Machines (RBM). Their proposed method was meant to classify the degradation patterns used for early diagnosis of a possible failing system and resulted in a more efficient manner than the existing methods for system degradation classification.

However, Zhang's method used a shallow convolutional neural network, meaning there was only one convolutional layer. Babu *et al.* proposed the first regressional deep learning network, having more than one convolutional layer, on the C-MAPPS dataset [14]. Instead of discretly classifying the degradation pattern, they attempted to estimate a mapping function to predict the RUL. Due to the complexity of accurately determining the system health status at each time, the paper proposed a piece-wise linear degradation model similar to the one shown in Fig. 2 [15].

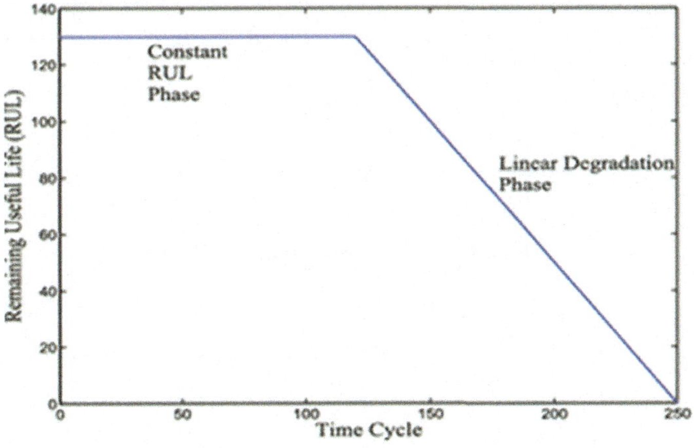

Fig. 2. Piecewise linear RUL target function [15]

Similarly, Li *et al.* proposed a deep convolutional neural network (DCNN), more commonly used for 2D image classification to estimate the RUL of the C-MAPSS data. Their goal was to extract useful features from the different sensor measurements [16]. They proposed to pre-process the data in a 2D format where one dimension is the feature number (the features obtained from the sensors on the subject under test) and the other is the time sequence of each feature. They also consider some important data techniques in their ML method by opting out of using pooling (an approach to summarize the presence of features). This is due to the small amount of initial features in the PHM and the idea that information can be down sampled during prognosis [16]. Furthermore, a dropout technique is used to prevent overfitting of the data by dropping different hidden neurons during training and completely dropped off during testing. Using dropout in this example to help prevent features from repairing more than once should be considered when implementing a model with a different dataset [16].

The performance of the DCNN method for most ML methods is calculated using the scoring function and the root mean square error (RMSE). Since the scoring function penalizes late prediction more than early prediction, its use is important in preventing catastrophic results. Li *et al.* showed the best performance with at mean of 12.61 and a score of 273.7 in their DCNN of a subset of NASA C-MAPSS and their accuracy is demonstrated in the graphs in Fig. 3. The time it takes to train a neural network should also be noted as a metric in performance evaluation of the methods used as indicated by the line in Fig. 4 [16].

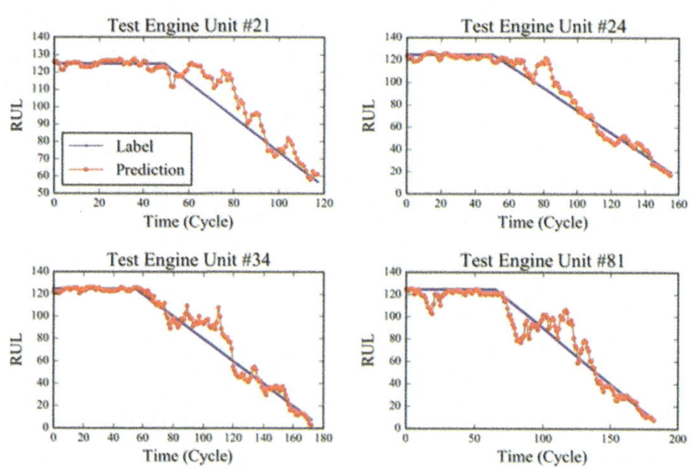

Fig. 3. DCNN model accuracy [16]

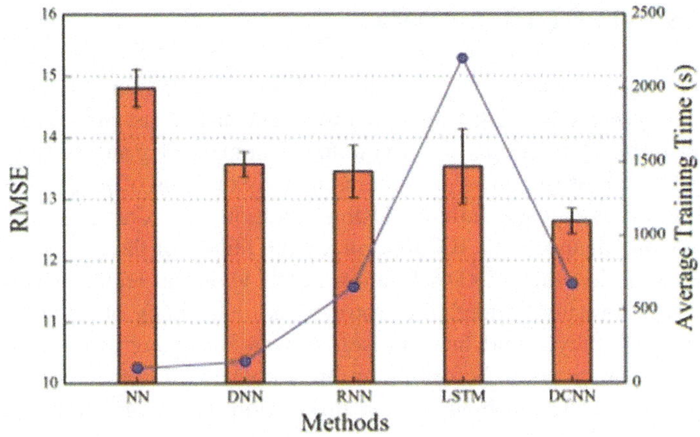

Fig. 4. Prognostic performance by different methods on FD001 (a subset of the NASA C-MAPSS) [16]

3 PRONOSTIA

The size of the data is not the only thing that matters. PRONOSTIA has also collected a popular dataset for RUL of rolling bearings that contains data of one sensor collecting the time-frequency domain of the raw machinery vibration for different rolling bearings at different speeds with different loads. Xaing Li *et al.* also explored data pre-processing using the short-time Fourier transform (STFT) and took the lower half of the frequency below half of the Nyquist frequency [17]. Figure 5 illustrates a flow chart of the process of taking the raw vibration data through their ML model to estimate the RUL. Since the data can be modeled in a 2D format, a Convolutional Neural Network (CNN) was used to estimate the RUL of the bearings. The performance measurements used were the mean absolute error (MAE), mean absolute percentage error (MAPE), and again the RMSE. Vibration signals will be a common measurement from sensors that may need to be converted first into a time-frequency domain to apply an appropriate ML model to.

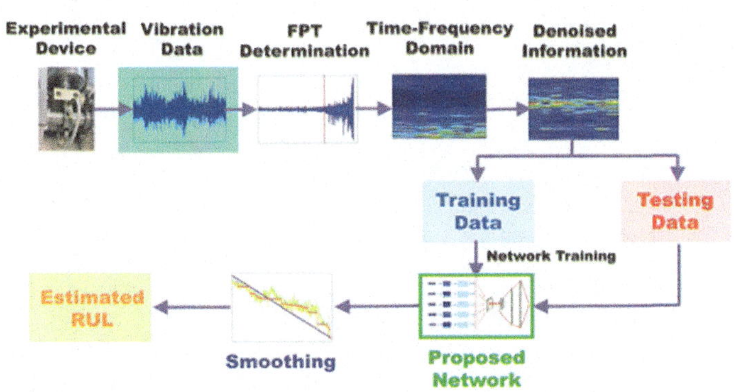

Fig. 5. Raw vibration ML method for prognostics [17]

4 SCADA Wind Turbine Data

On a different scale, a supervisory control and data acquisition (SCADA) computer system for gathering and analyzing real time data has amassed a large amount of labeled wind turbine vibration data containing examples of gearbox failures and the data leading up to those failures. It contains many examples of gearbox failures with the data that build up to these failures. Using vibration data, Carrol *et al.* were able to predict months in advance with 100% accuracy the failure of the gearbox [18]. Unlike the previous examples, the SCADA training inputs varied in features such as gear oil temperature, gear bearing temperature, rotor speed, and many other non-time related features. The work they found that one sensor was not enough to accurately predict failure. For example, a month before failure they noticed a lower temperature than the year before failure on the turbines power production. However, they concluded that power production, generator RPM, and rotor RPM were consequently required as well to predict failure [18]. SCADA used other known ML algorithms like artificial neural network (ANN), a support vector machine (SVM), and logistic regression instead of CNN as described before because their data is not easily translatable to a 2D format where DCNN can be used. This demonstrates the point that each dataset may require evaluation of different ML methods used.

5 PEMFC

New generations of fuel cells called proton exchange membrane fuel cells (PEMFC) have been studied to estimate its RUL [19]. Figure 6 highlights the complexity of a system and the importance of having a sensor data collected to implement intelligent PHM to maintain systems. Similar to the previous datasets, they take a power model over time as one of their features for estimating the RUL. Most importantly Xue *et al.* realized that their failure threshold would be determined when the cell system could no longer power to a certain value [19]. Their experiments concluded with .4633% error when predicting the RUL of the PEMFC. Once a fairly accurate neural network for estimating the RUL of a system has been determined, one should decide at what threshold would be the most cost effective and safest moment to replace or repair the system.

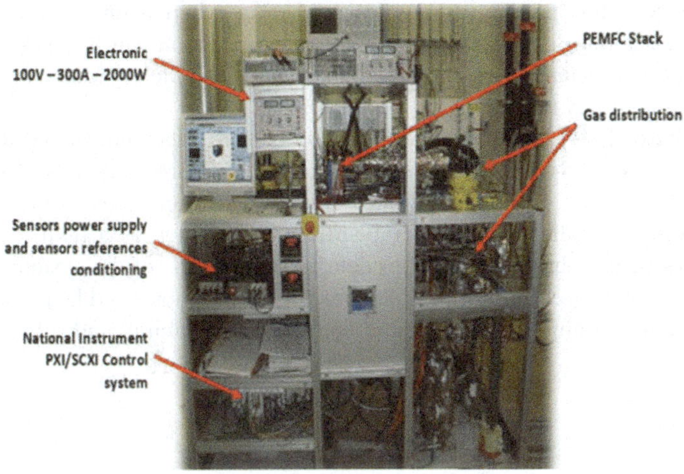

Fig. 6. PEMFC fuel cell test bench used to examine the remaining useful life [19]

6 Other Approaches

Not all data and approaches necessarily need to come directly from sensors on the system. Nash *et al.* developed a model to detect and plan for material deterioration, which would largely increase the reliability and safety of large systems [20]. The degradation of materials can be translated to the RUL of a system. Examples of deep learning in detecting degradation include its use in railway ties [21], fasteners [22], and in corrosion detection [23] using images instead of sensor data. In each case, the degradation in the system could be classified to a point that the system engineers could make an informed decision for when the system needed repairs thereby saving them money and increasing the overall reliability of the system.

7 Discussion

Using ML to classify and predict the RUL of systems in prognosis and health management requires a large amount of data. These datasets can vary in time, frequency, and other dimensions and come from many different types of systems that may require different neural network models, whether it comes from sensors or images. It has been shown in these studies that proper system planning for PHM should be highly considered at the beginning of a project to reduce the cost in the long run. When considering what ML method to use, first consider whether enough sensor data has been collected and whether that data has been labeled in order to use an easier supervised learning ML method. For data that varies in time or has two dimensions, a DCNN may be a more appropriate ML method. For data that has multiple dimensions, other methods such as ANN and SVM are more suitable. Since not all of the data will be useful, examine it for non-classifying data points such as certain frequencies such as in the rotating bearings experiments. After predicting when a unit is close to failure whether it be a month, week, or days an informed decision can be made to replace a crucial component potentially saving them the cost of completely replacing or repairing the system after it has failed.

The future of PHM is based on advancements in ML techniques and the design of new neural networks. However, there is still a need for large amounts of data to train more accurate models. Additionally, since a lot of data is collected using sensors, real time data could be collected on an embedded system, such as a field programmable array that has neural network trained to determine the RUL of the system. Ultimately, utilizing the right datasets and ML models intelligent prognosis and health management can largely reduce maintenance costs and increase the reliability and safety of systems.

References

1. Orsagh, R., Brown, D., Kalgren, P., Byington, C., Hess, A., Dabney, T.: Prognostic health management for avionic systems. In: 2006 IEEE Aerospace Conference (2006)
2. Sankavaram, C., Pattipati, B., Kodali, A., Pattipati, K., Azam, M., Kumar, S., Pecht, M.: Model-based and data-driven prognosis of automotive and electronic systems. In: 2009 IEEE International Conference on Automation Science and Engineering (2009)
3. Muller, A., Suhner, M.-C., Iung, B.: Formalisation of a new prognosis model for supporting proactive maintenance implementation on industrial system. Reliab. Eng. Syst. Saf. **93**(2), 234–253 (2008)
4. Dumargue, T., Pougeon, J.-R., Massé, J.-R.: An approach to designing PHM systems with systems engineering, June 2016
5. Yan, J.: Machinery Prognostics and Prognosis Oriented Maintenance Management. Wiley-Blackwell, Chichester (2015)
6. Barton, P.H., Ogden, R.: Stirling cryocooler prognostics and health management (PHM). In: 2009 IEEE AUTOTESTCON (2009). https://doi.org/10.1109/autest.2009.5314024
7. Sutharssan, T., Stoyanov, S., Bailey, C., Yin, C.: Prognostic and health management for engineering systems: a review of the data-driven approach and algorithms. J. Eng. **2015**(7), 215–222 (2015)
8. Benner, C.L., Peterson, R.A., Russell, B.D.: Application of DFA technology for ımproved reliability and operations. In: 2017 IEEE Rural Electric Power Conference (REPC) (2017)
9. Data Catalog: Data.gov. https://catalog.data.gov/dataset/c-mapss-aircraft-engine-simulator-data-d8ed0. Accessed 27 Nov 2019
10. Xu, J., Wang, Y., Xu, L.: PHM-oriented integrated fusion prognostics for aircraft engines based on sensor data. IEEE Sens. J. **14**(4), 1124–1132 (2014)
11. Rumelhart, D.E., Hinton, G.E., Williams, R.J.: Learning representations by back-propagating errors. In: Anderson, J.A., Rosenfeld, E. (eds.) Neurocomputing: Foundations of Research, pp. 696–699. MIT Press, Cambridge (1988)
12. Bektas, O., Marshall, J., Jones, J.A.: Comparison of computational prognostic methods for complex systems under dynamic regimes: a review of perspectives. In: Archives of Computational Methods in Engineering, July 2019
13. Zhang, C., Sun, J.H., Tan, K.C.: Deep belief networks ensemble with multi-objective optimization for failure diagnosis. In: 2015 IEEE International Conference on Systems, Man, and Cybernetics (2015)

14. Babu, G.S., Zhao, P., Li, X.-L.: Deep convolutional neural network based regression approach for estimation of remaining useful life. In: Database Systems for Advanced Applications Lecture Notes in Computer Science, pp. 214–228 (2016)
15. Heimes, F.O.: Recurrent neural networks for remaining useful life estimation. In: 2008 International Conference on Prognostics and Health Management (2008)
16. Li, X., Ding, Q., Sun, J.-Q.: Remaining useful life estimation in prognostics using deep convolution neural networks. Reliab. Eng. Syst. Saf. **172**, 1–1 (2018)
17. Li, X., Zhang, W., Ding, Q.: Deep learning-based remaining useful life estimation of bearings using multi-scale feature extraction. Reliab. Eng. Syst. Saf. **182**, 208–218 (2019)
18. Carroll, J., Koukoura, S., Mcdonald, A., Charalambous, A., Weiss, S., Mcarthur, S.: Wind turbine gearbox failure and remaining useful life prediction using machine learning techniques. Wind Energy **22**(3), 360–375 (2018)
19. Xue, X., Hu, Y., Qi, S.: Remaining useful life estimation for proton exchange membrane fuel cell based on extreme learning machine. In: 2016 31st Youth Academic Annual Conference of Chinese Association of Automation (YAC) (2016)
20. Nash, W., Drummond, T., Birbilis, N.: A review of deep learning in the study of materials degradation. npj Mater. Degrad. **2**(1) (2018)
21. Gibert, X., Patel, V.M., Chellappa, R.: Material classification and semantic segmentation of railway track images with deep convolutional neural networks. In: 2015 IEEE International Conference on Image Processing (ICIP) (2015)
22. Gibert, X., Patel, V.M., Chellappa, R.: Deep multitask learning for railway track inspection. IEEE Trans. Intell. Transp. Syst. **18**(1), 153–164 (2017)
23. Petricca, L., Moss, T., Figueroa, G., Broen, S.: Corrosion detection using A.I: a comparison of standard computer vision techniques and deep learning model. In: Computer Science & Information Technology (CS & IT) (2016)

Machine Learning and Analytics

Random Generation of Visibility Aware Polygons

Laxmi Gewali(✉) and Bibek Subedi

Department of Computer Science, University of Nevada Las Vegas, Las Vegas, NV, USA
{Laxmi.Gewali,Bibek.Subedi}@unlv.edu

Abstract. Problems dealing with the generation of random polygons has important applications for evaluating the performance of algorithms on polygonal domain. We review existing algorithms for generating random polygons. We present an algorithm for generating polygons admitting visibility properties. In particular, we propose an algorithm for generating random polygons admitting large size kernels.

Keywords: Visibility · Polygon decomposition · Kernel

1 Introduction

Development of efficient algorithms for generating simple polygonal shapes has wealth of applications in applied computational geometry. Any algorithm that takes input a simple polygon should be tested for its performance by executing on several polygons with different numbers of vertices and shapes. What is really needed is a set of randomly generated polygons of various shapes and structures. While there are several algorithms for generating pseudo-random integers [8], papers and research outlets addressing the issue of generating random polygons are very rare. Only a few papers on generating simple polygons randomly have been reported [11, 13]. Since no polynomial time algorithm for generating random polygon is known, researchers either use heuristics [1] or generate restricted classes of polygons [11, 13]. In this paper, we consider the problem of generating random simple polygons that satisfy certain visibility properties. Specifically, we examine the problem of generating random polygons that are star-shaped. In a star-shaped polygon P, there is a connected set k called *kernel* such that any point inside P is visible to all points inside k.

The paper is organized as follows. In Sect. 2, we present a condensed review of algorithms for generating simple polygons. We also review several results dealing with the computation of kernel and techniques for searching kernels. We next review algorithms for constructing kernels under stair-case visibility which has applications in VLSI design.

In Sect. 3, we present the main contribution of the thesis. We introduce the problem of generating random simple polygons that admit large size kernel. We present a space decomposition approach for searching and constructing polygons that are guaranteed to admit a kernel. Our technique is based on applying quad-tree decomposition for

H. Selvaraj et al. (Eds.): ICSEng 2020, LNNS 182, pp. 151–159, 2021.
https://doi.org/10.1007/978-3-030-65796-3_14

searching the candidate kernel points. The approach can be used to generate pseudo-random star-shaped polygons.

Finally, in Sect. 4, we examine several scopes extending and generalizing the algorithms proposed in this research.

2 Review of Existing Algorithms

2.1 Preliminaries

Generating two-dimensional shapes and structures of a given set of points is a central problem area in computational geometry. One such problem is connecting points to form a simple polygon. A few algorithms have been reported for constructing a simple polygon from a given set of points [1, 13]. It is remarked that a simple polygon consists of a set of line segments enclosing a connected area that partition the plane into three parts (i) polygon boundary (ii) unbounded exterior, and (iii) bounded connected interior [10]. A straightforward way for generating a simple polygon from a given set of points is to angularly sort and connect. This is illustrated in Fig. 1. To sort the points angularly, an axis point is picked inside the convex hull of input points.

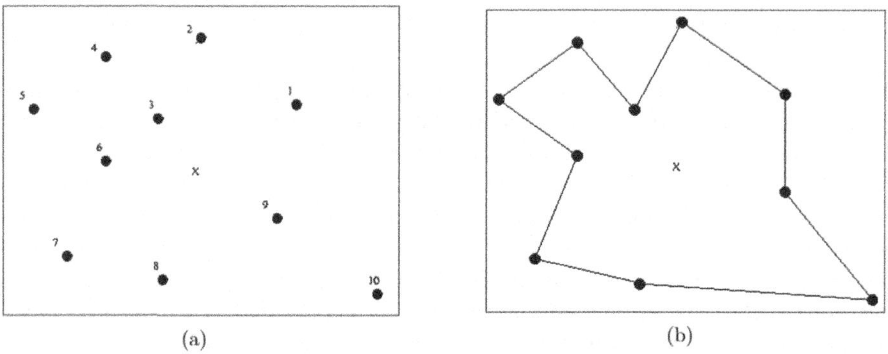

Fig. 1. Generating simple polygon by angular sort

In the left side of the Fig. 1, a given input points are shown. The axis point is marked as 'x'. The integers adjacent to input points represent the ordering of the angularly sorted (about x) list. In the right part, the polygon implied by the sorted list is displayed. Since sorting is used in this procedure, this approach of constructing polygon needs *O(n log n)* time.

2.2 Generating Random Polygons

An efficient algorithm for generating x-monotone polygon uniformly at random has been proposed in [13]. It is remarked that a simple polygon is called *monotone* with respect to a given direction *d*, if the polygon boundary consists of two chains *top chain* and *bottom chain* both of which are monotone with respect to *d*. Without loss of generality, we assume that direction *d* is *x*-axis. Alternatively, an x-monotone polygon is such that its intersection with any vertical line is either empty or one-line segment as shown in Fig. 2. The top chain and bottom chain of this polygon are vertices 1, 2, 3, 5, 7, 8 and 1, 4, 6, 8 respectively.

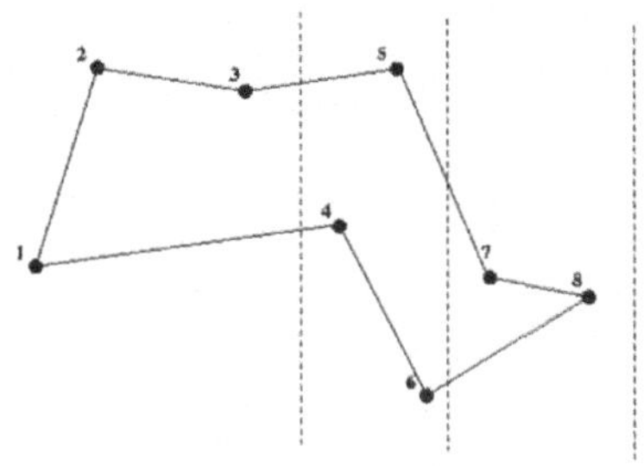

Fig. 2. Illustrating x-monotone polygon

In [1, 13], it is argued that if one can count the number of monotone polygons that can be formed (for a given set of vertices $S_k = \{s_1, s_2, ..., s_k\}$ with $k > 2$) then a formal method for random generation is applicable. The counting is done by observing that a given input point can be present either in the top chain or in the bottom chain. The authors arrive at a relationship for counting the number of monotone polygons in terms of a recurrence relation given below.

$$TN(k) = \sum_{j \in V_T(k)} BN(j+1)$$

$$BN(k) = \sum_{j \in V_B(k)} TN(j+1)$$

Where, $TN(k)$ is the number of *x*-monotone polygons with vertex set S_k that have edge (s_{k-1}, s_k) on their top chain. $BN(k)$ is the number of *x*-monotone polygons with vertex set s_k that have edge (s_{k-1}, s_k) on their bottom chain. $V_T(k)$ is the set of points that are above visible from the point s_k and $V_B(k)$ is the set of points that are below visible from the point s_k. The details are in [13].

The time complexity of the algorithm is bound by $O(n^2)$, where *n* is the number of input vertices. The recursive relation counts the number of monotone polygons with the given set of vertices S_k by scanning the vertices in the forward direction. A random number between 1 and the total count is picked and the *x*-monotone polygon corresponding to the selected number is constructed by scanning the vertices in the backward direction.

An algorithm for generating random star-shaped polygons has been given in [11]. Auer and Held [1] gave an algorithm to enumerate all the star-shaped polygons from a given set of points *S*. The algorithm takes $O(n^4)$ time and generates polygons uniformly

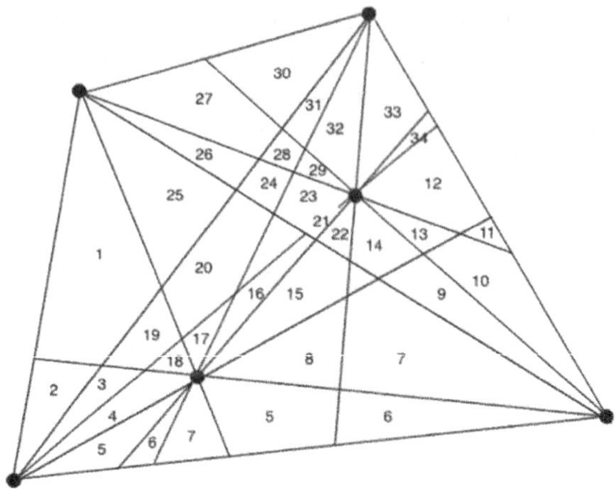

Fig. 3. Illustrating quadratic number of cells

at random. The authors use the fact that a star-shaped polygon is fixed once the kernel of the polygon has been specified. That means each star-shaped polygon has a unique kernel. Once we find the kernel of the polygon, we can sort the vertices angularly around the kernel and construct the polygon in $O(n \log n)$ time. Auer and Held [1] showed that for point sites of size n, there can be at most $O(n^4)$ kernels. This can be verified by drawing the lines through each pair of points and counting the number of regions induced by the lines that lie inside the convex hull of the point sites S. This is illustrated in Fig. 3. The total number of such lines is in the order of ${}^{n}C_2 = O(n^2)$. Therefore, the total number of regions (cells) can be at most $O(n^4)$. In other words, there can be at most $O(n^4)$ kernels. One possible arrangement of points that achieve this bound can be obtained by placing $n - 2$ points on the parabola $y = x^2$ equally in either side of y-axis and placing the remaining 2 points at $(-\infty, -\infty)$ and (∞, ∞).

2.3 Visibility Properties of Polygons

Understanding the visibility properties of polygonal shapes is perhaps the most widely investigated area of computational geometry [10]. While exploring the visibility properties of simple polygons, the boundary of the polygon is taken as an opaque wall: a polygon is viewed as a room with walls as a boundary. Under the standard definition of visibility, two points p and q inside the polygon are mutually *visible* if the straight line segment connecting p and q does not intersect with the boundary. Under this definition, many practical algorithms having application in robotics and computer-aided manufacturing have been developed. One such algorithm is the computation of *visibility polygon* induced by a point q inside the polygon. The visibility polygon for an interior point q is the set of points visible from q. Visibility polygons can be computed in $O(n)$ time for simple polygons and for polygons with holes, algorithms for time complexity $O(n \log n)$ is known [2, 4, 7, 9].

Another useful visibility property of a simple polygon is the concept of Kernel. The *Kernel* of a simple polygon is the set of points from which the polygon is visible. Not every simple polygon admits kernel. Those simple polygons that admit kernel are usually called *star-shaped* polygons. Fig. 4 illustrates an example of a simple polygon admitting kernel.

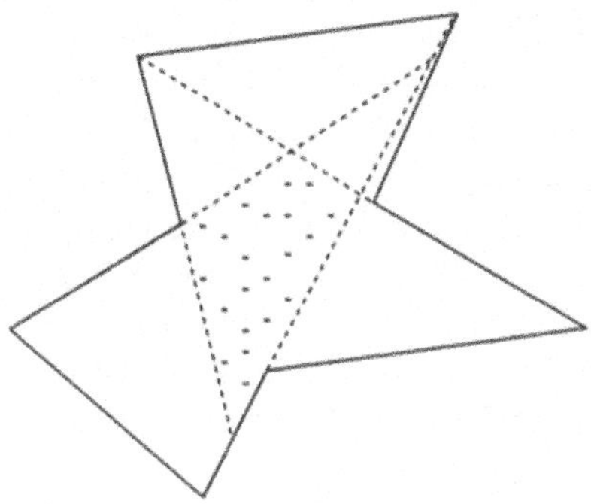

Fig. 4. Illustrating kernel of a simple polygon

Lee and Preparata [9] gave an algorithm that runs in linear time in the number of the edges for finding the kernel. The authors took advantage of the order of the edges in the polygons to find the linear time algorithm.

The notion of visibility has been generalized [6, 11, 12]. One such generalization is the introduction of *stair-case* visibility [6, 11, 12]. Two points p and q inside an orthogonal polygon are called *s-visible* if p and q can be connected by a *stair-case* path. It is remarked that in an orthogonal polygon the boundary edges are parallel to x- and y-axis. A path p' is a *stair-case* path if its edges are parallel to x- or y- axis and is monotone with respect to one of the co-ordinate axes. Intuitively a stair-case path does not back-up. Under *s-visibility*, the problem of determining and recognizing s-star have been considered [6]. An algorithm for computing *s-kernel* in an orthogonal polygon is reported in [11, 12]. The time complexity of this algorithm is $O(n^2)$, where n is the number of vertices in the polygon. It is interesting to note that while polygons with holes never admit kernel under the standard notion of visibility, there could be s-kernels for s-star polygons with holes [6]. Even though a s-kernel could have $\Omega(n^2)$ components, such polygon can be recognized in $O(n)$ time [6]. It is known that the problem of computing visibility polygon under stair-case visibility inside a polygon containing holes has a lower bound of $\Omega(n \log n)$ [6]. This result is extended to orthogonal polygons with holes in [6]. This result is established by reducing sorting problem to the problem of computing s-visibility polygon [6].

3 Star Polygons with Large Kernels

In this section, we study the development of an efficient algorithm for generating polygons (for given point sites) that tend to have a large area kernel.

3.1 Preliminaries

Given a set S of n points $\{v_0, v_1, \ldots, v_{n-1}\}$, our objective is to construct a star-shaped polygon whose vertices are exactly the points in S. A very simple method for generating a star-shaped polygon is to use the technique described in Sect. 2. Specifically, a reference point r_0 is picked inside the convex hull of points in S. All points are angularly sorted to obtain an order list L'. The vertices are connected in the order implied by L' to obtain polygon Q. The star-shaped polygon so constructed admits a region called *kernel* k_r such that any point inside Q is visible from all points in k_r. It is straightforward to observe that the shape of the kernel k_r is convex.

One interesting question arises for picking the reference point. If we pick another reference point, the admitted kernel will be much larger. These observations motivate us to look for an algorithm that generates a star-shaped polygon that maximizes the area of the implied kernel. The problem can be formally stated as follows.

Max Kernel Polygon (MK) Problem
Given: A set of points $S = \{v_0, v_1, \ldots, v_{n-1}\}$.
Question: Construct a star-shaped polygon with vertices in S that maximizes the implied kernel.

3.2 Quad-Tree Based Kernel Search

We propose to use the quad-tree technique introduced by Raphel Finkel and J.L. Bently [5]. In this technique, a two-dimensional bounded rectangular region R is orthogonally partitioned into four ortho-rectangular sub-regions which we refer as *quadruplets*. The process of partitioning into quadruplets proceeds recursively to arrive at a small *region of interest*. The implied result of this recursive partitioning is the formation of a quad-tree where each internal node has four children.

In our kernel searching context, the point of interest occurs when a small region is found that results in a large kernel. We illustrate this quad-tree based kernel search with a running example. Fig. 5a shows randomly generated seven-point sites with the smallest enclosing axis parallel rectangle (drawn by the dashed line). The edges of the resulting polygon implied by the angular sorting of the point sites are drawn by the solid lines. The corresponding kernel by selecting the axis point at the center of the bounding rectangle is shown shaded. The algorithm next proceeds to search for the other kernels in each of the quadruplets inside the current region R. This process is continued recursively until the region of interest is reached. A sequence of resulting kernels found in this way are shown in Fig. 5a–i.

More specifically, the smallest ortho-rectangle R_0 enclosing all input vertices is determined. The first candidate kernel point is the center c(R_0) of R_0. The star-shaped polygon and corresponding kernel k(R_0) is computed. Next, R_0 is partitioned into four parts (NE, NW, WS, SE) of equal areas. Star-shaped polygon construction, and the corresponding kernel construction is repeated recursively on each of the four NE-, NW-

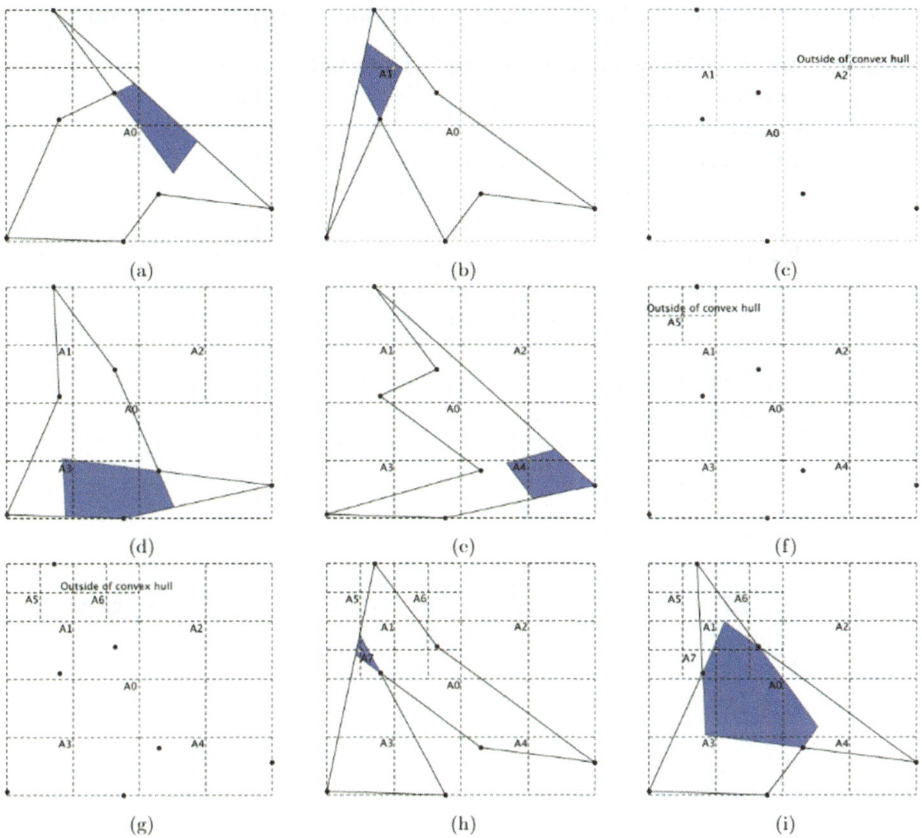

Fig. 5. Illustrating kernel search

, WS-, and SE-parts. The recursive partitioning stops whenever conditions (listed below) are satisfied.

Condition A1: The height of the quad-tree is no more than the pre-determined integer value.

Condition A2: Area of running max kernel must be at least twice the area of the next candidate node of the quad-tree.

A formal sketch of the algorithm based on above description is listed as Algorithm 1. The complexity analysis of Algorithm 1 can be done in a straightforward manner. In the while loop, the dominant step is finding the kernel [9]. One execution of Step 12 can be done in $O(n)$ time by using the algorithm given in [9]. The while loop exe-cutes t times, where t is the total count of nodes in the quad-tree constructed during the recursive partitioning in Step 10. Hence the execution time of algorithm 1 is $O(tn)$.

A formal sketch of the kernel searching algorithm is listed as **Algorithm 1**.

Algorithm 1: Kernel Searching Algorithm

```
Input:  Point sites P = p0, p1, ..., pn-1
Output: Star polygon S with large kernel
Compute enclosing ortho-rectangle R0
Find kernel k0 corresponding to R0
Q ← Empty Queue
Push R0 into Q
GlobalMaxKer ← k0
while Q is not empty do {
   Ri ← Pop node from Q
   Partition Ri into 4 children:  Ri1, Ri2, Ri3, Ri4
   Push those children to Q that satisfy
      condition A1 and condition A2
   Compute kernels ki1, ki2, ki3 ,and ki4
      corresponding to four children
   MaxKer ← Kernel with maximum area among
             ki1, ki2, ki3 ,and ki4
   if area(M axKer) > area(GlobalM axKer) then
       GlobalMaxKer = MaxKer
}
S ← Vertices corresponding to GlobalMaxKer
```

4 Discussion

We summarized a critical review of existing algorithms for random generation of polygonal shapes. We also reviewed algorithms related to visibility properties (particularly kernel counting and detection) of simple polygons. We proposed an algorithm for generating random polygons that tend to have large size kernels. The algorithm for generating such polygons is based on quad-tree searching.

The proposed quad-tree based algorithm stops after reaching a certain value for the height of the implied quad-tree (Condition A1). How to pick the bound for such height is an important question in itself? In our experimental investigation, the algorithm stops after reaching the height in the range 6–8. How to determine this stopping height range is a very critical question which also needs further exploration. An important related question is the selection of a subset Q of input node set S such that the corresponding polygon has large size kernel. One possible way of selecting such a sub-set is using the convex layers [14] of S. We recursively find the convex hull of. We can use the following greedy strategy to remove the layers (i). Remove the innermost layer l_1 from S and compute the largest kernel K_1 on the remaining points. (ii) Put l_1 back in S. Remove the next layer in an outward direction and repeat the same process. (iii). Remove the layer l_i corresponding to the maximum value of Ki from S to get Q. We can use the above strategy multiple times to remove more than one layer. It would be worth to experimentally explore this strategy.

References

1. Auer, T., Held, M.: RPG – heuristics for the generation of random polygons. In: Proceedings of 8th Canadian Conference on Computational Geometry, pp. 38–44 (1998)
2. Asano, T.: An efficient algorithm for finding the visibility polygons for a polygonal region with holes. Trans. IECE Jpn. **E-68**, 557–559 (1985)
3. Barber, C.B., Dobkin, D.P., Huhdanpaa, H.: The quick hull algorithm for convex hulls. ACM Trans. Math. Softw. **22**, 469–483 (1996)
4. ElGindy, H.A., Avis, D.: A linear algorithm for computing the visibility polygon from a point. J. Algorithms **2**(1981), 186–197 (1981)
5. Finkel, R.A., Bently, J.L.: Quad trees, a data structure for retrieval on composite keys. Acta Informatica **4**(1), 1–9 (1974)
6. Gewali, L.P.: Recognizing s-star polygons. Pattern Recogn. **28**(7), 1019–1032 (1995)
7. Lee, D.: Visibility of a simple polygon. Comput. Vis. Graph. Image Process. **22**, 207–221 (1983)
8. Lehn, J.: Pseudorandom number generators. Oper. Res. **91**, 9–13 (1992)
9. Lee, D.T., Preparata, F.P.: An optimal algorithm for finding the kernel of a polygon. J. ACM (JACM) **26**, 415–421 (1979)
10. O'Rourke, J.: Art Gallery Theorems and Algorithms, vol. 57. Oxford University Press, Oxford (1987)
11. Sohler, C.: Generating random star-shaped polygons. In: Proceedings of 8th Canadian Conference on Computational Geometry, pp. 174–179 (1999)
12. Vassilev, T.S.: Visibility: finding the staircase kernel in orthogonal polygons (2015)
13. Zhu, C., Sundaram, G., Snoeyink, J., Mitchell, J.S.: Generating random polygons with given vertices. Comput. Geometry **6**(5), 277–290 (1996)
14. Chazelle, B.: On the convex layers of convex set. IEEE Trans. Inf. Theory **IT-31**(4), 509–517 (1985)
15. Subedi, B.: Generating kernel aware polygons. MS thesis, Department of Computer Science, University of Nevada, Las Vegas (2019)

Assessment of Groundwater Quality and Quantity Using Supervised Machine Learning

Ashika Mettu(✉) and Shahram Latifi

Electrical and Computer Engineering Department, UNLV, 4505 S Maryland Pkwy, Las Vegas, NV 89154, USA
mettu@unlv.nevada.edu, Shahram.latifi@unlv.edu

Abstract. With the urban development of Las Vegas region in the past 3 decades, the water usage has increased rapidly and, accordingly, there has been a growing interest in assessment of water quality and quantity in this region. The availability of water in Nevada is not commensurate with the water usage. Water quality assessment involves expensive and time-consuming laboratory tests and in-depth analysis of many parameters. This study implements a supervised machine learning algorithm that uses cost-free methods to predict groundwater quality characteristics, thereby helping to reduce the cost of extraction and laboratory analysis. This study employs five input parameters, namely, Temperature, Humidity, Precipitation, Soil moisture (pH), and Well depth. The study uses the "Gradient Boosting Trees" and "Gradient Descent" methods to train the data and to predict the output. The results of this study provide real-time predictions of groundwater quality and quantity parameters, including water levels, nitrate concentrations. The Mean Absolute Error (MAE) values range from 16.7 to 39.48. A limited set of input parameters is used to achieve a cost-effective solution.

Keywords: Groundwater predictions · Gradient boosting · Supervised machine learning · Water quality

1 Introduction

Nevada is composed of four districts with a geographic area of 110,567 square miles. 80% of the geographic area is desert, and the rest of Nevada accounts for about 60% of the geographic area of the state. Groundwater is a vital resource. The groundwater development is comparatively low. The average annual groundwater consumption is more than the average annual groundwater supply. This substantial dependency on groundwater for both domestic water and irrigation purposes is right now moving towards the breaking point. After a long period of abuse, the amount of water reached an unreasonable level, which posed a serious threat to the amount of groundwater.

Out of concern for potential water quality, the Las Vegas Valley Water District has been sampling and analyzing water from many water sources and water delivery points in the area and the neighboring city of North Las Vegas. The valley-wide networks

H. Selvaraj et al. (Eds.): ICSEng 2020, LNNS 182, pp. 160–171, 2021.
https://doi.org/10.1007/978-3-030-65796-3_15

collect water samples from all major wells and frequent sampling from each site as many as 34 water-quality characteristics.

For decades, many studies performed on local water-quality issues of Las Vegas Valley for a better understanding. The wells sampled in the current program are not enough in number to detail the spatial variability of chemical concentrations. The key issue related to groundwater quality is the pollution of chemical substances, such as fluoride, nitrate and chloride salts. These chemicals will increase the overtime content of groundwater in the well, posing a health hazard. Ground Water contain Nitrate related substances concentration and is untraceable without performing a deep lab analysis due to Nitrate unique characteristic (No color, No odor, and no taste). The main danger of drinking groundwater is with the nitrate-nitrogen content present.

In 1962, the US Public Health Administration set the recommended limit for nitrate nitrogen to 10 milligrams per liter. Since then, the USA Environmental Protection Agency (EPA) has adopted the 10 milligrams per liter standard as the extreme contaminant levels (MCL) for controlling nitrate nitrogen and One milligram per liter as the maximum pollutant level for controlling nitrate related contents in drinking/irrigation/etc. water systems. Past USA Geological Department surveys and tests on private wells observed over 9% of wells surpassed nitrate concentrates standards. Health authorities forced to adhere to measures and standard examinations fluctuating from one year to three years to guarantee that a critical increment in nitrogen concentrations doesn't increment and shows the chances of reoccurrence.

Sources of nitrogen (NO3-N) in groundwater include sewage treatment, fertilizing crops, parks, golf courses, gardens, and natural nitrogen sources. Appropriate choice of the site assist spares with costing of building these wells.

The main problem with water analysis is the cost of the water extraction process. It has been found that the cost of propelling water from deeper surfaces increases. Greater power and high-power engine motors and different assets are utilized. Extract water from shallower surfaces is likewise expensive. After extraction, if there is no drinking water, we will waste money and time. The rest of the paper is organized as follows. Section 2 states the purpose and motivation behind the work while Sect. 3 describes the data characteristics used in our experiments. Section 4 describes the machine learning methods used in this work. This section also includes a brief review of neural networks. Section 5 presents the results while Sect. 6 depicts on future work and possible extension of this work.

2 Purpose

The purpose of this work is to implement a 360-degree predictive monitoring system to monitor the quality and quantity of groundwater with fewer parameters than what is used in current systems. This system provides time-series water quality (nitrate) data for predicting water quality problems at sampling sites. This project can be expanded to predict the sources of β,α radiations and the sources of chemical substances in uncertain shallow zones, and conduct continuous monitoring (1) to verify whether the concentration of chemical substances in the area has increased (2) identify and quantify the source of chemical substances.

3 Water Data Preprocessing

The data was obtained from USGS, discussed in this section.

3.1 Data Collection

Each algorithm was trained and tested on a data set collected from USGS containing 300 samples from seven different sources of Nevada State, namely Clark, Eureka, Lyon, Nye, Churchill, Lincoln and Carson City Counties collected throughout 1995 to 2018. Each District contains about 40 samples (Fig. 1).

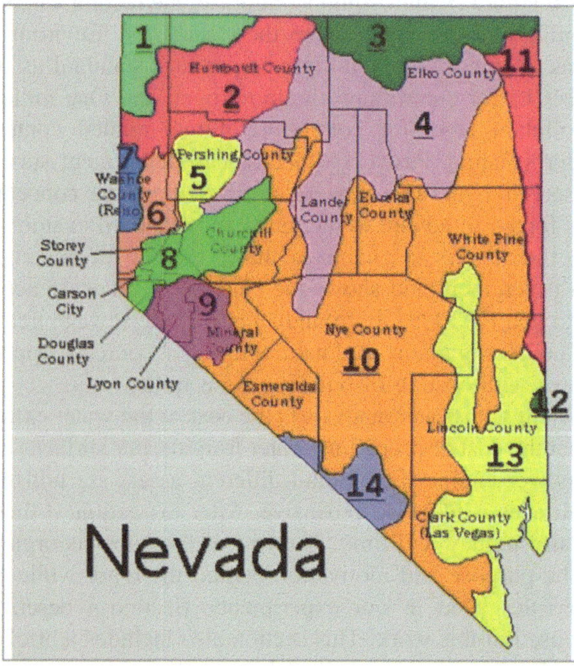

Fig. 1. The picture below provides the location of underground wells in Nevada.

3.2 Water Quality Index (WQI)

WQI is calculated using six available parameters. WQI model predicts almost 99.5% of the variation in the data. WQI is a measure of water quality, calculated using available parameters, including temperature, pH, turbidity, CaCO3, conductivity, and sum of dissolved solids. For this work, WQI uses the two water quality parameters pH and temperature in the data set for calculation. As seen from the equation below, WQI is obtained by multiplying q value with weight, adding up, and dividing the result by the sum of weights (Table 1):

Table 1. Weight factor and weight of each parameter

Parameter	Value
pH	0.11
Temperature	0.10

$$\mathrm{WQI} = \frac{\sum q_{value} \times \omega\_factor}{\sum \omega\_factor} \tag{1}$$

q-value = ranges from 0–100
w_f actor = the weight of a respective parameter.

3.3 Water Quantity Class (WQC)

A ground water well is designed to produce 5 gallons per minute. according to the Nevada State ground water resources. Based on that and data collected, the water Quantity Class is divided into four class (Table 2).

Table 2. Water quality classifications

Water in the well (gallons per day)	Water quantity class
0–50	Bad
50–200	Medium
200–400	Good
400+	Excellent

4 Analysis

Data analysis is performed using correlation and data splitting. Here, we train input data and machine learning algorithms to predict output using nominal parameters.

4.1 Data Collected

OUTPUT:

- Groundwater levels
- Nitrate Concentration

INPUT:

- Temperature - Help to predict Nitrate Concentration
- Humidity - Help to predict Ground Water Level
- Precipitation - help to predict the Ground Water Level
- Soil/Water Ph around the area – to predict groundwater level
- Well Depth - what type of compositions might be mixed in groundwater.

4.2 Correlation Analysis

Correlation Analysis is performed to extract correlations between various parameters. We implemented deep analysis algorithms to obtain accurate results using effective correlation method called Pearson correlation.

Using all available parameters to evaluate water prediction is a long and expensive process. We conducted a relationship study to isolate possible relationships between parameters. w used the Pearson method to perform correlation analysis and defined perfect correlation between parameters.

The correlation analysis indicates:

- Temperature is highly correlated with Soil/Water Ph loosely with Well depth, humidity
- Humidity is associated with Precipitation and Temperature, Soil/Water pH
- Precipitation is highly correlated with Well depth, Humidity and loosely with correlated with pH
- Soil/Water Ph is highly correlated with Temperature, Well depth and loosely with Humidity
- Well depth is highly correlated with Precipitation, Soil/Water Ph and loosely with temperature, and humidity
- Nitrates correlated with pH and Temperature
- Ground-Water Levels is highly correlated with Well depth, Temperature and Precipitation.

4.3 Data Split

The first steps are to test the data and derive the predictive model performance by measuring the tested data. Cross-Validation method is used to perform data splitting of the parameters.

The data is divided into k sub-datasets and iterated on all datasets. This method ensures data split is performed accurately. This technique requires additional methods of putting resources based on the number of iterations and matrix datasets. We used minimal water quality characteristics and few cross-validations. We split the data into five subsets (k), perform cross-validations. The final data contained 300 samples, and we ensured that 40 samples were used.

5 Machine Learning Algorithms

This Section provides information related to Supervised Machine learning Algorithms utilized here. This project uses Gradient Boosting Trees and Neural Networks algorithms towards obtaining an estimate of the Water Quality and Water Quantity.

5.1 Gradient Boosting Trees

The Gradient Boosting Trees method uses values (y) and approximate value of G(x), sum of functions h_i(x). class 'H' data is weak learners.

Gradient boosting Algorithm is the most prevailing method to build predictive models, perform supervised machine learning tasks, sorting and regression. Boosting is a method of converting weak learners into strong learners. This project implements regression method.

Gradient boosting involves three elements (Fig. 2):

- A loss function to enhance.
- A weak learner for predictions.
- An additive model to add weak learners to minimize the loss function.

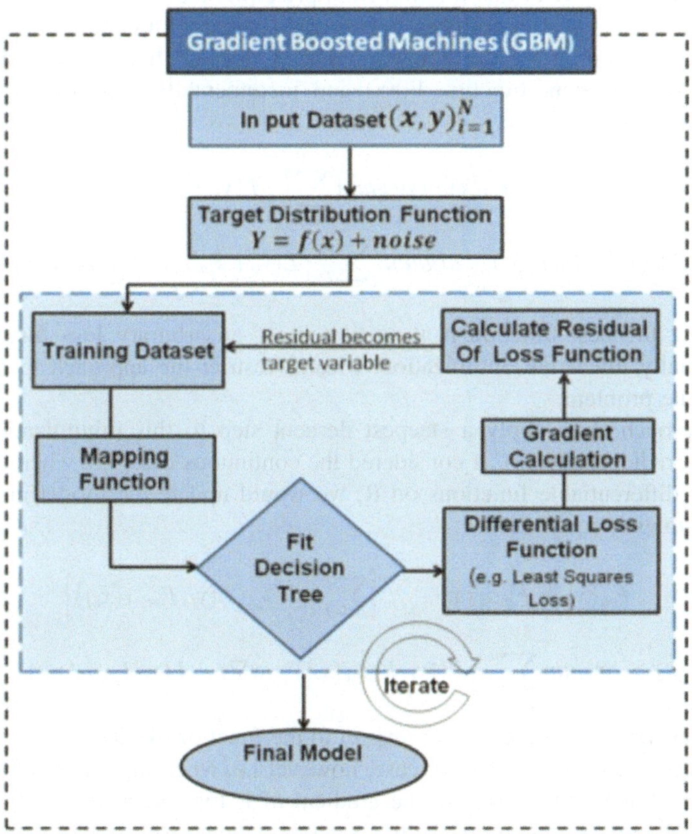

Fig. 2. Picture is a flow chart of Gradient Boosted Machines (GBM).

5.2 Process

The Supervised learning process possesses parameters y and x as output and input respectively. These parameters are defined using a joint probability distribution P (x, y). Using a trained data set {(x1, y1), (x2, y2), …. (xn, yn)} of values of x and equivalent values of y, to find an approximation G(x) to a function F(x) that minimizes the expected value of some specified loss function L(y, F(x))

$$G = \arg\min \ Ex, y \ [L(y, F(x)] \tag{2}$$

The gradient boosting method adopts a value of y and an approximate of G(x) in the form of a weighted sum of functions h_i(x) from some class 'H' called weak learners:

$$G(x) = \sum_{i=1}^{M} \gamma_i h_i + Constant \tag{3}$$

Based on the empirical risk minimization principle, the method attempts to find an approximation G(x) that minimizes the average value of the loss function on the training set, i.e., minimizes the empirical risk. It does so by starting with a model, consisting of a constant function F0(x) and incrementally expands it in a greedy fashion:

$$F_0(x) = argmin \sum_{i=1}^{n} L(y_i, \gamma) \tag{4}$$

$$F_m(x) = F_{m-1}(x) + argmin \left[\sum_{i=1}^{n} L(y_i + F_{m-1}(x_i) + h_m(x_i)) \right] \tag{5}$$

Choosing the best function h at each step for an arbitrary loss function L is a computationally infeasible optimization. Means, restrict the approach to a simplified version of the problem.

The approach is to apply a steepest descent step to this minimization problem (functional gradient descent). If considered the continuous case, i.e. where H is the set of arbitrary differentiable functions on R, we would update the model in accordance with the following equations

$$F_m(x) = F_{m-1}(x) - \gamma_m \left[\sum_{i=1}^{n} \nabla_{F_{m-1}} L(y_i, F_{m-1}(x_i)) \right] \tag{6}$$

$$\gamma_m = argmin \sum_{i=1}^{n} L((y_i, F_{m-1}(x_i)) - \gamma \nabla_{F_{m-1}} L(y_i, F_{m-1}(x_i)))] \tag{7}$$

where the derivatives are taken with respect to the functions Fi for i = {1, 2... m} and γ_m is the step length. In the discrete case, however i.e. when the set H is finite, chosen the candidate function h closest to the gradient of L for which the coefficient γ may then be calculated with the aid of the line search on the above equations. In addition to the methods described here, we also used Neural Networks to obtain a basis for results comparison.

5.3 Neural Networks

Neural Network contains inputs, outputs, and hidden layers. Neural networks algorithm can provide continuous predictions at the output; This method is quite effective for output estimation and prediction. This Neural Networks is very best analyzing numerous variables at a time. Solution is very complex; Neural Network gives a very good data model (Fig. 3).

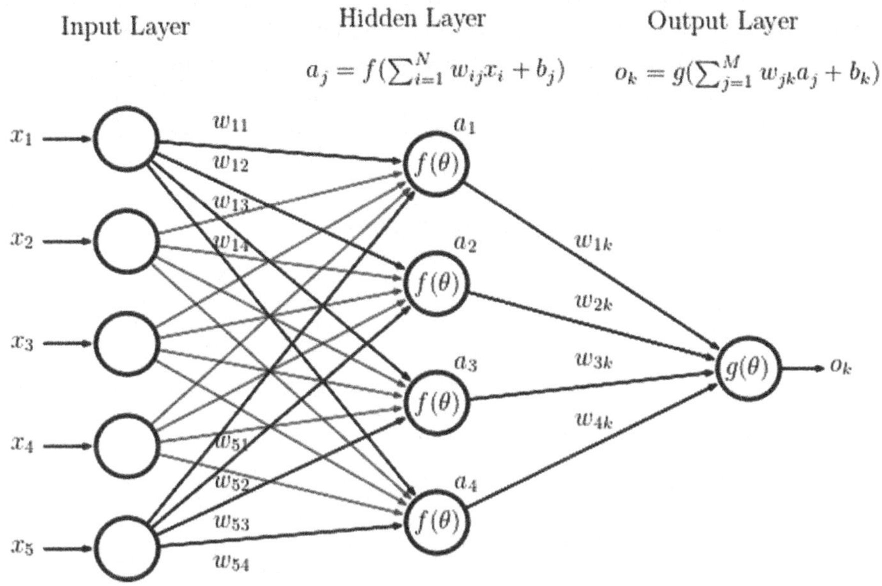

Fig. 3. Neural Network layers

For performance comparison, the Neural Networks model is also tested to deliver Water Quality and Quantity Predictions. For the Neural Network Model, we used historical data including pH and temperature. The data is co-related, and this model finds the relationship between the changes, t bua large amount of data is required to get better results. Successful application of neural network model is highly dependent on the right input data selection.

6 Results

This section evaluates the accuracy of the machine learning algorithms that were implemented.

The Gradient Boost method ran at approximately 60 to 120 s to predict, it tended to have more accurate prediction rates for nearly all factors.

The Gradient Boost method predicts dissolved nitrate concentration (Quality) and ground water levels (Quantity); however, prediction of the other factors also considered. Each prediction accuracy score is measured ten times, then the highest score, lowest score and the mean values for each factor prediction are considered. Dissolved nitrate was also successful with a 70% better prediction. The rest of the factors ranged in prediction accuracy from 80% to 90%. Most stayed within one or two degrees with minimal variance.

The Neural Network algorithm ran at approximately two minutes for each prediction where the weaknesses of neural networks are as follows, defining architectural parameters, extensive process to train the data and difficulties in data cleansing, correlation interpretation methods. Neural Network works well on static data, images and text. Not good for small data sets. Not suitable for time-series prediction problems. Water quality parameters at a fixed time interval and did not provide expected output predictions.

6.1 Graphs (Gradient Boosting Regression Trees)

See Figs. 4 and 5.

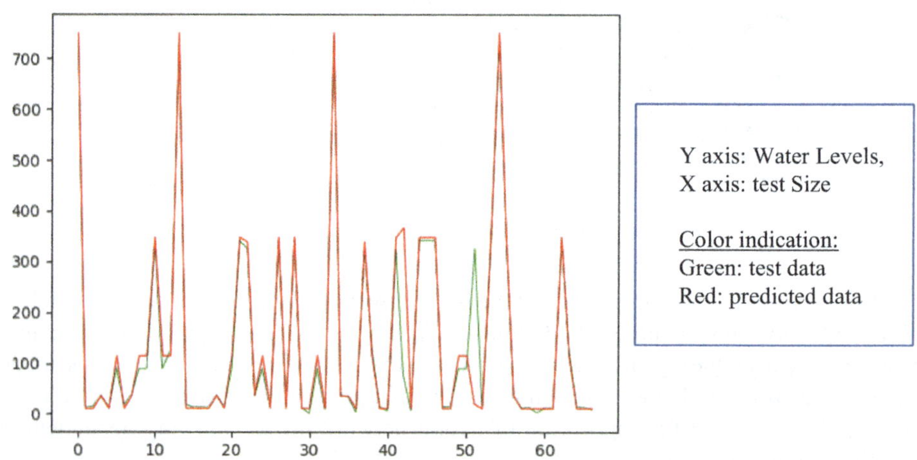

Fig. 4. Water levels

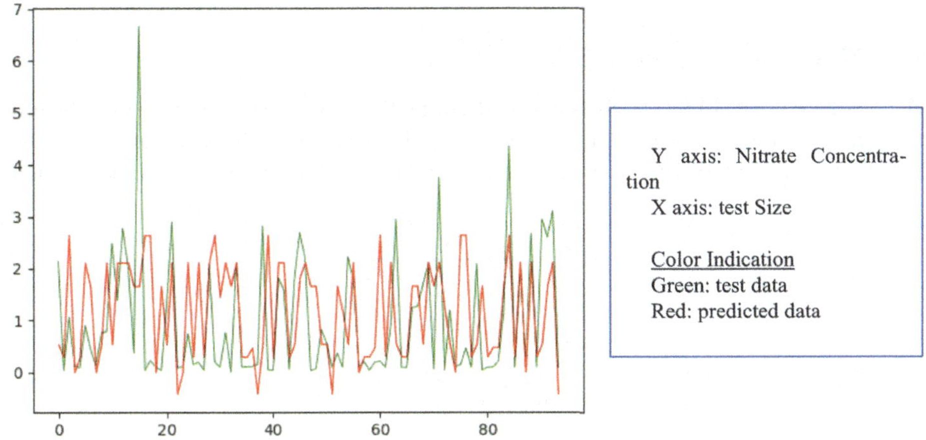

Fig. 5. Nitrate concentration

6.2 Accuracy Measures

Error Analysis according to the Gradient Boosting Regression is shown below in the figure. It explains how an increase in number of decreases the error (Fig. 6).

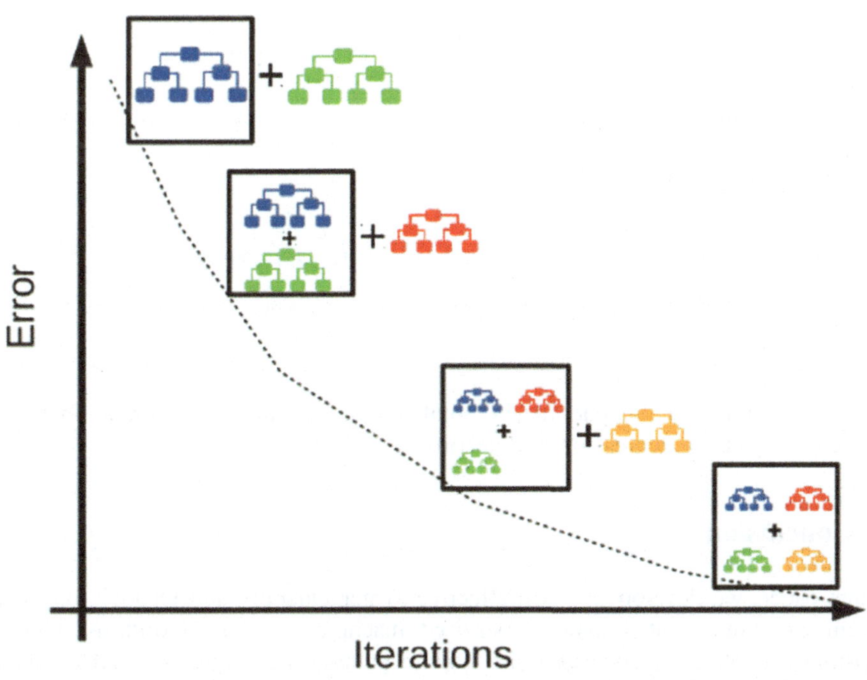

Fig. 6. Error analysis - gradient boosting regression

This project includes two accuracy measures listed as:

6.3 Mean Square Error (MSE)

Mean square error (MSE) is obtained by dividing the sum of squared errors by the total number of predicted values. These features can be used for errors with larger weights. This formula is useful when there are large weights to solve large errors in the problem.

$$\text{MSE} = \frac{\sum \left(x_{ops} - x_{pred}\right)^2}{n} \quad (8)$$

x-obs = actual value, x-pred is the predicted value
n = total number of samples collected

6.4 Mean Absolute Error (MAE)

Mean absolute error (MAE) is a measure of accuracy for regression. It sums up the absolute values of errors and divides them by the total number of values. It gives equal weight to each error value.

The formula of MAE is shown below.

$$\text{MAE} = \frac{\sum \left(\left\|x_{ops} - x_{pred}\right\|\right)}{n} \quad (9)$$

xobs = actual value
xpred = predicted value
n = total number of samples

```
Gradient Regression Water-Level Error:  16.790440997393983

Gradient Regression Nitrate Error:  39.859831597222225
```

Due to minimal input parameters, small set of data and less correlation factors, resulted in Gradient Regression Nitrate Error as 39.859.

7 Conclusion

In this paper, we developed a cost-effective Water Quantity and Quality forecasting system and implemented using supervised machine learning algorithms (Gradient Boosting and Neural Networks) with limited parameters to measure and obtain the desired 360-degree view of water characteristics. This project uses Five water quality parameters to implement the algorithms, predict the water volume with MAE of 16.7, and water volume with accuracy of 83%. The results laid a solid foundation for a cheap real-time water quality and Quantity 360-degree prediction system. This project also researched various Machine Learning Algorithms.

This project addresses research on existing water prediction frameworks and practices. The predicaments of these frameworks are: requirement of human and laboratory resources, the need for an in-depth analysis of 15+ water parameters to be used in traditional Machine Learning Algorithms, difficulty in determining correlation/ irrelevant factors, inability to eliminate the redundant data, and the need for extensive manual data purging.

We discovered machine learning algorithms that use no more than five parameters to predict water quality. We extracted the sample data from USGS (300 samples from seven districts in Nevada). This project additionally demonstrates that the Gradient Boosting Algorithm can deliver a fairly accurate output.

Finally, this work can be extended to design a real-time IoT-360 prediction and monitoring water framework. The implemented algorithms are sufficient for large-scale systems with minimal required parameters and integrate with real-time sensors. The algorithms can be extended to employ the parameters from real-time sensor systems and use-case scenarios (irrigation, human, Medical, etc.) and can take readings from other chemical sensors. This IoT System predicts water quality quickly and sends alerts on consumption readiness to the respective sources. Furthermore, this 360-prediction system can identify water based traumatic diseases like typhoid and diarrhea.

References

1. Data Obtained from https://nwis.waterdata.usgs.gov
2. Willmott, C.J., Matsuura, K.: Advantages of the mean absolute error (MAE) over the root mean square error (RMSE) in assessing average model performance. Clim. Res. **30**, 79–82 (2005)
3. Sakizadeh, M.: Artificial intelligence for the prediction of water quality index in groundwater systems. Model. Earth Syst. Environ. **2**(1), 1–9 (2015). https://doi.org/10.1007/s40808-015-0063-9
4. Friedman, J.H.: Stochastic gradient boosting. Comput. Stat. Data Anal. **38**, 367–378 (2002)
5. Srivastava, G., Kumar, P.: Water quality index with missing parameters. Int. J. Res. Eng. Technol. **2**, 609–614 (2013)
6. Ranković, V., Radulović, J., Radojević, I., Ostojić, A., Comić, L.: Neural network modeling of dissolved oxygen in the Gruža reservoir, Serbia. Ecol. Model. **221**, 1239–1244 (2010)

An Efficient Approach for Selecting QoS-Based Web Service Machine Learning Models Using Topsis

Miguel Angel Quiroz Martinez(✉), Josue Leonardo Moncayo Redin, Erick David Alvarado Castillo, and Luis Andy Briones Peñafiel

Department Computer Science, Salesian Polytechnic University, Guayaquil, Ecuador
mquiroz@ups.edu.ec,
{jmoncayor,ealvaradoc3,lbrionesp}@est.ups.edu.ec

Abstract. With the advancement of Service Oriented Architecture (SOA), web services have gained great popularity playing a vital role in performing daily transactions and information exchange based on the interaction of different applications within or outside through communication protocols, allowing to support the business requirements and data consolidation of any company. With the increase in the number of web services with the same functionalities, the problem that arises is that not all of them are efficient, which makes it difficult to make a decision to select the best ones that meet all the user's requirements. The problem can be solved by considering the quality of web services to distinguish web services with similar functionality. The objective of this paper proposes several automatic learning models to classify web services in categories according to QoS attributes using a refined data set, then select the best model based on performance criteria through the TOPSIS method. The deductive method and exploratory research technique were applied to study the QWS dataset.

Keywords: Web services · QoS · Machine learning · TOPSIS · QWS dataset

1 Introduction

Today, the growing popularity of Web services technologies on the World Wide Web has demonstrated the great utility and inherent capabilities in the development and deployment of comprehensive applications used for both business-to-business and business-to-customer integration over the Web [1]. Web services have been widely accepted as the best standards-based way to implement Service Oriented Architecture (SOA) to support the business requirements and data consolidation of any enterprise [2]. Service-Oriented Architecture is a powerful ally when it comes to managing large volumes of data (cloud data and data hierarchies). Web services are part of the following three technical elements, as SOAP, i.e. ("Simple Object Access Protocol"), also UDDI, i.e. ("Universal Description Discovery and Integration"), and finally WSDL, i.e. ("Web Service Description Language"). To obtain data or generate operations on that data in all possible formats, such as usually XML and JSON, which are platform-

H. Selvaraj et al. (Eds.): ICSEng 2020, LNNS 182, pp. 172–182, 2021.
https://doi.org/10.1007/978-3-030-65796-3_16

independent languages that allow the exchange of data between applications that transport large amounts of information. JSON is more natural to analyze, while XML is more complex [3]. However, the problem that arises at the time of discovery and selection, not all web services are efficient, and the adoption rate remains slow because of multiple challenges causing conflict in decision-making [4]. Most of these web services provide similar functional attributes, while their non-functional attributes such as response time, performance, reliability, security, availability, robustness, interoperability, and scalability may be quite different. These non-functional attributes have a direct impact on the quality of service (QoS) and user satisfaction [5].

Machine learning and data mining are the scientific fields in which the concept of classification and prediction has been exploited and explored. Different machine learning models of classification have been used to achieve better results in measuring prediction performance. To overcome the mentioned problems, this paper proposes to classify web services according to their QoS, and we use several well-known machine learning models such as Logistic Regression (LN), Naïve Bayes (NB), k-Nearest Neighbors (kNN), Random Forest (RF), Neural Network (NN), Support Vector Machines (SVM) to predict and classify according to QoS automatically attributes a web service in 4 categories from a data set called QWS. This real dataset allows us to take advantage of historical invocation experiences made by other users in similar contexts, through data mining by extracting QoS patterns (non-functional attributes) from web services [6]. Cross-validation was performed ten times to train and test to evaluate the results of the different models and determine the most efficient and accurate model to ensure optimal performance.

Therefore, the selection of an accurate classification model based on intelligent techniques supports the classification of web services by category: bronze, silver, gold, platinum according to their QoS attributes. It is an essential factor that helps in the tasks of identifying which new web services are efficient, achieving to satisfy the restrictions and QoS requirements of the user [7]. The objective of this paper is to select effectively and efficiently among various models the best candidate classification model using criteria such as performance metrics obtained in the classification of services based on QoS properties (non-functional properties). This is achieved by comparing and ordering the models according to their efficiency by means of an appropriate measure (MCDM) [8].

In this context, the technique of order of preference by similarity to the ideal solution (called TOPSIS) was used. TOPSIS is an effective method of selection for decision-making with multiple objectives (MCDM) that is applied to the classification or order of a set of alternatives. This method is applied in many fields because of its objectivity and completeness of object evaluation [9].

This method adopts the measure of the closeness of the final result to the ideal solution and the distance to the ideal negative solution of the evaluation rates. When the alternative solution is much closer to the ideal solution and much farther from the ideal negative solution, it is more favorable.

The rest of this paper is organized as follows: In Sect. 2, we review the preliminaries. In Sect. 3, we show the case study with its respective results. Finally, in Sect. 4, we present a conclusion and outlines of our future work.

2 Preliminary

In this section, you will find each of the important elements that make up the research work, as the first part presents a brief summary of multi-criteria decision-making (MCDM), then explains step by step the algorithm of the TOPSIS method, finally describes the QWS data set, the models of automatic learning classifiers and performance metrics.

2.1 Multi-criteria Decision-Making (MCDM)

The concept of multi-criteria decision-making (MCDM) is described as the process of finding the best selection alternative among all available alternatives, in which multiple qualitative criteria are used to assign value to each of the alternatives in order to compare and select the best option. MCDM are structured methods that have been used and highly valued for more than three decades in aspects of interdisciplinary nature such as political, economic, socio-cultural, technological, environmental, and legal (PESTEL) to the presence of different criteria in real problems in everyday life [10].

It was first developed in the early 1970s and rapidly spread around the world, attracting more and more researchers to the subject, and is known as the effective tool to study the identification and choice of alternatives according to the values and preferences of the decision-maker, has now been researched and refined. The general principles of the MCDM methods are: (i) alternatives, (ii) criteria, (iii) criteria weights, (iv) aggregation methods [11]. Hwang and Yoon in 1981 devised two groups of MCDM methods: as the first group i) MADM (Multiple Attribute Decision-Making) is used to solve discrete problems: on which the criteria are setting analysis is based, as the second group ii) MODM (Multiple Objective Decision-Making) is used to solve continuous problems: on which the analysis is based on multiple objectives and competition [12].

There are several methods for selecting the MCDM problem, and each has its specific characteristics. The following are some of the different MCDM methods that exist: as is AHP i.e. (Analytical Hierarchy Process), also PROMETHEE i.e. (Preference Classification Method for Assessment Enrichment), another is TOPSIS i.e. (Technique for Order of Preference for Similarity to Ideal Solution), also ELECTRE i.e. (Reality of Choice Elimination and Translation), and finally SAW i. e. (Simple Additive Weighting), among other methods that are currently used in various fields and that contribute to selecting the best alternatives by analyzing the weightings according to the criteria and the different proximity coefficients [13].

2.2 Topsis Method Algorithm

Originally the method of Discrete Multicriteria Decision TOPSIS (Technique for Order Preference by Similarity to Ideal Solution) was proposed by Hwang, C.L. and Yoon, K. This technique is based on the concept that it is desirable that the chosen alternative of the ideal solution be at the smallest geometric distance from the positive ideal solution (PIS) and at the largest geometric distance from the negative ideal solution (NIS) [14]. TOPSIS provides the best optimal candidate should have the least relative proximity to the PIS and the most relative proximity to the NIS. TOPSIS requires a performance score of each A_i alternative on each C_j criterion [15, 16]. Each of the steps within the proposed TOPSIS algorithm is described below:

The previous knowledge that initiates the topsis method is taking the decision matrix $D = [a_{ij}]$ as seen in the expression (1):

$$D = \begin{bmatrix} a_{11} & a_{12} & \cdots & a_{1n} \\ a_{21} & a_{22} & \cdots & a_{2n} \\ \cdots & \cdots & \ddots & \cdots \\ a_{m1} & a_{m2} & \cdots & a_{mn} \end{bmatrix} \tag{1}$$

Where a_{ij} represents the assessment of alternative A_i with respect to criterion C_j and, $W_j = [w_1, w_2, \ldots, w_n]$ is the weight vector associated with the criteria.

Step 1. The standardized decision matrix $R = [r_{ij}]$ is constructed. The initial decision matrix D is normalized; this allows the comparison between criteria by determining each r_{ij} element of the matrix used in the expression (2):

$$r_{ij} = \frac{a_{ij}}{\sqrt{\sum_{i=1}^{m} a_{ij}^2}}, (j = 1, 2, \ldots, n), (i = 1, 2, \ldots, m) \tag{2}$$

Step 2. The standardized weighted decision matrix $V = [v_{ij}]$ is constructed. Each r_{ij} element is multiplied by its weight associated with each w_j criterion with the expression (3).

$$v_{ij} = r_{ij} \cdot w_j, (j = 1, 2, \ldots, n), (i = 1, 2, \ldots, m) \tag{3}$$

Step 3. Determine the geometric distance of the slope of the positive ideal solution (PIS) and the distance of the slope of the negative ideal solution (NIS). The positive ideal values (A^+) determine the maximum benefit for each criterion through the expression (4) and the negative ideal values (A^-) de-terminate the minimum benefit for each criterion through the expression (5):

$$A^+ = \left\{\left(\max_{i=1}^{n} |j \in I^+|\right), \left(\min_{i=1}^{n} |j \in I^-|\right)\right\} = \left[v_1^+, v_2^+, \ldots, v_n^+\right] \tag{4}$$

$$A^- = \left\{\left(\min_{i=1}^{n} |j \in I^+|\right), \left(\max_{i=1}^{n} |j \in I^-|\right)\right\} = \left[v_1^-, v_2^-, \ldots, v_n^-\right] \tag{5}$$

So that I^+ and I^- make up the set of cost and benefit type criteria, respectively.

Step 4. Calculate the distance measurement based on Euclidean distance. The distance of each alternative to the PIS by means of the expression (6); Otherwise, the distance of each alternative to the NIS by means of the expression (7):

$$d_i^+ = \sqrt{\sum_{j=1}^{m} \left(v_{ij} - v_j^+\right)^2}, (i = 1, 2, \ldots, m) \tag{6}$$

$$d_i^- = \sqrt{\Sigma_{j=1}^m \left(v_{ij} - v_j^-\right)^2}, \ (i = 1, 2, \ldots, m \tag{7}$$

Step 5. Calculate the relative proximity for each alternative to the PIS.
The closeness coefficients are obtained with the following expression (8):

$$C_i^+ = \frac{d_i^-}{d_i^+ + d_i^-}, 0 \leq C_i^+ \leq 1, (\ i = 1, 2, \ldots, m). \tag{8}$$

Step 6. The last step of the algorithm classifies the alternatives in order of agreement, starting with those whose proximity coefficient is closest to the ideal solution (greater relative proximity). The best alternatives will be those with the highest C_i^+ value, i.e., closest to 1.0.

2.3 QWS Dataset

The Quality Web Services (QWS) data set is the first to capture the quality of services (QoS) of actual web services, and was conducted in 2007 as part of Eyhab Al-Masri's doctoral thesis work using the Web Service Tracking Engine (WSCE) to collect the services, most of which were obtained from public web sources [17]. This QWS dataset is available in an online repository at no cost for educational and other non-commercial purposes. This collection of web services is intended to provide a valid basis for web services researchers and has been applied to a wide variety of research/thesis projects in the areas covered by service computing such as web service quality prediction, service-oriented analysis, web service discovery, web service resource management, web service modeling, etc. It consists of different rows of data from more than 5000 web services, of which the public data set only consists of 365 web services that have been randomly chosen, and their measurements used 13 QWS (Web Service Quality) [18]. The QWS attributes represent the average values of the measurements collected over that period. Web services in the QWS data set are classified into four categories, such as (i) Platinum (high quality), (ii) Gold (medium quality), (iii) Silver (low quality), and (iv) Bronze (low quality) [19, 20].

2.4 Machine Learning Classifier Models

This section provides an overview of the fundamental self-learning supervised classification models used for data analysis and generating a validated model, with the goal of predicting categorical classifications for new instances, based on past behavior, experiences, or characteristics seen in historical data to solve a current problem. To do this, a learning process is developed from a set of data containing the characteristics, which is divided into two that are called training and test set, which allows the generation of a model based on rules or relationships to be able to classify and predict new instances [21]. The classification allows us to search and extract the patterns correctly through a special field called objective(category). There are two types of classification: binary classification and multiclass classification [22]. The main classifiers used are described in Table 1 below.

Table 1. Models of automatic learning classifiers

Classifiers	Description
Logistic Regression (LR)	It is an efficient model that results in binary classification or multiclass classification; the probability of an instance is predicted using the task logistics function
Naïve Bayes (NB)	It is a very fast model, it is based on Bayes' theorem, the assumptions are independence among the predictors, the probability that the presence of each attribute does not influence others
k-Nearest Neighbors (kNN)	It is an essential model that is used to try to seed groups of similar patterns with sample data sets to find the distance between neighbors
Random Forest (RF)	It is an outstanding model based on a decision tree at the time of learning and works by proximity search, then averaging its predictions
Neural Network (NN)	It is an optimized model, consisting of a set of (artificial) units connected and transmitting signals inspired by the functioning of the brain using the Multilayer Perceptron Sense
Support Vector Machine (SVM)	It is a model based on the minimization of structural risk; the decision function of classification by compressing the raw data into the set of support vectors (hyperplane as decision plane)

2.5 Performance Metrics

Performance metrics were used to assess the performance of service qualification models: Accuracy, recall, specificity, and F-measurement, which are widely used in any classification task [23]. These metrics are calculated based on the assigned class values of true positive (TP), false positive (FP), true negative (TN) and false-negative (FN), using the confusion matrix to describe the performance of a classifier, where the correct and incorrect predictions made by the model are defined in relation to the actual classifications of the test data; where the rows correspond to the current classes of an instance and the columns represent the predicted or estimated classes [24, 25]. In general, the metrics used are shown in Table 2, where description and the respective formula for their calculation are detailed for each.

Table 2. Performance metrics

Metrics	Description	Formula
Accuracy (A)	It measures the percentage that was correctly classified as TP and TN over a total number of classifications	$A = \frac{TP+TN}{TP+TN+FP+FN}$
Precision (P)	Measures the percentage of correct positive ratings overall positive ratings	$P = \frac{TP}{TP+FP}$
Recall (R)	Measures the percentage of correct positive ratings of correct overall ratings	$R = \frac{TP}{TP+FN}$
Specificity (SP)	Measures the percentage of negative ratings correctly on negative ratings	$SP = \frac{TN}{TN+FP}$
F-Measure (FM)	Measures the harmonic mean between the precision metric and the recall metric	$FM = 2 \cdot \frac{P \cdot R}{P+R}$

3 Case Study

In the case study, the well-known data set called QWS was used. By using the Orange Data Mining tool in a regular use of the widget, this dataset is taken to evaluate and test the machine learning models where we classified the web services in four categories according to the QoS attributes, using 10 times cross validation we observed their performance in Test & Score. Cross-validation is applied 10 times to train the models and then test them. In this process the data set is divided into 10 parts with a homogeneous distribution of equally randomly sized classes. Of the 10 parts, 9 are used as training data, while the other part is retained as validation data for testing the model. This process is repeated continuously until all 10 different parts are completed. This evaluation was carried out using an Intel Core i7 3.40 GHz and 8 GB RAM with Windows 10 64bit.

The alternatives take into account will be denoted by A_i (i = 1, 2, …,m). Where A_1 is Logistic Regression, A_2 is Naïve Bayes, A_3 is k-Nearest Neighbors, A_4 is Random Forest, A_5 is Neural Network, A_6 is Support Vector Machine. While the criteria used to evaluate the performance of the alternatives are C_j (j = 1, 2, …,n). Where C_1 is Accuracy, C_2 is Precision, C_3 is Recall, C_4 is Specificity, C_5 is F-Measure. Based on the results obtained from the classification of the previous parameters, the TOPSIS method is used to select the best model. First, a decision matrix is created, i.e., the matrix m x n (6 x 5) Starting from m alternatives A_i, i = 1, 2,…, m which will be evaluated based on the criteria C_j, j = 1, 2,…, n whose generic element is aij, Table 3.

Table 3. Decision matrix D = [aij]

	C1	C2	C3	C4	C_5
A_1	0,7520	0,7510	0,7510	0,7520	0,9110
A_2	0,7930	0,7960	0,9030	0,7930	0,9290
A_3	0,6910	0,6010	0,6120	0,6010	0,8320
A_4	0,9710	0,9780	0,9750	0,9010	0,9820
A_5	0,9090	0,9290	0,9100	0,9190	0,9740
A_6	0,9340	0,9340	0,9840	0,9340	0,9730

The values are normalized by (2). This normalization, which is necessary due to the presence of different units of measurement in matrix D, is carried out by determining the Rij of each element of matrix R, as shown in Table 4.

Table 4. Standardized decision matrix R = [rij]

	C1	C2	C3	C4	C_5
A_1	0,3621	0,3643	0,3609	0,3721	0,3978
A_2	0,3818	0,3831	0,3859	0,3924	0,4056
A_3	0,3327	0,2915	0,2941	0,2974	0,3633
A_4	0,4675	0,4744	0,4685	0,4458	0,4288
A_5	0,4377	0,4506	0,4373	0,4547	0,4253
A_6	0,4497	0,4530	0,4729	0,4622	0,4249

Then the weighting vector is then incorporated into the normalized decision matrix using Eq. (3) to achieve the weighted normalized decision matrix shown in Table 5.

Table 5. Weighted standardized decision matrix V = [vij]

	C1	C2	C3	C4	C_5
A_1	0,0724	0,0729	0,0722	0,0744	0,0796
A_2	0,0764	0,0772	0,0772	0,0785	0,0811
A_3	0,0665	0,0583	0,0588	0,0595	0,0727
A_4	0,0934	0,0949	0,0937	0,0892	0,0858
A_5	0,0875	0,0901	0,0875	0,0909	0,0851
A_6	0,0899	0,0906	0,0906	0,0924	0,0850

The ideal solutions A^+ and A^- are constructed based on normalized evaluations and weighted in matrix V is determined using (4) and (5), respectively, used in Table 6.

Table 6. Ideal solutions A^+ y A^-

	C1	C2	C3	C4	C_5
A+	0,0935	0,0949	0,0946	0,0924	0,0858
A^-	0,0665	0,0583	0,0588	0,0595	0,0727

Using the ideal solutions A + and A-, the Euclidean distances for each parallel PIS and NIS alternative are calculated using (6) and (7). For each of the alternatives the relative closeness is calculated with (8). The alternatives are then classified based on the value of Ci +. The higher the value of C_i^+, the better the performance of alternative A_i, shown in Table 7.

Finally, among six models of learning machines with respect to five criteria, after using this method. The classification obtained is clearly the following A4 > A6 > A5 > A2 > A1 > A3. Therefore, the best machine learning model to classify web services is Random Forest.

Table 7. Euclidean distances d_i^+ and d_i^-, Relative proximity and ranking $^-$

	d_i+	d_i^-	C_i^+	Rank
A_1	0,0424	0,0264	0,3837	5
A_2	0,0335	0,0350	0,5106	4
A_3	0,0678	0,0000	0,0000	6
A_4	0,0034	0,0658	0,9512	1
A_5	0,0106	0,0585	0,8470	3
A_6	0,0056	0,0641	0,9194	2

4 Conclusion and Future Work

On the basis of the article presented, it is concluded that with the increase in web services that have the same functional but different non-functional attributes, i.e., quality of service, and therefore selecting the best becomes too complex, it is demonstrated that the most viable solution is to classify web services according to their quality of service using the machine learning classifier models. In our approach, we propose several models and then compare them and select the most efficient model to classify and predict the web services in this case from the QWS data set. The results of the cross-validation are encouraging, and some of the models obtained a favorable score above 0.90. For multi-criteria decision making, the Topsis method was used to compare the performance between the algorithms in terms of mean values and standard deviations according to performance metrics. This method allowed us to find the best model, which is the Random Forest model, and the worst model, which is the Nearest Neighbor model. It is a very simple calculation procedure, which will encourage researchers from different areas of knowledge to use it. It is important to note that TOPSIS is a well-structured and reliable method and has been shown to serve as decision support for solving multi-criteria decision problems.

Future work aims to further investigate the relationships or rules between each attribute of service quality, to conduct the experiment with a data set that has many more instances, and to test with other classifiers not used in this article. It is suggested to develop a hybrid approach using the method of Topsis and entropy to obtain more accurate results in the selection.

Acknowledgments. This work has been supported by the GIIAR research group and the Salesian Polytechnic University of Guayaquil.

References

1. Ghadge, T., Bogiri, N.: Framework for web service composition and invocation. In: International Conference on Automatic Control and Dynamic Optimization Techniques, ICACDOT 2016, pp. 1030–1035 (2017). https://doi.org/10.1109/ICACDOT.2016.7877743

2. Bagtharia, P., Bohra, M.H.: An optimal approach for web service selection. In: ACM International Conference Proceeding Series, 21–24 September 2016, pp. 121–125 (2016). https://doi.org/10.1145/2983402.2983436
3. Yuan, Y., Zhang, W., Zhang, X., Zhai, H.: Dynamic service selection based on adaptive global QoS constraints decomposition. Symmetry (Basel) **11** (2019). https://doi.org/10.3390/sym11030403
4. Yang, Y., Ke, W., Wang, W., Zhao, Y.: Deep learning for web services classification. In: Proceedings - 2019 IEEE International Conference on Web Services, ICWS 2019 - Part 2019 IEEE World Congress on Services, pp. 440–442 (2019). https://doi.org/10.1109/ICWS.2019.00079
5. Ouadah, A., Benouaret, K., Hadjali, A., Nader, F.: Combining skyline and multi-criteria decision methods to enhance web services selection. In: 12th International Symposium on Programming and Systems, ISPS 2015, pp. 124–131 (2015). https://doi.org/10.1109/ISPS.2015.7244975
6. Chippa, M., Priyadarshini, A., Mohanty, R.: Application of machine learning techniques to classify web services. In: IEEE International Conference on Intelligent Techniques in Control, Optimization and Signal Processing, INCOS 2019 (2019). https://doi.org/10.1109/INCOS45849.2019.8951339
7. Ouadah, A., Hadjali, A., Nader, F.: A hybrid MCDM framework for efficient web services selection based on QoS. In: Proceedings of the 2018 International Conference on Applied Smart Systems, ICASS 2018, pp. 24–25 (2019). https://doi.org/10.1109/ICASS.2018.8652037
8. Swami Das, M., Govardhan, A., Vijaya Lakshmi, D.: Web services classification across cloud-based applications (2019). https://doi.org/10.1007/978-981-13-0589-4_23
9. Kazar, O., Rezeg, K., Belouaar, H.: Web service selection based on TOPSIS algorithm. In: Proceedings of the 2017 International Conference on Mathematics and Information Technology, ICMIT 2017, January 2018, pp. 177–182 (2017). https://doi.org/10.1109/MATHIT.2017.8259713
10. Agwa-Ejon, J.F., Mulaba-Bafubiandi, A., Pretorius, J.H.: The multi-criteria decision making, a tool for the promotion of technological inter-connected world through digital decision making. In: PICMET 2017 - Portland International Conference on Management of Engineering and Technology, Technology Management for the Interconnected World, Proceedings 2017, January 2017, pp. 1–7 (2017). https://doi.org/10.23919/PICMET.2017.8125301
11. Ouadah, A., Hadjali, A., Nader, F., Benouaret, K.: SEFAP: an efficient approach for ranking skyline web services. J. Ambient Intell. Humaniz. Comput. **10**, 709–725 (2019). https://doi.org/10.1007/s12652-018-0721-7
12. Kumar, R.R., Mishra, S., Kumar, C.: A novel framework for cloud service evaluation and selection using hybrid MCDM methods. Arab. J. Sci. Eng. **43**, 7015–7030 (2018). https://doi.org/10.1007/s13369-017-2975-3
13. Hosseinzadeh, M., Hama, H.K., Ghafour, M.Y., Masdari, M., Ahmed, O.H., Khezri, H.: Service selection using multi-criteria decision making: a comprehensive overview. J. Netw. Syst. Manag. (2020). https://doi.org/10.1007/s10922-020-09553-w
14. Yang, Y., Yu, N., Chen, Y.: Trusted cloud service selection algorithm based on lightweight intuitionistic fuzzy numbers. IEEE Access **8**, 97748–97756 (2020). https://doi.org/10.1109/ACCESS.2020.2996197
15. Regunathan, R., Murugaiyan, A., Lavanya, K.: A QoS-aware hybrid TOPSIS–plurality method for multi-criteria decision model in mobile cloud service selection (2019). https://doi.org/10.1007/978-981-13-1610-4_50

16. Ikram, M.A., Hussain, F.K.: Software as a Service (SaaS) service selection based on measuring the shortest distance to the consumer's preferences. Lecture Notes on Data Engineering and Communications Technologies, pp. 403–415 (2018). https://doi.org/10.1007/978-3-319-75928-9_36.
17. Vaishali, Kumar, R., Solanki, S.: Web service ranking and selection based on QoS. Springer, Singapore (2019). https://doi.org/10.1007/978-981-13-5802-9_28
18. Al-Masri, E., Mahmoud, Q.H.: Investigating web services on the world wide web. In: Proceeding 17th International Conference on World Wide Web 2008, WWW 2008, pp. 795–804 (2008)
19. Nivethitha, S., Gauthama Raman, M.R., Gireesha, O., Kannan, K., Shankar Sriram, V.S.: An improved rough set approach for optimal trust measure parameter selection in cloud environments. Soft Comput. **23**, 11979–11999 (2019). https://doi.org/10.1007/s00500-018-03753-y
20. Dilli, R., Argou, A., Reiser, R., Yamin, A.: IoT resources ranking: decision making under uncertainty combining machine learning and fuzzy logic (2018). https://doi.org/10.1007/978-3-319-95312-0_11
21. Shafi, S., Qamar, U.: Web services classification using an improved text mining technique. In: Proceedings - IEEE 11th Conference on Service-Oriented Computing and Applications, SOCA 2018, pp. 210–215 (2019). https://doi.org/10.1109/SOCA.2018.00037
22. Vazquezl, M.Y.L., Peñafiel, L.A.B., Muñoz, S.X.S., Martinez, M.A.Q.: A framework for selecting machine learning models using TOPSIS (2021)
23. Own, H.S., Yahyaoui, H.: Rough set based classification of real world web services. Inf. Syst. Front. **17**, 1301–1311 (2015). https://doi.org/10.1007/s10796-014-9496-3
24. Wang, Z., Tong, V.J.C., Chin, H.C.: Enhancing machine-learning methods for sentiment classification of web data. Lecture Notes in Computer Science (including subseries Lecture Notes in Artificial Intelligence (LNAI) and Lecture Notes in Bioinformatics), vol. 8870, pp. 394–405 (2014)
25. Kamath, S.S., Ananthanarayana, V.S.: Semantics-based Web service classification using morphological analysis and ensemble learning techniques. Int. J. Data Sci. Anal. **2**, 61–74 (2016). https://doi.org/10.1007/s41060-016-0026-x

Automated Agronomy: Evaluation of Fruits Ripeness Using Machine Learning Approach

Grzegorz Chmaj(✉), Saugat Sharma, and Henry Selvaraj

Department of Electrical and Computer Engineering, University of Nevada, Las Vegas, USA
{grzegorz.chmaj,henry.selvaraj}@unlv.edu, sharms8@unlv.nevada.edu

Abstract. Fruit orchards require a lot of tasks, monitoring the current state of fruits and anticipated time of harvest. Precise estimation of harvest time is a key knowledge for the supply chain and timely delivery to grocery stores. The current process relies on human agronomists, who visit orchards often and do the visual evaluation of number of fruits, their ripeness and expected time of harvest. In this paper, we propose a preliminary work for automation of the fruits' evaluation process, using machine learning algorithms to evaluate pictures of the trees. For our current system we have used pictures – one picture per tree, to constitute the base for evaluation of each tree using multiple pictures and video captured by drones. In this paper, Convolutional Neural Network (CNN) is used for fruit image classification based on ripeness stage of the fruits. Fruits classification was based on appropriate surface color and shape features and CNN was used to extract these features for classification. The model showed 96.43% accuracy. The ultimate goal of this work is to fully automate the process of orchard trees evaluation, including estimation of number of fruits on the tree (including non-visible ones), their ripeness and time of harvest to address the commercial market delivery planning requirements.

Keywords: Machine learning · Agronomy · Drones · UAV

1 Introduction

The delivery of grocery products to the stores is a sophisticated and time sensitive process. Perishable products such as fruits and vegetables are not only important elements of grocery store operation, but also time sensitive as they need to be harvested right in time and then delivered to stores in a timely manner. Orchard trees fructify multiple times a year and not synchronously in locations with most sunny and warm days over the year. Therefore, fruits are harvested at different times from different trees and visual/tactile evaluation is needed periodically, to determine the ripeness of fruits and the number of fruits that will be ripe and ready for harvest on a specified, future date. The knowledge of the quantity of fruits possible to harvest at specified date is a key information for grocery stores and their supply planning process. The evaluation process is currently done by agronomists who visit orchards in person. The industry demand is to introduce automation of this process by using drones to deliver pictures

H. Selvaraj et al. (Eds.): ICSEng 2020, LNNS 182, pp. 183–191, 2021.
https://doi.org/10.1007/978-3-030-65796-3_17

and videos of orchard trees that would then be delivered to computational system, which would do the estimation of number of fruits on the tree, ripeness and timing evaluation automatically.

In this paper, we present a partial element for such an orchard evaluation system. We focus on algorithms for evaluating pictures of trees. In future papers we will describe the remaining components of the system: drone (UAV) operation, UAV-to-computation multimedia delivery, UAV path planning and more advanced features of machine learning based algorithm: estimating number of fruits on the tree based on partial images/videos, categorized evaluation of ripeness, estimating the total number of deliverable fruits from an orchard for any given day in the future. Due to CoViD-19, current access to the testing orchard is limited, thus training of algorithms for the results in this paper is limited to the number of pictures we already have. Once orchards reopen, presented algorithms will be retrained and results will be included in future papers. The target system is presented in Fig. 1.

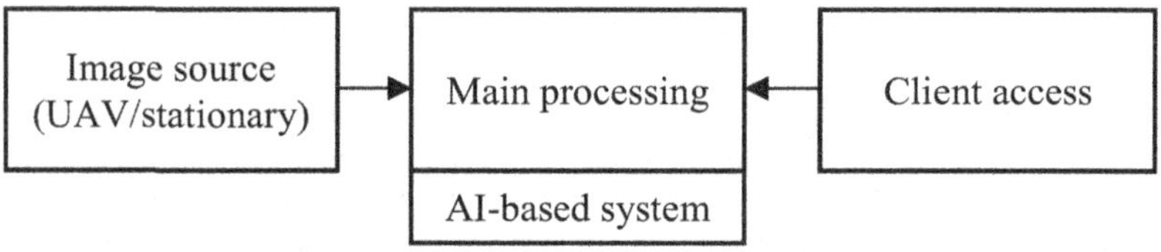

Fig. 1. General structure of the target system

2 Literature Review

The research work that is essential for the proposed system falls into the following categories: a) agricultural publications related to state of fruits and ways of their ripeness evaluation b) image processing for object recognition and features evaluation and c) machine learning for feature extraction and recognition.

The method of ripeness estimation based on video sourced from aerial vehicle was presented in [1] – authors used color features and genetic algorithms to determine four classes of apple ripeness (unripe, half-ripe, ripe, overripe). The algorithm detects apples and determines their ripeness – artificial neural networks are used in conjunction with genetic algorithms. Training was done using static images, authors assumed that in their future work such images would be delivered from aerial vehicle. In [2], the vision based robotic system for ripeness evaluation for sweet pepper is described. It includes detection, ripeness plus quantity evaluation and tracking. Fruits are tracked to avoid duplicate counting of each identified fruit. Vision system delivers 30 frames per second image input, coming from slowly moving robotics platform. Another method for apple orchards was presented in [3]. Instead of vision input, the ethylene sensors are used to determine the apple ripeness. The efficiency of the system was determined using simulated environment, including distribution of ethylene in the orchard environment and the impact of the wind generated by UAV motors, as the ethylene sensor would be embedded on an unmanned aerial vehicle. The results show that this method of measurement is sensitive to wind, and that a UAV can sense ethylene at a maximum of 10% detection probability and at less than 20 feet from the ground. Authors of [4] use the smartphone-based spectrometer for ripeness testing. The proposed solution includes light source, spectrometer, filters, microcontroller, wireless circuits and software application running on the smartphone directly. The method uses the spectral data analysis acquired as fluorescence from chlorophyll on fruit skin exposed to ultraviolet light. Next method, using another type of data input, has been presented in [5]. Ripeness of pears is determined using acoustical vibrations – based on time-course changes in the elasticity index and texture index. The vibrational method is non-destructive and uses a laser Doppler vibrometer. Results show the correlation of mechanical attributes (hardness, crunchiness, and thickness) with elasticity index and small or no correlation of chemical attributes (sweetness, juiciness, and acidity).

In [6] convolution neural network (CNN) is used for the strawberry image ripeness classification. In [6] the input image is converted into matrix using the 3 * 3 convolutional layer, then batch normalization is performed. After that, it has used 2 * 2 Max-pooling where final data is flattened and fed into the Dense layer to generate the 3-classification output. In [6] training is performed with a total of 240 images and testing is done on 60 images, with an overall accuracy of 91.6%. Similarly, in [7], it shows the CNN implementation for the CIFAR-10 Image classification performed in multithread GPU. The CIFAR-10 has RGB images similar to images we used. The architecture used in [7] has used 3 convolutional layers, where the depth of layers is 16,16 and 32 respectively. The final images were applied to a fully connected layer and finally categorized into 10 different classes. In [8], the grapefruit ripeness classification accuracy is compared between CNN and SVM methods. Color feature and shape of the grapes [8] were chosen as features for classification where CNN model showed 79.49% accuracy, whereas SVM showed 68% classification accuracy, concluding that CNN is better for image classification than SVM. Similarly, in [9], CNN is used to classify the ripeness of the pineapple fruit. Color is used as features for classification [9]. In this research, image data are collected by taking pictures through 3 webcams tricked from Arduino Controller and labeled as unripe, partially ripe and fully ripe. This research has used already trained MobileNet CNN architecture and has predicted 98.38% and 90.77% of training and test accuracy [9].

3 Methods

For the classification of ripeness of Orange fruit, we divide fruits into three categories: 1. *Green Unripe* 2. *Partial ripening (yellow color stage)* 3. *Ripe*. The labeling of the fruits is done providing unique index to each of the stage. Index "0" is labeled as green unripe fruits, index "1" as partial ripe fruits and index "2" as ripe fruits (Fig. 2). Indexing is done randomly—the index number to any fruits category does not matter and any number can be assumed to any class but it must start from "0" as the label file is processed in python "List" and python "List" index starts with "0". The images of the oranges are all stored in a folder and label is stored in a separate script file – each orange is given a tag either "0", "1" or "2" depending upon there categories they belong to. This constitutes input data set.

a) Green unripe: label 0 | b) Partial ripe: label 1 | c) Ripe: label 2

Fig. 2. Input dataset – orange types

The dataset used in experimentation can be characterized as follows (Table 1):

Table 1. Input dataset characteristics

Category of fruits	Number of images	Label
Green unripe	44, ids: 1–45	0
Partial unripe	40, ids: 46–86	1
Ripe	50, ids: 87–137	2

Convolutional Neural Networks (CNN) architecture was used to train and classify the Orange fruit stage. Before feeding into the CNN, all the images of different forms are first reshaped into size of 224 * 224 pixels and is stored in the list along with its label values. The array of labeled data (0 to 2) is converted into one-hot vector. The whole set of images is split into training and test sets, where 20% of images is training set. Our images consist of RGB components whose pixel values ranges from 0–255. The picture pixel values are kept as matrix form. For simplicity, these pixel values are normalized or rescaled between range 0–1 by a rescaling factor of 1/255. Thus, this will make mathematical computation easier with small values between 0–1 than the higher values between 0–255.

3.1 Creating Model

The model type used is "Sequential" built in Keras. Keras is high level neural network API [10]. It was originally written in Python and it allows very fast experimentation [10]. Keras runs on top of TensorFlow, CNTK, or Theano [11]. TensorFlow is deep neural network library [11]. It is used for numerical computational and machine intelligence [11]. We have used Keras on top of TensorFlow for our research. "Sequential Model" allows us to build the model layer by layer [11]. There are fully connected layers, max pool layers and activation layers built with Keras in Sequential model [11]. Thus, these layers make the architecture of the CNN. At first our image is converted into pixel values in matrix format. The input shape of our image is 224 * 224 * 3—Height,

Width and number of channels. The convolutional layer attracts the feature of our image by using filter. Filter moves over the input image and produces the feature map. We have used 3 * 3 size convolutional layer. Then, the activation layer is used. There are three types of most commonly used activation function: 1) sigmoid function 2) hyperbolic tangent function 3) Rectified Linear Unit (ReLu) [11]. We have used ReLu layer for our research.

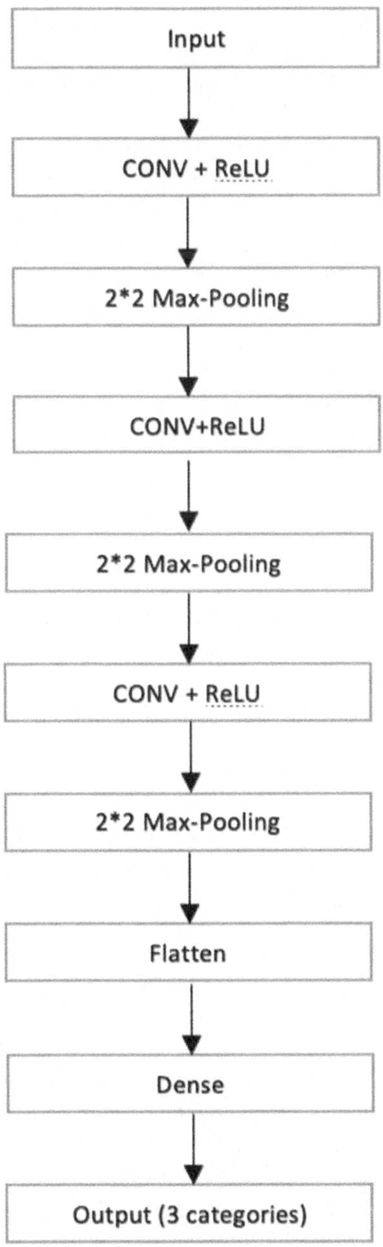

Fig. 3. Flowchart showing the CNN model.

ReLu layer converts all the negative pixel values in our feature map to zero keeping the positive values untouched [11]. After the ReLu layer, pooling operation is done by pooling layer. Pooling layer is used to reduce the dimension of the feature map [11]. We have used Max-Pooling layer for our research. Max-Pooling layer computes the maximum of a local feature map [12]. We have used 2 * 2 as size of the Max-Pooling layer. Hence, it reduces the dimension of feature map. The Final layer is the Fully Connected layer, which performs the classification process. Before feeding into the fully connected layer, the final output from the Max-Pooling layer is flattened. Flatten layers converts the Pooled Feature Map into a single column and feeds into a fully connected layer/Dense layer [11]. The Dense layer converts every input to every output by a weight [11].

In our research, the final output from the Fully Connected layer is categorized into 3-classes: 1) green unripe 2) partially ripe 3) ripe. For final classification after the Fully Connected layer, we have used "SoftMax" activation function. "SoftMax" activation function squashes the final output values into probabilities by normalizing the prediction [13]. We have used "Adam" as optimizer and "categorical cross-entropy" as a loss function. Optimizer controls the learning rate [13]. The total parameters are 2,797,795.

Figure 3 shows the CNN Model where the input (RGB) image is applied to Convolutional, ReLu and Max-Pooling layer in series—these steps are executed 3 times in series - and finally it is flattened and passed to Dense layer whose output generates the final output classifying the image.

4 Results

The model was run for 3-epoches. An epoch is the number of times a complete data set is passed for training in the learning process of the learning machine. Thus, we have sent the complete data set 3 times for learning process. Accuracy is obtained by evaluating the number of correctly categorized images in our test data. Our research achieved an accuracy of 96.43% (Table 2).

Table 2. The accuracy and losses in each epoch.

Epoch	Train loss (%)	Train accuracy (%)	Validation loss (%)	Validation accuracy (%)
1	88.10	47.22	52.932	75
2	57.49	75	41.88	64.29
3	52.13	69.44	28.93	96.43

Graph in Fig. 4. shows the result obtained after completing the training in each epoch. The training accuracy is represented by blue line and validation accuracy by green line. The X-axis represents the number of epochs (total of 3 in our research) and y-axis represents the accuracy (0 to 1 or 0% to 100% accuracy). It shows the plot for accuracy in each epoch for training and validation. The figure shows the training accuracy being 0.4 in the first epoch, 0.75 in the second epoch and 0.69 in our final epoch. Similarly, validation accuracy was 0.75, 0.64 and 0.96 in first, second and third epoch respectively.

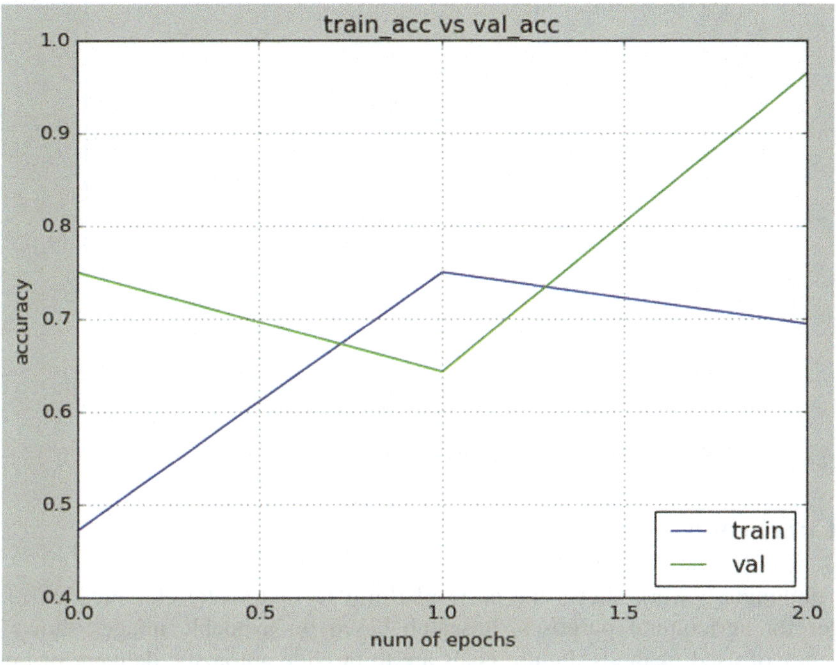

Fig. 4. Graph showing training accuracy vs validation accuracy

Figure 5 shows the plot for train loss and validation loss. Train loss is represented by blue line and validation loss by green line. The figure shows the loss for both train and validation decreasing in every epoch. The training loss decreased from 0.88 to 0.57 and finally to 0.52 in first, second and third epoch respectively. Similarly, validation loss also decreased from 0.52 to 0.41 and finally to 0.28 in first, second and third epoch respectively. Hence, it is proved that loss is minimized on each successive iteration (epoch).

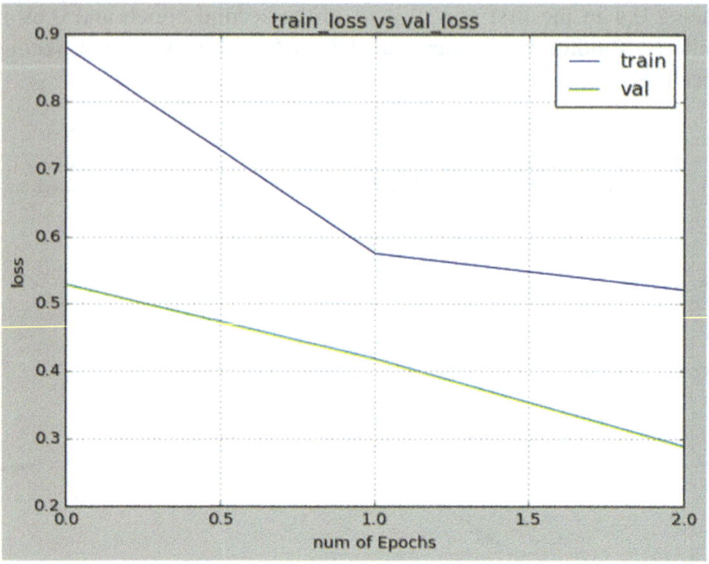

Fig. 5. Graph showing training accuracy vs validation accuracy

5 Conclusion

This preliminary work shows the basic platform for automatic classification of fruit images for agricultural purposes. Research based on available images shows high efficiency of used methods. Future goals are to provide automatic delivery of images, estimating number of ripe fruits on the tree and predicting number of ripe fruits for anticipated date. For this, we can categorize the set of fruit images into ripeness categories and depending upon the prediction done by CNN: if the ripeness predicted fall in one category, then we can say what is the approximate time it will take to mature into next categories. In addition, the exact labeling of the fruit images manually is a must to determine in which stage of ripeness the fruit falls. Thus, we can apply appropriate object detection algorithm in order to detect and count the number of fruits in a tree and apply algorithm to extract those detected images and predict its stages of ripeness and estimate its ripeness time (if is not ripened) with the algorithm developed in this research paper.

Work prepared with the cooperation with U.S. entity ATACT (Academy Of Threat Assessments, Countermeasures and Technology).

References

1. Sabzi, S., et al.: An automatic non-destructive method for the classification of the ripeness stage of red delicious apples in orchards using aerial video. Agronomy **9**(2), 84 (2019). https://doi.org/10.3390/agronomy9020084

2. Halstead, M., McCool, C., Denman, S., Perez, T., Fookes, C.: Fruit quantity and ripeness estimation using a robotic vision system. IEEE Robot. Autom. Lett. **3**(4), 2995–3002 (2018). https://doi.org/10.1109/lra.2018.2849514
3. Valente, J., Almeida, R., Kooistra, L.: A comprehensive study of the potential application of flying ethylene-sensitive sensors for ripeness detection in apple orchards. Sensors **19**(2), 372 (2019). https://doi.org/10.3390/s19020372
4. Das, A.J., Wahi, A., Kothari, I., Raskar, R.: Ultra-portable, wireless smartphone spectrometer for rapid, non-destructive testing of fruit ripeness. Nat. Sci. Rep. **6**. Article number 32504. https://doi.org/10.1038/srep32504
5. Taniwakia, M., Hanadab, T., Tohroc, M., Sakurai, N.: Non-destructive determination of the optimum eating ripeness of pears and their texture measurements using acoustical vibration techniques. J. Postharvest Biol. Technol. **51**, 305–310 (2009). https://doi.org/10.1016/j.postharvbio.2008.08.004
6. Thakur, R., Suryawanshi, G., Patel, H., Sangoi, J.: An innovative approach for fruit ripeness classification. In: International Conference on Intelligent Computing and Control Systems (ICICCS). IEEE
7. Han, S.-H., Lee, K.-Y.: Implemetation of image classification CNN using multi thread GPU. In: International SoC Design Conference (ISOCC). IEEE
8. Kangune, K., Kulkarni, V., Kosamkar, P.: Grapes ripeness estimation using convolutional neural network and support vector machine. In: Global Conference for Advancement in Technology (GCAT). IEEE (2019)
9. Chaikaew, A., Thanavanich, T., Duangtang, P., Sriwanna, K., Jaikhang, W.: Convolutional neural network for pineapple ripeness classification machine. In: International Conference on Electrical Engineering/Electronics, Computer, Telecommunications and Information Technology (ECTI-CON). IEEE (2019)
10. Arnold, T.B.: kerasR: R interface to the Keras deep learning library. J. Open Source Softw. **51**, 305–310 (2017). https://doi.org/10.1016/j.postharvbio.2008.08.004
11. Sewak, M.: Practical convolutional Neural Network. Packt Publishing Ltd., Birmingham (2018)
12. Atienza, R.: Advanced Deep Learning with Keras. Packt Publishing Ltd., Birmingham (2018)
13. Priyanka, C.H., Rachel, V.S., Harshith, B., Moulisha, R.: Color recognition algorithm using a neural network model in determining the ripeness of a banana. J. Eng. Sci. **9**(2), 305–313 (2009)

Internet of Things

Adaptation-Enabled Architecture for Internet of Things Systems

Patryk Schauer(✉) and Łukasz Falas

Faculty of Computer Science and Management, Wrocław University of Science and Technology, Wybrzeże Wyspiańskiego 27, 50-370 Wrocław, Poland
{patryk.schauer,lukasz.falas}@pwr.edu.pl

Abstract. Internet of Things (IoT) systems development faces numerous challenges related to assurance of high level of availability, reliability and maintainability. Many of those challenges are directly correlated with the operating environment of such systems and limitations resulting from them, which can impact system components' connectivity and operability. Due to this fact, IoT systems need to account those challenges and offer mechanisms which enable effortless reconfiguration of services and distributed processes conducted in such systems. In this paper, a novel architecture enabling easy adaptation of IoT services and reconfiguration of system processes is presented. The article discusses dependability related challenges which where consider during the development of the architecture, proposes an architecture and tools improving the overall dependability of the system, by providing solutions enabling easy recomposition and migration of containerized services between cloud and edge devices and presents a IoT system development methodology, which utilizes all of the features and supporting solutions of the proposed architecture.

Keywords: Internet of Things · IoT systems dependability · IoT systems architecture · IoT systems adaptability

1 Introduction

Internet of Things systems are quickly becoming an inherent part of modern society's everyday life. The transition from a quickly evolving concept to technology, which becomes adopted in many real life fields like economy, e-commerce, infrastructure, industry and healthcare was very rapid. Currently, there are many IoT systems which function in a direct connection with real world and real people. For instance, there are numerous Smart City application which focus on people and traffic monitoring, which is a basis for intelligent methods utilized for proper traffic management aimed at reduction of traffic congestion. There are also many pilot application of e-Health solutions that are responsible for remote patient monitoring and detection of life critical health conditions. Similarly automation methods in many industrial processes are becoming more complex and IoT oriented, hence the production systems are constantly moving to a state in which workers are collaborating with machines instead of operating them. These leads, again, to a situation in which proper machine environment

H. Selvaraj et al. (Eds.): ICSEng 2020, LNNS 182, pp. 195–204, 2021.
https://doi.org/10.1007/978-3-030-65796-3_18

monitoring and reliable decision making is crucial for health and wellbeing of people interacting with such systems.

All of these examples show that modern IoT systems function in environments which introduce a lot of new concerns for the development of new systems for such critical applications. In a more classical ICT applications, system failures could lead at most to reduced availability of certain services, denial of service for end users and in general, failures which could generate financial costs, but in most cases would not affect human health. However, in IoT applications in which there is often a direct connection to a life critical systems, failures can lead to harm or even threaten life of people involved in interaction with the system. It can be easily imagined that a failure in traffic management system can lead to car accidents and any failures in production systems can lead to health threatening situations for workers involved in the production process.

Due to this characteristics of those operating conditions, IoT systems require solutions which can guarantee a high level of dependability, which incorporates aspect such as, i.a., service availability, reliability, maintainability, security and failure recovery. However, many research works in the field of IoT focus either on the general concept of IoT and high level architectural solution, or on specific applications of IoT systems in various fields like e-Health, Smart City, etc., while the topic of long term dependability is often not the core theme of such research works.

In contrary, in this paper a more practical approach to the topic of IoT architecture is presented. The discussed approach and solutions are aimed at providing IoT researchers, as well as developers and practitioners, with architectural patterns and guidelines which enable development of IoT systems, designed with migration, reconfiguration and optimization solutions, which aim at increase of overall dependability such systems. The main focus of the proposed architecture is to facilitate the process of adaptation of computing services and process conducted in cloud environment and on edge devices, in order to maintain system operability even if the environmental factors introduce device to cloud connectivity issues or in case of failure of data processing components deployed in the system. This idea is supported by the fact, that the proposed architecture was developed with consideration of Quality of Service aware service composition methods, which can be utilized for automated reconfiguration of processes conducted in IoT systems in order to maintain they operability, in line with defined QoS requirements, in case of failure detection.

In the following sections of this paper a brief summary of related research works will be presented, which will be followed by a discussion of dependability challenges occurring in typical IoT systems. The main part of the article will present IoT systems building methodology, which can be utilized to facilitate development of reliable and migration enabled computing service and composition of adaptable data processing scenarios. The description of the proposed methodology will be followed by presentation of a practical IoT system's architecture incorporating failure detection mechanisms, composition methods and migration and reconfiguration supporting tools, which can be utilized to provide a high level of overall systems dependability.

2 Related Works

One of the main advantages of service based IoT systems is free and open approach to architecture. Support for few protocols such as HTTP REST is enough to preserve possibility of any component's integration. In relation to IOT-A [4], one of the first attempts of formalizing IoT architectures, the Internet of Things Service Systems Architecture was proposed [1]. This approach aggregates flexibility of service based systems, generic architecture and unlimited possibilities of IoT. In that paper, Universal Hardware Platform (UHP) was proposed as a base architecture model for any IoT devices. That model is composed of eight layers where data between internal communication interfaces such as RS232, 1-Wire, etc. and external communication interfaces like IP based networks are processed.

Next step of service based IoT architecture evolution was the practical implementation of Internet of Things Service Systems Architecture presentation [2]. In that paper IoT service based systems challenges were presented and above mentioned UHP model was used as generic a model for every IoT system component, not only devices, but also software based services. Moreover, the practical implementation utilized Docker containerization as a base mechanism of computing resources provider and publish-subscribe as the main communication mechanism. In that work the proof of proper efficiency, based on measurements, was presented.

A kind of extension of two previously mentioned articles was paper about IoT service based systems building methodology [3]. Basing on previous experiences in that artcle the main focus was aimed at describing computing and communication services as well as IoT service based systems management. Especially that work was focused on IoT services building process and its methodology.

The main motivation for this article was need to overcome challenges described in the next section. It is hard to find a solution to these problems in IoT research papers. Despite curious problem and promising way of conducting research in paper discussing migration in IoT cloud [7] utilized methods bases only on state machines and results are under any reasonable values. In modern application time values over hours are not acceptable. In addition the development methodology for system which can utilize those methods is not clearly stated in the mentioned paper.

There are also many works discussing the need for migrating IoT services to the Edge, in relation to not only cloud but also edge computing approach [5]. This kind of solutions can be a helpful for some communication problems such as transmission delay from end to end. However, this is not enough if we want to satisfy all requirements and not waste computing and communication resources.

There are also promising articles where workload based IoT service migration is presented [6]. Unfortunately that work focused on possibility of migration between IoT devices and from/to cloud, aside from any other aspects like service orchestration, resources allocation, etc.

3 Dependability Related Challenges in IoT Systems

In computer systems theory, dependability is defined as a trustworthiness of a computer system such that reliance can justifiably be placed on the service it delivers [8]. In practical scenarios and assessments, system's overall dependability can be defined and measured via Quality of Service attributes such as, i.a, availability, reliability, safety, integrity, maintainability, fault prevention, fault tolerance and fault recovery.

As it was stated, dependability of IoT Systems is an important feature due to the connection with real world and high potential of direct influence on people involved in processes conducted in such systems. Hence, high level of overall dependability of an IoT systems should be one of the crucial requirements for life critical IoT systems like traffic management systems, city security systems, e-Health systems, etc. However, providing required level of dependability in such systems is not a trivial task, due to distributed nature of such systems, involving numerous computing nodes and interconnection points, which potential all of them can become a point of failure disturbing the execution of processes conducted in the IoT system.

On the basis of the analysis of potential failure points in IoT systems, a set of three key groups of dependability related challenges involving communicational, computational and environmental threats were identified, which were described in the following subsections of this article.

3.1 Connectivity Related Challenges

One of the key characteristics of an IoT system is its distributed architecture, which involves clouds servers, IoT gateways, edge devices, sensors and actuators. In many applications those components are not only distributed in terms of logical architecture, but they are also dispersed geographically. Due to this fact, connectivity in IoT systems is often a mix of local and personal area networks involving low energy wireless protocols and wide area networks utilizing often wireless (e.g. LTE) connections which are prone to failures.

Due to those facts, IoT systems can often suffer from wireless communication links failures, data loss during wireless transmissions, failures of data transfer methods and protocols due to link instability and similar wireless networking related problems. Many of those problems are difficult to solve on the data transfer protocol level hence, one of the key challenges for IoT systems is to introduce architectural solutions, that will be able to detect network failures and reactively perform reconfiguration of an a data processing, and in many cases decision making, processes in order to prevent data loss and system inoperability. For instance, in case of a security systems which are monitoring valuable assets e.g. petrol during transport, it would be desirable that in case of connectivity failure with sensor data services deployed in the cloud, secondary services responsible for such analysis (which are able to operate with limited functionality, on limited resources) should be deployed on IoT gateway or edge device itself operating in a local area network to maintain desired monitoring functionality. Similarly, if a severe packet loss is detected during the data transfer, additional services responsible for data storage should be automatically migrated and deployed on edge device in order to store the data until it can be reliably resent.

3.2 Computing Nodes Dependability Related Challenges

In many IoT applications distributed components of a system have their specified tasks and role in the general architecture of the system. In many scenarios, such distribution of task is a desired trait of a system, however without proper mechanisms supporting failure recovery such solutions can lead to reduction of systems functionality or even system-wide failure in case of key components. For instance, in case of an IoT gateway failure which hosts data aggregation services and services responsible for data transfer to cloud functionality of the whole system would be greatly diminished. Similar situation can occur when a service executed by one of computing nodes becomes obsolete (e.g. image recognition model in one of deployed services can be vulnerable for specific data which introduces classification errors). In such situation, IoT system should be able to quickly reconfigure itself by deploying a new version of a service to such node or, if the new version of a service requires more computing power, by deploying it to a completely new node which features required resources, while managing the connectivity between services appropriately.

In such cases IoT system should be capable to detect such kind of failure and to perform a reconfiguration of deployed business process, which can be conducted by performing a new composition of services and their deployment on different components of the system, which will enable system to maintain its functionality.

3.3 System Environment Dependability Related Challenges

Due to direct connection with real life IoT system are also susceptible to changes in the environment. Such changes may be related to different aspects of reality which is monitored and influenced by such systems. For instance, in cases of rapidly increased volume of data that should be processed (e.g. greatly increased car traffic can lead to an increased need for processing power required to identify and analyze all of the objects captured by cameras) a reconfiguration of the data analysis process may be required in order to maintain desired QoS parameters of the offered functionality. In many cases such reconfiguration may only require multiplication of certain services and nodes, but in some scenarios it may require a change of the analytics function (replacement by a more efficient, but maybe less accurate method) that will enable system to maintain required dependability. Similarly to the previous challenges, in order to guarantee a proper level of dependability in such situations, IoT system's architecture should provide migration, reconfiguration and composability capabilities that can be easily integrated with monitoring and watchdog systems responsible for failure detection and execution of failure recovery procedures.

4 IoT System Development Methodology

To overcome challenges described in previous section, the IoT system development methodology was developed. The main feature of this methodology is its division into three areas of interests. First is related to atomic services designing, developing, deployment and execution. Second one is related to IoT systems, as a complex system based on service based systems architecture. Third area concerns computing and communication resources.

4.1 IoT Services

As it is generally known from service based systems IoT services should have several features like reusability, composability, loose coupling, autonomy and design by contract. In assumed methodology Process related to atomic service consist of service designing and developing. Artefact of this actions is Docker container image with implemented service. Moreover description of this image in repository should have enough information to allow composition process to choose this or other service.

The most important action in the following process is the service deployment. To handle previous mentioned requirements it is necessary to deploy in repository as many versions (compute architecture mean) of services as different types of computing resources are available in considered infrastructure.

Services repository should provide two services. One responsible for offering services list with proper descriptions. Second one provides docker images ready to use on different architectures.

The last action related to atomic services. In assumed architecture it is based on Docker images form repository and controlled by orchestration process.

4.2 IoT Complex Systems

Main goal of building computer systems is handling requirements. On their basis, system designer prepares IoT system design and optional requirements for atomic services, if it is necessary. One of the most important action is service composition, which is based on that design and results in a composite service plan that can execute the designed process. This process is responsible for meeting the functional requirements. Service composition could also be executed when composite service update is needed. For example, when system does not work properly.

Next step is the resource allocation. In opposition to the composition, resource allocation is responsible for meeting the nonfunctional (QoS) requirements, like memory amount, network interface bandwidth, etc. Similarly, to previous this step can also be executed to update the complex service when amount of resources on several nodes are not enough for proper service work.

Results of the resource allocation is passing to the Docker orchestration (Fig. 1).

4.3 Resources

In the proposed methodology resources can be divided into two groups, computing and communication resources. First group consists of computing nodes which allows to run Docker containers. It can be general purpose servers, virtual machines, but also specialized IoT devices. Second group are simply all software and hardware mechanisms using for communication between services.

Strongly related to resources are orchestration, monitoring and failure detection. The simplest function of resources monitoring is measuring resources utilization.

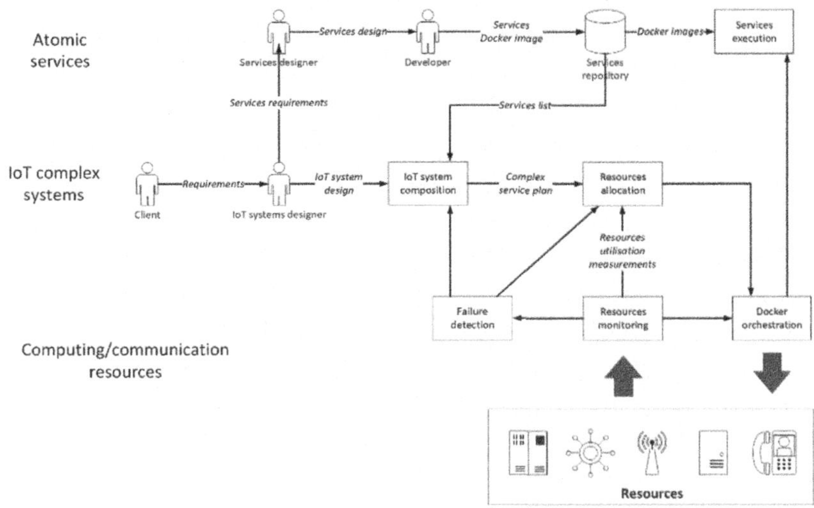

Fig. 1. IoT system development methodology

This and other functions are responsible for providing necessary information to failure detection and orchestration processes.

The failure detection is making decision support system which can trigger IoT system update. This decision can be made not only in relation with information about resources as well as services work. For simplification it is settled in the computing/ communication resources section, because the most valuable information of problems usually came from resources. When service is not responding, it usually not sending other information as well.

Last but not least element is orchestration. It relays on monitoring data and this process can control services execution requested by resources allocation.

5 Adaptation-Enabled Architecture for IoT Systems

The main goal of the proposed IoT system's architecture is to support the adaptability capabilities of such systems, which in effect can enable high increase in their overall dependability. The base assumption for the proposed architecture was that, automated web service composition methods for functional composition and resource allocation methods for ensuring optimized resource allocation and required level of QoS, integrated with monitoring and failure detection methods can effectively increase the overall dependability of such system. The detailed schema of the architecture was depicted in the Fig. 2.

The two main components of the architecture are providing features enabling increased dependability are the IoT system builder and the IoT system supervisor. The IoT system builder implement is responsible for the automated IoT system composition process, which involves selection and composition of IoT services that are required for proper execution of business process conducted in the IoT system. The service

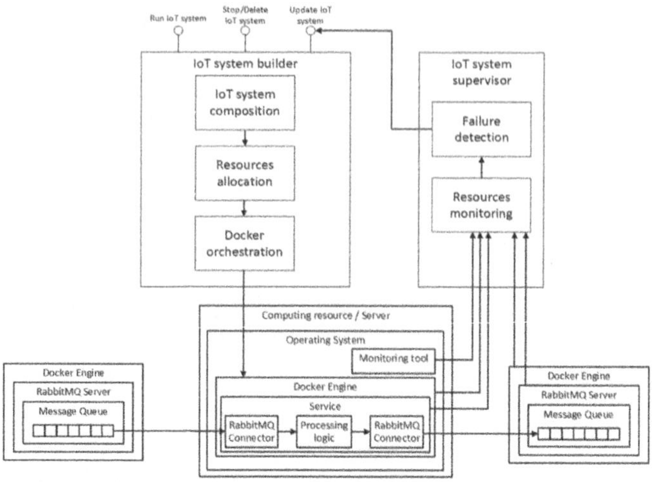

Fig. 2. IoT system development methodology

composition process is followed by the resource allocation process which is responsible for optimized deployment of services between various nodes of the system in such a way that ensures compliance with defined QoS parameters for the business process conducted in the IoT system. The final element of this component is the Docker orchestration module which is responsible for integration of computing nodes connected to the system and for the deployment of services to computing nodes according to the result of service composition and resource allocation.

The IoT system supervisor component is responsible for constant monitoring of the infrastructure (cloud servers, IoT gateways, edge devices) connected to the IoT system. The resource monitoring module of the supervisor is connected with dedicated monitoring tool gathering detailed data about the state of the computing node and, additionally it utilizes monitoring APIs of the Docker Engine, RabbitMQ Server and deployed services to gather additional data of the overall state of the system. The data gathered by this module is processed by the failure detection module which utilizes anomaly detection methods that detect failures on the basis of comparison between current values parameters of the deployed services and process and their expected, as well as historical, values of these parameters. The detection of any failures or anomalies in system's operations results in generation of a new request to the IoT system builder in order to updated the select IoT system's business processes structures and services in a way that will ensure recover to proper operability and compliance with the defined QoS and functional requirements.

All of the process and related services that execute them are deployed on the computing resources available in the system (i.e. cloud servers, IoT gateways, edge devices). Main assumption of the proposed architecture was that the computing resource should be generic and system agnostic, so it could be easily repurposed and utilized in a different way in different process without need for any extensive reconfiguration of the resource itself. Due to this fact all of the services deployed in the system must be containerized, because Docker Engine was selected as an execution environment for them due to support for container isolation it offers and due to efficiency and low image storage footprint. Also, such solutions enable deployment of multiple services that participate in different business processes on the same computing node without affecting the configuration of the computing resource that is hosting them. The only component installed on the computing resource that is architecturally specific is the monitoring tool that is responsible for gathering data about the computing resource that are sent to the supervisor. Additionally, this component is also responsible for simple failure recovery actions in critical situations when connection with the supervisor and IoT builder is unavailable.

The communication between services deployed in the system is realized in a publish-subscribe model utilizing the RabbitMQ message broker. The broker is also containerized and can be easily migrated between different computing resources in order to maintain standardized communication mechanisms between services and to provide a high level of reliability for communication mechanisms, i.a. due to possibility to multiply message broker instances and to deploy them both in cloud on or edge devices (to maintain messages when edge device is disconnected from the cloud servers). Those methods and capabilities were in depth discussed in research works described in related, previously published article [2].

All of the presented components of the architecture as a whole provide an environment for IoT systems and their business process that emphasize real time system's health monitoring and automated reconfiguration capabilities, which are inherent for adaptability, resulting in increased dependability of such systems.

6 Conclusions

The concept of IoT has quickly became adopted on a large scale in various field of everyday life like economy, e-commerce, infrastructure, industry and healthcare. Due to this rapid adoption and the fact that IoT systems are often directly connected with the real world and that they can influence people involved in their business process and their health, the dependability and adaptability issues related to those systems are highly important.

In this paper an adaptable architecture for IoT systems was proposed, that aims in providing a high level of adaptability and dependability in order to ensure proper operability of IoT system in compliance with functional and QoS requirement in various system states and in case of failure occurrence. The proposed architecture was supported by a description of system development methodology that can support researchers and practitioners in the process of building and improvement of adaptable IoT systems.

Future research works will focus on further research on this architecture and the possibilities for its practical utilization. Current research works validated the operability and legitimacy of the proposed architecture in terms of support for monitoring, adaptability and service migration between cloud server, IoT gateways and edge devices. The research works that are planned for the future will focus on in depth experimentation focusing of verification of efficiency of the adaptability and migration mechanisms. Especially they will focus on verifying the composition and resource allocation methods effectiveness and accuracy in the context of the complete system as well as in the measurement of the total time required for system reconfiguration in the event of failure occurrence. Future research works will also focus on improvement of currently utilized web service composition and resource allocation methods that are currently utilized in the proposed architecture.

Acknowledgment. This work was cofinanced by Polish Ministry of Science and Higher Education within LIDER programme coordinated by The National Centre for Research and Development.

References

1. Schauer P., Debita G.: Internet of Things service systems architecture. In: Barbucha, D., Nguyen, N., Batubara, J. (eds.) New Trends in Intelligent Information and Database Systems. Studies in Computational Intelligence, vol. 598. Springer, Cham (2015)
2. Falas, Ł., Schauer, P., Świątek, P., Trzaska, R.: Practical implementation of Internet of Things service systems architecture. In: 25th International Conference on Systems Engineering, Las Vegas, NV, pp. 291–298 (2017)
3. Schauer, P., Falas, Ł., Świątek, P.: IoT service based systems building methodology. In: 26th International Conference on Systems Engineering, Sydney, Australia, pp. 1–9 (2018)
4. Carrez, F., Bauer, M., Boussard, M., Bui, N., Carrez, F., Jardak, C., Loof, J.D., Magerkurth, C., Meissner, S., Nettstrater, A., Olivereau, A., Thoma, M.,Walewski, J.W., Stefa, J., Salinas, A.: Deliverable D1.5 – Final architectural reference model for the IoT v3.0 (2013)
5. Oteafy, S.M.A., Hassanein, H.S.: IoT in the fog: a roadmap for data-centric IoT development. IEEE Commun. Mag. **56**(3), 157–163 (2018)
6. Dupont, C., Giaffreda, R., Capra, L.: Edge computing in IoT context: horizontal and vertical Linux container migration. In: Global Internet of Things Summit (GIoTS), Geneva, pp. 1–4 (2017)
7. Paulraj, G.J.L., Francis, S.A.J., Peter, J.D., Jebadurai, I.J.: Resource-aware virtual machine migration in IoT cloud. Future Gener. Comput. Syst. **85**, 173–183 (2018)
8. Laprie, J.C.: Dependability: basic concepts and terminology. In: Laprie, J.C. (eds.) Dependability: Basic Concepts and Terminology. Dependable Computing and Fault-Tolerant Systems, vol. 5. Springer, Vienna (1992)

Internet of Things Systems as Complex Services

Patryk Schauer(✉), Łukasz Falas, and Arkadiusz Warzyński

Faculty of Computer Science and Management, Wrocław University of Science and Technology, Wybrzeże Wyspiańskiego 27, 50-370 Wrocław, Poland
{patryk.schauer, lukasz.falas, arkadiusz.warzynski}@pwr.edu.pl

Abstract. Popularity of Internet of Things (IoT) systems are growing and becoming important role in everyday life. Development of hardware, network infrastructure and software solutions provide an opportunity to take on more responsibility and perform more complex tasks by these systems. Performing complex operations requires developing more advanced systems, which leads to natural requirement to use an architecture that allows easy extension and maintenance. Additionally, systems of this type must contain may different nodes, such as databases, communication services, user interfaces, etc., which requires easy system modularization. This document presents the possibilities of using service-oriented architecture to build Internet of Things system as a complex service, which ensures that the requirements for modern IoT systems are fulfilled.

Keywords: Internet of Things · IoT systems · System as a service

1 Introduction

Internet of Things solutions are gaining ever increasing popularity in modern society, resulting in a vast range of applications, ranging from small scale environments dedicated for personal usage, e.g. smart home solutions, to large scale deployments utilized in industrial and social environments, e.g. Industry 4.0 and Smart City solutions. Such systems often have a complex architecture, consisting of multiple components responsible providing required functions, which are often distributed considering both, logical architecture and geographical location. Moreover, many IoT systems rely on integrations and communication with external services or even other IoT systems in order to deliver the required functionality.

In conjunction with the increased industrial and social demand for new IoT applications, this briefly described characteristics of IoT systems clearly indicates the complexity of their architecture and resulting challenges related to their development. In spite of this, one of the most important challenge for the future development of IoT based systems, is the development of architecture patterns that will support development of composable, adaptable and reliable IoT systems, while decreasing the overall system complexity and development time.

H. Selvaraj et al. (Eds.): ICSEng 2020, LNNS 182, pp. 205–212, 2021.
https://doi.org/10.1007/978-3-030-65796-3_19

This article discusses the problem of IoT systems architecture and proposes a novel, composite service-based approach for development of such systems. Due to the fact that many complex functions conducted by IoT systems can be modeled as processes with strictly distinguished atomic services, with well-defined interfaces it is reasonable to assume that complex functions conducted by IoT systems can be implemented as composite services. Moreover, such approach also simplifies the interoperability between different systems, due to the fact that such integrations can be accomplished via inter process communication.

While such approach might have been discussed in various research papers, the main discriminator of the proposed approach is the fact, that the service-based approach enabling composition of such process is conducted on both, data processing and data transfer level. In the proposed solution network connections between data processing services are abstracted as communication services, which offer additional QoS enabling mechanisms like queue-based throughput control and reliability mechanisms enabling easy and implementation independent reconnection and migration of data processing services connected via abstracted communication service.

The presented research works are an extension of research works discussed in previous articles [1, 2]. This article discusses a service-based architecture and supporting solution enabling implementation of IoT systems in the form of composite services. The following discussion presents a generic solution that it should allow development of vast array of different Internet of Things systems and applications. All of the solutions presented in this paper utilize standardized mechanisms for building computing services and communication, hence they can be easily implemented by engineers and IoT systems architects.

2 Related Works

Today, almost all electronic devices can be connected to the Internet. Starting from personal networks enabling the connection of sensors monitoring the vital parameters of people, through networks connecting elements of road infrastructure in order to better manage traffic and systems based on sensors monitoring the environment, ending with huge communication systems used in banking and finance [3, 4]. The differences between the classic Internet and the Internet of Things have many grounds, including differences in the design and implementation of these systems, as well as differences in the construction of the devices used. For over 10 years, Internet of Things systems have been associated and built on the basis of object identification mechanisms, one of the most important publications in this area is [5].

In addition to focusing on the topics related to the techniques of close communication with devices, the authors of that publication also noted the need to build systems of this type based on relations with reality. They also suggested using the Service Oriented Architecture (SOA) paradigm as the underlying architecture for the software layer.

Although the basis for the operation of the Internet of Things systems are devices that enable interaction with the outside world, the development of this field aims at integration with this reality, and especially with humans. Several years after the publication of this article, other researchers referred to the 2010 vision when discussing the implementation of one of the exemplary scenarios [6].

The use of the service composition known from the service paradigm (SOA) [7, 8] enables the construction of decision support systems regarding the operation of things, changes in the state of objects or data processing. The above-mentioned decision making is based on information obtained from the environment and its description formalized in the form of reality models built on the basis of domain knowledge. Due to the variety and multitude of things and computing services, it is necessary to develop a universal architecture on the basis of which it will be possible to design and build Internet of Things systems.

The process of designing the Internet of Things systems is inextricably linked with the description of reality, e.g. in the form of a domain ontology, as well as with modeling decision-making processes that occur in reality. Some analogies are known from the SOA paradigm [9].

3 IoT System Components

The main assumption for the proposed architecture, is the fact the in a composite service-based Internet of Things system several generic component classes can be distinguished, which enable the utilization of a process-based architecture in IoT systems development. The following component type should enable development of majority of IoT systems and applications according to the proposed architecture:

- Computing services – components responsible for data processing and implementation of required atomic business functions that can be composed into complex processes.
- Communication services – abstract services responsible for ensuring proper and reliable connectivity between services, enabling QoS functions and recomposition of services without data loss.
- Databases – data access services that store IoT data and which can be utilized in composite service workflows.
- Internet of Things devices – IoT devices utilizing interfaces compatible with the abstract communication services.
- Mediators – service responsible for transferring data to and from databases, devices, etc. via standardized interfaces.
- User interfaces and mechanisms responsible for data visualization – applications offering graphical interfaces for users to interact with the processes conducted in the IoT services.
- External systems – services and systems that are a part of the IoT systems environment, which can be accessed via mediators in order to ensure proper, reliable and standardized connectivity.

The identified components that were listed above should enabled development of any typical IoT system, focusing on data processing and decision-making supporting or process steering.

4 Computing Services

It is clear that the main element of Internet of Things systems are computational services. They are responsible for data processing and perform an executive role in the process of supporting decision making. From the point of view of Internet of Things systems, it is important that computing services are equipped with at least one input and output interface that receives or sends data using communication mechanisms. Computing services are equipped with interfaces that use communication mechanisms to transfer data to other services.

Services communicating directly with devices, databases, mechanisms responsible for data visualization, etc. will be called mediators. However, mediators apart from having only an input or output interface (instead of both at the same time) can and usually will have the same structure as computational services.

The assumption derived from the SOA paradigm is the statelessness of services. This also implies that the service instance performs tasks (processes data) only within the composite service for which it was launched. This does not exclude a situation in which a composite service can be decomposed into smaller components, and then a single instance of a computing service can perform tasks within several composite services.

Instances of computing services are created and launched in the process of launching a composite service. They use the computing resources available in the computing clouds, Internet of Things devices, terminals and users' computers. The architecture of computational resources is irrelevant as long as the formats and types of data are compatible and the communication interfaces support the communication services provided by the composite service.

One of the solutions designed to fulfil the requirements of the Internet of Things systems is the use of resources, in particular computing, delivered in the cloud computing. The main advantage of such a solution is high flexibility and scalability of resource use and ensuring high availability and security. Currently, an increasingly frequent model of resource sharing is container technology (in particular, realizable by the Docker platform [11, 12]), displacing classical virtualization. The advantage of this is the fact that new instances of computing services launched in the form of containers are created very quickly, the time to launch a container is in the order of several hundred milliseconds, without losing the flexibility of portability. While it is not possible to run the container, with a computing service, if it has not been prepared in accordance with the hardware architecture of the computing resource, adapting it to the new architecture should not be a problem. In the case of subtitling services that are inherently portable between architectures, such as Python, adapting to the new architecture will only change the base image. The mechanisms of the container platform will be able to prepare a new image of the container with a computing service for the Internet of Things in a virtually automatically way.

5 Mechanisms of Communication

The communication mechanisms, also called communication services, are responsible for the data transfer between the interfaces of the computing services. The communication service has multiple inputs and multiple outputs. Data are sent to the communication service after they have been created (processed by the sending service) and received by the next service when it is ready to receive the next portion of data. The communication service performs the functions of data transmission channels and queuing system. The use of the communication service requires an appropriate communication protocol.

Data sent to a single instance of the communication service, from multiple senders, are sorted according to the sequence of arrival, and it is only possible to distinguish their sources if they contain information about the source themselves. The data can be sent to a single queue or duplicated to multiple queues. The configuration of queues may include the following parameters:

- Queue size (number of messages or sum of message sizes),
- Validity time of the message in the queue,
- Message transactionality (whether downloading/processing the message must be confirmed for it to be deleted from the queue).

Queueing systems operating as part of communication services, like computing services, can be deployed on resources regardless of their architecture. The only criterion will be resource availability and obtained values of non-functional (qualitative) parameters.

As mentioned above, the communication service is a queuing system with input and output communication channels. The characteristics of communication channels depend not only on the characteristics of the queuing system, but also on the characteristics of the computing services instance with which they connect a given queuing system. Obviously, if the queuing system uses the same resources as the service sending data, for example they will be run within a single virtual machine, the communication channel between them will have very low latency.

There are many implementations of Message Oriented Middleware systems [13]. One of the most popular solutions is RabbitMQ. It is an open source software with AMQP and MQTT communication protocols. Due to the extended functionality, as well as the availability of documentation and libraries for most of the popular programming languages, this work uses the specification of this software as an example for a queueing system. Moreover, the ability to run the RabbitMQ server in the form of a Docker container ideally matches the proposed model of the Internet of Things system, in which the freedom of choice of resources on which computing services and queuing systems are run is a key issue.

6 Internet of Things Devices, Databases and Other Elements of Internet of Things Systems

Besides the computational and communication services described above, which are key elements of the Internet of Things systems, the sources and targets are equally important. They are usually implemented in the form of Things Internet devices, databases, user interfaces and visualization mechanisms. All these elements provide data for computing services or receive from them.

Databases are the most universal because their role is only to store data. Their proper use and integration with the whole system of the Internet of Things requires the use of the service as a mediator. The mediator has information about the structure of the database and is equipped with an appropriate communication interface compatible with the protocols used by communication services. Mediators will usually be services with relatively low requirements for computing resources. The additional role of mediators may be to convert formats or data types. A mediator can be treated as a computing service and usually its implementation will be compatible with it. The difference lies in the use of a specific communication protocol to the database, on the input or output of such a service.

Similarly to databases, devices may also require mediators to integrate them into the Internet of Things system. The use of increasingly available devices, built on the basis of computational architectures such as ARM, allows much more freedom in the design, construction and operation of Internet of Things systems. In such cases, the device is no longer simply a data provider or actuator implementing the decision. The ability to run computing services, queuing systems, as well as databases on the Things Internet device increases the flexibility of the system, its security and availability.

Other Internet of Things elements such as user interfaces or data visualization mechanisms as well as connections to other ICT systems can be realized in various ways. It is necessary for them to be compatible with protocols used by communication services and with data types and formats for their integration. The most desirable situation will be one in which all elements of the Internet of Things systems will be implemented on the basis of a common universal model, which will significantly improve building and extending the capabilities of the Internet of Things systems.

7 Internet of Things System Example

The above describes the individual elements of Internet of Things systems. Figure 1 shows a typical functional diagram of such a system implemented in the form of a complex service with additional elements.

In the presented system there are seven communication services (queuing systems). They allow data transfer between all elements of this system and external systems. In this system, data comes from the database, two devices and the internal system. Then they are processed by four computing services.

Finally, they are transferred to two external systems, a device and a database, and are passed on to the user interface for visualization. It should be noted that the

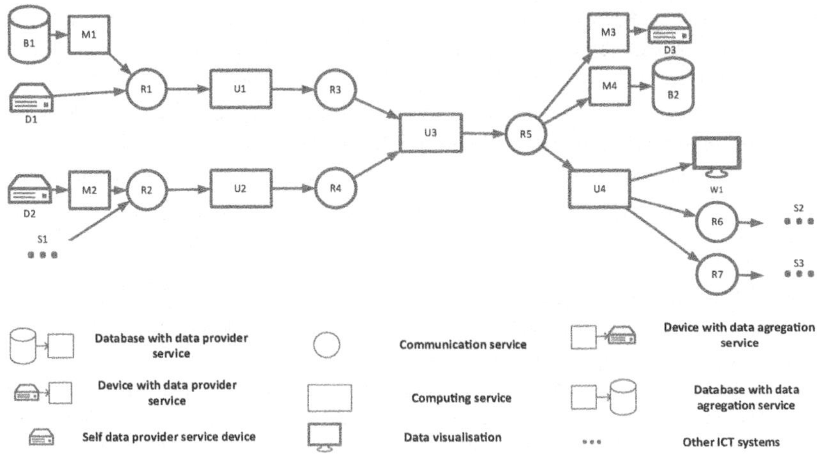

Fig. 1. Przykładowy schemat funkcjonalny usługi złożonej tworzącej system Internetu Rzeczy.

composite service provided in this case has a both sequential and parallel structures. Some services have only one input and output. Others receive data from two sources and send it through a single output interface. The opposite situation is also possible, when a service accepts only one data stream and generates three different ones. The important thing is that if the data coming out of the same type of service and are to reach different recipients, the function of their duplication is taken over by the communication service. There, successive portions of data generated by the service are transferred and then wait for the recipients to download. This allows to distribute the task of data processing among different providers.

8 Conclusions

This article discussed the topic of composite service based architecture for Internet of Things systems. The article presented a general concept for such architecture, focusing utilization of mechanisms and patterns, which enable practical utilization of the proposed solution by software engineers and system architects. The proposed approach emphasizes the composability IoT of systems in order to increase their reliability, adaptability and ease of development. In spite of other research works in this field, one of the discriminators of the propose approach is the introduction of communication mechanisms enabling abstraction of communication service which simplifies inter-service connectivity, increases communication reliability and introduces capabilities enabling application-layer QoS management for IoT connections between services. Preliminary verification of the proposed approach on the presented Internet of Things example have shown the viability of the proposed approach.

The results of the proposed research can be easily adapted by practitioners due to utilization of common tools and standards, which additionally enables potential for improvement of already developed and deployed IoT systems. According to presented research, utilization of such approach should shorten new feature development time for such systems and increase their maintainability and reliability. Further research works in this field will focus on research on experimental research on proposed abstract communication services, especially in terms of their capabilities and performance related to QoS management in QoS reliant IoT systems.

References

1. Patryk, J.S., Łukasz, L.F., Paweł, R.Ś.: IoT service based systems building methodology. In: 26th International Conference on Systems Engineering, ICSEng 2018, Sydney, NSW, Australia, 18–20 December 2018, pp. 1–9 (2018)
2. Łukasz, L.F., Paweł, R.Ś., Patryk, J.S., Rafał, T.: Practical implementation of Internet of Things service systems architecture. In: ICSEng 2017: 25th International Conference on Systems Engineering, Las Vegas, Nevada, 22–23 August 2017, pp. 291–298 (2017)
3. Lewandowski, J., Arochena, H., Naguib, R., Chao, K.-M., Garcia-Perez, A.: Logiccentered architecture for ubiquitous health monitoring. IEEE J. Biomed. Health Inform. **18**, 1525–1532 (2014)
4. Brzostowski, K., Drapała, J., Grzech, A., Świątek, P.: Adaptive decision support system for automatic physical effort plan generation-data-driven approach. Cybern. Syst. **44**(2–3), 204–221 (2013)
5. Atzori, L., Iera, A., Morabito, G.: The Internet of Things: a survey. Comput. Netw. **54**(15), 2787–2805 (2010)
6. Soro, A., Brereton, M., Roe, P.: The messaging kettle: It's iotea time. Aarhus Ser. Human Center. Comput. **1**, 2 (2015)
7. Świątek, P., Stelmach, P., Prusiewicz, A., Juszczyszyn, K.: Service composition in knowledge-based SOA systems. New Gener. Comput. **30**(2–3), 165–188 (2012)
8. Falas, L., Stelmach, P.: Web service composition with uncertain non-functional parameters. In: Camarinha-Matos, L.M., Tomic, S., Graça, P. (eds.) DoCEIS, vol. 394 of IFIP Advances in Information and Communication Technology, pp. 45–52. Springer (2013)
9. Swiatek, P., Stelmach, P., Falas, ł., Schauer, P.: Enabling automatic composition of stream processing services through ontology-based standardization. In: 2014 IEEE 16th International Conference on e-Health Networking, Applications and Services (Healthcom), pp. 88–93, October 2014
10. Stelmach, P., Swiatek, P., Schauer, P.: Communication protocol negotiation in a composite data stream processing service. In: Swiatek, J., Grzech, A., Swiatek, P., Tomczak, J.M. (eds.) Advances in Systems Science, pp. 681–690. Springer, Cham (2014)
11. Kaur, K., Dhand, T., Kumar, N., Zeadally, S.: Container-as-a-service at the edge: trade-off between energy efficiency and service availability at fog nano data centers. IEEE Wirel. Commun. **24**, 48–56 (2017)
12. Turnbull, J.: The Docker Book: Containerization is the New Virtualization. JamesTurnbull (2014)
13. Ding, H., Zhang, C., Chen, X., Shi, J., Wang, W.: Cloud-mom: a content-based real-time message-oriented middleware for cloud. In: 2018 IEEE 20th International Conference on High Performance Computing and Communications; IEEE 16th International Conference on Smart City; IEEE 4th International Conference on Data Science and Systems, pp. 750–757, June 2018

Communication Services in the Internet of Things

Patryk Schauer(✉), Łukasz Falas, and Arkadiusz Warzyński

Faculty of Computer Science and Management, Wrocław University of Science and Technology, Wybrzeże Wyspiańskiego 27, 50-370 Wrocław, Poland
{patryk.schauer, lukasz.falas, arkadiusz.warzynski}@pwr.edu.pl

Abstract. In recent years the development of complex Internet of Things (IoT) systems and their increased use has resulted in the growing importance of communication services, which task has been to provide uninterrupted and undisturbed communication between nodes with different interfaces. This paper presents the capabilities of communication between Internet of Things nodes including application of various communication protocols as well as software and architectural solutions ensuring adequate quality of connections. It also discusses the differences in various types of networks and quality assurance methods.

Keywords: Internet of Things · IoT systems · Communication services

1 Introduction

Network communication is one of fundamental elements of Internet of Things systems. In many applications, IoT systems gather large sets of data from numerous distributed sensors and data preprocessing services, which need to be transfer between edge IoT devices and data processing services via various networks and transmission mediums. In order to provide a reliable and efficient connectivity between various types of services utilized in IoT systems, a proper network management mechanism need to be utilized, that enable connectivity quality assurance and Quality of Service management for network connections, enabling compliance with requirements defined for the IoT system's connectivity.

This article discusses various challenges in the field of IoT systems network connectivity. Especially, it discusses various communication models in context of connectivity quality assurance. In many studies, researchers focused only on a single type of network, usually related to communication between Things Internet devices. This paper introduces a broader approach to this problem and identifies communication scenarios that occur in the Internet of Things systems. On this basis, the types of networks where communication occurs were characterized and the most common transport and applications layers protocols used in them were described. It also introduces the topic of management in ICT networks by describing measures of communication service quality and mechanisms of its assurance. The summary of this paper is discusses of the Internet Things communication scenario based on the

H. Selvaraj et al. (Eds.): ICSEng 2020, LNNS 182, pp. 213–222, 2021.
https://doi.org/10.1007/978-3-030-65796-3_20

proposed model utilizing application-layer queuing mechanisms for QoS management of IoT systems connectivity.

2 Related Works

Much of the work on IoT communication focuses on specific aspects, such as wireless communication and low energy protocols that can be used in edge devices.

The article [1] presents an overview of communication protocols used and recommended for use in the Internet of Things. However, this article only covers the wireless communication aspect and related protocols.In the [2] work attention was drawn to the need for a cross-sectional approach to the analysis of communication in the Internet of Things, but this article focuses on issues related to the software that is used for communication, treating the protocols-related issues as secondary to the discussed issues.

An interesting approach, seemingly complete, is presented in [3]. It covers both the issues related to the IoT hardware layer, wireless communication, as well as the problem of computing resources provided within the framework of cloud computing. Unfortunately, it does not include communication between the two areas. On the other hand, the work [4] addresses the issues related to the efficiency of transport protocols used in the Internet of Things and presents the results of empirical research conducted with the use of a real test environment.

Similar research, enriched with the topic of virtualization on IoT devices, is presented in [5]. Similarly, in [6] the problems related to dispersion and limited resources were addressed, drawing the attention of developers to the need to select appropriate protocols for various applications. Last but not least, the works [7, 8] dealt with the issue of communication security in terms of transport protocols.

In order to fill the missing gap and combine the usually considered separately topics of wireless communication and network and transport layer protocols, this article presents an end-to-end overview of issues related to Internet of Things communication, considering both communication with IoT devices, as well as software components typically implemented as web services.

3 Communication Scenarios

Communication services in Internet of Things systems provide for the transmission of data between Internet of Things devices and data-processing services. The transmission process includes links from different networks implemented in various technologies with different parameters.

Figure 1 presents scenarios of communication services in Things Internet systems. Components marked with numbers from 1 to 8 are starting or ending nodes.

Nodes marked with numbers from 1 to 5 are sensors (data generators) or actuators (receivers). The numbers from 6 to 8 represent services that process and store data. Any pair of different numbers from this set represents a communication scenario in the Internet of Things system. The connection between basic services within a complex service that is an implementation of the Internet of Things system is always one of such scenarios.

The different scenarios can be distinguished by type or location. Different types of nodes result directly from the functions performed by basic services. This function determines the choice of the service instance, and therefore the computing resource on which this service is to be launched (in other words, what its "location" is to be). They are independent of communication resources and concern data sources (and purposes) and computational resources. Services with the same functional parameters can be used interchangeably, which will affect the achievable values of the communication service parameters.

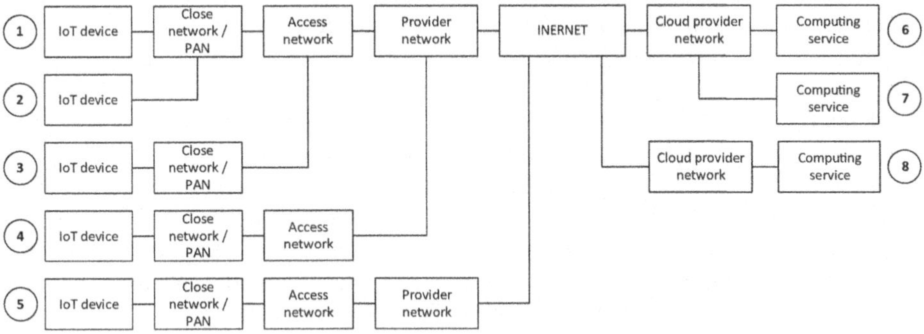

Fig. 1. Internet of Things communication scenarios.

As already mentioned, location depends on the function provided by the service, it has a direct impact on communication. As can be seen in Fig. 1, the nodes may be in the same network segment. However, nothing prevents services located in networks of different operators from communicating with each other. Furthermore, one could be a telecommunications operator, where the node would be a mobile phone and the other would be an operator providing computing resources, and the node is a data processing service. Moreover, there may not be a direct point of contact between the networks of both operators. They may connect through the networks of other operators.

The scenario described above is most realistic and known from general practice. However, it has an impact on resource availability. Such complexity of communication limits the possibilities of traffic management and therefore ensuring the Internet of Things quality of services.

4 Internet of Things Network Types

This section focuses on the complexity of the problem of managing communication resources in Internet of Things networks resulting from the heterogeneity of these resources. In particular, two features of these networks are important. The first one is the differences resulting from the types of these networks and the variety of protocols used. The second feature is related to the nature of the Internet itself, which is a combination of networks of different operators, or in other words, owners. Those who manage the Internet's parts, which are used to establish even single connections between services, may have different management objectives or strategies.

The characteristics of the two main segments of the network occurring in communication between services of the Internet of Things are described below. These are local area networks (or even smaller networks such as personal area networks) and links that form backbone networks of operators and between operators.

4.1 "Near" Networks

The topic of the Internet Things is inseparably connected with communication mechanisms and protocols designed to connect devices in the nearest area of the monitored or controlled facility. Originally, the concept of the Internet of Things required only the ability to identify the object, later, along with the technological and logical development, the ability to transmit the status of a given object became a standard. The communication mechanisms developed for these functions assumed one-way communication to the nearest node with increased functionality, usually acting as a hub. Devices included in the network of this type usually have very limited resources. Due to the need for miniaturization, they cannot process data in a complex way and limited access to electricity forces the use of very limited communication protocols.

Apart from the previously mentioned, one of the most important features of "near" network nodes is wireless communication. The values of parameters of connections established by these nodes do not allow for free communication. Moreover, the use of ISM (Industrial, Scientific, Medical) frequencies additionally limits the possibilities of this type of equipment.

The "near" networks in the Internet of Things are characterized in particular by very short ranges (<10 m), low bandwidth (<1 Mbps, e.g. 250 kbps for ZigBee), high latency due to wireless communication and high latency variability because of the use of common radio frequencies by more and more devices.

At the edge of "near" networks there are usually devices with communication interfaces allowing to connect such network to the Internet. Frequently, these devices will provide access to the resources of the "near" network in the form of services (e.g. using REST or SOAP protocols). Alternatively, these devices may be directly responsible for transferring data from devices within the network and vice versa. Communication with the "near" network may be organized with the use of radio and cable connections. Depending on the communication technique used, it is natural that such a connection will have different parameter values. The development of technologies used by wireless networks (e.g. LTE, 5G, etc.) moves their characteristics closer to cable networks, or at least makes it possible to provide services at the quality level reserved only for wired networks.

Specific network layer protocols such as 6LoWPAN [9] have been developed for "near" networks in the Internet of Things, prepared for low-energy devices, while allowing the integration of many advanced services.

4.2 Inter-operator Networks

One of the main postulates of the Internet of Things concept is the possibility of combining many seemingly unrelated elements within one system in order to extend functionality that would not be possible without such integration. Often the elements connected within a single system can be geographically distant. Therefore, connections of telecommunication operators are an inseparable element of the Internet of Things network architecture. The links organized in a single operator's network as well as links between operators are important. Usually they will be characterized by high values of quality parameters, i.e. high throughputs or low delays brought by intermediate nodes.

Unfortunately, in this situation the user, provider or designer of Internet of Things services does not have any possibilities to manage communication resources and control traffic on the connections between individual atomic services.

4.3 Internet of Things Network Protocols

The concept of the Internet of Things is not strictly defined. Some attempts to systematize it, such as the report of the European IOT-A working group [10] only gave the concept a general outline and, by indicating important elements, allowed for flexibility in the implementation. Nevertheless, the solutions developed by practitioners allow to formulate the principles of the Internet of Things, based on the protocols used. Below are some examples of protocols [11], together with an indication of their differences and applications for which they are best suited.

The most popular protocol used for communication within service systems is Hypertext Transfer Protocol (HTTP). In the higher layers, it is usually enriched with a Simple Object Access Protocol (SOAP) or a much less demanding Representational State Transfer (REST) mechanism. Due to its great popularity and, in the case of REST, great simplicity of implementation, HTTP is a very popular protocol used in service interfaces. It is a text-based protocol, which additionally facilitates its analysis. Moreover, the Hypertext Transfer Protocol Secure (HTTPS) version, which allows to prevent the communication from being suspected by third parties, is a trusted method of communication on the Internet. Communication with the use of this protocol is based on the concept of client-server, where one side provides a service interface and the other connects to it. In this way it is possible, among other things, to retrieve data, update, delete it, as well as initiate processing. In the case of communication between equivalent nodes, this may be a drawback, because such a model forces the need to implement the server on at least one side.

The protocols used for streaming data are the Realtime Transport Protocol (RTP) and the more extensive Real Time Streaming Protocol (RTSP). Unlike the state HTTP, RTP and RTSP is designed to control the streaming session. These protocols control the sequence of data transmission, and can adapt to the prevailing network conditions. Unfortunately, they are not suitable for all data transmissions. Mainly due to their complexity, resulting from specialization in multimedia data transmission.

The answer to the needs of communication in the Internet of Things are the protocols implementing communication in the publish - subscribe model. This type of communication does not require the devices or services to implement service interfaces. A broker is used for communication. The server hosting this function provides interfaces to which interested parties connect and transmit data to it or receive data that are stored in it. The broker is organized in the form of queues and data exchange points. When data is sent to the broker, it goes to the exchange point declared by the sender. The queues are connected to this exchange point, from which the receivers retrieve data. As a rule, two recipients connected to one queue will "steal" each other's messages, thanks to which it is possible to realize a scenario in which two services alternately process successive portions of data. If the recipients are connected to different queues, they will have a copy of the same data, which makes it possible to realize a scenario in which the same data is to be multiplied. The most popular protocols of this type are Advanced Message Queuing Protocol AMQP) and Message Queuing Telemetry Transport (MQTT), which is much less demanding and therefore very popular in Internet systems. These protocols can be used to transmit measurement data and expectation values through actuators as well as, due to the high performance of brokers, streaming data. Moreover, broker mechanisms can successfully replace, still very rarely implemented in networks, data link layer protocols for group communication (multicast). Such application is particularly advantageous in the case of data streams, when the advantages of group communication, related to bandwidth savings, are greatest.

5 Measures of Communication Service Quality

The main parameters used to assess quality in ICT networks are:

- bandwidth,
- the delay and its fluctuation,
- error rate.

The values of all these parameters are interrelated and changes in one of them usually lead to changes in the other. With regard to the Internet of Things systems the above parameters will be of varying importance.

Considering the commonly used communication protocols in the Internet of Things, the error rate at application layer level will always be zero. In most cases, these protocols use TCP in the transport layer, which provides retransmissions. Decreasing the value of this parameter in the lower layers will obviously affect the quality of connections. However, this will be reflected in the values of other parameters. The impact of this parameter on the quality of connections is particularly important in wireless networks, which are most vulnerable to packet losses. Moreover, some traffic control mechanisms work on the principle of packet rejection. In this situation, the value of available bandwidth in the application layer will decrease and the delay will increase.

The vulnerability of Internet of Things systems to capacity deterioration and delays are related to the architecture of these systems. Internet of Things systems, in particular those based on the service paradigm and described as a complex service, use multiple connections between atomic services. Each pair of services which need to establish communication will require an appropriate value of bandwidth. Failure to meet this requirement may lead to further connection degradation due to overload. Particularly, it will cause an increase in delay and eventually packet rejection. As already mentioned, because of the protocols used, this will be compensated by retransmissions, which will further reduce the value of available bandwidth. This general scenario is very dangerous because it may lead to a situation where the link becomes useless. Therefore, knowledge about service requirements for the bandwidth of the links between them should be matched in such a way as to satisfy them.

With regard to Internet of Things systems, the value of delay is important not only as a characteristic of single connection, but of the entire complex service. The delay on individual links of a complex service is cumulative in relation to the entire complex service. In serial structures of the service graph, the delays will add up, and in parallel, the highest value will be important.

The management of communication services is only possible when the value of parameters characterizing the links is known. The information about them may come from the operator or from measurements taken.

5.1 Quality Assurance Methods

The communication quality management can be conducted on different layers (assuming a layer model e.g. ISO/OSI [12]). Moreover, quality management mechanisms should be located on all layers. It is not always possible to access and manage information on all, especially the lower layers. For the designer, supplier, or user of Internet systems, these mechanisms will usually not be accessible. Due to the limited possibilities of managing the nodes involved in communication between individual services and the assumption that the purpose of communication is to provide services, and "(…) the only important criterium of quality of network services is the quality of services provided in the application layer (…)". [13], the methods of communication resources management considered in this work concern the highest layer, i.e. the one in which communication takes place between services.

The basic mechanisms of quality management in ICT networks are flow control and route management. In general, flow control is designed to optimize packet delay when it is greater than the assumed critical value. Route management, on the other hand, allows for the selection of links that additionally meet the criteria for available capacity.

The literature describes the communication quality aspects, which are defined by two concepts of ICT traffic management, in networks with quality assurance [14, 15]. They are:

- Integrated Services (IntServ),
- Differentiated Services (DiffServ).

The first one assumes guaranteeing quality by reserving network resources for each flow. The second consists in limiting the amount of traffic entering the network and assigning it to classes with different service quality levels. Unfortunately, both concepts, to be fully implemented, require integration in the network layer. This makes them available only to the entities managing the networks, and not to the services using them.

There are also concepts of network services without quality assurance. They can be classified into two groups:

- systems with losses,
- systems with delays.

The first is based on rejecting traffic that exceeds the capabilities of the network. The second one assumes the possibility of compensating for temporary lack of resources by buffering traffic. Although traffic buffering increases the delay on a given connection, within certain limits the capacity does not degrade. As mentioned earlier, Internet of Things communication protocols use retransmission mechanisms to provide a guarantee of data delivery. This packet rejecting mechanism will also delay the transmission as a result. However, unlike the data buffer in network nodes, the retransmitted data will have to travel the entire path between services again. As a result, the effective bandwidth available on a given connection will decrease, especially in the higher layers.

6 Internet of Things Communication Scenarios

As already mentioned, the communication mechanisms in the Internet Things are defined by the protocols used. Application of publication and subscription-based communication protocols requires a message broker. Introducing an additional element into the communication certainly results in increased use of resources. However, the benefits of this solution may confirm its usefulness. With regard to traffic management and quality assurance, a message broker based on a queuing system offers additional possibilities.

Due to the flexibility of such an intermediary mechanism and the possibility to deploy it on any computing resource, it becomes available to manage routes in the network by selecting appropriate instances of this mechanism. Moreover, it becomes possible to control traffic through its configuration. The use of a queuing mechanism permits the implementation of a redundant traffic compensation mechanism by stopping data in queues. However, the most important is the fact that these mechanisms are implemented in the application layer. This is crucial because of the lack of possibility to manage communication resources on the lower layers, because there is no access from the level of services to the network management mechanisms of the operators that intermediate in communication. While there are mechanisms in the protocol specifications of the network layer, these mechanisms are usually not available to end customers.

Inter-service Communication

The simplest and most common type of communication in Internet of Things systems is the one-way communication with the use of queuing mechanism shown in Fig. 2. The communication model described in this way concerns the application layer.

In this situation, Service 1 sends the data to the queuing mechanism, and service 2 retrieves them at any time, appropriate for itself. Such a mechanism allows the recipient to control the traffic, which is beneficial because in the case of temporary unavailability

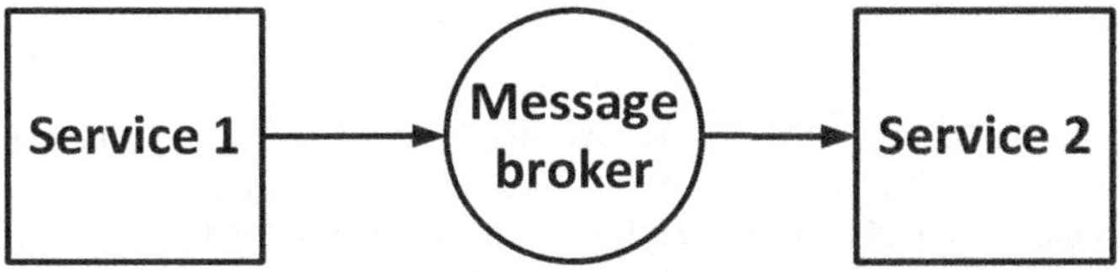

Fig. 2. Scheme of communication between a pair of services using a queuing mechanism.

of the recipient, no communication resources are wasted (for example for pointless retransmissions).

All three nodes involved in communication are located on computing resources. In such organized communication model, two links from the computing service to the queuing mechanism and from there to the next service can be specified. The exact values of the link parameters between the computing services will depend on many factors, but it is still possible to define limit values. The value of the capacity of such a link will not be greater than the smaller value of the capacity of the links from the computing service to the queuing mechanism and from it to the next computing service. In turn, the value of delay will be the sum of the delay values of individual lines.

In addition, the queuing mechanism has a capacity that limits the possibility of data buffering. In fact, this capacity will be greater than that of queues available on devices that are network nodes of lower layers. An additional advantage of such a mechanism is the possibility to multiply traffic on demand. If it is necessary to send the same data to two recipients, they will be sent to two queues, assigned to particular recipients. Such a mechanism of group communication (multicast) does not require additional support from the network. Moreover, the recipient is responsible for controlling the data transmission.

7 Conclusions

This paper presented the methods of communication it the Internet of Things systems using various communications protocols and models. It also describes how to assess the quality of communication using typical IoT elements. Creating new systems and developing solutions for the Internet of Things, attention should always be paid to the differences resulting from the size of the network and protocols used. In order to ensure quality, it is important to consider the advantages and disadvantages of the adopted model and the location of the entire Internet.

Currently, research works focusing on experimental verification of the application-layer queue-based QoS management, implemented according to the presented scenario utilizing message broker intermediary, are being undertaken. The results of the experimental verification of the proposed approach will be discussed in following articles related to the topic of quality assurance in IoT systems connectivity.

References

1. Al-Sarawi, S., Anbar, M., Alieyan, K., Alzubaidi, M.: Internet of Things (IoT) communication protocols: review. In: 2017 8th International Conference on Information Technology (ICIT), Amman, pp. 685–690 (2017). https://doi.org/10.1109/ICITECH.2017.8079928
2. Zhiliang, W., Yi, Y., Lu, W., Wei, W.: A SOA based IOT communication middleware. In: 2011 International Conference on Mechatronic Science, Electric Engineering and Computer (MEC), Jilin, pp. 2555–2558 (2011). https://doi.org/10.1109/MEC.2011.6026014
3. Tayeb, S., Latifi, S., Kim, Y.: A survey on IoT communication and computation frameworks: an industrial perspective. In: 2017 IEEE 7th Annual Computing and Communication Workshop and Conference (CCWC), Las Vegas, NV, pp. 1–6 (2017). https://doi.org/10.1109/CCWC.2017.7868354
4. Chen, Y., Kunz, T.: Performance evaluation of IoT protocols under a constrained wireless access network. In: 2016 International Conference on Selected Topics in Mobile & Wireless Networking (MoWNeT), Cairo, pp. 1–7 (2016). https://doi.org/10.1109/MoWNet.2016.7496622
5. Falas, Ł., Schauer, P., Świątek, P., Trzaska, R.: Practical implementation of Internet of Things service systems architecture. In: 2017 25th International Conference on Systems Engineering (ICSEng), Las Vegas, NV, pp. 291–298 (2017). https://doi.org/10.1109/ICSEng.2017.57
6. Mun, D., Dinh, M.L., Kwon, Y.: An assessment of Internet of Things protocols for resource-constrained applications. In: 2016 IEEE 40th Annual Computer Software and Applications Conference (COMPSAC), Atlanta, GA, pp. 555–560 (2016). https://doi.org/10.1109/COMPSAC.2016.51
7. Rahman, R.A., Shah, B.: Security analysis of IoT protocols: a focus in CoAP. In: 2016 3rd MEC International Conference on Big Data and Smart City (ICBDSC), Muscat, pp. 1–7 (2016). https://doi.org/10.1109/ICBDSC.2016.7460363
8. Zamfir, S., Balan, T., Iliescu, I., Sandu, F.: A security analysis on standard IoT protocols. In: 2016 International Conference on Applied and Theoretical Electricity (ICATE), Craiova, pp. 1–6 (2016). https://doi.org/10.1109/ICATE.2016.7754665
9. Mulligan, G.: The 6lowpan architecture. In: Proceedings of the 4th Workshop on Embedded Networked Sensors, EmNets 2007, New York, NY, USA, pp. 78–82. ACM (2007)
10. Carrez, F., Bauer, M., Boussard, M., Bui, N., Carrez, F., Jardak, C., Loof, J.D., Magerkurth, C., Meissner, S., Nettstrater, A., Olivereau, A., Thoma, M., Walewski, J.W., Stefa, J., Salinas, A.: Deliverable D1.5 – Final architectural reference model for the IoT v3.0, July 2013
11. Karagiannis, V., Chatzimisios, P., Vazquez-Gallego, F., Alonso-Zarate, J.: A survey on application layer protocols for the internet of things. Trans. IoT Cloud Comput. **3**(1), 11–17 (2015)
12. Chodorek, A., Chodorek, R., Pach, A.R.: Dystrybucja danych w sieci Internet. Wydawnictwa Komunikacji i Łacznosci (2007)
13. Grzech, A.: Sterowanie ruchem w sieciach teleinformatycznych. Oficyna Wydawnicza Politechniki Wrocławskiej (2002)
14. Papir, Z.: Ruch telekomunikacyjny i przeciazenia sieci pakietowych. Wydawnictwa Komunikacji i Łacznosci (2001)
15. Gozdecki, J., Jajszczyk, A., Stankiewicz, R.: Quality of service terminology in ipnetworks. IEEE Commun. Mag. **41**, 153–159 (2003)

A Smart Environment Monitoring Application for Mobile Internet of Things

Samia Tasnim[1(✉)], Ahmon Ferguson[1], Brooke Gordon[1], Chad Gordon[1], Kishwar Ahmed[2], and Idongesit Mkpong-Ruffin[1]

[1] Florida A&M University, Tallahassee, FL, USA
{samia.tasnim,ahmon1.ferguson,brooke2.gordon,chad1.gordon, idongesit.ruffin}@famu.edu

[2] University of South Carolina Beaufort, Bluffton, SC, USA
ahmedk@uscb.edu

Abstract. With the advent of better wireless technology and an increase in smartphone usage, a new mode of data collection (a.k.a. mobile IoT) has emerged. Smartphones are equipped with different types of sensors, which aid in the collection of heterogeneous real-time mobile IoT data. In this paper, we develop a smart mobile application to assist in overall air quality monitoring, and consequently to improve public health. We design a distributed environment and air quality monitoring application leveraging mobile IoT devices. The proposed mobile application allows users to submit and view real-time environment (precipitation) and air quality (particle matter and ozone) information in their vicinity. Furthermore, the application provides users opportunities to view historical reports of any location of interest. We perform experiments using our developed application while considering various real-life scenarios.

Keywords: Internet of Things (IoT) · Mobile crowdsensing · Environment monitoring · Air quality monitoring · Public health · Smart cities

1 Introduction

Internet of things (IoT) is a system that connects people and devices (e.g., sensors, smartphones, wearables) anytime and anywhere. With the advent of better wireless technology and an increase in smartphone usage, a new mode of data collection (a.k.a. mobile IoT) has emerged [3]. Nowadays, smartphones are equipped with different types of sensors (e.g., temperature sensors, humidity sensors, accelerometers, gyroscope). Therefore, these smartphones can be used to collect various real-time mobile IoT data. Such measurement data help us on various aspects, such as, improving the transportation systems, monitoring the air quality and the consequently reducing spread of diseases, responding to natural disasters, among others.

Recently, air pollution has become a global major public health crisis [1,14]. According to a recent report [13] published by World Health Organization

H. Selvaraj et al. (Eds.): ICSEng 2020, LNNS 182, pp. 223–233, 2021.
https://doi.org/10.1007/978-3-030-65796-3_21

(WHO), 4.2 million premature death occurs annually due to air pollution. High concentrations of particulate matter with a diameter less than 2.5 μm (PM2.5) in the pollutant air causes various serious illness, such as cardiovascular or respiratory diseases, cancers, among others. Unfortunately, 91% of the global population inhabit areas where the air pollution levels exceed WHO defined limits [12].

In this paper, we develop a smart mobile application (i.e., app) to assist in overall air quality monitoring, and consequently to improve public health. We present an intelligent and economical system that enables the collection of heterogeneous environmental and air pollution real-time data. As such, people will be able to monitor the environment and take appropriate measures, consequently being able to prevent serious diseases caused by air pollution. The application we develop enables participants to input data and create reports based on weather conditions and/or air quality.

Our contributions in the paper can be summarized as follows. We design a distributed environment and air quality monitoring application leveraging mobile IoT devices. The novelty of our application is that it has both the provision of factual data submission of user's locality, as well as alternative option of viewing the real-time and historical data reports submitted by other users. Moreover, we display different environmental and air quality features (e.g., ozone, particle matter, and precipitation) on a single application. In state-of-the-art mobile applications, only one type of environmental parameters (e.g., precipitation or particle) is being displayed [2,11]. In general, these applications show data from fixed weather stations, hence, the data is sparse. In contrast, our method allows participants to submit real-world data using the application interface. Consequently, we can cover the environmental and air quality data of a fine-grained bigger region.

The rest of this paper is organized as follows. In Sect. 2, we present work in related areas. In Sect. 3, we present the system architecture, and describe the framework of our methodology. In Sect. 4, we outline different components of our designed application, as well as present the experimental results. Finally, we conclude the paper in Sect. 5.

2 Related Work

In this section, we describe the research works that are most pertinent to our work.

Researchers have been developing mobile applications that can be used to monitor air pollution. However, most of these applications allow users to view data only. None of the apps have been designed for user participation in the real-time data collection. Plume labs [11] has created an app "Air Quality by Plume Labs" to monitor the air quality. The way this data is collected is by stations set up in areas to observe and study the air's quality. Even though the reports are accurate, the data is quite sparse. There is no provision for the users to contribute in the air quality data. Because of the expensive land-based measurements with satellite imagery, the map could show data of only a few cities around the globe.

mPING is an application that allows user to submit precipitation data based on user's region [2]. Specialists use the data to confirm that their systems are working properly. Moreover, researchers developed an app that can be used to detect the variation in temperature. It requires to be in physical contact with the object being sensed and use conduction to monitor changes in temperature.

Various interesting applications have been developed for traffic monitoring (e.g., [8,17,21]). Wazirali et al. designed an image sharing app embedding messages into the image [22]. Another application of IoT is aiding the food supply chain with the use of smart sensing [15]. It monitors the freshness of the food while transporting from one location to another.

Researchers have also been working on developing algorithms for environmental parameter estimations for mobile crowdsensing data stream [4,9,16,18–20]. Furthermore, Chen et al. [5] designed an air pollution prediction method that is guided by the physics of the spread of gas. In contrast, Genc et al. [6] developed an air pollution forecasting model based on the data of Ankara, Turkey using air pollution index. Kumar et al. developed an ARIMA-based data prediction mechanism considering temporal relationships [10]. Moving forward, Zhang et al. [23] and Gill et al. [7] developed data cleaning methods for environmental data streams. However, these methods are not much scalable due to the requirements on fixed sensor set-up.

3 System Architecture

In this section, we present the high-level system architecture and discuss details of different system modules. Figure 1 shows the overall architecture of our developed system. The system contains the modules for: (1) mobile IoT-based data collection, (2) data processing, (3) real-time report view, and (4) historical report view. We describe each of the modules and associated tasks next.

3.1 Mobile IoT-Based Data Collection

The mobile IoT platform is the module responsible for crowd-based data collection. Due to wide-spread ubiquity of smartphones, any participant can submit the precipitation and air quality information of his/her surroundings through our application. First, the smartphone's GPS sensor detects the location of the participant who might be interested in data submission. This ensures that no participant can intentionally provide incorrect location information. The location data (that is represented by latitude and longitude) provided by the GPS sensor is incorporated with the participant-reported data, such as, precipitation, ozone, and particle information. Since the application is used in a smartphone, the participant can be stationary or moving, while contributing to the data collection procedure.

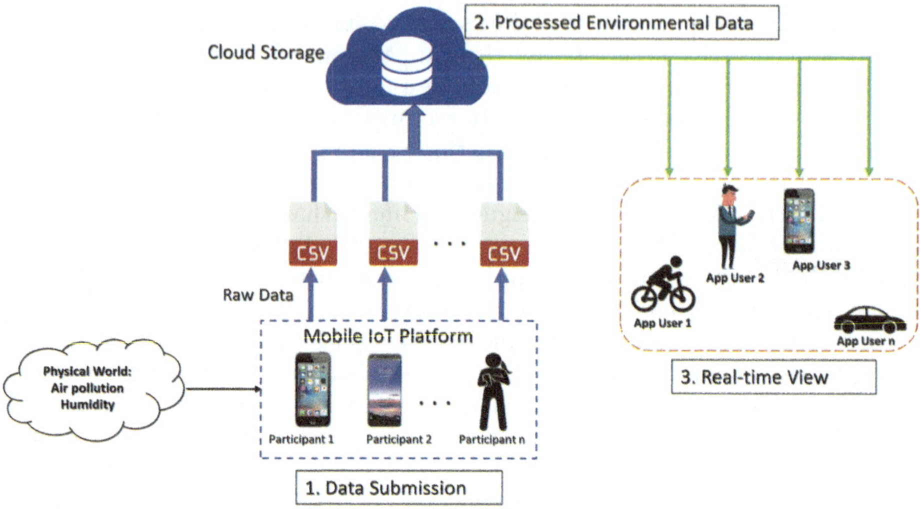

Fig. 1. Overall system architecture.

3.2 Data Processing

Our system includes a secure cloud database for storing and processing the data. Whenever a participant submits environmental data using wireless communication (such as WiFi or LTE), the data along with time and location information is stored in a CSV file. Data reports from multiple participants are sent to the cloud database where data cleaning (e.g., duplicate elimination, formatting) is performed. The processed data is finally stored in the cloud storage module.

3.3 Real-Time Report View

Using our system, users can observe the environment and the air quality conditions in real time. Whenever an application user selects the environment properties of their interest, the pertinent data is fetched from the cloud storage and displayed on the user's smartphone. In step 3 of Fig. 1, a user can be on motion (e.g., riding bicycle, driving car, walking), while viewing the environment condition of their surrounding in real-time. The GPS sensor detects the location of the smartphone user and provides the appropriate report of user location.

3.4 Historical Report View

Our system allows a user to view historical environmental and air pollution data of a chosen location. Figure 2 depicts the details of the historical report view flow. First, an application user selects the desired date and time on the calendar. Next, s/he needs to provide the zip code of the location of their interest. We have two different types of reports: one type of report for precipitation data,

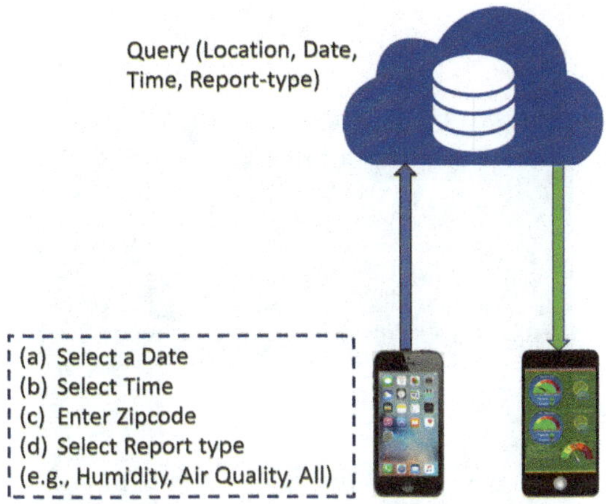

Fig. 2. Historical report view

and another type of report that displays the air pollutant (e.g., ozone, particle) measurements. The choice of report type is the input that the user provides. After choice submission, a query including the user-provided choices is transferred to the cloud storage from the smartphone. We assume that the cloud storage is secure and has abundant storage and processing capabilities. Finally, the queried information is fetched and transferred to the requested smartphone user. The whole process is reasonably fast. The users can see the detailed report on their application screen.

4 Experiment Results

In this section, we present experiment study to show usage of our developed mobile application.

4.1 Home Screen

Our ENVAir mobile application runs in the user smartphone. It is developed for the android operating system. The homepage of our application is presented in Fig. 3. The home screen features the two buttons, as a link for users to go directly to the next screen to collect information or to observe. A user can select the option of either air quality or precipitation monitoring.

Fig. 3. Home screen of the ENVAir App

4.2 Precipitation

Figure 4(a) presents the precipitation search screen. After the user selects the precipitation option, s/he is redirected to this page. A brief description of the application is shown at the beginning, so that users unfamiliar with the application will have the necessary background information. Next, the user can search for any location by zip code and can receive a report of that location. The date and time picker functionalities allow the user to choose the desired date and time. Finally, when the user clicks the "See results" button, a query is sent to the cloud database to fetch the requested report. As shown in Fig. 5(b), the precipitation report shows rain in Tallahassee, FL. Various legends along with their representations are shown in Fig. 5(a). It helps participants to understand and submit precipitation reports accordingly. The "Go Back" Button allows one to return to the search reports page.

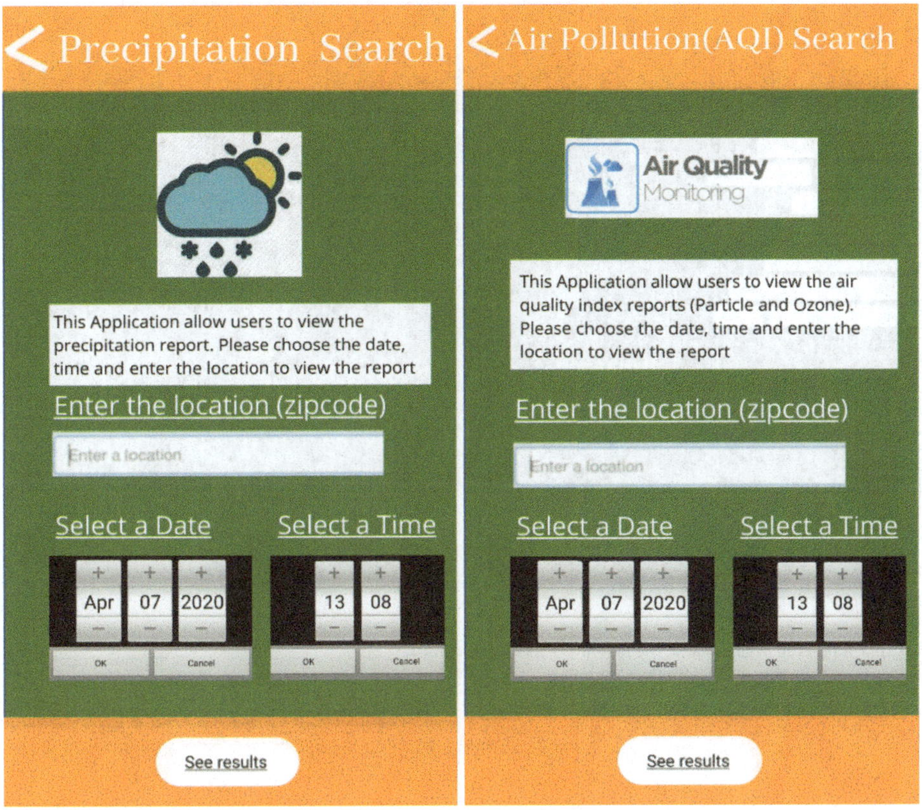

(a) (b)

Fig. 4. (a) Precipitation report search, and (b) Air quality report search

4.3 Air Quality

Figure 4(b) presents the air pollution search screen. A brief description of the application is shown at the beginning, so that users unfamiliar with the application will have the necessary background information. Similar to the precipitation search, a user can enter the location information, as well as can choose a date and time to view the air quality index (AQI) reports. Figure 6 depicts the air quality index for a searched area. Furthermore, the key description shows the AQI classification of healthy, moderate, unhealthy and hazardous air based on the pollutant range. This data allow the user to make logical observations (e.g., select a less polluted area for a walk) among the environment.

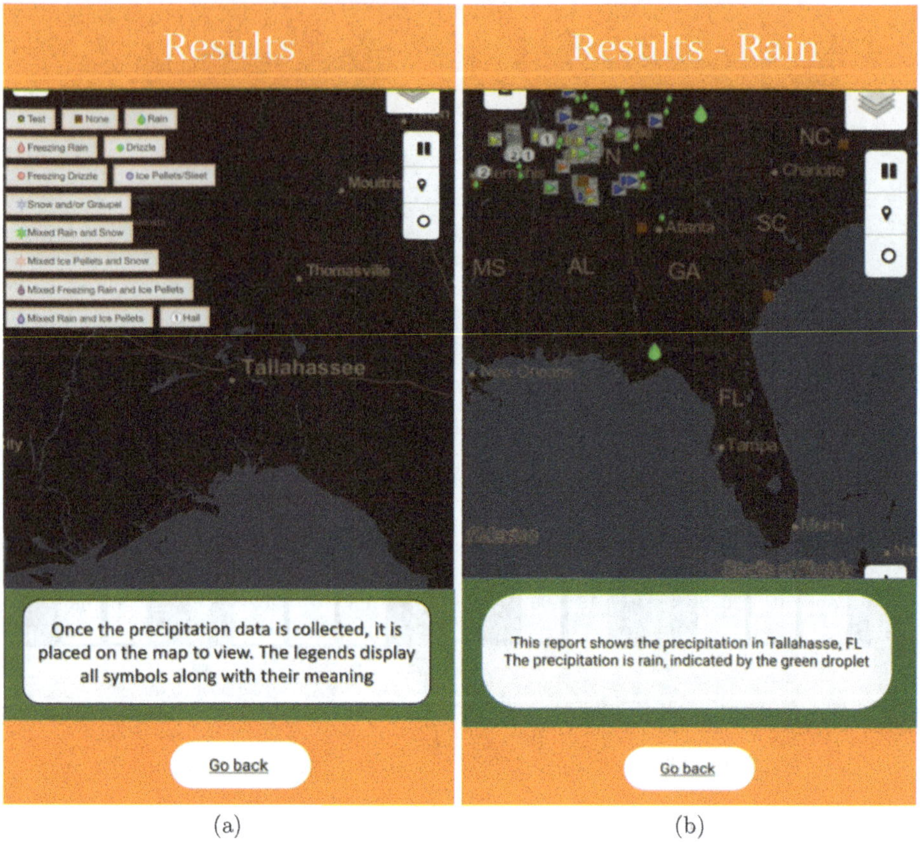

Fig. 5. (a) Key description of precipitation classification, and (b) A rain report example.

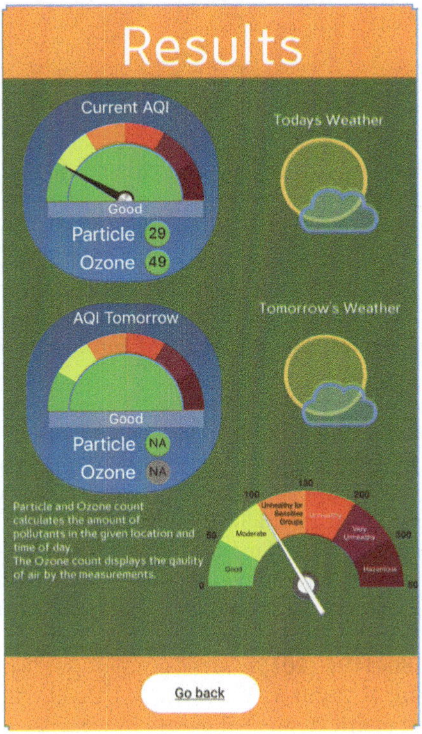

Fig. 6. Key description of air quality AQI (Particle & Ozone)

5 Conclusions

The fast-growing IoT system has influenced and materialized many industries, such as smart cities, smart grid, smart healthcare, and so on. The ubiquitous smartphone has significantly increased the usage of mobile IoT data. In this paper, we present a smart environment monitoring application for IoT system. The success of our application lies in providing the users the opportunity of submitting as well as viewing multiple environment and air quality properties in real time in a cost-efficient manner. We performed various real-life experiment studies using our proposed system. In this work, we considered only honest participation from the users. In our future investigation, we will consider malicious participation scenarios from users.

Acknowledgment. The authors would like to thank Marquis A. Bryant and Marleak J. Barriner for their contribution. This work was supported in part by NSF grant OIA-1655740 and an ASPIRE grant from the Office of the Vice President for Research at the University of South Carolina.

References

1. Air pollution may have killed 30,000 people in a single year, study says, July 2019. https://www.cnn.com/2019/07/23/health/air-pollution-us-deaths-study/index.html
2. mping crowdsourcing weather reports (2020). https://mping.nssl.noaa.gov/. Accessed 14 July 2020
3. Alnahdi, A., Liu, S.-H.: Mobile Internet of Things (MIOT) and its applications for smart environments: a positional overview. In: 2017 IEEE International Congress on Internet of Things (ICIOT), pp. 151–154. IEEE (2017)
4. Chen, L., Ding, Y., Lyu, D., Liu, X., Long, H.: Deep multi-task learning based urban air quality index modelling. Proc. ACM Interact. Mob. Wear. Ubiquit. Technol. **3**(1), 2 (2019)
5. Chen, X., Xu, X., Liu, X., Pan, S., He, J., Young Noh, H., Zhang, L., Zhang, P.: PGA: physics guided and adaptive approach for mobile fine-grained air pollution estimation. In: Proceedings of the 2018 ACM International Joint Conference and 2018 International Symposium on Pervasive and Ubiquitous Computing and Wearable Computers, pp. 1321–1330 (2018)
6. Deniz Genc, D., Yesilyurt, C., Tuncel, G.: Air pollution forecasting in ankara, turkey using air pollution index and its relation to assimilative capacity of the atmosphere. Environ. Monit. Assess. **166**(1–4), 11–27 (2010)
7. Gill, S., Lee, B.: A framework for distributed cleaning of data streams. Procedia Comput. Sci. **52**, 1186–1191 (2015)
8. Kishino, Y., Takeuchi, K., Shirai, Y., Naya, F., Ueda, N.: Datafying city: detecting and accumulating spatio-temporal events by vehicle-mounted sensors. In: 2017 IEEE International Conference on Big Data (Big Data), pp. 4098–4104. IEEE (2017)
9. Koukoutsidis, I.: Estimating spatial averages of environmental parameters based on mobile crowdsensing. ACM Trans. Sensor Netw. (TOSN) **14**(1), 2 (2018)
10. Kumar, U., Jain, V.K.: Arima forecasting of ambient air pollutants (o 3, no, no 2 and co). Stoch. Env. Res. Risk Assess. **24**(5), 751–760 (2010)
11. Plume Labs: Lets do something about air pollution, May 2020. https://plumelabs.com/en/
12. BBC News: Air pollution: what are the effects on humans?, April 2019. https://www.bbc.com/news/av/science-environment-47831610/air-pollution-what-are-the-effects-on-humans
13. World Health Organization: 9 out of 10 people worldwide breathe polluted air, May 2018. https://www.who.int/news-room/air-pollution
14. World Health Organization: Ambient air pollution - a major threat to health and climate, May 2018. http://www.who.int/airpollution/ambient/en/
15. Pal, A., Kant, K.: Smart sensing communication and control in perishable food supply chain. In: 2019, submitted to ACM TOSN (2019)
16. Restuccia, F., Ghosh, N., Bhattacharjee, S., Das, S.K., Melodia, T.: Quality of information in mobile crowdsensing: survey and research challenges. ACM Trans. Sensor Netw. (TOSN) **13**(4), 34 (2017)
17. Tasnim, S., Caldas, J., Pissinou, N., Iyengar, S.S., Ding, Z.: Semantic-aware clustering-based approach of trajectory data stream mining. In: 2018 International Conference on Computing, Networking and Communications (ICNC), pp. 88–92. IEEE (2018)

18. Tasnim, S., Ataur Rahman Chowdhury, M., Ahmed, K., Pissinou, N., Sitharama Iyengar, S.: Location aware code offloading on mobile cloud with qos constraint. In: 2014 IEEE 11th Consumer Communications and Networking Conference (CCNC), pp. 74–79. IEEE (2014)
19. Tasnim, S., Pissinou, N., Iyengar, S.S.: A novel cleaning approach of environmental sensing data streams. In: 2017 14th IEEE Annual Consumer Communications & Networking Conference (CCNC), pp. 632–633. IEEE (2017)
20. Tasnim, S., Pissinou,N., Iyengar, S.S., Shahid, A., et al.: Reputation-aware data fusion and malicious participant detection in mobile crowdsensing. In: 2018 IEEE International Conference on Big Data (Big Data), pp. 4820–4828. IEEE (2018)
21. Tayeb, S., Pirouz, M., Latifi, S.: A raspberry-Pi prototype of smart transportation. In: 2017 25th International Conference on Systems Engineering (ICSEng), pp. 176–182. IEEE (2017)
22. Wazirali, R., Chaczko, Z., Gibbon, J.: Steganographic image sharing app. In: 2017 25th International Conference on Systems Engineering (ICSEng), pp. 494–499. IEEE (2017)
23. Zhang, Y., Szabo, C., Sheng, Q.Z.: Cleaning environmental sensing data streams based on individual sensor reliability. In: International Conference on Web Information Systems Engineering, pp. 405–414. Springer (2014)

Special Session: Applied Media Informatics and Technology

Interaction Design for Email Client: User Movements Reflected in Gesture-based Operation

Ziran Fan and Takayuki Fujimoto(✉)

Graduate School of Information Sciences and Arts, Toyo University, Tokyo, Japan
swterc@gmail.com, fujimoto@toyo.jp

Abstract. This study looks into users' gesture-based operations to be performed on smartphone interface, and focuses on the factor of realistic experience each operation produces. Then we devise a new interface design concept by incorporating the element of reality in the design theory of Natural User Interface (NUI). The element of reality constitutes the user experience of digital content, and NUI forms the sensory reality of operating the digital content. Proposed here is an interaction system in which the realistic experience of digital content use is emphasized using the two characteristics of gesture operations: "expressiveness" and "problem-solving ability based on spatial thinking". This study takes as an example the emailing tasks which anyone today has had experience with, and attempts to solve the everyday issues of incorrect email transmission, based on the design concepts the authors propose. Focusing on the action of sending emails, the study is aimed at raising users' awareness of emailing practice by creating a realistic experience of "sending something to a recipient", using three-dimensional gesture operations on the app's interface. On top of this, a practical email transmission/interaction system will be built by integrating the incorrect transmission prevention mechanism into the realistic experience of "sending things", which was represented by gesture operations.

Keywords: Interface design · Interaction design · NUI · Gesture · Information media

1 Introduction

Gestures, along with gaze and facial expressions, are important components of non-verbal communication. Gestures play an important role in daily communication. Today, gestures are considered crucial not just in interpersonal communication but also in the "communication with machines", or human-machine interaction [1]. As digital technologies advance, particularly the field of user interface design is evolving rapidly to facilitate communication between human and computer.

Computer interface started with Character User Interface (CUI) and Command Line Interface (CLI), which enabled users to interact with computer by issuing commands in text form. It then developed into Graphical User Interface (GUI) in 1990s, which is characterized by the use of mouse and icons [2]. In recent years, an interface that is

H. Selvaraj et al. (Eds.): ICSEng 2020, LNNS 182, pp. 237–246, 2021.
https://doi.org/10.1007/978-3-030-65796-3_22

closer to actual human than GUI is being developed. For example, Natural User Interface (NUI) is drawing public attention as the next- generation interface design which attempts to realize more human-like communication by using gestures, voice commands and touchscreen [3]. Different systems have been proposed using gesture-based interfaces to meet diverse user practices and environments [4].

This study focuses on the element of realistic experience to propose and design an interface with which users can control digital devices with human-like movements, using gestures. In this paper, a prototype design of interaction system is proposed as an email client for daily emailing tasks.

2 Background

According to the "2018 White Paper on Information and Communications in Japan" published by Ministry of Internal Affairs and Communications of Japan, email usage accounts for the largest part of all the functions and uses of the internet [14]. As the report published by ITmedia Inc. in 2014 shows, there are three challenges in using e-mail: (1) Increase in unsolicited emails including spoofing and spams; (2) Information leakage due to incorrect emails transmission; and (3) Increase in operating cost [15].

For increases in junk emails and operating cost, the system environment by service providers is responsible whereas the problems of misaddressed emails and incorrect attachments/e-mail content are caused by users' carelessness. Problems related to users' carelessness are difficult to solve compared to those taken care of by the email service providers through system improvement or cost reduction because the system or environment is not the only factor causing problems.

At the moment, solutions to the problems caused by users' carelessness are limited. These include double-checking the content before sending the emails, typing all the recipients' addresses in the BCC field when sending a group email and encrypting attachments. However, all these well-practiced solutions just "encourage users to be more careful", and would not help them overcome their carelessness. Therefore, it is difficult to completely prevent them from making careless mistakes. In this study, we propose an interface design that helps users become aware of their mistakes in the emails by enhancing the sensory reality of mail transmission, using gesture operations.

3 Purpose

3.1 Related Studies

Various interface designs have been proposed so far to allow users to interact with computers using gestures [5]. There are a large number of such designs, including those that did not turn out to be a success. Operating on smartphone touchscreen is a typical example. Also, gesture-based computer operations are not always performed by actually touching the device.

For example, some recent studies used an illuminance sensor and a distance sensor - not smartphone screen - to measure the gestures detected near the smartphone so that the measurements will be reflected in smartphone operations [6, 7]. These interfaces deliver a more realistic operational experience by solving the problem of a small device screen, which hindered users from grabbing or throwing real objects. As mentioned above, these recent interface designs use gestures as a motif, and are based on the concept of Natural User Interface (NUI). NUI emphasizes the "intuition" element, which certainly enhanced the usability of devices.

However, all of the afore-mentioned examples are lacking in "reality" of using things. For example, when smartphone measures gestures with both luminance sensor and distance sensor, and has the measurements reflected in the device control, a divergence may occur between actual gestures and the movements displayed on the screen because users do not actually touch the object. For this reason, operations that require precise movements can hardly be performed. This is the biggest challenge with NUI [8, 9]. After taking all these into account, we devise in this study a NUI-based interface design method, which incorporates the element of "realistic experience" in the intuitive operation.

3.2 Examination of Reality Element

"Reality" is an essential factor to nurture attachment to a tool or its condition and to enhance its convenience. As Fujimoto (2018) mentions, people are fascinated by things that they can actually touch and feel, and getting physical sensations while using them not only improves users' motivation but also leads to greater work efficiency [10].

Many analog contrivances and machines have been digitized in the past few years with the advanced digital technologies. For use on smartphones, especially, many analog tools have been digitized, which is making people's lives dramatically convenient. However, digital devices such as smartphones provide virtual experience to users, and it is clear that they lack in sensory experience which their real-life, analog counterparts provide.

This phenomenon is seen not only in the digital content like smartphone applications but also in the interface of any digital devices or human-machine interaction. For example, advanced technologies are applied to methods such as voice control, eye-gaze device operation and mechanical control with brain waves, making it easier for people with disabilities to use computers. However, not all these interfaces come with highly realistic experience.

For example, "sound" is one instance that shows the importance of realistic experience in interface. Like the use of smartphone, when users type characters on the keyboard or scroll the picker to choose a date, or when an application notification is displayed, a corresponding sound is played for each particular operation. These sounds allow users to become conscious of the operation. In case no sound is heard in response to user actions, it might be considered as an indication of a broken device.

In addition to sounds, iPhone has a vibration feature, which stimulates users' sense of touch. Unlike auditory sensation, the feature gives sensory reality of operation when users hold the device. This tactile experience emphasizes the characteristics of the features according to the intensity of the vibration. These interface designs that deliver realistic experience in the interaction improve smartphone user experience. This helps users perform their actions more consciously. However, the element of reality often tends to be treated as a by-product of interface design. At present, it is merely one of the convenient features to enhance interface usability.

4 Proposal of the Design Method

4.1 Design Concept

Gesture operations excel in the intuitive feel compared to the interfaces such as button and pointer [11]. As the method of nonverbal communication, sometimes they also function as the representation of what certain terms mean. The reason why people find operations with physical sensations excellent in term of intuitive feel is because people tend to prefer the interface design, to which they can apply the feeling of physical actions as same as their "routine daily practices" or "accustomed manner". Relevantly, " →" might be recognized as "proceed" and " ←" might be perceived as "return." The reverse case would be also possible. Likewise, documents can be written either vertically or horizontally, and a book can be either right-opening or left-opening. In short, they are all customary ways and are "somehow" common sense.

The first concept of the proposed system is to grant the clear meaning to the interface operations, which are designed to be interpreted with feeling: "somehow." It goes without saying that not all of the people would not have the impression of "proceeding" to " →." Therefore, this study attempts to design the operations with physical sensation, which could be intuitively recognized by a majority, who have the general understanding, and then incorporates the operation with physical sensation that represents the activity of sending a mail. People have used mail client systems on smartphones or computers "somehow" by clicking, tapping and sliding on screens. However, this study considers the interface that requires the users' gesture operations, which enable users to have sensory experience of email transmission. The proposed interface is different from the conventional interfaces.

The second concept refers to the strength required for the gesture operation, which leads to the problem resolution. Body-motion technologies including gestures accelerate human spatial thinking and play the role to activate design thinking [12]. Specifically, the multidirectional examination of the relationship among the objects, which are placed in the 3-dimensional space, can lead to creativity for their unexpected usages or problem-solutions. In their production process, people operate, construct, connect and pile the objects with body motions.

The production process of body-motion technologies is not just suitable for the cases where people actually touch or operate physical objects, but also for functions on the virtual interfaces of digital devices. For example, the previous research studied the computer graphics representation of the two-string problem, which is a typical problem resolution model conceived by Maier (1931). The result shows that the gesture operations can simulate thinking activity and support the problem resolution in more creative way than the interface operations by the mouse [13].

Gestures are basically the operations performed in the 3-dimensional space so that they excel in providing the sensation of using objects with body motions. For that, users can better perceive the characteristics regarding what they are using and understand how to use it. It is the remarkable feature of spatial thinking, and the second point of the proposed concept: the advantage of gesture operations for interface.

This study focuses on the activity of "email transmission." "Email transmission," means, "sending a letter," if it is applied to the actual activity in our daily lives. People put letters in a mailbox when they send them. In short, supposing the receiver forward, we perform the specific action: "stretch out a letter" or "exert a letter." However, emails can be sent just by clicking the "send" button as for the conventional email client interfaces. The physical actions of "pass a letter forward to a receiver" or "put a letter in a mailbox," the activities we actually perform are not represented. The sensory reality of mail transmission is lacked because the gesture of "sending a mail" is not reflected to the interface design.

This lack of the realistic experience of sending mail may cause the problems like incorrect transmission. Moreover, sometimes the "convenient function "upon which people can send emails by one-click with the send button, would induce them to make mistakes regarding the checking tasks. It is often the case that people accidentally send emails by mistouch on the button, and also they send emails thoughtlessly even before they complete checking mail content.

The first concept aims to make users perceive the reality of sending the emails.

The second concept allows users to perceive the problem of incorrect email transmission by adopting the characteristic of spatial thinking by body motions and it aims to improve the users' consciousness toward the checking tasks to prevent email incorrect transmission.

The following is the image of the proposed email client interaction: Write a mail → "send" (first concept: improved interface operability with sensory reality) → spatial thinking by the gesture operations with physical sensation regarding "sending" (second concept: users can recognize if they have mistakes) → write (modify) a mail.

This study proposes a new email client interaction system based on those two design concepts: ① representing "sending an email" with physical sensation and ② to prevent (resolve) the mistakes (problems) by improving the realistic experience of email transmission based on the idea of spatial thinking regarding operations with physical sensations.

4.2 Design Method

This study targets iPhones provided by Apple Inc. And proposes an email application equipped with an interaction system.

Fig.1 shows the mechanism of application. The application interface is conscious of the original user movements for emailing and emphasizes the sensory reality of sending an email. It automatically checks the mistakes of email content and reflects the results to the strength required for the operation of email transmission. Thereby this application aims to construct the system to prevent the incorrect email transmission.

The development environment adopts the Xcode provided by Apple Inc. And programming language is Swift. "Mailer" function by MessageUI framework is used for the application. In regards of mailer function, the application does not construct an original email client system, instead it runs the mailer function, which is initially

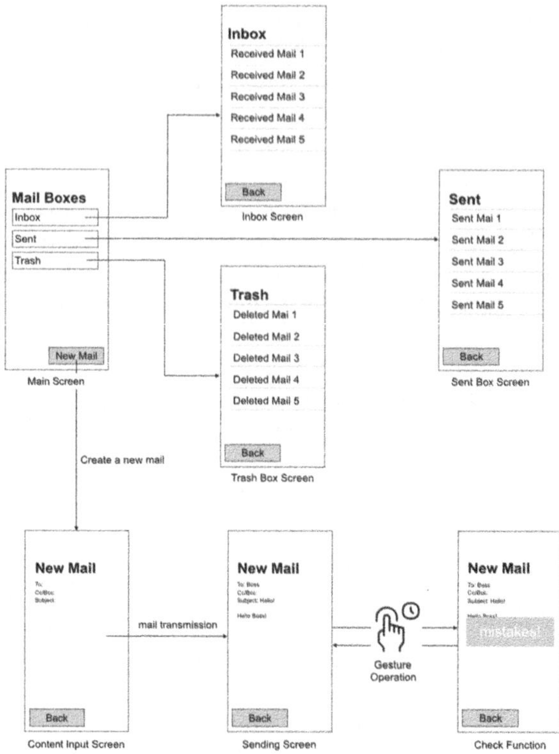

Fig. 1. Mechanism of application.

equipped in iPhone by default to perform the functions of email transmission and reception. For that, the specification of basic emailing functions such as transmission and reception, content input, draft, deleting(trash) is as same as the pre-installed application, "Mail" on an iPhone. We proposed and designed the new interface operations and email client interaction system.

The functions of the application are mainly classified into 2 parts: the interface operations of email transmission and the interaction system to prevent incorrect email transmission. The interface operations for smartphones are certainly unified regardless of, the variety of terminal models, usage environments, used applications (Fig. 2). One exception is 3D Touch: pressing operation. At present, only Apple's iPhone models, 6s and later, and Android models, OS10 and later support this operation. However, 3D Touch is the function, which has not been popularized yet. This study specifically designs the interface regarding mail transmission utilizing the function of this pressure-sensitive operation. As the word "pressing-sensitive" shows, this function detects the amount of the pressure with which users touch the smartphone screens and performs the particular functions depending on the touch strength. For example, regarding the iPhone, the pressing function (3D Touch) enables users to display the common actions of an application on the home screen, and to check the received message on an SNS application quickly, and to move the cursor freely for character entry.

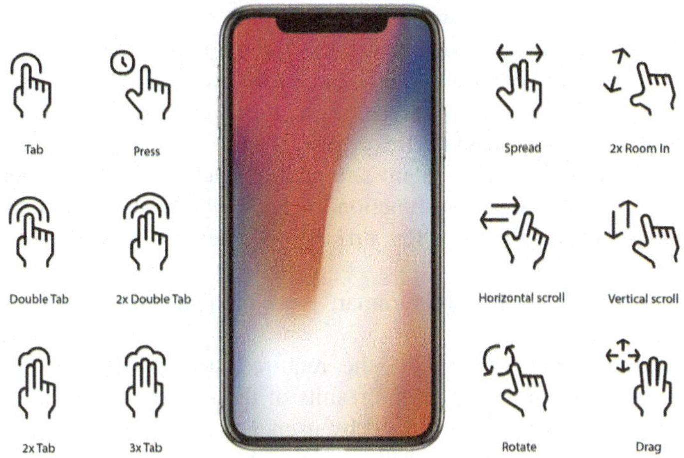

Fig. 2. List of smartphone interface operations.

The characteristics of 3D Touch are not only the measurement of pressing strength, rather 3-dimensional operational feel produced by the "push deeper" gesture. The interface operations performed on smartphone screens are almost planar movements and they cannot provide sensory reality of 3-dimensional operation because they are all completed in thin surfaces. The interaction proposed in this study aims to represent the movement of sending a mail, and the image of "stretching out a letter to the receiver supposing the user forward" is important. The "push deeper" operation of 3D Touch conjure the image of the movement, "stretch out a mail forward", and the sensory reality of sending email would be emphasized by pressing the screen hard (Fig. 3).

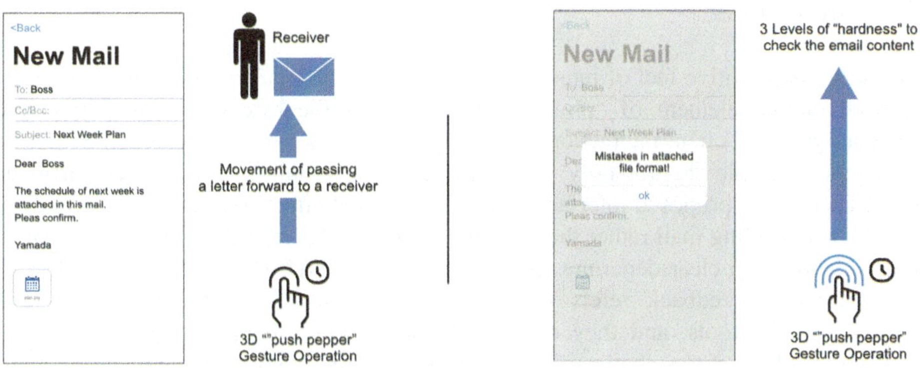

Fig. 3. Mail client interaction system based on 3D touch gesture operation.

Moreover, the application set the function to prevent incorrect email transmission related to the gesture operation of "pushing deeper". The problems of incorrect email transmission are mostly caused by unconsciousness of the mistakes and even though they notice the mistakes after the transmission and users can no longer cancel the messages from being sent. In short, to prevent the incorrect email transmission, users need to check the email content correctly before the transmission.

Thereby, regarding the application design, the following functions would work automatically before users' sending the emails.

① Check function for mail content and address including CC recipient.

② Collation with email history function to check whether users replied to the latest email for long email threads like "Re: title A" → "Re: Re: title A" → "Re: Re: Re: title A."

③ Proofreading function for grammar, typographical errors, and attached file's name.

Moreover, the application adjusts the required strength for the "push deeper" operation to send an email based on the results of these checking functions. There are 3 levels designed for hardness perceived by users. It changes as the following scales: "critical mistakes" (regarding mail address and attached file), "subtle mistakes" (regarding grammar and typographical errors), and "no mistakes." When you "press deeper" on the screen, the "hardness" (the tactile feedback to 3D Touch) produce sensory reality in different degrees. This design attempts to enable users to perceive the realistic experience of having mistakes in email by the difference of operational feel by based on the idea of spatial thinking with gestures.

If "critical irregularities" are detected users cannot complete "send" operation despite of their "push deeper" gesture. Emails can be sent if 1.5 you push deeper on the screen 1.5 times as strong as normal in the case of "subtle mistakes". If no mistakes are detected, the system reacts with the strength of "medium", which is originally set for 3D Touch by default. For any kind of mistakes, the application would display the notifications to request you to review the email by yourself.

5 Conclusions

Recently, the intuitive feel of interface design is attracting public attention. This study focused on the element of "physical sensation" as the reaction to this trend and designed the mail client interaction based on the gesture operations to provide realistic experience. People do not only prefer using tools in intuitive ways. The realistic experience of the proposed interface is the element that affects the whole experience of writing and sending mail rather than just the operational feedback in the use of content (mail). It makes a clear departure from the concept of NUI (Natural User Interface).

Operational feedback refers to the human impression to the operability and usability of the tools, and they consist of, the elements regarding the surface and appearance of interfaces; "voice input is easy to use" or "the function indication on this button is difficult to understand." On the other hand, the interaction with sensory reality proposed in this study refers to the "whole experience" produced by the physical action of using a tool. Prospective comments can be described as "it was fun to use it," "I am fascinated to this" or "I got motivated."

Thereupon, we incorporated this element of reality into sending the mails and proposed the interaction system to provide the realistic experience and intuitive operational feel. This attempt is the fusion of the advantages of the precedent design methods and the indication of the new possibility for content design. The primary future work is to evaluate and verify the effect of this new design method. There may be a claim that the deeper-push operation may not be necessary because the simple mail automatic check function is enough".

From the viewpoint of the result, the outputs with or without deeper-push may be the same, However whether users need to the send button or not, in the last step of the emailing procedure, can completely change the meaning of the experience of emailing. People tend to feel a sense of attachment for physical objects that can be touched. They think that the experiences provided by those objects deserve money and effort. For example, if digital content is presented together with real books or vinyl records, many people will choose the latter even though they are a little more expensive. In short, the enjoyment and pleasure are condensed in the physical actions, operations and sensations and they are the fundamental importance and value regarding content of the physical products.

This study proposes this old/new element of the sensory experience as the design concept not only for the interface but also for the whole digital content in this digital age and we would continue to consider the possibility to expand of this design concept.

Acknowledgement. This work was supported by JSPS KAKENHI Grant Number 17K00730.

References

1. Horioka, H., Okamoto, Y.: A study of gesture recognition using smartphone sensors. In: The 1st IEEE Global Conference on Consumer Eletronics (2012)
2. Liu, W.: Natural user interface - next mainstream product user interface. In: IEEE 11th International Conference on Computer-Aided Industrial Design & Conceptual Design (2010)
3. Zao, J.K., Lin, C.-T., Ko, L.-W., She, H.-C., Dung, L.-R., Chen, B.-Y.: Natural user interfaces: cyber-physical challenges and pervasive applications. In: IEEE International Conference on Internet of Things, and IEEE Green Computing and Communications and IEEE Cyber, Physical and Social Computing (2014)
4. Furui, N., Sakamoto, D., Sugiura, Y.: Gesture design and implementation for smart-home control by cushion interface. J. Inf. Process. Soc. **60**(10), 1859–1868 (2019)
5. Hatori, A., Kobayashi, H.: A preliminary study of IoT-device control using gestures recognition. In: 56th Annual Conference of the Society of Instrument and Control Engineers of Japan (2017)
6. Kato, K., Matsumura, K., Sugiura, Y.: Inputting method applying method performed outside smartphone. In: Proceedings of Entertainment Computing Symposium, pp. 87–92 (2019)
7. Yamashita, D., Aida, H., Yamamoto, T., Miki, M.: Gesture recognition using built-in illuminance sensor for one-hand operation of smartphone. Harris Sci. Rev. Doshisha University **58**(4), 196–201 (2018)
8. Mankar, S.M., Chhabria, S.A.: Review on hand gesture based mobile control application. In: International Conference on Pervasive Computing (2015)
9. Dawes, J., Brian, M., Bialek, H., Johnston, M.L.: Wireless smartphone control using electromyography and automated gesture recognition. In: 40th Annual International Conference of the IEEE Engineering in Medicine and Biology Science (2018)

10. Fujimoto, T.: Ideology of AoD: analog on digital -operating digitized objects and experiences with analog-like approach. In: Proceedings of the 7th International Congress on Advanced Applied Informatics (2018)
11. Dong, H., Danesh, A., Figueroa, N., El Saddik, A.: An elicitation study on gesture preferences and memorability toward a practical hand-gesture vocabulary for smart televisions. IEEE Access **3**, 543–555 (2015)
12. Thomas, L.E., Lleras, A.: Swinging into thought: directed movement guides insight in problem solving. Psychon. Bull. Rev. **16**(4), 719–723 (2009)
13. Fang, W., Lin, Y., Sheu, F., Chen, N.: Exploring problem solving performance through natural user interfaces. In: IEEE 13th International Conference on Advanced learning Technologies (2013)
14. Ministry of Internal Affairs and Communications of Japan: 2018 White Paper on Information and Communications in Japan (2018). https://www.soumu.go.jp/johotsusintokei/whitepaper/ja/h30/html/nd252120.html
15. ITmedia Inc. White Paper of the Solutions of the 3 subjects in the use of email (2014). https://www.itmedia.co.jp/enterprise/articles/1407/17/news003.html

A Support System for Habitual Journaling

Chie Shiraishi(✉) and Takayuki Fujimoto

Graduate School of Information Sciences and Arts,
Toyo University, Tokyo, Japan
Chie.shiraishi0123@gmail.com, fujimoto@toyo.jp

Abstract. What is 'Diary'? In Kojien, which is the most famous Japanese dictionary, says "A diary is a record of daily events, and report of an owner's feelings and thoughts."

A diary has existed from long time ago and there are many well-known diaries in history. These were for daily reports and it is clearly that people needed the diary from ancient times. Today, this role has not been lost and people, you can get a diary everywhere. In recent years, people use not only paper diaries but also diary applications or blogs for recording their daily events. There also have been appearing new types of diary formats such as Twitter, Facebook, Instagram, and etc. Diary forms are diversified now, and the presence of diary is still regarded as important. The following are three main merits of writing a diary. First, with diary, people can record what happened in daily lives. Second, keeping a diary brings their mental health to the owner. And finally, people can improve on their creativity by writing a diary. While some start writing diaries for evaluating these merits, a considerable number of people feel difficult to form a habit of keeping a diary every day for various reasons, even though they know these merits. There would be various reasons; "because It's bothersome." or "because I often forget to write an entry of the day. Therefore, we devised a support system for enabling people to build a habit of keeping a diary, aiming to provide merits of writing a diary as it is.

1 Diary Contents

A diary is originally a tool to keep track of an owner's daily life. We organized general diary contents focusing on the diary's original role. (We also understand that diary content cannot be completely summarized to only one diagram, and it is diverse depending on individuals.)

In the diagram below, the contents are divided into two types: external records, and internal records. In this study, we define "external record" as the record of the events that happened without the diary owner's interaction, and we define "internal record" as the record of the events that happened with the owner's will, felling and actions.

Specifically, external records refer to "date", "days of week", "weather", and "current news". In most of the cases, "weather" and "date" of the day are recorded to clearly show when a certain diary entry is written. Regarding "current news", it is helpful to recall the owner's memory of the day. Because the society and trend of the world in those days can be linked to it as triggers.

We subdivided "internal records" into four categories.

H. Selvaraj et al. (Eds.): ICSEng 2020, LNNS 182, pp. 247–257, 2021.
https://doi.org/10.1007/978-3-030-65796-3_23

First, there are the records of the owner's health and mood. Specifically, they are the records whether an owner's physical condition or mood is good or bad. They are records about the body and feelings. A diary owner sometimes writes in the health statement such as weight or blood pressure. By recording these kinds of information, the owner can objectively realize his or her mental and physical state of the body. Therefore, it can be a help to maintain good health. Second, there are the records of daily events. Owners can record of what happened at the day, such as what they did, where they went, what they watched and whom they met. By looking back the day, what made them happy or sad becomes clear. In many cases, most of them are associated with their feeling. It is also possible to objectively understand the causes of good or bad mood. Third, there are the records of what owners want to do. They can record what they want to try, what they want to eat, the books that they want to read, and etc. In short, they are records to remind the owners to make future opportunities for those "want to (do)". It is a quite practical way to remember what they are thinking about at a certain time and what they want to do. In addition, there is the merit of increasing the feasibility of the future projects not only by keeping information in one's head, but also by writing them in a diary. Finally, there are the records of what owners want to memorize. They are the records to memorize the particular words and thoughts that diary owners want to take root in their mind. Specifically, dreams, tomorrow's goals and cheering messages to themselves are recorded for the future. Thus, the general contents of diary consist of the previously mentioned two types and subcategories upon each. Sometimes owners write about health and body conditions, and other times they record the events of the day. Still other times, owners record what they want to memorize. It can be said that the contents of the diaries are diverse.

【Diary Contents】						
<External record>	<Internal record: about health and mood>	<Internal record: daily events (most of cases, they link with feeling)>		<Internal record: what you want to do>		<Internal record: what you want to memorize>
Date	Health	Event	Impressive Scenery	Food you want to eat	What you want to create	Words you want to memorize
Day of the week	Body condition	Food you ate	What you got praised with	Dishes you want to make	What you want to know	Thinking way you want to memorize
Weather	Mood	Something you made	What you got scolded with	Person you want to meet	What you want to research	Dream you dreamed
Current news	Feeling	Something you bought	What you got angry with	Place you want to go	What you want to draw	short-term goal
		Gift you got	What you have done your best	TV program or movie you want to watch		long-term goal
		people you met	What you felt attractive	Book you want to read		Cheering messages for yourself
		Kindness from others	What you challenged	What to express thankfulness for		Your ideal future
		Kindness to others	What you studied	What you want to overcome		Person you want to be like
		What you talked	What you thanked for	What you want to try		Your role models
		Place you have been to	What someone thanked you for	What you want to tell someone		What you want to improve
		TV program and movie you watched	What you did well	What you want to do		
		Books you read	What you did not do well	What you want to buy		

2 Currently Used Diaries

There are four types of diaries. paper, application, SNS and blog. We analyze these features, merits and demerits of each type of diary.

2.1 Paper Diary

The most commonly used diary is a paper-based diary. some paper diaries have a limit regarding the number of lines and what to write in, however most of them are designed for each user's free-style writing basically. The remarkable feature of the paper diaries is "handwriting" and their format that reflects owners' preference without so much restriction. The unique font of handwriting makes one's diary one and only. Writing with a hand brings mental stability to the diary owner and also there is the merit for ease to organize his or her feeling while writing. There is no restriction of character number and writing formats, therefore writing in paper diary can be quite comfortable for the people who like writing. Moreover, for the owners, it would be a good point of paper diaries to be "tangible objects" instead of digital data. There are also some demerits. For example, paper diaries may be inconvenient to carry with because they are tactile objects and they may deteriorate over the years. There are also weak points such as inability to lock with passcode and the limited clue (only dates) to look back on the diary easily.

2.2 Diary Application

Diary application can be used through a smartphone, a computer or a tablet. In the case of paper diaries, it was a demerit for the owners to take out the diaries for writing because they are tangible objects. However, people can use diary applications on their smartphones or computers, and they can make diary entries everywhere. Diary applications are more convenient than paper diaries in term of the users' convenience to make entries at any time and place, for example, even in an occasion such as on train, or waiting for subway. In addition, the users can utilize functions of inserting photos and location information to the diary entries.

The special feature of the diary applications is their use-openness to smartphones, computers and tablets. As long as people have these devices, they can record without a restriction of the location and time. In respect of that, diary applications are effective for time management and timesaving. They can record events before they forget, and they can easily review their diaries. There is also merits unique to diary applications, users can save photos and location information as a digital data. It is also possible to limit who can open the applications. Regarding all of devices, users' personal information is protected by passcode lock and fingerprint recognition. When looking back diary entries, they can easily find what they want by using keyword searches and photos. It is a demerit that the applications cannot produce each user's uniqueness than paper diaries. Although it is convenient to write in the application on which entries are recorded as data, but users cannot show their distinctiveness because of the limited fonts and writing formats.

2.3 Diary Blog

A Diary blog is a diary that publishes content written by individuals for an unspecified number of people. Blogger write about daily events, restaurants, their interests, and etc. Entries are open to the public on the web pages. Comparing with SNS, Blogs are free from the, limit of the number of letters, and bloggers can post longer articles. It is easy to find a certain old article by performing keyword search even on the Internet. The special feature of a blog diary is the posts from an individual user to the unspecified number of people. Whereas paper diaries and diary applications are recorded privately by default, blog posts are made to open to the public. In most cases, blogs have a system that bloggers can get income with advertisement on their posts. Recently, there is the people who write on blogs as a business. Now blogs are not just record of daily lives and people pay attention to their new value as a method to get profit. Regarding the blog as a diary, it is the merit that bloggers can make articles public. Therefore, it is possible to let others know what they have experienced and their thinking as a dispatcher of information. Regarding articles, they are stored in bloggers' account, and there is no problem of deterioration due to long term preservation. Another merit of blogging is that bloggers can get profits if their blogs have a lot of number of accesses. The disadvantage is that they cannot write in a short time. Bloggers need writing ability because articles are open to visitors. To those who originally do not like writing, or, who do not like to do something that takes a long time, writing diary blogs do not fit.

2.4 SNS as a Diary

SNS (Social Networking Services) as a diary is a diary that publishes contents written by individuals to an unspecified number of people like a diary blog. Originally, SNSs were media to connect people, but in these days, they have been also used as diaries. The special feature of the SNS is a diary for the public not private and there is limit the number letter to write. There are also functions for feedbacks from others such as "like". In particular, SNS is a much more conscious of others than blogs, and it can show other people's reactions directly by "like", comments or other functions. Other people's reactions. In terms of that, SNSs do not simply function as diaries: they are more of the posts to hook lots of "like" from others.

The merit of SNS is that users can directly know the feedbacks from others about the content that they posted. If users post photos and sentences about their recent trips, and the posts get lots of "like" from others, they would be happy. There is also a role that relates to users' mental health, and SNSs have been increasing the values as new forms of diaries. In addition, users can check diaries of others by hash tag or keyword searching, and they can develop interests in not only their own diaries but also in others' diaries. The demerit is that there is a possibility of their mental health may be harmed by feedback from others. Regarding the case in which a user's mental health is ruined because of others' evaluations, it cannot be said that the SNS diary takes a meaningful role as a diary. Another demerit is the limited number of words. users may not be able to write all of what they want to due to the limited number of words, and they have to cut certain amount of words.

2.5 Comparison and Analysis

We compared and analyzed the following four diaries; paper diary, diary application, diary blog and SNS as a diary. There are nine items for comparison; the purpose of writing a diary, writing method, font, the character limit, insertion of photo/location information, long-term preservation, the way of searching past content, lock function and others. We clarified the characteristics of these types of diaries thorough comparison and considered the features for the supporting system to enable the users to form a habit to keep a diary.

	[Paper diary]	[Diary application]	[Diary Blog]	[SNS as a diary]
The purpose of writing a diary	· To record daily events · Good mental health by writing · To record not to forget · To look back later	· To record daily events · Good mental health by writing · To record not to forget · To look back later	· To show to an unspecified number of people · To show happened events and your ideas to others · To record daily events · To get income by advertisement	· To show to others (especially friends) · To show happened events and your ideas to others · To get lots of "Like" · To record daily events
Writing method	Handwriting	Typing	Typing	Typing
Font	user's original	provided font	provided font	provided font
Character restriction	no limit basically, but it depends on diary size	no limit basically	no limit basically	limited
insertion of photo/location information	not available in digital data	available	available	available
Long-term preservation	it may deteriorate over time	saved to device	saved to account	saved to account
Past contents search	not searchable with the conditions	Searchable by keyword, photo or location information	Searchable by keyword, date or categories	Searchable by keyword, photo or hashtag
Lock functions	no lock function	locked by passcodelock or fingerprint recognition	ID and password required to log in to your account	ID and password required to log in to your account
others	customizable type and format			

In regard to purpose of writing diary, there are remarkable differences in respect of added values that accompany, with the basic role: recording daily events. As for diary blogs and SNS as a diary, users can get profits; incomes and outside feedbacks. Thus, the purpose of writing diary has been diversified. Regarding writing method and font, only paper diary has a distinctive feature. Handwriting and unique font are impressive to produce "one and only" diary, but regarding diary application, diary blog, and SNS as a diary, all the entries are input in the common fonts as digital data. With regards to the character limit, there are no limits except for SNS as a diary (but there are also some exceptions such as one-line diary and Q&A diary.) There are also differences in the character limits depending on which SNS is used as a diary, the range is from 280 characters on Twitter to 63,206 characters on Facebook. In respect of insertion of photo and location information, application, blog, and SNS have the applied function. Owners can adhere photos to the pages of paper diaries, but they cannot be inserted as digital data. As to the long-term preservation, digital data never deteriorate because it is stored in the cloud or on users' account. Although digital data is suitable for long-term preservation, users have to note that if they forget to save data or lose their accounts, they cannot recover the data. When turning back the past entries of diaries, it is convenient to quickly reach information that they want to know. Digital data enables users to do that utilizing words. They can easily find the necessary information by keyword searching, tag searching and location information searching. Concerning lock functions, it is indispensable function to protect diary owner's privacy. In the case of paper diary, there is no lock function as common system, and it is necessary to keep it in a particular place or lock it by key as object to protect them. As for application, it is possible to have access limit by passcode lock or fingerprint recognition. On the other hand, diary blog and SNS as a diary set access limit by ID and password to log in to user's account. Thus, there are various features regarding diaries, and the merits and demerits are different depending on each. Therefore, to build a habit to keep a diary, it would be a significant step to select and use diary that fits each user.

3 The Reasons Why People Cannot Form the Habit of Keeping a Diary and Remedies

What is the reason for that they cannot develop the habit of keeping a diary? We performed research on why people cannot keep their diary to enable the users to make the most of the proposed application in this study.

3.1 The Reasons Why People Cannot Form the Habit of Keeping a Diary

Possible reasons are as follows:

1. Not fully realizing the effects or benefits of keeping a diary
2. Not knowing what to write
3. Difficulty of putting thoughts into words
4. Getting tired of writing every day
5. Simply forgetting to write

The above five are the common things that stop one from keeping a diary. We will analyze the causes of these problems and discuss solutions in the next part.

3.2 Problems and Solutions

The first problem – not fully realizing the effects or benefits of keeping a diary – occurs when the benefits of journaling and developing the habit are not clear to people. In addition, the positive feeling as a benefit of journaling can be hard to get in a short period of time. Gradually, they can get less motivated, which makes it harder for them to continue. A solution to this problem would be to help users understand the benefits of keeping a diary and feel the effect more easily with visual means.

The second and third problems – not knowing what to write and the difficulty of putting thoughts into words can occur when people are not used to keeping diaries, or are not comfortable writing, or have no special events to write about. If they are not used to keeping diaries or are not comfortable writing, it would be harmful for them to choose not to write just for these reasons. They should learn to write better through experience but might get bored before they develop the habit. In order to prevent these from happening, creating some sample entries should help. Keeping a diary in Q&A format would also be helpful. This would also be valid when a diary owner had nothing special to write about for the day. Using these tricks to help writers come up with ideas is one of the ways to form their diary-keeping habit.

The fourth problem – getting tired of writing every day - occurs because they feel they are wasting time in keeping diaries, in addition to the things they feel mentioned in the other problems so far. Therefore, it is necessary to keep diaries that does not take too much of their time.

The fifth problem - simply forgetting to write - happens to people who have not developed the habit of keeping diaries yet. If they forget to write even for one day, they might quit keeping a diary pretty soon. It is important to write as continuously as possible. A visible diary that they can make entries whenever and wherever they wish to should be an effective solution to the problem of forgetting to write.

3.3 New System that We Propose

Here, we discuss the features of the support system that helps users build a habit of keeping a diary, based on what has been addressed in the previous paragraph.

The system we propose has the following five features:

① Diary application gives users more chances of checking the diary.
② Diagrams and charts including that of the mood tracker help users realize the effects and benefits of keeping a diary more easily.
③ Sample entries are provided.
④ Diary formats are provided.
⑤ Multiple choices allow easy and efficient use of the diary.

First, we choose to use a diary application for an environment that works regardless of location, date and time. Being visible and useful regardless of geographical location, the use of the application is meant to help users remember to write. We also use a mood tracker to make it easier for users to realize the benefits and effects of journaling. By visually obtaining information using diagrams and charts, it enables the users to objectively understand their own different mood patterns, and maximize the effect of keeping a diary on their mental health. In addition, it is also helpful that the application provides sample diary entries and formats. Showing examples of effective use of a diary and providing frequently-used phrases in the format make it possible to save time to write. Lastly, bothersome tasks are reduced by multiple-choice diary. The trouble in this context means taking the time to bother over what to write and putting thoughts into words. Multiple-choice diary should solve these problems.

4 Overviews of the Application We Propose

Based on the new system proposed in Sect. 3.3, we propose a new application.

4.1 Features of the Application We Propose

The diary application proposed in this study is different from other similar applications in three aspects.

The first one is that a mood tracker is used in it. The second is that sample entries and formats are provided. The last one is that it is a multiple-choice diary. The mood tracker captures the user's mood from the information obtained through the user's selections of mood and event. Furthermore, graphic representation of the information shows the changes in the mood, contributing to the user's good mental health.

Next, as mentioned earlier, providing sample entries and formats helps solve the problems of not knowing what to write and getting tired of writing every day. To help users continue keeping a diary, formats with simple sentences are needed.

Also, multiple-choice format is one option that enables users to write quickly and efficiently.

With these, we aim to bring users an efficient and practical diary application.

4.2 Transition Diagram Showing the Vision of System Design

In this paragraph, we show the system design for the new diary application in the transition diagram below.

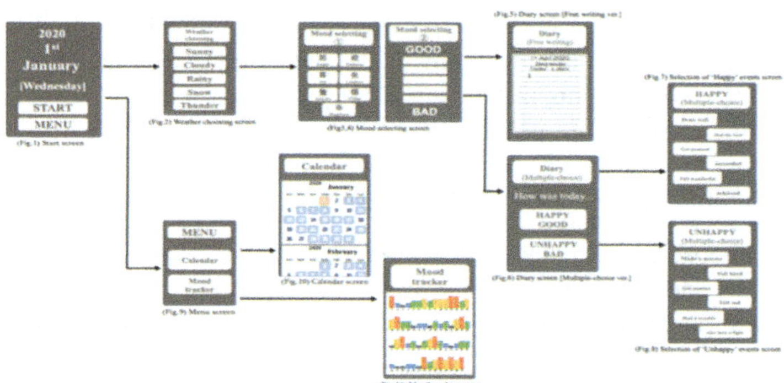

Start screen (Fig. 1). On the first screen, the date is displayed. After confirming that the year, date and day of the week match the diary date, select START or MENU to proceed to the next screen.

Weather selection screen (Fig. 2). After pressing START in the start screen, weather screen is displayed. In this screen, choose the weather each day from the following five options sunny,, cloudy, rainy, snowy, and thunderstorm.

Mood selection screen (Fig. 3, 4). Weather screen is followed by the mood screen. Here, input your mood in two different stages.

First, the entry starts with selection of one Kanji or Chinese character. Seven Kanji characters on the screen represent the following emotions: 怒(anger), 悲(sadness), 喜(joy), 笑, (laughter), 驚(surprise), 穏(calm), and 幸(happiness). Select the Kanji that represents your mood. Secondly, you enter your mood using the 5-point scale. As shown in Fig. 4, there is a scale that ranges from GOOD to BAD. If your choice is closer to GOOD, your mood would be recorded as good. If it is closer to BAD, your mood would be recorded as bad.

We use these two steps because we would like to make sure the record is accurate and specific. It would be difficult to understand exactly how the user is feeling if only one of them is used. For example, if we know the reason why the user is feeling good, it would help us understand the user's mood more accurately. The accurate and specific records help maintain the user's mental health, which is the original role of a diary.

The information from these two steps are shown in the mood tracker screen, and the record for a certain period of time is displayed in a chart.

Diary screen [Free form] (Fig. 5). The screen for mood selection is followed by the diary screen for recording events. There are two recording options: free form and multiple choice. In this screen, you can write freely just like in regular diaries. You can have your dream diary, using your own text, a font of your choice and so forth (Fig. 5).

As the diary application proposed in this study provides sample text and multiple-choice format, the originality of the user's diary may be lost. In order to retain the user's uniqueness, we have adopted free form option with which users could show their originality in their own choice of words or wording.

Diary screen [Multiple-choice ver.] (Figure 6). Unlike the free form diary, the other diary screen is where you can record events by choosing from multiple options. This would save time and improve efficiency. As shown in Fig. 6, in the first screen of the multiple-choice diary, you start by choosing HAPPY/GOOD or UNHAPPY/BAD. This selection process is for reflecting on the events of the day and for recalling each of them objectively and thoroughly. The diary will be created based on these details.

'Happy' events selection screen (Fig. 7). Fig. 7 shows the screen, which is displayed after selecting HAPPY/GOOD. Here, you choose the events that made you happy and record them. The contents are classified into 6 categories, which are done well, did the best, got praised, succeeded, felt wonderful and achieved. We plan to increase selections in the future. The user chooses the applicable content, which replaces the sentence provided in the format.

For example, if you want to write that you did the best at work, you can build the sentence below by selecting HAPPY/GOOD→did the best→work .

"Today's <u>HAPPY</u> event is that I <u>did the best</u> at <u>work</u>"

The content of the underlined part changes as the user makes a choice. The contents of the other parts than the underlined one are prefilled in each format. There are many different formats.

Also, you do not have to choose only one event because additional events can be entered.

'Unhappy' events selection screen (Fig. 8). Fig. 8 shows the screen, which is displayed after selecting UNHAPPY/BAD. Here, you choose the events that made you unhappy. As in the previous screen, the contents are classified into 6 categories: made a mistake, felt hard, got, scolded, felt sad, had a trouble, and got into a fight. We plan to increase selections in the future. The user chooses the applicable content, which replaces in the sentence provided in the format.

For example, if you want to write that you made a mistake at work,

you can build the sentence below by selecting UNHAPPY/BAD→made a mistake→work.

"Today's <u>UNHAPPY</u> event is that I <u>made a mistake</u> at <u>work</u>"

The content of the underlined part changes as the user makes a choice. The contents of the other parts than the underlined one are prefilled in each format. There are many different formats.

The mood tracker reflects the choice between HAPPY and UNHAPPY that the user makes in the multiple-choice diary screen, just as it does on the mood screen.

Menu screen (Fig. 9). Pressing MENU in the start screen leads you to the menu screen. Here, you can choose CALENDAR or MOOD TRACKER.

Calendar screen (Fig. 10). In the calendar screen, you can see the date on which the diary was written. Aiming at a diary that users can easily maintain, we adopt a calendar, which is easy to use and visually appealing to them. As shown in Fig. 10, the numbers in blue squares are the dates the diary was written, and the number in red square is the date you have selected and want to edit. By scrolling up and down, you can move to the previous or the next month.

Mood tracker screen (Fig. 11). In the mood tracker screen, the information on the user's recorded mood can be viewed in a chart. You can grasp the changes in your mood during a certain time period quickly and objectively to maintain your good mental health. There are three pieces of information that are reflected in the mood tracker. The first is the information from Kanji selection (Fig. 3), the second is the one from 5-point scale (Fig. 4), and the last is the one from the diary screen (Fig. 6). All of these results are reflected in the mood tracker.

We take these three pieces of information into account to prevent users from making uncertain choices based on impulsive judgement. In this system, even if you consider your mood to be neutral, the mood tracker may display results differently because of two other criteria. For example, even if you choose neutral in the 5-point scale, your choices of 笑(laughter) in Kanji selection and HAPPY/GOOD in diary screen make the mood tracker show a result that is better than neutral.

We are planning to think of the specific ways to make a comprehensive decision using the information from these three sources. One method we are considering at this point is to digitize each piece of information and calculate the average value of these three.

5 Conclusion

We have been conducting a research on a support system to help people develop a habit of keeping a diary. People have kept a diary to record their daily activities since ancient times. Why do diaries still play the same role today? It is because a diary not just works as a means of journaling, but it also offers other positive effects. Keeping a diary provides people with various effects such as good mental health and development of creativity.

As we researched existing diaries such as paper, diary, application, blog and SNS, we found that each had its own characteristics, merits and demerits, and realized that users need to select one that suits themselves. Also, diaries can have a diverse audience. Depending on whether they are meant for themselves or their friends, or even other groups of people, different styles of writing were used for different objectives. Next, we found the reasons why some people cannot make a habit of writing a diary, and proposed specific solutions. Tracing the causes helped us identify the underlying problems, which led to the solutions to obstacles that hinder habitual journaling. There are three major differences between the application proposed in this study and existing diaries. The first one is that the mood tracker function contributes to the users' mental health. The second one is that they can write efficiently with the help of the sample contents and format provided in the application. The third one is that it offers an efficient multiple-choice diary to save time. All of these are functions to enable users to develop a habit of keeping a diary. We adopted functions that cannot be found in other types of diary in order to make it one that diarists can use comfortably. To get the full benefit from the functions, users need to be informed of their merits and effects. We would also like to show them how to effectively use the functions at the same time. For the application of this study, an after-development user survey needs to be conducted. We are planning to have the target users use the application for a certain period of time and answer some questions. The questions will ask whether the users were able to feel effects or write every day, and what they have found good or bad. We would then like to improve the application in the future based on the feedback received.

Acknowledgement. This work was supported by JSPS KAKENHI Grant Number 17K00730.

References

1. Caldeira, C., Chen, Y., Chan, L., Pham, V., Chen, Y., Zheng, K.: Mobile apps for mood tracking: an analysis of features and user reviews. In: AMIA Annual Symposium Proceedings, April 2018
2. Bush, N.E., Gary Ouellette, B.A., Kinn, J.: Utility of the T2 mood tracker mobile application among army warrior transition Unit Service Members. Military Med. **179**(12), 1453–1457 (2014)
3. Kumiyo, S., Toshihiro, K.: Effect of 'Reflecting Diary' on Mental Health" Psychologist: bulletin of the Graduate School of Professional Clinical Psychology, Kansai University (2018)
4. Katsura, R., Sakamoto, A.: The effect of the amount of web diary writing on psychological health: a panel study examining the mediating effects Jpn. J. Appl. Psychol., 177–185 (2015)
5. Hayashi, K.: The Japanese Journal of home care nursing. Houmonkango to kaigo, vol. 20(7) (2015)
6. Cho, C.-H., Lee, T., Kim, M.-G.: Mood prediction of patients with mood disorders by machine learning using passive digital phenotypes based on the circadian rhythm: prospective observational cohort study. J. Med. Internet Res. (2019)
7. Janssens, K.A.M., Bos, E.H., Rosmalen, J.G.M., Wichers, M.C., Riese, H.: A qualitative approach to guide choices for designing a diary study. BMC Med. Res. Methodol. (2018)
8. Asakura, T., Kakiuchi, S.: A Study on Periodic Patterns of Subjective Fatigue Symptons among Elementery School Children. Bulletin of Tokyo Gakugei University, pp. 145–155 (1994)
9. Justine Dowd, A., Jackson, C., Tang, K.T.Y.: MyHealthy Gut: development of a theory based self-regulatory app to effectively manage celiac disease. Mhealth (2018)
10. Kajiwara, A.: Development and Verification of the Self-help `mental `health Application Intervention for `young Workers. Waseda J. Hum. Sci. **32** (2019)

“Ready-Made Computing“ for Cost-Effective Service Performance

Kazuya Murata(✉) and Takayuki Fujimoto

Graduate School of Information Sciences and Arts, Tokyo, Japan
muratatoyo0702@gmail.com, fujimoto@toyo.jp

Abstract. In recent years, computer-related products have been greatly developed. Today, very high-performance computers, networks and systems continue to appear and are evolving every day. Along with this, it is becoming a social standard that anyone can easily use very high-performance computers or systems. We assume that it is possible to construct devices and systems that have the same level of performance as those of extremely expensive commercial devices and systems by utilizing such high-performance devices and systems. Therefore, we propose the method to construct devices and systems with performance comparable to high-end devices and systems at low cost by utilizing resources produced for mass use. In this research, this method is defined as “Ready-Made Computing”, following the terminology of the field of fashion and art. “Ready-made” in this research does not only mean mass-produced products. By reconfiguring or combining ready-made products and making most of them, it is possible to give ready-made resources new meanings and roles, and this is a concept that can be further evolved. This research aims to construct the system at the same level as existing business services or systems by combining mass-produced “ready made” resources.

Keywords: Ready-Made · Low cost · Remote class · Remote class environment

1 Introduction

In recent years, computer-related products have been greatly developed. Today, very high-performance computers, networks and systems continue to appear and are evolving every day. In other words, it is becoming a social standard that anyone can easily use very high-performance computers or systems. For example, the devices such as the smartphones like iPhone and Google Pixel, are available to anyone who pays for. As for a personal computer, anyone can obtain a computer equipped with a high-performance CPU and OS. Regarding the network, anyone can easily and unconsciously use a network capable of high-speed communication such as optical line or 4G/LTE. In this research, we focus on the social standard upon which such high-performance devices and networks can be easily reached by anyone casually. We assume that it is possible to construct devices and systems that have the same level of performance as those of extremely expensive commercial devices and systems by utilizing such high-performance devices and systems. Because of the use of commonly

H. Selvaraj et al. (Eds.): ICSEng 2020, LNNS 182, pp. 258–266, 2021.
https://doi.org/10.1007/978-3-030-65796-3_24

available equipment, we can build and operate the system at lower costs than commercial equipment. From this, we assumed that the combination use of the existing resources enables us to achieve a system that has the functions comparable to high-performance devices or systems while keeping costs low.

Therefore, we propose the method to construct devices and systems with performance comparable to high-end devices and systems at low cost by utilizing resources produced for mass use. In this research, this method is defined as "Ready-Made Computing", following the terminology of the field of fashion and art.

2 Definition of "Ready-Made Computing"

As the origin of "Ready-Made Computing", we borrow a technical term related to fashion and art. In the field of fashion and art, clothes can be roughly classified into three levels according to the degree of perfection.

First, they can be classified as "Haute Couture" as the highest-grade perfection. "Haute couture" is French. "Haute" means "luxury" and "Couture" means "sewing", the word means "luxury and custom tailoring". For example, it is used for a single piece of clothing worn by the nobles or royal family. Unlike ready-made clothes, it is a one-of-a-kind item that is created upon an order and is made only for the client. For this reason, the degree of perfection is so high as to be incomparable to ordinary clothes, and it is extremely expensive. For example, Marie Antoinette's costumes in the age of the French Revolution were in excess of about 1 million dollars. "Haute couture" now refers to tailored clothing that generally costs more than 100 thousand dollars. The number of clients for "Haute couture" is said to be hundreds worldwide.

Next level is the term, "prêt-à-porter". The term has become widely recognized for its use in the advertisement by a French ready-made clothes manufacturer in 1949. In French, it means "ready to wear." In the past, "prêt-à-porter" was used to refer to general clothes for the mass, but in recent years, high-quality off the rack clothes are called "prêt-à-porter". Specifically, it is a Chanel suit. Chanel suits are high-status ready-made clothes that cost around 5 thousand dollars to 10 thousand dollars. This is a kind of fashion product of "prêt-à-porter."

Finally, there is a term, "ready-made". "Ready-made" is the special term introduced by a French artist, Marcel Duchamp. Originally, "ready-made" is a word that means mass products. For example, Marcel Duchamp's first ready-made work is a "Bicycle Wheel" in Fig. 1 [1]. It is the work based on the fact that in 1913, Marcel Duchamp mounted a bicycle wheel on the seat of a kitchen stool. Since then, Marcel Duchamp has devised several works of art that was created by tweaking to existing objects or defined the existing objects as an opus. In another example, a Japanese, Sen no Rikyu had a similar idea. Sen no Rikyu had the idea of "Mitate", regarding items that were not originally tea tools as tools for brewing tea. The word, "Mitate" means "to see things as something different, not as what it should be." One of Sen no Rikyu's most famous works is "Take Ichijugiri Hanaire (Flower bamboo vase with side opening [2])".

(Figure 2) There is an anecdote that this was not created especially as a flower case, and he used just a simple bamboo tube by calling it as a flower case. This is a perspective that other Japanese artists could not think of, at the time. This is also said to be one kind of "ready-made". "Ready-made" in this study does not only mean mass-produced products. By reconfiguring or combining ready-made products and making most of them, it is possible to give ready-made resources new meanings and roles, and this is a concept that can be further evolved. Figure 3 shows an overview of the three – layer hierarchy.

Fig. 1. "Ready-Made" Work No.1 Marcel Duchamp "Bicycle Wheel" (1913)

Fig. 2. "Ready-Made" Work no.2 Sen no Rikyu "Take Ichijugiri Hanaire: Flower bamboo base with side opening" (1590)

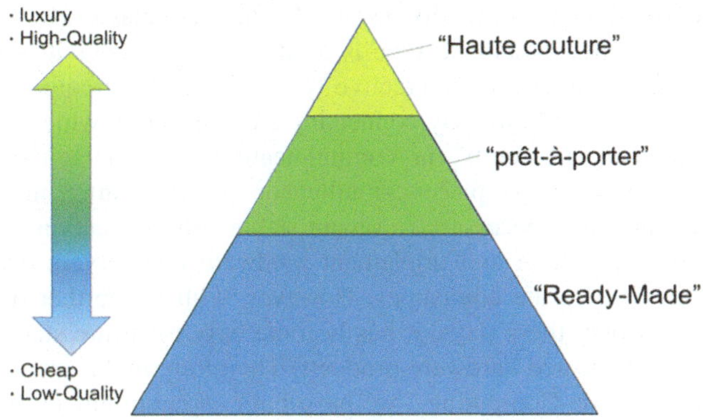

Fig. 3. Three-levels hierarchy of fashion and arts.

As an example of applying these three-layers hierarchy to the computer field, there is a supercomputer. "Haute Couture" corresponds to the supercomputer. Supercomputers are extremely expensive one-of-a-kind computers because of their required performance. For example, there is the Japanese supercomputer "K computer". It can be said that it is one and only computer in the world, namely, an Haute Couture computer. This is because K computer cannot be replaced with another computer. Next, "prêt-à-porter" corresponds to parallel computers. By connecting hundreds or thousands of computers in parallel, it becomes possible to show performance close to that of a supercomputer. The hardware to be paralleled and the software to control it are very special but they are not so expensive as supercomputers. However, it is still expensive from a general point of view and it is a kind of "prêt-à-porter" computer. Finally, "ready-made" corresponds to mass products that are widely distributed and, available to anyone. The computers that we casually use is "ready-made" computers. In other words, "Ready-Made Computing" is the meaning of "Ready-Made Resource Oriented Computing."

Based on the concept, this research aims to construct the system at the same level as existing business services or systems by combining mass-produced "ready made" resources. Therefore, the proposed method in this research is defined as "Ready-Made Computing".

3 Application Fields of "Ready-Made Computing"

In this research, we propose a new method, "Ready-Made Computing". This chapter describes the examples and fields that the "Ready-Made Computing" method is (or can be) applied to.

3.1 Remote Class Environment

First, the most remarkable example of our present research is "Remote Class." At present, remote classes at universities are divided into two types: "simultaneous interactive" and "on-demand." The on-demand type is a method in which there is no restriction on time and place, and the user learns by watching the recorded class videos at home. In this method, there is a high demand from adult students, disabled students and other students, who have difficulty physically attending classes. On the other hand, in most of the case, the on-demand type is used at online universities because of its characteristics. The simultaneous interactive type remote class requires a large classroom unlike the on-demand type. By connecting a classroom in which a class is being conducted to another classroom via communication, the interactive class can be practically facilitated in two places simultaneously. Regarding the simultaneous interactive type, since it is necessary to secure stable communications, in most cases, universities tend to purchase and implement hardware and servers dedicated to the remote class provided by the companies. However, with the further development of computers and networks these days, it has become easy to secure stable communications even without dedicated hardware products. Therefore, in this study, based on the concept of "Ready-Made Computing", we have built an environment that can provide the simultaneous interactive type remote classes with the same level performance as dedicated hardware offers, by combining general mass-produced resources that anyone can reach. Dedicated hardware and servers provided by the companies correspond to the "prêt-à-porter". The remote class environment that we build corresponds to "ready-made" because only mass-use resources are used. In this research, we are studying a remote class environment based on the concept of "Ready-Made Computing." In the field of distance education, various similar studies have been conducted [3–5]. However, there is no research on the distance learning environment based on the concept such as "Ready-Made Computing" proposed in this research, and it has high novelty.

3.2 Pear to Pear File Sharing

"Pear to Pear" is a method by which computers connected to a network directly communicate with each other. Before the development of this method, most of data communication and file sharing were performed by a "Client-Server" method by which communication between computers is always performed via a server. However, pear to pear is now being used in various fields due to the P2P technology development. One example is virtual currency. For virtual currency, P2P networks and blockchain technologies are used for the core technologies. In this method, customers are connected via a P2P network, and virtual currency exchange is shared via a blockchain. Another example is "LINE" application. Regarding LINE, user information is managed by a server. On the other hand, conversations, images, and telephone communications between users are all performed on a P2P network. "Ready-Made Computing" is a technology that enables communications among users' devices only by the network technology without using a server that requires high costs for operation and maintenance.

3.3 Physical Examination by Familiar Equipment

In order to understand our health conditions, we need to have regular medical examinations in healthcare facilities. Through a health check, we can obtain the information about not only our own health condition but also the presence or absence of diseases. On the other hand, in recent years, casual devices that can perform simple medical examinations have appeared. For example, the Apple Watches have the application that allows users to easily monitor their health. At present, it is still in the range of entertainment applications. However, if the technology is further developed in the future, by a system of a small device such as the Apple Watch that anyone can obtain, the health state will be able to be measured without using expensive medical equipment. In addition, it is also expected that doctors will obtain information about the patients' conditions in crisis directly by remotely grasping their health conditions from the information through small devices.

3.4 Video Distribution

Previously, delivering video has been a very expensive method. In the past, the only way to deliver video was "television." Therefore, if it is desired to distribute a video, tens of millions or hundreds of millions of enormous investments are required. Even in recent years, when video distribution via the Internet are rapidly spread, 4 K high-resolution video and switching in real time required extremely expensive equipment. In other words, television broadcasting is "haute couture", and high-spec Internet video distribution is "prêt-à-porter". However, recently, the technology of video distribution via a network is rapidly advancing. By using existing video distribution platforms and combining general shooting equipment products and computers, it is possible to construct a high-resolution and high-performance video distribution system via the Internet, comparable to television broadcasting. Ultimately, anyone with a single smartphone can easily shoot and post videos. We suppose that this is also a kind of "Ready-Made Computing".

3.5 Photography with a Camera

In the field of photography, there is an example to which "Ready-Made Resources Oriented Computing" is applied. That is relevant to the cameras' history in photography. Conventionally, camera equipment for business use was very expensive due to its needed performance, and there were times when huge costs of several million yen were required. However, today, smartphone camera functions have evolved tremendously. Regarding the iPhone 11, which is the "high-end model" of the current iPhone models, adopts a three-lens camera for the first time as a smartphone, and it can perform as good as other cameras such as a wide-angle camera and a telephoto camera. At present, even when taking commercial photographs, it is possible to take photographs at the same level as expensive cameras by the combination use of a smartphone, an off-the-shelf tripod and a stabilizer. This case can be also said as the example of "Ready-Made Resources Oriented Computing."

4 Construction of Remote Class Environment by "Ready-Made Computing"

As described in Sect. 3, we conducted research on the construction of a remote class environment by "Ready-Made Resources Oriented Computing". In this chapter, we briefly describe the remote class environment. In constructing the remote class environment for this research, we refer to similar research and previous research on remote class [6–10].

The remote class environment is practical representation of the concept of "Ready-Made Resources Oriented Computing". Therefore, all the equipment used for the environment construction is mass-produced resource that is available to anyone. Table 1 shows the equipment used for the remote class environment in this study. The combination of these devices has realized the remote class at the same quality level as the remote class by dedicated hardware.

Table 1. Equipment used for the remote class environment

No. ①	MacBook Pro (Main Computer)
No. ②	"Google Hangouts"
No. ③	Network Environemnt in Toyo University
No. ④	VGA or HDMI connector and Cable
No. ⑤	FIFAN USB Condenser Microphone
No. ⑥	3.5 mm Analog Sound Cable

Note that Table 1 lists particular name of equipment. However, this does not mean that the exact same devices are necessary, and any device having the same function can be a substitute. Figure 4 shows a remote class environment for which the devices were combined and actually used.

Fig. 4. Equipment combination for actual remote classes

Upon the built remote class environment, the actual remote classes were held at Toyo University with the equipment combination. As shown in Fig. 4, the remote class environment was constructed in two classrooms. In addition, a computer and a camera for each classroom were set up to show the classroom, and that enabled students to see each other's classroom while the remote classes were conducted. In regard to the practice of the remote class environment, Toyo University's "Sports and Nutrition" class has been held for two years from 2018 to 2020 by connecting two remote campuses. We were able to successfully deliver the classes up to end of the university curriculum without any major problems. As for image and sound qualities, the remote class environment could provide the same level qualities as the remote class environment by dedicated products. Figure 5 shows the remote class of "Sports and Nutrition." This photo shows classrooms where students are using the class support application "Respon" at two campuses at the same time.

Fig. 5. Remote class by "Ready-Made Computing"

5 Conclusion and Future Work

In this study, we have proposed and defined "Ready-Made Resources Oriented Computing" as a method of constructing a system with the same performance as business-use computers and systems by combining mass-produced resources. In this research, as the first step, we constructed a remote class environment utilizing only general mass-produced equipment. Practically, we have succeeded in operating remote classes at Toyo University using the built remote class environment.

As a future work, there is the examination of the applicability to other fields. At present, we are conducting research on a remote class environment using "Ready-Made Computing", but research on applicability to other fields is left as a future task. We believe that "Ready-Made Computing" can be applied to not only to the field of information science but also to, theoretically, every fields. In the future, we plan to explore fields where "Ready-Made Resources Computing" can be applied to and we will build new systems.

References

1. MoMA Learning Team: Marcel Duchamp. Bicycle Wheel, Education at MoMA. https://www.moma.org/learn/moma_learning/marcel-duchamp-bicycle-wheel-new-york-1951-third-version-after-lost-original-of-1913/. 10 December 2019

2. Tokyo National Museum: Take Ichijugiri Hanaire. https://www.tnm.jp/modules/r_collection/index.php?controller=dtl&colid=G4217. 10 December 2019
3. Kudo, M.: Study by distance lectures given using information and communication technology in an art class of junior high school in an isolated Island in Hokkaido. J. Assoc. Art Educ. (39), 113–125 (2018)
4. Ito, M., Ichimura, H., Munkh-Ochir, S.: 3E05 a case study of distance-learning system for physics class between Salesian Polytechnic and Mongol Kosen by Using The NTT Smart Video Conference Service. In: Proceedings of Annual Conference of Japanese Society for Engineering Education 2018(0), pp. 454–455 (2018)
5. Nakano, N.: Experiences of the interactive classes using the free software "web clivker". J. Faculty Staff Dev. High. Educ. (15), 121–126 (2017)
6. Aizawa, S., Kobayashi, Y.: A study of remedial education on the utilization of a remote education system for junior high school refusal students. Stud. Educ. Res. Traning, Center Educ. Res. Traning, Gifu SHotoku Gakuen Univ. **17**, 243–250 (2017)
7. Koide, T., Yamada, H., Kobayashi, S.: Practical approaches to interactive lectures for progress of enginnering design class. Research reports of Tokyo Metropolitan College of Industrial Tevhnology (10), pp. 82–88, March 2016
8. Nakano, K.: An attempt of Interactive Classes using free software "PingPong". Bulletin of the Faculty of Education Ehime University (62), pp. 135–142, October 2015
9. Izumi, F., Okabe, S.: Education environments for interactive distance learning. J. High. Educ. Lifelong Learn. (22), 75–81 (2015)
10. Kamata, M.: Problems and view of e-learning systems to support interactive learning in large classes. J. Chiba Univ. Commerce **52**(1), 87–101 (2014)
11. Murata, K., Fujita, K., Fujimoto, T.: Construction of remote class system that enables interactive communication and comparison existing remote class system. In: The 17th International Conference on E-Learning, e-Business, Enterpritse Info Systems, & e-Government (EEE 2018), Nevada, USA, August 2018
12. Murata, K., Fujimoto, T.: Design of interactive remote class environment by "infrastructure-based low-cost computing. In: The 2nd International Conference on Interaction Design and Digital Creation/Computing (IDDC 2019), 7–12 July 2019, pp. 948–953, Toyama, Japan (2019)
13. Murata, K., Fujimoto, T.: Construction of remote class environment by "low-cost computing" and introductin of class support application. In: The 17th International Conference on Scientific Computing (CSC'19), 29 July–01 August 2019, pp. 80-pp.84, Las Vegas, Nevada, USA

Design of Smartphone Software Keyboard for Sensory Experience

Yui Tanaka(✉) and Takayuki Fujimoto

Graduate School of Information Sciences and Arts, Toyo University, Tokyo, Japan
{s4B101810014, fujimoto}@toyo.jp

Abstract. In recent years, a movement in which various things are becoming mobile have been spreading. Keyboards are one example, not only battery-powered ones, but also foldable ones, and recently there appeared keyboards that can be connected to smartphones. However, it is troublesome to carry around a device additionally even though a keyboard size is minimized. Some keyboard products are projected on a desk with a laser system, but that does not give users the actual feeling of using a device. Therefore, we propose an application to turn a smartphone into a keyboard. Unlike the previous smartphone keyboard applications, it is more conscious of the "sensory experience" and proposes the implementation of physical sensation of touching. Furthermore, by using two smartphones together, it is possible to make them one full keyboard just like an actual keyboard.

Keywords: Human interface · Mobile device · Smartphone · Keyboard

1 Background

1.1 Spread of Mobile Office Environment

In recent years, many cafes, restaurants, and shopping malls have been equipped with wi-fi services, providing an environment that can be immediately connected to the Internet. Communication technology has been improved, and time-lag has become less of a concern. With the emergence of various applications and tools that are useful for sharing information, there are many companies that facilitate remote work and do not have offices by using online chat or communication functions. As work styles diversify, it is now common to see people using PCs and other devices in restaurants or cafes. In particular, among IT companies, Introduction of telecommuting and remote work system has increased productivity in some companies such as Six Apart Co., Ltd. and Recruit Management Solutions Co., Ltd. The mobile office environment is expected to expand in the future.

When working outside of offices, most people use laptops. Their processing is quick, and capacity is large. However, in recent years, the performance of tablet terminals has been remarkably improved so as to be comparable laptops. It does not take long time before the difference between a laptop and a tablet will be almost same. Software that only supported PCs has released application versions one after another.

H. Selvaraj et al. (Eds.): ICSEng 2020, LNNS 182, pp. 267–276, 2021.
https://doi.org/10.1007/978-3-030-65796-3_25

Unlike a PC, users can operate the screen by tapping the screen, which enables intuitive operation and has various advantages. With the use of an external device such as Apple Pencil, a user can do more things like input by handwriting. Using a tablet PC needs less space relatively, it is best suited to build a mobile office environment.

When building a mobile office environment using a tablet device, the first inconvenient matter is the keyboard. Tablet devices usually do not have keyboards, instead they have screen keyboards: keyboards are displayed directly on the screens. Therefore, the usable range of the screens become narrow. If one wants to use the entire screen of the tablet, he or she has to buy a commercial keyboard. However, that is costly, and that is very cumbersome, as he or she has to carry another device. If there is a keyboard that can take up less space and enables work without having to carry new external devices, the demand will be high.

1.2 Disappearing "Realistic Experience"

In recent years, the mobile service movement of various things has been spreading.

Because of that, "real feel" of operating devices has been disappearing. However, from the view that there are many bookstores in the streets and the popularity of paper diaries, it can be seen that the "sensory experience" that people have with their own feelings is important. The idea is already mainstream in the entertainment industry such as game industry, and it is the fact that "sensory experience" is a keyword regarding the interface that gives people fun and familiarity (reference [2, 3]). In addition, the "tactile sensation" accompanying the sensory experience is an important sensation for human beings, and is being proved to be important also in infant education. The purpose of this research is to apply the "sensory experience" including the tactile sensation to the office environment and propose a keyboard application with the "real feel" to make a mobile office environment more convenient.

In this paper, "sensory experience" includes "tactile stimulus" and "consciousness" due to the actual operation by the user.

2 Purpose

The purpose of this research is a proposal of keyboard application with an interface designed concerning "sensory experience." The user can use the keyboard without carrying an external device, and he or she can also use the keyboard quite easily because the system's look and the feeling of use is designed to be like a real keyboard. When inputting texts using a small device such as a smartphone, it is often the case that only the thumb is used. In this application, the interface enables the user to operate the system with both hands. We believe that this point will give the user a realistic experience of "operating the keyboard" and it will lead to easier texts input.

3 Precedent Case

3.1 A Mobile Keyboard in Recent Years

There appeared various keyboards by increasing awareness of the concept, "mobile office." Out of them, at present, there are three portable keyboard types:

(1) Wireless keyboard
(2) Sensor keyboard
(3) Smartphone application

3.2 Wireless Keyboard

Many wireless keyboards are used by connecting to the base unit via Bluetooth communication. In recent years, these support not only tablet PC, but also smartphones, and miniaturization and weight reduction are on progress. iClever Bluetooth keyboard (reference [4]) has a unique shape that can be folded in two vertically. Users do not need worry about possible damages when he or she carries that, because it can be carried as relatively a small object. It is rechargeable, and according to the official website, it can be used for 60 h on a single charge (Fig. 1).

Fig. 1. iClever Bluetooth Keyboard (from the official website)

Fig. 2. Jelly Comb Wireless keyboard (from the official website)

Some keyboards can connect to two devices at the same time. Figure 2 show a wireless keyboard produced by Jelly Comb in America (reference [5]). It can connect to any device with Bluetooth regardless of the installed OS. Two connected devices can be switched with a single button. It successfully covers the weakness of mobile terminals where the screen is quite small and multiple screens cannot be open at a time like a desktop PC.

It is certain that keyboards are reducing their weight and size. Obviously, people have to note that they have to carry extra keyboards in addition to the mobile device. The risk of loss or breakdown outside can be also predicted.

3.3 Sensor Keyboard

Users can input texts using a sensor keyboard projected with a laser. It is also called as virtual keyboard. Their looks are neo-futuristic and smart. Recently, various functions have been added. Figure 3 shows a laser keyboard, "Epic" by the Celluon company. This is an improved version of the previous model "Magic Cube". It is small and convenient to carry with; length 70 mm, width 35 mm, and depth 20 mm. It also has an interface like a trackpad.

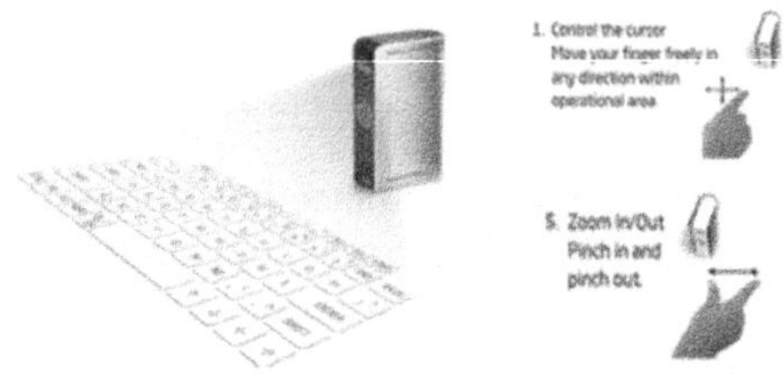

Fig. 3. Celluon Epic (from the official website)

Besides, Serafim's "Keybo" can be switched to a keyboard for each of seven languages, such as Arabic and Chinese, and it can be also changed to the piano's keyboard display. The device itself can be used as a smartphone stand and can be also used as a mobile battery (Fig. 4). These keyboards eventually require the users to carry them as external devices. However, "Selfi Type", by Samsung technology, which was released in January 2020, uses a smartphone sensor to detect users' finger movements. Even though keys are invisible to the user's eyes, characters can be input simply by moving his or her middle finger.

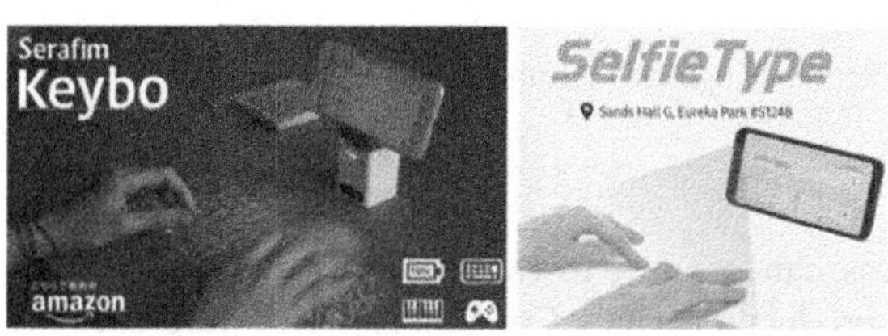

Fig. 4. Serafim "Keybo" (left), Samsung Electronics Co., Ltd. "Selfie Type" (right) (from the official website)

While these keyboards seem smart and convenient at first sight, it cannot be said that they are quite popular. The reasons are that the performance of the sensor is not qualified yet and the user cannot obtain "physical sensation" of operating the tool. Some people claim that keep on touching on an empty flat place will directly cause the impact to people's fingers, and it increases fatigue. When the physical sensation of "pressing" a key that has been familiar with people, is lost, they tend to feel uncomfortable.

3.4 Smartphone Application

Basically, it is provided as an application associated with the remote mouse application. Many of these applications have the concept that a smartphone can be either a mouse or a keyboard. Therefore, the input device serving as the parent device is almost always the desktop PC.

Many mouse applications have the concept that they can also input characters while operating the cursor on a smartphone. Yao Ruan's application, "Remote Mouse" is also available to Windows tablets (but not to iPad or Android tablets). The smartphone can be used not only as a mouse, but also a display of the keyboard and enables user to enter texts. However, it looks is the same as a smartphone's screen keyboard, and users have to tap a particular button to switch to another screen when they want to enter symbols or numbers.

"WiFi Mouse" by Shimeng wang is also designed to be as a mouse and a keyboard to input characters. In addition, it can be displayed like a game controller. Although it is closer to the actual keyboard than RemoteMouse. there is no sound when a key is pressed.

"iWritingPad Keyboard Mouse" provided by 2ConnectMe Limited will become more realistic with additional charges. Although appearance is realistic, sound or vibration at the time of operation is not mounted at all. In addition, it has a specification that the key is too small to tap the wrong place frequently (Figs. 5, 6, 7).

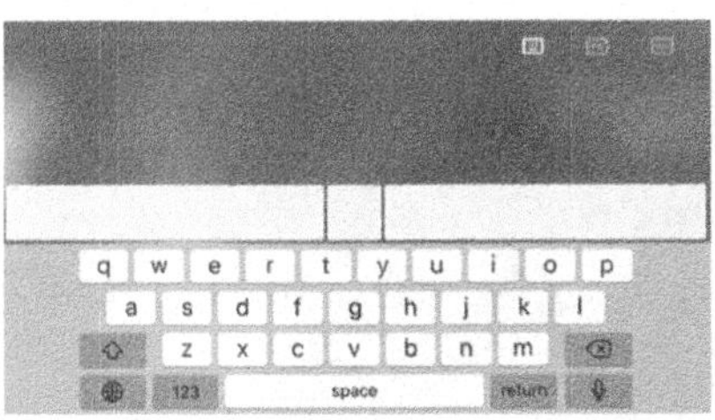

Fig. 5. Remote Mouse Texts input screen

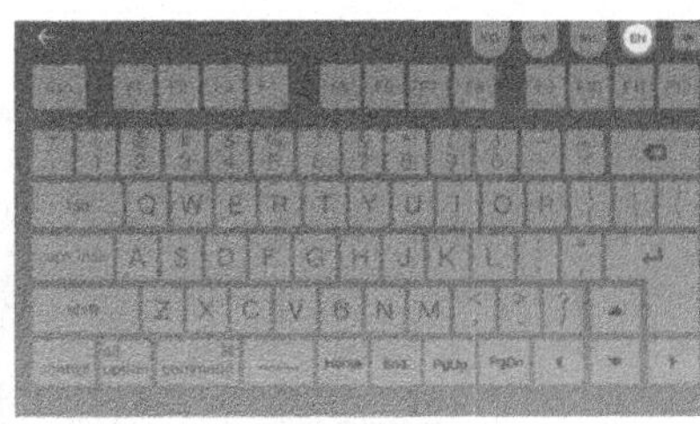

Fig. 6. WiFi Mouse Texts input screen

Fig. 7. iWritingPad Texts input screen

In this way, existing keyboard applications have a "reality in appearance", but lack of operational feedback when the keyboards are pressed. Some applications have sound effects that are similar to that of smartphone screen keyboards, but they are always constant, and the sounds are different from the actual keyboards'. It is not designed for the two-handed typing style in which users can move five fingers for each hand, like a normal keyboard, and it is general for the user to hold the smartphone with both hands and operate the keyboard by an application with his or her thumbs. Under such a condition, user would not get the feeling like using the real keyboard.

3.5 sMouse

We have previously proposed "sMouse" as an interface that is conscious of physical sensation (reference [1]). This is the application that allows users to use their smartphone as a mouse. Other than this application, there are applications that can turn a smartphone to a mouse, but all of them premise on the operation to slide the touch panel with a finger on the smartphone. However, sMouse enables a user to operate the cursor on the connected PC by holding his or her smartphone and moving it like a physical mouse.

Another feature is that there is some kind of feedback to the user, such as a "click" sound and vibration like "brrr" when the user clicks the mouse button. As a result, the user can obtain a "physical sensation" of operating the mouse (Fig. 8).

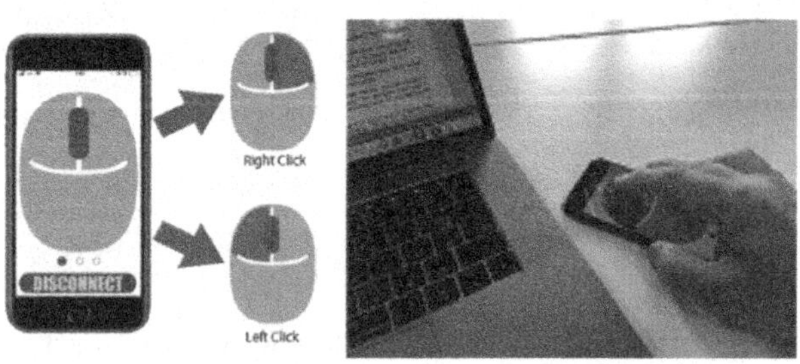

Fig. 8. iWritingPad Texts input screen

In this research, a keyboard application is created with the same concept as sMouse, taking into account the "sensory experience".

The goal is to create a more useful and convenient mobile office environment by making the application close to the physical keyboard interface.

4 Application of Proposal

In this section, the proposed keyboard interface that is conscious of "sensory experience" will be described in detail. Existing wireless keyboards can give a user "tangible feeling," but the user has to carry an external device. There is a great demand for a smartphone to be a keyboard because any businessman would have a smartphone nowadays. Sensor keyboards, which are smaller than wireless keyboards, are more portable, but lack the "user's sensory experience" of pressing keys. If a smartphone substitutes with a keyboard, certainly we can get the feeling that we are operating a physical object.

Existing keyboard applications have the same specifications as screen keyboards displayed on smartphones.

Even if they look like a physical keyboard visually, these are too small to operate. In addition, in most of the cases, they have no reaction when a key is pressed by a user. It can be said that they lack sensory experience.

Therefore, we implement the operational feedback similar to an actual keyboard's to the application proposed in this research.

4.1 Overview

Figure 9 shows the start screen that opens when this application is started. This application premise on the landscape orientation to display keys as large as possible. In this paper, a connected device to the keyboard is called a "Base unit".

Fig. 9. Application startup screen image

By pressing "A", you can set the connection between a smartphone and a tablet. The proposed application premise on, basically two base units' simultaneous connections. The same mechanism as the existing mobile mouse applications' is used.

By connecting to the same wi-fi and matching passwords, you can connect base units to a specified server, and terminal control (keyboard operation) of base units becomes possible.

Upon "B", you can select the type of the keyboard. For the proposed application, we prepared US keyboard, EU keyboard, and JS keyboard used in Japan. You can select your favorite keyboard.

With "C", users can select the use with one smartphone or two smartphones. This function will be described in the latter part.

By tapping D, you can start to use a smartphone as a keyboard with the specified settings on the screen.

4.2 A Screen of Keyboard

This application can be used with either of one or two smartphones. People who have only one smartphone can use this application, but if two smartphones are connected as base units, they can be used as one keyboard. This enables the display of keys larger than existing applications', and it leads to the easier use of the keyboard. In addition, typing is not limited to the operation only by your thumb, and you can operate the keyboard with both hands. As a total experience, typing with the application is closer to the operational method for a physical keyboard.

Regarding the use with one smartphone, they can select whether the full screen keyboard display or common smartphone screen keyboard display. Users can switch it by the upper right button.

When two base units are connected, one keyboard is divided into two: a part operated by the right hand and the other part operated by the left hand following the method of the touch-typing and each will be accordingly displayed on each smartphone. The right or left assignment can be switched with a single tap. Users can type with both hands, just like a normal keyboard. It is also possible to allocate the role of the numeric keypad to one smartphone (Fig. 10).

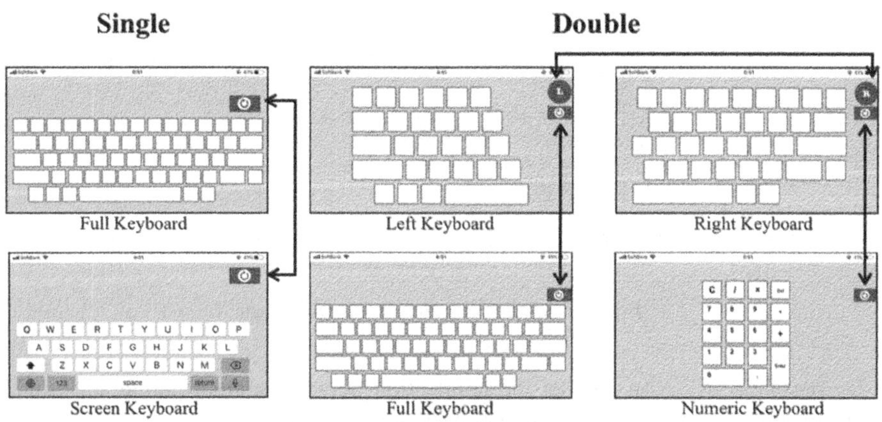

Fig. 10. Application startup screen image

4.3 Implementation of Sensory Experience

To provide the feeling of a real keyboard, "sound" and "vibration" are utilized in addition to "need to use both hands".

Regarding some existing applications, the keys are set to sound when they are tapped, but all the keys sound in the same way constantly just like a smartphone's on-screen keyboard. Users can get some sensory feeling of using a keyboard better than no response, but it is far from physical sensation of the real experience. Speaking of a real keyboard, a sound should change depending on how hard the key is pressed.

Therefore, in this application, the sound changes depending on tapping hardness. We set a small sound for soft tap and a loud sound for hard tap.

Vibration effect is also added. Soft vibration is given when tapping softly, and strong vibration is given when tapping hard. However, since these operations are performed on the smartphone on the desk, too strong vibrations may sound unpleasant. Therefore, adjustment of intensity is possible upon the settings (Fig. 11).

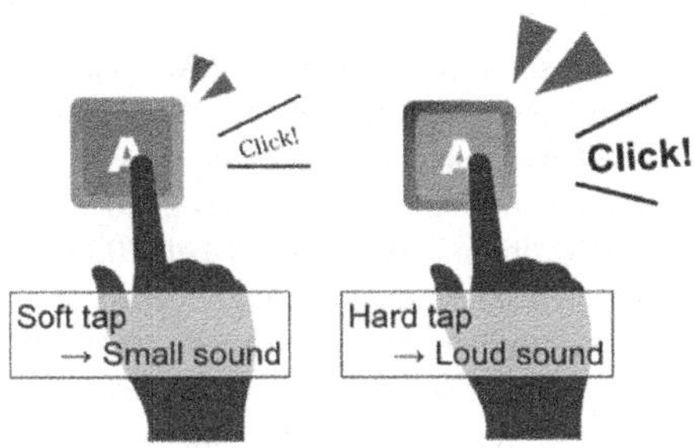

Fig. 11. Application startup screen image

5 Conclusion and Future Works

A In this study, we proposed a keyboard application with the interface that is conscious of "sensory experience." In recent years, various mobile keyboards have appeared from the concept of the mobile office, but it is very troublesome for users because they have to carry an external device in addition to the main device. Sensor keyboards have also appeared, but their performance is still on the way of developing, and they do not provide users with physical sensation of "pressing". The operation of existing keyboard applications is totally different from the real ones'. By using the proposed application, a user can use their smartphone as a keyboard. Furthermore, by connecting two devices, two-handed touch typing is possible, and the operation can be much closer to a real keyboard operational method. By adding "sound" and "vibration" effects, concerning "physical sensation" at the time of user's operation, we could have the application provide the feedback like the real keyboard and realized easier use.

As a future task, it is necessary to develop a prototype and conduct an evaluation experiment. It is required to thoroughly examine which combinations of the tap pressure and sound is comfortable for users.

This application adopts wi-fi as the connection method. and security measures are also left as an issue. Since the concept of mobile office premise on the use wherever users are, there is also a need for a mechanism that can secure safe use under shared wi-fi such as a cafe. In counter to the tendency upon which various office devices become mobile and lose tangibility, we believe that the interface design that allows users to have "sensory experience" of using a tool has certain needs.

In the future, we will continuously devise interfaces that are conscious of "sensory experience" from various points of views.

Acknowledgement. This work was supported by JSPS KAKENHI Grant Number 17K00730.

References

1. Tanaka, Y., Fujimoto, T.: sMouse: a development of the analog operability of computer mouse. In: 7th International Congress of Advanced Applied Informatics, p. 11. IEEE (2018)
2. BANDAI Submarine Rescue. https://bandainamcotec.co.jp/product/007/. Accessed 09 Feb 2020
3. NatsumeAtari GunArena. https://www.gunarena.jp/. Accessed 09 Feb 2020
4. "iClever". https://www.iclever.co.jp/. Accessed 08 Feb 2020
5. "Jelly Comb". https://www.jellycomb.com. Accessed 08 Feb 2020
6. "CELLUON". https://celluon.com/. Accessed 08 Feb 2020
7. "Serafim". http://serafim-tech.com/ja/. Accessed 08 Feb 2020
8. "C-Lab Inside". https://research.samsung.com/culture/ces2020. Accessed 08 Feb 2020
9. "remote Mouse". http://www.remotemouse.net/. Accessed 05 Feb 2020
10. Fujimoto, T.: Therefore the designer is flaming. Chu-okoron-shinsha Inc, 2016. [in Japanese]
11. Fujimoto, T.: The problem of content design in the Internet age: how do theft and plagiarism happen? In: 7th International Conference on Innovative practices in Business, Social Sciences and Humanities Research (2017)
12. Fujimoto, T.: A design that makes information easy to understand: what is information design and what is not it? J. Inf. Sci. Technol. Assoc. **65**(11), 450–456. Information Science and Technology Association, Japan (2015). [in Japanese]
13. Tanaka, J., Katsuma, R.: Efficiency improvement of HIRAGANA typing for a finger mounted device. In: 2019 IPSJ Kansai Chapter Conference. Information Processing Society of Japan (2019). [in Japanese]
14. Ochi, T., Kurahashi, M., Dekihara, H., Imai, M., Kuramoto, A.: A proposal for constructing multi-dimensional measurement environment for input interfaces: how do the university students in the smartphone generation use them. Information Processing Society of Japan (2019)

Problems with the Corporate Information System for Work Efficiency Improvement

Wangjie Xu(✉), Takayuki Fujimoto, and Ziran Fan

Graduate School of Information Sciences and Arts, Toyo University, Tokyo, Japan
Xuwangjie2018@gmail.com, swterc@gmail.com, fujimoto@toyo.jp

Abstract. This paper investigates and clarifies the current situation of work efficiency improvement in business operations. From companies operating in Japan, authors extract cases based on company size. We consider the information systems used in their offices and sort out their strengths and weaknesses. In particular, we focus on the systems used by many companies for business communication. In addition, we also concentrate on basic tasks common to all companies, such as attendance records and approval requesting procedures for documents, and clarify issues and problems related to work efficiency. Based on the results, we analyze the requirements for further work efficiency improvement of the companies.

Keywords: Interface design · Interaction design · Information media · Information communication environment · Work efficiency

1 Purpose

In recent years, various services and systems have been proposed to efficiently maintain and share business information in companies and organizations. For example, many companies have introduced ICT to their business, such as digitized attendance records and in-house communication with dedicated tools. On the other hand, implementing new systems and technologies changes the operational environment. As new systems' functions are improved, many of them require more complex operations and users' IT knowledge so that users have to take great time and effort to get used to the new work environment. Depending on the usage environment, convenience is rather reduced in many cases. In this study, we examine the companies' measures for work efficiency improvement and clarify the current situation. Then, we consider the problems and issues on those countermeasures and analyze the requirements for realizing further work efficiency.

2 Background

Work efficiency on business operations is now a major concern for companies. There are many companies where employees have to spend extra time and effort or have to go through complicated business procedures due to the lack of in-house environment to

H. Selvaraj et al. (Eds.): ICSEng 2020, LNNS 182, pp. 277–286, 2021.
https://doi.org/10.1007/978-3-030-65796-3_26

manage and share information. For example, the document processing to request approval is particularly complicated. In Japan, the requesting procedure for paperwork requires the approval by the heads of several departments. In each department, a document is checked, and after a certain period of processing, the applicant goes to the department to receive the document and this step has to be repeated. In the meantime, all the applicant can do is only waiting for the contact, and he or she often cannot know the processing status of the document and predict the completion time.

Many companies introduce and operate in-house information systems to improve work efficiency, but building a new in-house environment requires enormous funds and advanced technology, which is not easy for many SMEs (Small and medium-sized enterprises). As a result, the operation environment may not be improved, the work efficiency may be reduced, and the company's business performance may become increasingly sluggish, leading to a vicious cycle.

In order to avoid that, two measures have been widespread among SMEs.

One is the use of existing analog tools. For example, there is a method of recording attendance using a paper time card. Paper time card is easier to use than digital attendance systems and it is also easier to introduce it. However, using a paper time card requires inserting the card into a dedicated device every time. If there are more than 20 employees in the company, and all reaches to the office at about the same time period, it takes a long time to record their attendance (with cards), and that leads to inefficiency of the work.

The other is to incorporate commercially available and inexpensive information systems into internal operations. For example, there are cases where the email service, 'Gmail' developed and provided free by google is used as an in-house communication tool, and the use of free communication tool, 'Slack' to share business information related to schedule management, project confirmation and etc. Is done. However, these methods cannot specialize in their own business and cannot construct an optimized business environment. Concurrently, using a plurality of applications and systems together may cause problems such as insufficient information management or complicated information. Despite the introduction, there are many cases where the users leave the service or tool in the end.

In today's business where everything is complicated, pursuing work efficiency is indispensable for companies. However, many companies have been unable to introduce the best information systems due to financial or technical constraints and also the limited effects of existing tools.

3 Investigation of Preceding Cases

3.1 Current Status of Work Efficiency Method in Japanese Companies

Table 1 shows the criteria for determining the company size. According to the Small and Medium Enterprise Agency, the conditions for SME certification vary by industry. First, there are four industry classification conditions: ① wholesale, ② retail, ③ service and manufacturing, ④ construction, transportation, and others. In addition, there are specific conditions such as 'capital or total investment does not exceed X yen' and 'the number of regular workforce does not exceed X', and if an entity falls under either one, it is determined as a SME. In other words, companies that do not meet these conditions are considered to be major companies.

Table 1. Criteria for determining company size (Source: The Small and Medium Enterprise Agency).

Type of Industry	SMEs		Small Business
	The amount of stated capital or total amount of contribution	Number of employees always used	Number of employees always used
①Manufacturing industries, construction industries, transport industries and other industries (excluding ②~④)	300 million yen or less	300 or less	20 or less
②Unloading industry	100 million yen or less	100 or less	5 or less
③Service industry	Fifty million yen or less	100 or less	5 or less
④Retail	Fifty million yen or less	50 or less	5 or less

Table 2 shows the information systems used for business operations in three companies each of which represents a different company size category based on the above criteria.

Table 2. Survey results on corporate information systems and tools.

	large enterprise	Medium enterprise	Small enterprise
Email	Outlook	Gmail	Homepage email
Sugejuru	Outlook	Google	None
Attendance	Record Time	Kintaro	Excel
Internal communication	Skype For business	Email	None
Application submission	Lisa (paper)	paper	paper

The authors interviewed employees of the three companies to discover how information systems and tools are used in each organization.

Nissan Motor Co., Ltd., a large-sized company, has introduced advanced information tools and systems in all areas of business operations such as business communication, attendance management and information sharing.

Sonton Food Industry Co., Ltd., a medium-sized company, normally uses a commercial information system, and more noticeably, employs free services.

The small-sized Meililife Corporation rarely uses any information systems for internal business operations, and tends to rely on analog tools such as time cards.

It was found that, as the company size gets smaller, specialized information systems are less utilized and building of a corporate environment for information sharing becomes less likely.

3.2 Information Tools and Systems.

3.2.1 Outlook

MS Outlook is an email service developed by Microsoft. It is designed for computer operating systems, namely Apple's OS, Microsoft's Windows and Linux. Of them, it is most widely used with Windows (Fig. 1).

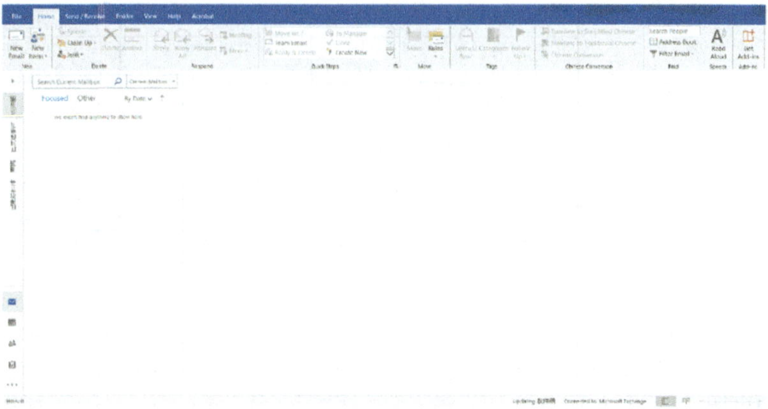

Fig. 1. Outlook user screen

Outlook is cloud-enabled, allowing users to manage multiple devices with a single account and share email data. Its email function is intended for business use, with a simple interface design and easy-to-understand email features.

On the other hand, Outlook supports only one domain name - @outlook, and does not display a company-specific address such as '@nissan'.

For this reason, the recipients cannot identify the sender's organization when they receive an email from a business partner, which may lead to confusion in some cases.

3.2.2 Skype for Business

Skype for Business, provided by Microsoft like Outlook, is a cloud-based computing service, which supports multi-participant video conferencing, real-time messaging and Internet calling. It is part of the services offered under Microsoft Office 365 brand, and is compatible with other services of the same brand (Fig. 2).

For example, the contacts in Outlook are also stored in Microsoft Exchange Server groupware. Thus the data can be easily imported to Skype for Business. You can also share these data with people within the group who use the same service.

The communication function of Skype for Business is SIP-based, and uses a client communication protocol. It also supports TLS and SRTP encryption, and is well-regarded for its excellent information security.

Skype for Business offers not only video calling services but also an environment for users to collaborate by sharing text, pictures, and graphic annotations in real time through whiteboard documents. This has helped the group's participants come up with a new way of using the tool as a 'collaborative team of presentation material creators', in addition to the traditional roles of 'presenters' and 'viewers'. Use cases include online voting and online surveys.

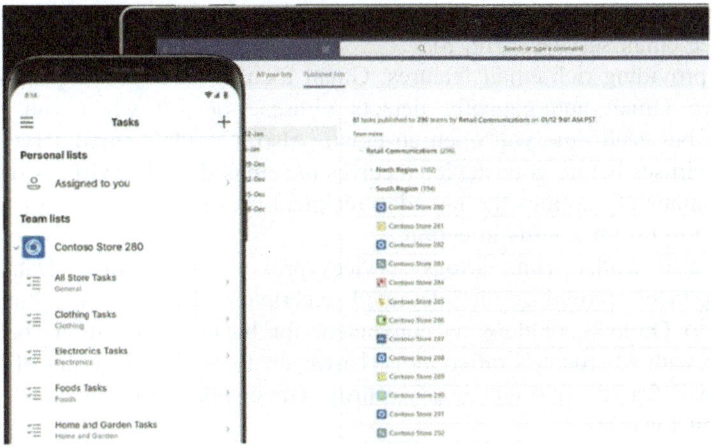

Fig. 2. Skype for Business user screen

Skype for Business has problems with information management. Users typically use the tool along with other services provided by Microsoft. This improves convenience and ease of use of the entire system of services, but also has the drawback that a single service does not complete tasks by itself. For example, Skype for Business specializes in instant communication with multiple users. In other words, completing real-time tasks other than video conferencing needs another service. In addition to basic data such as contacts stored in the cloud, users need to re-integrate other business-related information as well, and the resulting labor and inadequate information management become major issues.

3.2.3 Gmail

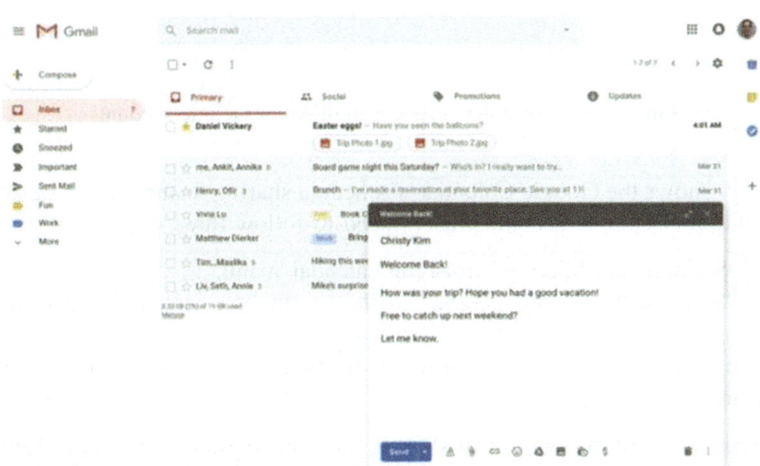

Fig. 3. Gmail screen

Gmail is an email service provided by Google. It boasts the highest usage of all the existing free email services (Fig. 3).

While providing rich email features, Gmail focuses on improving the security of attachments. Gmail automatically detects viruses not just when you receive an attachment but each time you open an email. Also, Google's email server scans the emails for viruses before your device receives the email data. If a virus is detected, the server automatically notifies the intended recipient of the email and stops transferring email data to prevent a virus infection.

Gmail also works with various services provided by Google. Data is shared between services, providing an integrated experience. That is why Gmail is often compared to Outlook. Outlook is convenient for business operations because it is compatible with Microsoft's office tools. However, from the viewpoint of the system such as email features and enhanced security, Gmail offers a better email messaging environment for users.

3.2.4 Google Calendar

Google Calendar's greatest strength is in information sharing. In addition to schedule sharing between devices managed in the same account, the tool also supports information sharing with other Google account users.

Fig. 4. Sharing function between different Google accounts

Figure 4 shows the Google Calendar's schedule sharing feature. To share schedules between different Google accounts, users need to follow three steps.

1. Select "Settings and Sharing" from the calendar menu.
2. Select "Share with specific people" in the destination screen and click on "Add people".
3. Set a level of permission in sharing with another user and send the schedule / calendar.

The four levels of permission in sharing the schedule are as follows (Table 3).

Table 3. Setting a level of permission.

·See only free/busy(hide details): Only "Available" visible
·See all event details: View the details of events (cannot be changed)
·Make change to events: You can see the contents of the schedule and can change the schedule
·Make changes and manage sharing: Can see the contents of the event and can change the event and sharing destination

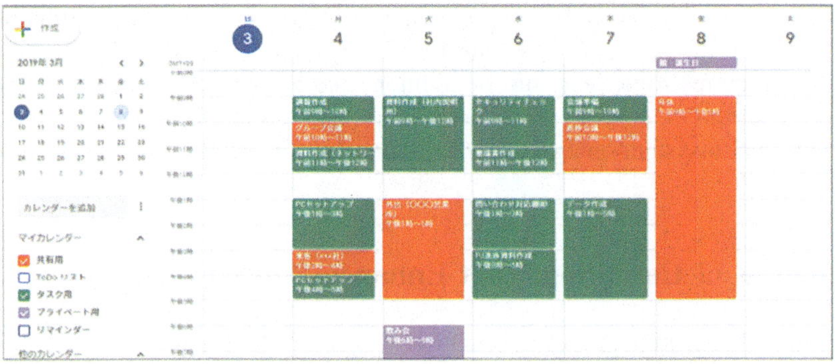

Fig. 5. Schedule management function

Besides sharing schedule, users can not only create new calendars using Google Calendar but also manage multiple calendars in one account. Google Calendar also has 'Tasks' function, which color-codes tasks using the colors specified by the user. You can easily tell which events are for work and which are private. The information added in Tasks also supports the schedule sharing function (Fig. 5).

However, Google Calendar is inconvenient in checking previous tasks because they are automatically deleted from the system once completed and cannot be viewed. Thus, while Google Calendar may be suitable for personal use, it still has a room for improvement in information management for company operations.

3.2.5 Record Time

Record Time is computer software for recording and managing employee attendance. By installing Record Time on the office computer, Record Time automatically begins operation as employees arrive at office and start working at computer each day. Daily work hour recording on Record Time starts when the computer is turned on and ends when it is turned off.

The problem with the software is that it only counts the hours of computer work in the company office. It does not record the hours of work done without using the office computer, e.g. offsite sales negotiations with clients, meetings and work from home. Attendance cannot be recorded accurately with Record Time alone, and employees are required to report their work hours by themselves.

3.2.6 Kintaro

Kintaro is a software program that records employee attendance. The software has not been updated for about 5 years, and its interface design and features have many problems including poor usability, difficulty understanding the features, and the limited number of supported systems. Its use is not widespread as it is used only by a small number of Japanese companies.

3.2.7 Lisa

Lisa is an information system for requesting various company paperwork and certificates. For example, submitting a form through Lisa to request an employment verification letter is much more convenient than the traditional method of submitting a paper form. However, the cost of introducing Lisa is high, and the system is complicated to use, which requires some learning. In addition, due to the factors such as inability to support the latest OS, costs to repair deficiencies and errors, or the impact on business operations due to system outages, very few companies have introduced Lisa system so far.

4 Analysis of Conditions that Enhance Work Efficiency

As shown in the examination of information systems/tools currently used in offices, the environment for sharing information within the company is split into multiple systems regardless of the size of the company, and is hardly an efficient one to support business operations. Also from employees' point of view, using multiple systems in work process leads to excessive work time, which may cause inefficiency.

In addition, many of the information systems currently used in the offices are legacy systems, and neither their features nor interfaces are compatible with the latest technologies. This may result in frequent errors, often hindering business operations. Above all, it fails to satisfy the needs of today's users such as 'multi-language support', 'simple interface' and 'natural operating environment'. In particular, many small- and medium-sized enterprises do not have an internal information system due to lack of funds. In that case, employees have to put much time and effort to do the work that a system would process instantaneously.

For example, in Japan, when making a document processing request, a worker is required to obtain approval from multiple supervisors, starting with his or her section chief, section manager and then department manager. If the request involves different departments, there would be additional work. Moreover, if email is to be used as the sole communication tool within the company, quick responses would become a challenge. With commercial communication apps like LINE, though, information may not be secured.

In order to solve these work efficiency problems, the cost of introducing an information system needs to be reduced in the first place. Furthermore, instead of processing business information with multiple systems and tools, an integrated information system should be used to cover company tasks including information sharing, attendance management and approval requests for documents.

'Desirable' systems should be easy to use and meet users' needs with the latest technologies, which enable the use of multimedia on smartphones, tablets and computers.

5 Future Work

Through this paper, we investigated how companies are looking to improve work efficiency, reviewed the information systems used in organizations to support business operations, and analyzed problems and solutions. Future research will propose and design a new information system that delivers efficiency in business operations based on these results. To this end, our future work will first aim to establish the concept of system design. In particular, a system mechanism is devised based on the theory of information design. We will then conduct on-site company surveys, accurately understand the needs of users, and reflect them in the system design.

References

1. Information Design Associates: Information Design: Designing for easy understanding. Graphic Inc. (2002)
2. Fujimoto, T.: "Design that makes information easy to understand: what is information design?" What is "not what" (Special feature: Design that makes information easy to understand). The Journal of Information Science and Technology Association, pp.450–456 (2015)
3. Fujimoto, T.: "Toward information design 3.0: designing information to" transmit "Building maintenance & management", pp. 41–46 (2013)
4. Sunaga, G., Nagai, Y.: Information design-give shape to information. Inf. Process. **41**(11), 1258–1263 (2000)
5. Ministry of Health, Labor and Welfare: Work Style Reform Support Handbook Ministry of Health, Labor and Welfare SME Agency (2019). (https://www.city.hamada.shimane.jp/www/contents/1534817309311/simple/hanndobukku.pdf). Acquired 25 Aug 2019
6. Takahashi, T.: Examples of ICT utilization in private companies (No.40) Promoting transformation into an integrated infrastructure service company by in-house development using agile methods: Maeda Construction Industry Co., Ltd. Administration & Information Systems, pp. 66–70 (2020)
7. About the definition of SMEs (www.chusho.meti.go.jp/faq/faq/faq01_teigi.htm#q1). Acquired 25 Oct 2019
8. Yamazaki, B.: "Producing results for both companies and individuals! Introduction of" work style reform "and points of practice (13) Let's liven up work style reform (Part 1) Overcoming wall ICT edition "Wage circumstances", pp. 74–77 (2019)
9. Yamada, K.: Environmental environment where employees can continue working and productivity improvement by introducing ICT (Special feature: Creating a welfare workplace in the future with diverse people and working styles). Monthly welfare, pp. 25–29 (2019)

10. Yamazaki, S.: Japan ICT Trend, IT product introduction status/needs of domestic mid-sized companies and small and medium-sized enterprises. InfoCom T & S world trend report: Information magazine of information and communication services in the world/Information and Communications Research Laboratory ICT creation Research Division, pp. 10–13 (2020)
11. Kawamata, K., Moriguchi, T.: Customer retention activities using IoT data:—a case of Ricoh @ Remote—. Mark. J. 125–136 (2020)

Re-experiencing the 'Process of Perception': Consideration of 'Memory' Reproduction with Data Mapping Structure of Brain Cortex

Ning Zhou(✉) and Takayuki Fujimoto

Graduate School of Information Sciences and Arts, Toyo University, Tokyo, Japan
zhouwuqi646@gmail.com, fujimoto@toyo.jp

Abstract. In this paper, we focus on 'memory', which is a human perceptual experience, and examine a method of recreating memory based on AI. Recent years saw a series of AI research on the theme of 'recreation of information'. This includes computation of neural information with AI based on the theory of cognitive neuroscience, and colorization of black-and-white photographs using human memory. This research focuses on the structure of the human brain cortex, and proposes Ultra Dot Matrix (UDM), a concept of transferring visual information data based on neural pathways. This is achieved by abstracting the cortex from the viewpoint of brain cortex topology, not by high-precision computation that simulates the whole cranial nerves, a method adopted in the previous research projects. This reproduces a more diverse 'memory-association' perceptual experience, which could not have been achieved by previous research. Re-experiencing perception process can be applied to a wide range of fields, e.g. solving social problems, new uses based on deep learning, and robot development.

Keywords: Cerebral cortex · Topology · Perceptual experience · Memory · AI · Brain science · Cognitive neuroscience

1 Introduction

The long history of brain research is said to have started with Hippocrates, known as the Father of Medicine, around 450 BC in ancient Greece. It can be said that the history of brain research is synonymous with the history of medicine. While there has been significant progress in brain science, less is known about human brain. In particular, it is still hard to say that there are long-accepted theories of memory, consciousness and emotion. However, successes through interdisciplinary approach to various research areas such as brain science, anatomy, and information science have been reported in recent years, and brain functions have gradually been revealed. For example, it is in the field of analyses such as 'factors affecting memory' and the 'contex of sensory organs' data transmission'. The latest research results in cognitive neuroscience have had a great impact on various fields such as clinical medicine, psychology, and artificial intelligence. In this paper, we propose a unique memory-association recreation system,

H. Selvaraj et al. (Eds.): ICSEng 2020, LNNS 182, pp. 287–296, 2021.
https://doi.org/10.1007/978-3-030-65796-3_27

UDM (Ultra Dot Matrix), with the motif of the brain data mapping structure, which stores daily information. The purposes of this paper are as follows.

1. Shed light on the meaning of this research by describing the research background such as cognitive neuroscience and artificial intelligence;
2. Explain the principle and basic structure of UDM proposed in this research; and.3. Clarify the achievements we have had so far and agenda items to work on.

2 Background

Today, it is said that artificial intelligence (AI) is having the third boom, and AI-related consumption and services such as Watson, Siri, Pepper, autonomous driving and Denou-sen, or human vs. Computer Shogi match, have attracted public attention. In recent years, the technology of deep learning, which evolved from the neural network technology and became a hot topic during the second artificial intelligence boom in the 1980s, has been attracting attention and is advancing remarkably in both image recognition and speech recognition. Deep learning is a technology that has been realized through accumulation of big data and a vast increase in computer processing power. It also suggests a new possibility.

It is a system of learning about representation of information. Deep learning is considered one of the methods of representation learning, and the subject of 'expression' is a theme that has long been discussed and studied in the field of artificial intelligence [1]. Yann LeCun, Yoshua Bengio, and Geoffrey Hinton, the leaders in the global deep learning community, told Nature magazine about the future of deep learning.

1. Unsupervised learning will become more important. Humans and animals do not need a teacher to learn from. We always discover the structure of the world by observing it.
2. By combining CNNs (Convolutional Neural Networks) and RNNs (Recurrent Neural Networks), a mechanism that determines where to look using reinforcement learning will emerge, which would outperform conventional methods.
3. Deep learning will be applied in the field of natural language processing in the future. With RNNs, understanding of words or whole documents will improve.

3 Representation of Memory

In recent years, the question of 'whether creativity can be realized by programming' has been a hot topic in the field of AI research (Matsubara [12]). According to Zhu (2018), 'intelligence' requires a 'causal relationship', which is based on the laws of physics and the laws of living things. Examples of the studies include the Cell Census Network (BICCN) project by National Institutes of Health (NIH), the Neuro-FAST project by Defense Advanced Research Projects Agency (DARPA), and the MICrONS project by The Intelligence Advanced Research Projects Activity (IARPA) in the United States.

The BICCN Project is intended to conduct research on mouse, human, and non-human primate brains from molecular, anatomical, and functional perspectives, and to collect, analyze, and share relevant data. Using the analyzed data, large data sets and terabytes of archives are built for brain data, and data storage/analysis tools are developed and enhanced. Research findings are applied to support researchers and clinicians engaged in the treatment of neurological diseases.

The Neuro-FAS project aims to visualize and decode brain functions through multiple approaches including data processing, mathematical modeling, and new interfaces. The project is conducted as part of the BRAIN Initiative, and the research findings are used to restore functions in the brains of US military personnel who have had brain injuries.

Lastly, the MICrONS project is aimed at understanding the algorithms the brain uses when it mathematically abstracts sensory data, learns, reproduces and transforms them, and finally applies them to the development of algorithms for machine learning. Currently, these data can be recognized by human brain, but state-of-the-art machine learning systems are still not capable of processing them.

On the other hand, this study aims at reproducing 'association' and 'imagining', which are higher-order functions of the brain. As we intend to achieve only the visual representation of the memory rather than to simulate the entire brain nerves through 'high-precision computation' as targeted in IARPA project, we will reproduce the mechanism of 'data transmission' in the neural pathways between brain cortices to build data maps. We believe that this excerpting technique provides more visual representation with less computational resources.

Also, 'memory'-related research projects which uses AI technology includes 'Rebooting Memories' by Watanabe et al. (2019) of The University of Tokyo. 'Rebooting Memories' is an application based on their study of colorization of black-and-white photographs with AI technology. The application associates photographs with 'memory', and visualizes it by mapping. This research not only archives the information on their past lives, but also gives people a better understanding of the ways of living and societies of the past. With this, the research demonstrates the importance of 'passing down the past memories that we must not forget'.

However, not all the tasks of 'Rebooting Memories' are performed by AI. It only uses AI technology for auxiliary processing, and this is not necessarily a novel idea in the areas of renovation and archive production of old content.

On the other hand, in this research, we aim to reproduce 'memory' with AI based on a completely new concept and method. In all the visualization processes, a unique, AI technology-based mechanism is used. Specifically, we propose a method to reproduce the process of 'association' by incorporating the data mapping structure of a part of the brain cortex.

For example, 'Rebooting Memories' recreates past situations or people with image information, i.e. photographs whereas this research allows people to experience the 'perception process' of how people remember things. As a result, a more vivid perception experience can be created and shared.

This study particularly aims at 'association' and 'imagining' which are higher-order functions of the brain. Instead of 'high-precision computation', which simulates the brain mechanism as sought in the IARPA project, we develop a tool, which focuses on emphasizing visual expressions. For this purpose, the first step we take is to study the existing research on brain science for 'data transmission' mechanism in the neural pathways between brain cortices, and see how so-called cortical data are connected, or the topology of these data. We then incorporate the findings in the construction of our unique memory-association reproduction system UDM (Ultra Dot Matrix) proposed in this research.

1. Production of media content that incorporates expressions of human memory, such as games, videos, and art works.
2. Engineering research such as development of learning aid materials for deep learning and robots.
3. We expect the results of this research can be applied in the following fields.

3.1 UDM Framework

Summary
This chapter details the unique memory and associative reproduction system, UDM proposed in this research.

Figure 1 shows an example of 'association' processing in UDM.

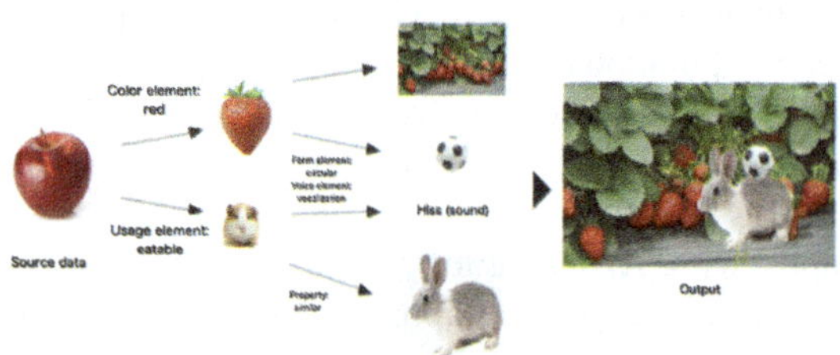

Fig. 1. UDM's process

Figure 1 shows the UDM's process to make a picture of an apple an 'association source' and recall and imagine other things based on the characteristics of color, shape and nature.

Figure 2 shows the mechanism of data analysis of 'association' in UDM, and Fig. 3 shows the image of brain cortex.

The UDM framework decomposes data in a specific way, packages it, and stores the packaged data in a virtual cortical layer. By constructing a 'bridge' that transmits the diffused signal in the virtual cortical layer, it is possible to freely diffuse the signal deprived from the original data and to output data through the bridge.

The connection rate between microcolumns of the cerebral cortex of about 2 × 108 is about 6 × 104 (Johansson, 2006), and in the case of random coupling, the combinations of the majority of columns is connected in about two steps. On the other hand,

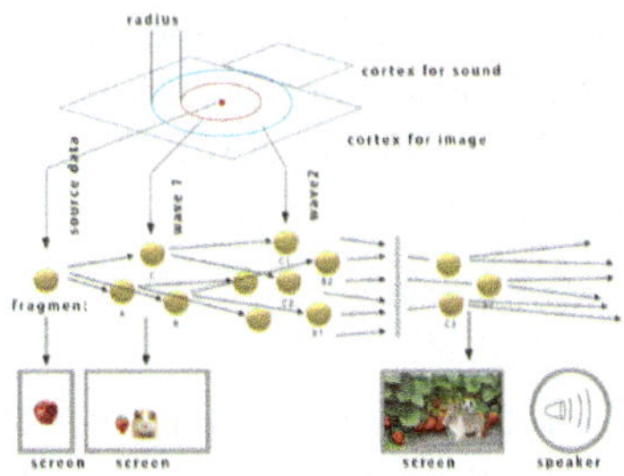

Fig. 2. ‘Association’ in UDM

the brain consists of anatomically independent units at various levels: organ level (neocortex, cerebellum, hippocampus, and basal ganglia): territory level (the primary visual cortex, the secondary visual cortex, the association area, and the somatosensory cortex within the neocortex): the level of organs such as CA3 and CA3.

The hypothesis that the brain is constructed as a set of computationally independent modules is called the ‘modular hypothesis’. The cerebral cortex is not only subdivided into smaller parts, but also multilayered. In this study, we consider the following three elements to abstract the cortex from the viewpoint of cortical topology, not the ‘biological characteristics’ of the cortex.

1. The number of layers of the cortex and the physical elements of each layer such as length and width characteristics of signal transmission between internal neurons.
2. Connection between the cortex and other cortices (red area in Fig. 3).
3. Features of external data writing and data output of the cortex.

In the ‘UDM Framework’, this research proposes the concept of ‘virtual cortical layer’ (POOL) for the abstraction of cortex with these three elements. POOL stores ‘data fragments’ of sensory data broken down into blocks in a specific way.

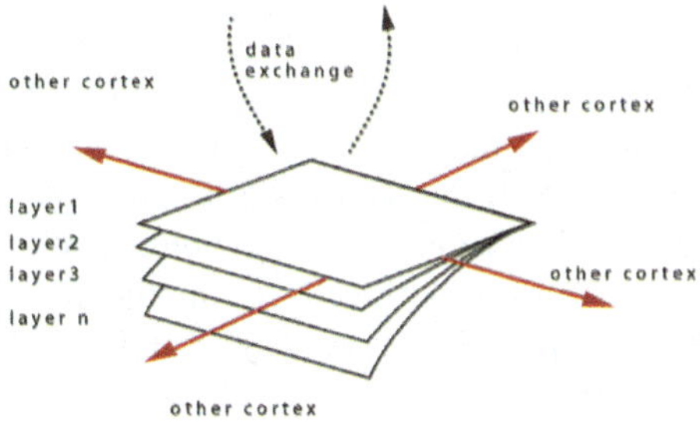

Fig. 3. Connection

Dot

DOT is a ‘data fragment’ of sensory data based on visual or auditory sense, which is broken down into blocks in a specific way.

The first step of data analysis in the 'UDM Framework' is to decompose and analyze data. Taking visual information as an example, when visual information such as photos and images are introduced to the 'UDM framework', the data is first subdivided, and useful data is extracted, and then transferred to the 'virtual cortex'.

3.2 Virtual Cortex

In the 'UDM framework', in order to send DOT to POOL (the virtual cortical layer), the data is first analyzed in the 'primitive feature test layer' and transferred to the 'virtual cortex' by the feature test method. For example, regarding a 'virtual cortex' that stores data relating to lightness, the data is sent to corresponding coordinates in the virtual cortex based on its own lightness information.

The subdivided data has a 'diffusion range' of the signal. In other words, it is a range that is connected to other DOTs. For example, in the 7×7 two-dimensional Pool (virtual cortex), as shown in Fig. 4, 'DOT A' is the center, and its 3×3 (red portion) is the 'diffusion range'.

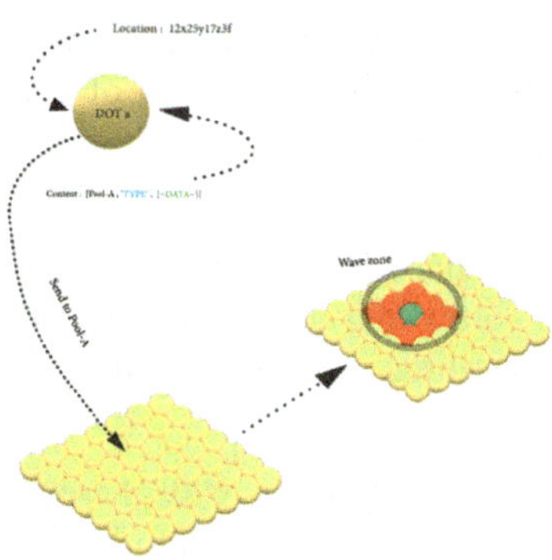

Fig. 4. Diffusion range.

'Diffusion range' means that other data in this range may be 'activated' at the next moment. The number of links in the 'diffusion range' is shown as an example in the one-dimensional space of Fig. 5. If the radius is 2, the number of links is 4. Under the same condition with the two-dimensional case, the number of links is 24. With the case of three dimensions, it becomes 124. Finally, with the four dimensional case, the number of links is 374. Four dimensions are x, y, z and 'depth.'

The following parameters are set in the 'virtual cortex' of the 'UDM framework' (Fig. 6).

1. Length: x
2. Width: y
3. Height: z
4. The fourth-dimension: f
5. Radius of 'diffusion range': Activity
6. Number of diffusions in the 'diffusion range': Amount

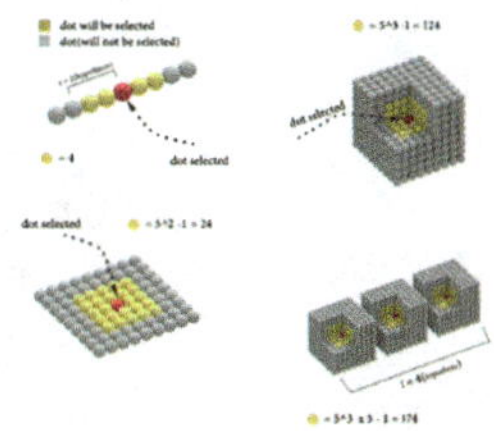

Fig. 5. Diffusion range.

7. Maximum number of simultaneous activities inside the cortex: Out-Amount
8. Parallel connection with other cortex: LINK
9. Vertical connection with other cortex: LINK
10. Cover that changes the characteristics of signal transmission inside the cortical layer: MASK

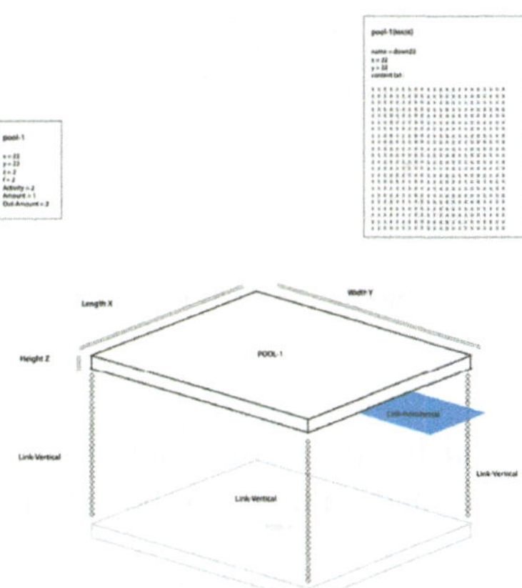

Fig. 6. MASK

Mask

The UDM framework adopts the concept of MASK in order to realize a function that specializes the transmission of internal signals. In the 'virtual cortical layer' to which MASK is added, within the 'diffusion range,' the internal signal transmission is not arbitrary, and it is transmitted to the direction specified by MASK.

3.3 Bridge

In order to spread signals between multiple 'virtual cortices' corresponding to the 'signal diffusion' described above, 'bridges' are required. There are two types of bridges (Fig. 7). One is a 'vertical bridge', which is the jointing surface of the two

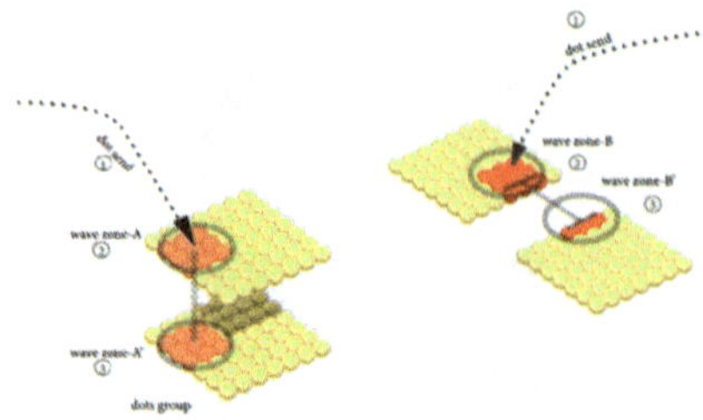

Fig. 7. Bridges.

'virtual cortices' that are in contact with each other, and signal diffusion between them passes without barrier. The other is a 'parallel connection bridge', which stretches from side to side between two 'virtual cortices' through which signals spread.

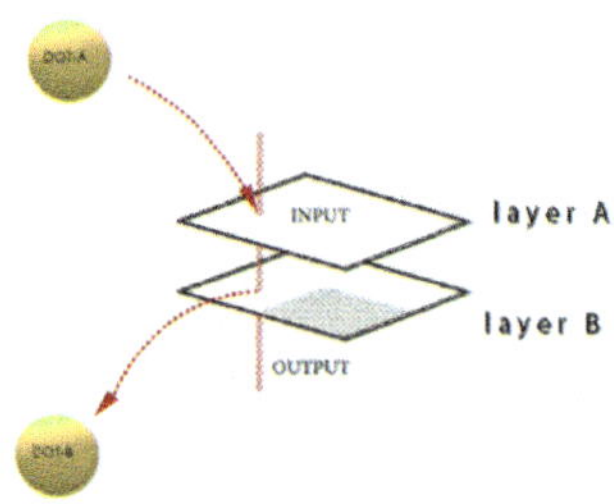

Fig. 8. Vertical bridge.

Through the 'vertical bridge', the two 'virtual cortices' are in a 'vertical relationship' as shown in Fig. 8, and the data inside are combined as 'pairs' by x and y coordinates, and simple 'input' and 'output' can be executed.

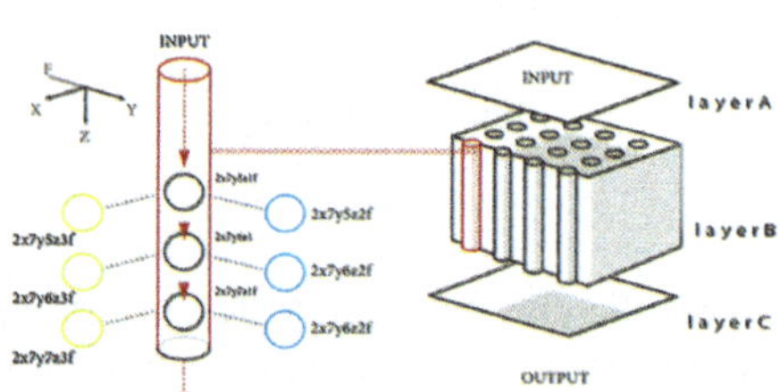

Fig. 9. Structure.

Figure 9 shows the structure for storing continuous data. 'Cortical layer A' and 'Cortical layer C' are thin and are responsible for 'input' and 'output' respectively. On the other hand, 'cortical layer B' is thicker and uses 'MASK' to store continuous data. The 'MASK' has the function of moving the spread signal vertically downward. A series of data passed along the way by the signal is made 'active' one by one.

3.4 Conclusions

In this paper, we examined a method of reproducing the 'association' process of human memory, specifically, regarding video and audio in memory, at an angle of visual expressions. We also devised a unique memory: association reproduction system, UDM. This research aims at 'reproduction of 'association' and 'imagination' of higher-order functions of the brain. Instead of 'precision calculation' that imitates the brain mechanism like the IARPA project, specifically only visual expressions were focused.

At present, the development of the part that stores the database is almost completed, and the next step is the development of 'data decomposition' (subdivision of data). Specifically, it is 'decomposition and analysis of information regarding sensory organs'.

The following are the future tasks.

1. We will collect more cases that can be applied to this study from existing brain research results (particularly research on cortical data mapping), and explore the possibility of reflecting them to UDM.
2. Develop a plug-in that decomposes 'visual' and 'auditory' data, and conduct research on 'mutual conversion between sound and visual' in the UDM Framework. Using the results, we will start applying the conversion to information design.

When the development of UDM is completed, we provide the system to users by making it public. UDM project is open to those who are interested in the system and want to participate in improving the 'UDM framework'.

Acknowledgment. This work was supported by JSPS KAKENHI Grant Number 17K00730.

References

1. Yutaka, M.: The future of artificial intelligence and new possibilities for data sharing. Inf. Manag. **58**(8) (2015)
2. MICrONS project by IARPA. https://www.quantamagazine.org/mapping-the-brain-to-build-better-ma-chines-20160406/
3. Research and Development Trends in Brain Information Technology in the 3. United States. https://www.nict.go.jp/global/lde9n2000000bmum-att/a1525652348170.pdf
4. Fujimoto, T.: 'Design to make information easy to understand: what is information design' what and what is not. Inf. Sci. Technol. **65**(11) (2015)
5. Tada, T.: Information design: a new attempt to express changing things. J. Jpn. Fuzzy Soc. **13**(3) (2001)
6. Yasushi Harada homepage. https://www.geijutsu.tsukuba.ac.jp/yhard/
7. Zhu, S.: Visual Search (2017). https://mp.weixin.qq.com/s/-wSYLu-XvOrsST8_KEUa-Q
8. Sunaga, T.: 'Expansion of Artifacts Supporting' Words 'and Information Design'. Special Issue on Design Studies, pp. 22–26 (2002)
9. So, W.C., Ching, T.H.W., Lim, P.E., Cheng, X., Ip, K.Y.: Producing gestures facilitates route learning. PLoS ONE **9**, e112543 (2014)
10. Cook, S.W., Mitchell, Z., Goldin-Meadow, S.: Gesturing makes learning last. Cognition **106** (2), 1047–1058 (2008)

11. Matsubara, H., Nakajima, H., et al.: The author of Future University Hakodate Kimagure Artificial Intelligence Project (2012). https://www.fun.ac.jp/~kimagure_ai. Accessed 21 Mar 2015
12. Watanabe, H., Niwata, A.: 'Decompression of memory: generation of' flow 'and succession of memory based on colorized photographs.' J. Digit. Arch. **3**(3), 317–323 (2019)
13. Johansson, C., Lansner, A.: Towards cortex sized artificial neural systems. Neural Netw. (2006)
14. Takahashi, K., Itaya, K., Nakamura, M., Koizumi, M., Arakawa, N., Tomita, M., Yamakawa, H.: RIKEN Center for Life Sciences, Keio University Graduate School of Media and Governance Shu. The 29th National Convention (2015)

Prototyping and Evaluation of Drawing Software to Provide Analog-Like Feeling by Sound

Yulana Watanabe(✉), Yui Tanaka, and Takayuki Fujimoto

Graduate School of Information Sciences and Arts, Toyo University, Tokyo, Japan
{s4B102000049,s4B101810014,fujimoto}@toyo.jp

Abstract. Today, most areas of our everyday life have been digitized. In this research, we focused on one example, which is drawing software. While various types of software are being developed, very few give a realistic experience when drawing, though they are highly capable of drawing faithful reproduction of the original. In this study, this sensation that users get when actually using the drawing material is referred to as "analog-like feeling". We have conducted research, focusing on the importance of this sensation users feel during work. The focus was also on benefits of "analog-like feeling". With it, users may get attached to the tools such as drawing materials, which help improve work efficiency and quality of the work. In addition, we examined different contributing factors that create "analog-like feeling". In this paper, we focused on the factor of sounds, and developed a prototype in which they have been digitally incorporated. This is a system that generates the sounds of three different drawing materials: brush, pencil, and marker when they are used for digital drawing. In addition, we asked testers to actually use the system for verification to check if the sounds have produced or enhanced such feeling. As a result, about 90% of the participants answered that they had had better drawing experience when the sounds were generated. This proves the importance of adding an analog-like feeling in digital drawing.

Keywords: Interface design · Drawing software · Analog · Digital · Information media

1 Background

1.1 Recent Developments of Digital Drawing and the Drawing Sensation

In recent years, digitization has been prevailing in many areas of our daily life. Drawing software, which is the subject of this research, is one of them. There are various types of software with which anyone can draw with simple operations regardless of age, from children to the elderly. Tilt Brush is a typical example. This is a software that draws in 3D space with dedicated equipment such as a head-mounted display and a controller. There are also drawing software for professionals and creators who need highly specialized knowledge and skills. Gravity Sketch is a software that

H. Selvaraj et al. (Eds.): ICSEng 2020, LNNS 182, pp. 297–306, 2021.
https://doi.org/10.1007/978-3-030-65796-3_28

focuses on "modeling", while working in 3D space using dedicated equipment such as a head-mounted display just like the above-mentioned Tilt Brush. In design and layout, it is easier to scale up and down than with analog approach, and the tool can be used on multiple devices such as iPads as well as dedicated devices. Different types of software such as Inkscape and Verve Painter are also good examples. They come with so many different functions that, with these software, users can draw with particular attention to color and details. In addition, users can make subtle changes to the color with changing amount of water or adjust bumps and glossiness. However, having too many features leads to complexity, which makes it difficult to acquire new users.

In this way, drawing software has been developed in many different approaches. Some are designed to produce analog-like experience, but they merely reproduce appearance. Very few of them focus on the actual sensation users feel during work. Whatever interface is used, the sensation that users get is the same. On the other hand, analog drawing starts by selecting drawing materials, or what to draw with, and base materials, or what to draw on. We consider the drawing sensation can differ depending on the selected tool. In this research, these different drawing sensations are incorporated into the digital tool. By combining the real feel of drawing, which is one of the charms of analog drawing, and the convenience of digital drawing, we propose a new software, which is totally different from precedent packages.

1.2 The Charms of Analog Objects

Today, while most areas of our life are being digitized, some people choose to go analog. Book is one example.

Today, many books have been digitized for people to subscribe to through their own personal devices anytime, anywhere. However, bookstores and second-hand bookstores are still around, and paper books are still selling online in huge numbers. It may be because people find appeals of books not just in the content printed in the form of characters or sentences. When we read physical books, we feel its weight and smell. In addition, when carrying them, there is a possibility that the pages may get folded by hitting other belongings. If your hands were dirty, it may leave marks on the books as you turn pages. These seemingly troublesome factors contribute, in fact, to the feeling of continued reading process.

The same can be said about music. While being able to listen to a variety of songs on a personal digital device with almost no restrictions just by paying a low monthly fee, many people still rent or purchase CDs. Also, when Sony released vinyl records a few years ago, many bought them. However, few people today have the equipment such as turntables or CD players to listen to music on the media they have bought. In many cases, the music is imported to a device like computer, and then transferred to a portable player. Also, Sony's vinyl records have a QR code on the package. Reading the code makes it possible to play the music. In other words, though the sound is no different from the downloaded music, users buy the vinyls because they find appeals in the size and appearance of the medium as an object.

It is clear that digital approach is far more convenient either in books or music. Nevertheless, people who choose to go analog seek appeals rather than convenience in the analog approach. Beyond the original purpose of delivering the content of books or music, owning the medium as an object gives a real-life experience of feeling the joy of ownership or of feeling the aging effect of the object. We assume that many people consider having this feeling as the appeal analog approach has and choose to go analog rather than digital.

1.3 Effects of Sound

There are various types of "sounds" we hear when working on something. This includes not only the sounds generated during work but also the noises such as the ambient sounds and the voice of people around. Many people play music for concentration while working. According to The Effect of Music Stimulation on Autonomic Nervous System Activities (2005), it has been proved that there is a relaxing effect on the listeners who listen to their favorite music, though the result may vary depending on each individual's personality or preferences such as music genres. Similarly, in A Basic Study of Human Engineering on Different Music Environments for Work Efficiency (2005), the comparison of the results of experiments performed using three types of music proved that the work time and the rate of right answers tend to improve when the work was done in a relaxing music environment. As Mamoru. I.'s study (2008) revealed, it has also been proved that reduced stress leads to fewer typing errors and improvement in work efficiency. Through our research, we have become fully aware of the relaxing effect of sound, which enhances work efficiency and concentration. This has also been verified in previous research. Also, for example, when drawing in art lessons or classes, we can hear the sharp noise of sharp-pointed pencils rubbed against paper in a quiet space. You can tell that someone is drawing something just by hearing the sound in that space. When the sound stops, you know that no one is moving and can imagine that the student might be lost for the next line. In this way, even when you are not drawing, you can get the realistic experience of drawing from these sounds. On the contrary, if you were there and hear nothing, it would not be easy for you to fully realize that someone is drawing only through the visual information.

In this research, we paid particular attention to the impact of realistic experience of working with analog tools and focused on the "sound", which is a dominant and influential factor out of many. Of all the sounds we have in our life, we focused on the one generated by work, thinking that incorporating the sound into digital tools would lead to improved work efficiency and high quality of work.

2 Purpose

Today, various drawing software packages have been released. They have been devised to be more enjoyable or easy to use for various purposes and uses. Because there are many types, it is difficult to summarize them all. Despite the development of a variety of drawing software, there is little that focuses on and provide the "drawing sensation." There are developed software packages that make the work output's appearance closer to analog, such as software that changes the line on the screen depending on the drawing tool and software packages that make subtle color differences depending on water content of paint used for watercolor painting. However, even if the reproducibility of the appearance is high, those software packages have no big differences in the digital interfaces that users actually operate, and it is difficult to obtain physical sensation of using the drawing tool. When people are working on drawing in analog, it is always accompanied with the physical sensation of drawing. We thought that this would increase the people's feeling of attachment to the tools used for the work, and it would lead to the improvement of the work efficiency and the quality. We propose drawing software that is not just an outcome of unilateral digitization, and also the system enables users to reaffirm the appeal of analog by adding the "realistic experience" to the trend of the development and spread of digital technology these days.

3 System Development

We developed a prototype of the proposed drawing software in this study. There are many possible factors to provide the analog-like feeling, but we specifically implemented sound combinations of plain paper (as the base material) and a brush, marker or pencil (as the drawing tool), instead of implementing all the factors mentioned in this paper. The sounds for implementation are based on the edited version of the free sound source on the Internet with reference to the characteristics and changes in waveforms by the difference of the base material and the drawing tool combinations, which was clarified in "A Drawing Software that Changes User's Realistic Experience by Sounds Generated by the Combination of Tools and Materials (2019)."

The sound for the brush was set as the sound with slightly rough impression. Reflecting the characteristic that there is no significant difference between the sound when people start to draw and the sound when the line is drawn continuously, the sound was edited so that it can be consistent and the volume is constant to some extent. The sound for marker was set as a slightly tingly sound, supposing the marker type that has a large tip surface, which comes in contact with the base material. Initially, for the sound at the beginning of drawing, we implemented a responsive sound generated when the marker touches the desk. However, as we proceeded with the development and discussion, we came to the conclusion that some people might find the sound uncomfortable, and this time we did not adopt the sound itself, instead, simply took over the characteristic of the raised volume of drawing starting sound. The pencil sound was edited to make it easier for the users to capture the difference between the sound that generates when people start drawing and the sound that generates while they are drawing. The users would hear the rather concrete sound when starting drawing because the marker tip comes in contact with the desk surface, but while they are drawing, they would hear the smoother sound like rustling. By making the difference between them obvious, the clicked sound and the drawing sound with the continuous click are clearly distinguished.

The loudness was set as similar to what you would hear when working on drawing in reality. The brush sound was set to a slight sound, and the pencil and marker sounds were set so that they could be heard rather clearly, even if it is not quiet around the user.

Regarding this prototype, we carried out the experiment only with the sound changes upon drawing so that the drawing sensation is not affected by the reproducibility of appearance as with the conventional drawing software. The thickness and density of the drawn lines were set as the same by default.

After the prototype use, in which the sounds are actually implemented, the questionnaire survey was conducted regarding the difference in the presence/absence of drawing sensation and the work efficiency compared with the conventional drawing software. Participants were 38 university students who casually use computers, and they had some knowledge and experience of drawing software.

As a method of the experiment, participants first worked on drawing in the conventional style in which the drawing sounds do not generate by the drawing operations and only the click sounds of the mouse were heard so that they could get used to the operations. Next, participants worked on drawing with the brush, marker, and pencil, in this order, with the sound effect of each drawing tool. Participants were required to write letters as much as possible so that the content to be drawn would be similar to those of drawing and writing in the real world, and they were asked to draw freely. Then they answered the questionnaire. There are two main questions.

First question: Regarding the state with/without sound, which state did you find better for the drawing sensation? Select one from the three options "With sound/Neither/Without sound." Second question: Regarding the state with sound, was the drawing sensation enhanced? Select one from five levels of "Extremely well/Very well/Neither/Not well/Not at all well". These two questions were answered regarding each drawing tool. In addition, by the open-ended question, the participants were asked to report their impressions and improvements regarding the use of the drawing tools.

4 Summary of Questionnaire Results

4.1 Brush

Upon the first question: regarding the state with/without sound, which state did you find better for drawing sensation?, 34 out of 38 respondents answered that they had found the state with sound better for the drawing sensation. Two participants said that the absence of sound had given them drawing sensation better, and two selected "Neither." Participants who reported that they had found the state with sound better for drawing sensation, also commented that the sound's very slight volume set by default had reminded them that when working on drawing in the real world, the surrounding environment should be quiet. The Participants who answered Neither or had found that the state without sound was better, also commented that even when working on drawing in the real world, the sound of the brush was too slight for them and they had never paid attention, and that was why the sound had not led to the drawing sensation.

For this reason, the response of the participants to the brush sound effect was not so positive compared to other drawing tools, because even if the slight sound of the brush was digitally provided, it was rather different from the impression of the actual sound. Some respondents claimed that they would like to move to a quieter location to hear the sound effect, because they knew that there should be the sound, and we changed the experimental environment by the request from the participants. As a result, many participants answered that they had had drawing sensation and besides they had been able to improve their concentration.

As a result of the questionnaire, 90% of the participants reported that the drawing sensation had been enhanced in the state with sound. Many of the participants who selected “Extremely well” wrote a single kanji character with a brush in the same way as they do in the real world. On the other hand, many of the participants who selected Not well/Neither tended to draw a spiral lines, which is rarely expected to be drawn with a brush in reality. Some respondents reported that although the state with sound had been better than the state without sound, the sound of the brush itself had been not so obvious that it had not led to the drawing sensation.”

Figure 1 shows a result of a question regarding the drawing sensation when drawing is performed in a state with a brush sound.

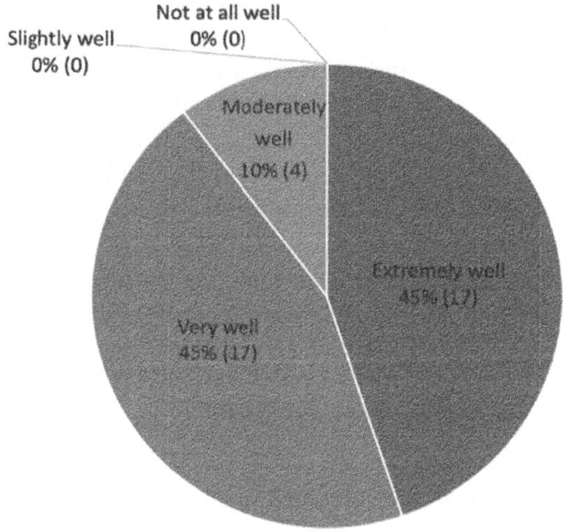

Fig. 1. Questionnaire results on the drawing sensation: Brush

4.2 Marker

Upon the first question: regarding the state with/without sound, which state did you find better for a drawing sensation?, 32 out of 38 respondents answered that they had found the state with sound better for the drawing sensation. Four participants said that the absence of sound had given them a drawing sensation better, and two selected “Neither.” The sound of the marker is distinctive compared to other drawing tools. Some participants reported that the sound of the marker had been so uncomfortable that they had not wanted to listen to it. In addition, as an explanation of this experiment, only the term, “marker” was given to the participants, but in fact, the term can indicate various types, which are on the market.

Through the free description of the questionnaire, it became clear that the drawing tool assumed by the word, “marker” differed depending on each participant. Some imagined a small pen that fits in a pencil case and is used to write a owner's name in a textbook, others imagined a highlighter pen to emphasizes important parts, and the others imagined a thick and large pen used for drawing on big base materials such as cardboard and thick paper. Some respondents said that the difference between the sound imagined by the term, "marker" and the sound actually heard had not led them to a drawing sensation.

As a result of the questionnaire on whether a drawing sensation was enhanced by the marker sound, nearly 90% of the participants answered that the drawing sensation had been enhanced. On the other hand, some respondents reported that the heard sound had been different from their image of a marker sound, and that it had taken a long time to connect it to a drawing sensation of a marker. It was characteristic that many of the participants who answered Extremely well/Very well pointed out their likeness for the marker sound as a reason. On the other hand, many of the participants who answered Not well/Neither, were uncomfortable with the sound. Regarding content drawn freely, there were various types of characters and lines, but there were no participants' report that there was a difference in the drawing sensation depending on the content.

Figure 2 shows the results on a question regarding the drawing sensation when they draw with a marker sound.

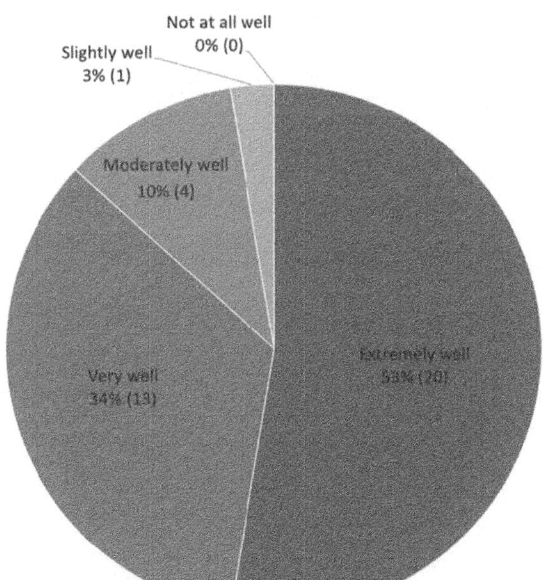

Fig. 2. Questionnaire results on the drawing sensation: Marker

4.3 Pencil

Upon the first question: regarding the state with/without sound, which state did you find better for a drawing sensation?, 38 out of 37 respondents answered that they had found the state with sound better for the drawing sensation. Only one participant answered, “Neither”, but none said that they had felt the drawing sensation better in the state without sound. Compared with other drawing tools, pencil’s sound obtained the most positive response from the participants in the experiment. Statements; “easy to use” and “fun” were made not only in the free description of the questionnaire, but also in many participants’ direct comments during the use in the experiment. Many people answered that the hard sound, which generates when the pencil contacts the desk at the beginning of drawing, had led to the drawing sensation, rather than the sound generated while drawing lines. On the other hand, some respondents who answered “Neither” commented that although the drawing sensation and work efficiency had been increased in the state with sound comparing with the state without sound, the sound impression had been as hard as that by the combination of a chalk and a blackboard, which had led to the different feeling from the actual drawing sensation of using a pencil.”

As a result of the questionnaire, 95% of the respondents in total answered that the drawing sensation had been enhanced by the pencil sound, and it indicates that the sound effect was extremely effective. There were also many positive feedbacks that were interpreted from the free description. On the other hand, there were reports that some people had found the pencil sound effect different from the real one and they had been able to notice that because the pencil sound was quite familiar with them.

Figure 3 shows a result on a question regarding the drawing sensation when they draw with a pencil sound.

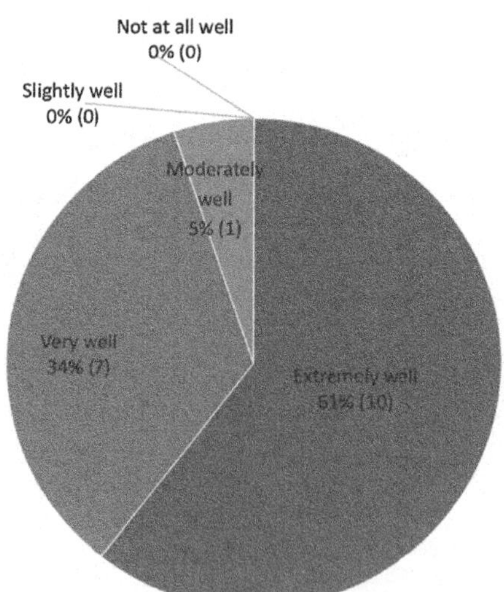

Fig. 3. Questionnaire results on the drawing sensation: pencil

5 Conclusion and Future Work

5.1 Conclusion

In this research, we reconfirmed the appeal of analog and considered the new digital way besides the convenience of digital by reproducing analog in digital. We specifically focused on drawing software, which has been remarkably developing today, and thought that providing the sensation of use would lead to improvement of work efficiency and quality of work. There are various factors to provide a realistic experience of drawing, and out of them, this research focused on 'sound', which is considered to take up a large proportion.

In this paper, we developed a prototype that incorporates sounds resembling the sounds heard when we are working on actual drawing in analog, into the system. Upon the system, the sound effect is provided by the movement of drawing in digital. Through the experiment and the questionnaire survey, it became clear that the participants felt that the drawing sensation had been enhanced in the state with a 'sound.' The novelty and usefulness of this prototype system was also proved.

5.2 Future Work

The development and experiments carried out in this study are only one example. As shown by the questionnaire results, many people obtained the drawing sensation, but also drew improvement points. We believe that it is necessary to clarify the detailed conditions for the users' recognition and the drawing sensation regarding each sound. In addition, to the drawing software, we will also propose to add the physical sensations that are gained when we are working on analog drawing and have not yet been digitized. For example, when working in analog drawing, it can be assumed that a difference occurs in sound due to individual characteristics such as pen pressure and speed, and it is expected that the sensation accompanied with the sound effects changes accordingly. These features will be reflected to the system as functions in the future. Furthermore, we aim to further improve the quality of the software based on the results of the survey. In addition to the drawing tools mentioned in this paper, we will provide crayon and chalk, which are generally used as drawing tools, and blackboards and whiteboards, which are widely used as base materials.

Acknowledgement. This work was supported by JSPS KAKENHI Grant Number 17K00730.

References

1. Watanabe, Y., Fan, Z., Fujimoto, T.: A mechanism of drawing software that provides a "real feel. Appl. Inform. Media Des., 59–62 (2019)
2. Watanabe, Y., Fujimoto, T.: Adding analog-like reality to drawing software with work sound effects. Int. J. Comput. Sci. Netw. Secur. **19**(9), 132–138 (2019)
3. Tanaka, Y., Fujimoto, T.: A design of application to turn a smartphone into a computer mouse and possibility of preventing from being copied. In: 1st International Conference on Interaction Design and Digital Creation/Computing (IDDC 2018) (2018)

4. Fan, Z., Fujimoto, T.: Proposal of a design method to apply the analog features to digital media. In: The 2nd International Conference on Applied Cognitive Computing (2018)
5. Fan, Z., Fujimoto, T.: Proposal of a scheduling app utilizing time-perception-reality in analog clocks. In: 1st International Conference on Interaction Design and Digital Creation/Computing (IDDC 2018) (2018)
6. Fan, Z., Fujimoto, T.: Implementation and evaluation of document production support system "DLO-Editor" with obsession mechanism. In: 25th International Conference on Systems Engineering (2017)
7. Fan, Z., Fujimoto, T.: Implementation of document production support system with obsession mechanism. In: 4th International Conference on Computational Science/ Intelligence & Applied Informatics (CSII2017) (2017)
8. Fan, Z., Fujimoto, T.: Proposal of a digital book application that offers analog-like usability. In: The 15th IEEE Transdisciplinary-Oriented Workshop for Emerging Researchers, p. 26 (2018)
9. Fan, Z., Fujimoto, T.: Method to control children's smartphone use based on the motif of analog fuel system. In: ICSTR Bangkok – International Conference on Science & Technology Research (2018)
10. Fujimoto, T.: Toward information design 3.0: the information design for 'communicate.' Build. Maint. Manag. **34**, 42–46 (2013)
11. Fujimoto, T.: Understandability design: what is 'information design'? J. Inf. Sci. Technol. Assoc. **65**, 450–456 (2015)
12. Fujimoto, T.: Ideology of AoD: analog on digital-operating digitized objects and experiences with analog-like approach. In: 1st International Conference on Interaction Design and Digital Creation/Computing (IDDC 2018) (2018)
13. Soma, Y.., Matsunaga, T., Soga, J., Uchiyama, H., Fukumoto, I.: A basic study of human engineering on different music environment for work efficiency. IEICE Technical Report 105(304), pp. 43–46 (2005)
14. Iwaki, M., Arakawa, S., Kiryu, T.: Influence on biosignal and working efficiency of sound environment in typewriting, The Institute of Electronics, Information and Communication Engineers (IEICE) technical report. ME and bio cybernetics 108(52), pp. 19–24 (2008)
15. Ohisa, N., Yoshida, K., Yanbe, T., Kaku, M.: Effect of autonomic nervous system activity while listening to music. Auton. Nerv. Syst. **42**(3), 265–269 (2005)
16. Karashima, M., Nishiguchi, H.: Effectiveness of listening to music before simple repetitive tasks: copying task and mental rotation task (theory and methodology). J. Jpn. Ind. Manag. Assoc. **63**(2), 29–40 (2012)

Facial Features in Stylized Images Typically Used for Face Recognition and Comparison of Impression Between Stylized Images and AI Images

Miki Sunakawa(✉) and Takayuki Fujimoto

Graduate School of Information Sciences and Arts,
Toyo University, Tokyo, Japan
{s3b101900035, fujimoto}@toyo.jp

Abstract. In a caricature, a person's character is often exaggerated. While the stylized picture may not look exactly like the model, it can make one feel familiar and attached to it. In this research, we compare our impressions between stylized images and the images created with AI. In addition, the tendency people have in recognizing the characteristics of a human face to create a stylized image is also investigated.

Keywords: IT · Caricature · Stylized image

1 Background

It is said that we are currently having the third AI boom. AI is roughly divided into two types: "Specialized AI" that works in specific areas, and "Versatile AI" with ability to generalize that is similar to or exceeds human performance levels. Currently, the development effort is focused on specialized AI and it is already in use. One of the typical applications of specialized AI is image recognition technology. Image recognition is an ability to recognize the features of and patterns in an image and specify what is in it.

2 Purpose

Image recognition technology is supported by "deep learning". In deep learning, computers learn from a huge amount of image data to label those images, indicating what they represent. As learning progresses, computers become capable of recognizing objects in a new image and identifying what they are by matching the database to the objects. By making full use of this technology, AI-led creation of realistic images and videos is also made possible. On the other hand, some say such a creation does not look like human and feels wrong no matter how elaborately it is made.

There is a way of depicting human faces, which is caricature. In a caricature, a person's character is often exaggerated. While the stylized picture may not look exactly

H. Selvaraj et al. (Eds.): ICSEng 2020, LNNS 182, pp. 307–313, 2021.
https://doi.org/10.1007/978-3-030-65796-3_29

like the model, it can make one feel familiar and attached to it. Stylized caricatures can be more memorable than those created precisely by artificial intelligence. In this research, we compare our impressions between stylized images and the images created with AI. In addition, the tendency people have in recognizing the characteristics of a human face to create a stylized image is also investigated.

3 Examples of AI Images

In this research, a realistic image created by artificial intelligence is called an "AI image."

3.1 Image Correction

Many people use an image correction technique, which brightens up photos taken at night. The technique analyzes images and matches them to the photographic data used in previous learning. Then, the artificial intelligence detects and removes from the image the noise that was produced when the photo was taken in a dark place. In addition, colorizing monochrome photos is also considered to be an image correction technique. By learning from a large amount of data on colors and corresponding labels necessary for image classification, color extraction can be performed more effectively (Fig. 1).

Fig. 1. Colorized monochrome photograph

3.2 Composite Image

AI has long been applied to image recognition to distinguish human faces with high accuracy. Now, not only can it identify the faces, but it can also generate human face images from the data used for identification. This technology is famous for the system developed by NVIDIA, an American semiconductor manufacturer. The system is based on a technology called GAN (Generative Adversarial Network), which was developed in Canada in 2014. This technology builds "AI for distinguishing faces" and "AI for generating faces", and continues training until the "generated face" cannot be distinguished from the "real face." One network generates AI images and mixes them into the image data while the other tries to detect the fake AI images. After the repeated adversarial trainings, the image generator improves its creation techniques and the accuracy of the AI image increases proportionately (Fig. 2).

Fig. 2. Non-existent person or object created by AI

4 Examples of AI Images

Even if an image is created to look like a very natural human, humans sometimes instantly recognize that it is an AI image. It is said that people feel discomfort, spookiness, and have a feeling like fear to AI images. This is considered to be a kind of phenomenon called "Uncanny valley." Uncanny Valley was proposed in 1970 by Masahiro Mori, a leading robotics engineer. "Uncanny valley" was proved to occur, but the fundamental cause was left unknown. Therefore, an experiment was conducted with 80 images composed of human photographs and doll photographs in various ratios. Then the examinees were asked to classify as human or doll. Furthermore, "whether they had a sense of unease to the images" was also asked. As a result, it was found that the images caused spooky feelings upon two conditions: "long time required for examinees' classification" and "Judgement difficulty whether the image is human's or doll's." Robots and androids were used in the "Uncanny Valley" experiment, but the conclusion is expected to be applicable to other artificial beings (Fig. 3).

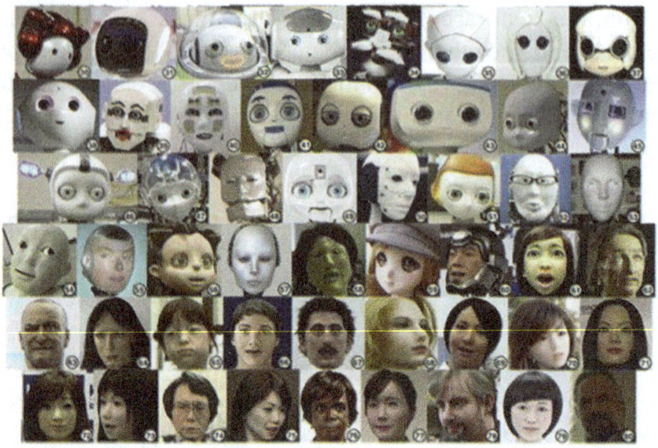

Fig. 3. Figure 3 80 images used in the "Uncanny Valley" experiment

5 Examples of Stylized Images

For comparison with AI images, this section gives examples of stylized images that can be created using electronic devices such as personal computers and smartphones.

5.1 Avatar

An avatar is the graphical representation of one's self-figure that is used on the Internet. This functions for smooth communication among users mainly on SNSs (social networking services) and social-network games. A user can select the face parts and create a character similar to himself or herself depending on the combination. A research has shown that avatars actually make a good impression on Internet communication (Fig. 4).

Fig. 4. Avatar example

5.2 Sticker

A sticker is illustrations used in messenger applications with which the users can expresses their emotions and moods intuitively. Google has developed the messaging application called "Allo", on which the users can create unique stickers based on selfie photos. A sticker is created by combining the data of the accumulated artist works and the features extracted from the users' selfies. A sticker is like a simple illustration with exaggeration. This simplicity generates a lovely impression. Google decided to limit image analysis for stamp creation to the parts needed for illustrations. Therefore, it was possible to make the function reach the operable level with relatively little learning (Fig. 5).

Fig. 5. Sticker created based on selfie

5.3 Caricature-like Portrait

Microsoft has researched and developed the technology with which the users can create images with stylization, including exaggeration by hand-drawn portraits. The research focuses on exaggerating shapes and stylizing the appearance to create a caricature that people can identify whose face it is even with stylized characteristics. In addition, the

Fig. 6. Caricature-like portrait by AI

system was trained with more than 8000 portraits on the Internet to master the exaggeration of the unique features of portraits. Although some caricature-like portraits have a high degree of perfection, since the system specializes in identifying facial parts, the output work does not provide ears or hairstyle expressions (Fig. 6).

6 Future Work

A research on creating caricature-like portraits with artificial intelligence is ongoing. It means that stylized images are regarded as attractive. Stylized images may be far from actual models due to the nature of exaggerating features. However, it is still possible to figure out who is stylized. Sometimes the stylized image can cause attachment or closeness feelings to people. In this research, we carry out a comparative survey with a stylized image and an AI image, created from the same person's face data. First, to create a stylized image, participants are asked to answer the questionnaire for the features of the faces. Specifically, they answer the number of the section that they found most distinctive regarding the face photos divided into nine sections so that the face parts can be distributed.

The section, which the most participants answered as the distinctive part can be said to be the face part, which is easily recognized as the characteristic of the person by others. We expect that we can identify the remarkable facial features to stylize for each face. Also, we will clarify the people's tendency in recognizing features when they look at another person's face. We will create stylized images based on the results of this questionnaire. We will also have AI create images from the same face photo data. Regarding the method, we are currently studying the possibility. Finally, we plan to evaluate the impressions by a questionnaire on which the completed stylized images and AI images are compared (Fig. 7).

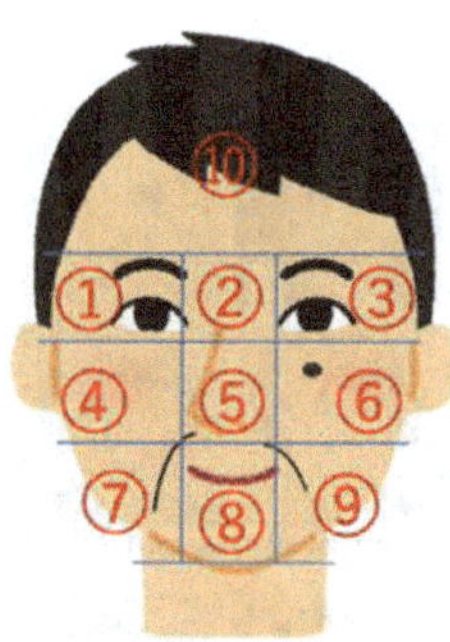

Fig. 7. Example of response to face photo

References

1. Beckett, J.: Generative Adversarial Network. NVIDIA Japan, 21 June 2017. https://blogs.nvidia.co.jp/2017/06/21/generative-adversarial-network/. Accessed 23 Jan 2020

2. Stone Washer's Journal, Artificial intelligence synthesizes images and changes movies and photos, 1 November 2017. https://stonewashersjournal.com/2017/11/01/gan/. Accessed 20 Jan 2020
3. WIRED Japanese Edition, Can you tell the real thing? Amazing accuracy of "fake face photo" created by AI, 24 July 2019. https://wired.jp/2019/07/24/artificial-intelligence-fake-faces/. Accessed 17 Jan 2020
4. WIRED Japanese Edition, Robots prove "Uncanny valley" that humans hate: research results, 6 November 2015. https://wired.jp/2015/11/06/uncanny-valley-creepy-robot/. Accessed 23 Jan 2020
5. Kyoshiro, S.: The eerie nature of android is anxious about the unknown, 20 November 2017, Academist Journal. https://academist-cf.com/journal/?p=6416. Accessed 23 Jan 2020
6. Hidehiko, N., Woonho, K., Kosuke, M.: Avatar effect at online community: case study of Japan-Korea Avatar sites. Ritsumeikan Bus. J. 17–36 (2010). Accessed 19 Jan 2020
7. Snow Company. ZEPTO. https://buildit.zepeto.me/?lang=en#. Accessed 19 Jan 2020
8. ITmedia. Google, neural network function to create original stamp from face photo in "Allo", 12 May 2017. https://www.itmedia.co.jp/news/articles/1705/12/news108.html. 21 Jan 2020
9. Cao, K., Liao, J., Yuan, L.: CariGANs: unpaired photo-to-caricature translation. ACM Trans. Graph. (Proceedings of Siggraph Asia 2018) (2018). https://cari-gan.github.io/. Accessed 20 Jan 2020
10. Iizuka, S., Simo-Serra, E., Ishikawa, H.: Let there be color!: joint end-to-end learning of global and local image priors for automatic image colorization with simultaneous classification. ACM Trans. Graph. (Proc. SIGGRAPH 2016), 110:1–110:11 (2016). http://iizuka.cs.tsukuba.ac.jp/projects/colorization/en/. Accessed 19 Jan 2020
11 Karras, T., Laine, S., Aila, T.: A style-based generator architecture for generative adversarial networks. arXiv:1812.04948v3 [cs.NE], 29 March 2019. https://arxiv.org/abs/1812.04948. Accessed 19 Jan 2020

Examination of Tourism Promotion Using Online Videos

Shunsuke Aoki(✉) and Takayuki Fujimoto

Graduate School of Information Sciences and Arts, Toyo University, Tokyo, Japan
{s3B102000015, fujimoto}@toyo.jp

Abstract. Inbound tourism business has grown all over the world in recent years, and the competition to attract tourists is becoming more intense. Each country is putting a lot of effort into tourism promotion. Videos, in particular, are often used for these tourism marketing campaigns. Promotional videos from these countries are delivered on the internet. While some of these countries have succeeded in attracting tourists from abroad with effective promotional videos showing their countries' charms, others are not so successful in drawing foreign tourists to countries due to poor video promotion despite a huge amount of money spent on the production. In this thesis, we analyze and compare two types of promotional videos: the ones from countries successfully promoting tourism with them, and the others from countries with poor promotional results.

Keywords: Tourism · Inbound · Promotional video · Foreign tourists · Advertising

1 Introduction

Since 2010, both international tourism revenue and global GDP have been increasing. The total number of international tourist arrivals in 2018 was 1.4 billion, and the total exports from international tourism for the same year was as much as $ 1.7 trillion. By region, South Asia saw the fastest growth with annual average growth rate of 10.5% in the period from 2010 to 2018. There are some factors that contribute to the worldwide growth in inbound tourism such as newly-launched low cost carriers (LCC) and the increase in their flights, diversified travel style, lower travel cost and digitalization in tourism. Many countries are striving to promote inbound tourism, because it leads to job creation, revitalization of rural areas, and economic development and growth.

Different promotional methods are used to attract foreign tourists. These methods include paper media such as guidebooks and brochures, and websites and blogs of travel agencies. Of them, video is a more effective medium than others, and is especially popular as it shows the charms of travel destinations by appealing to both visual and auditory senses of the audience. Promotional videos are thus created all over the world. Furthermore, with the popularity of video sharing websites like YouTube, it has become common to watch videos as a source of information. This is also making videos a popular medium for promotion. As tourism departments of national governments are actively working to promote inbound tourism, different countries create

H. Selvaraj et al. (Eds.): ICSEng 2020, LNNS 182, pp. 314–323, 2021.
https://doi.org/10.1007/978-3-030-65796-3_30

different styles of promotional videos that vary in structure or contents. Also, not few countries make promotional videos to attract repeat visitors to maximize the economic effect of tourism and to attract visitors to rural areas, rather than to draw new foreign tourists.

Amid the current promotion video boom, some countries have been successfully attracting foreign tourists using their videos whereas others attempted to do the same with no luck, failing to convey the region's charms despite a huge amount of money put into the video production.

2 Purpose and Background

With a surge in global tourism business, many countries are running promotional campaigns using video content distribution services on the internet. It would be no exaggeration to say that online PR video is today's most popular way of inbound tourism promotion. Tourism departments of national governments, travel agencies and local governments have created promotional videos, which can be viewed easily on the internet. Many countries and regions have succeeded in attracting foreign tourists with promotional videos. In the past, tourism promotion and publicity used to take a massive budget and a large staff. Now, however, PR brings a greater result with a small budget and little effort by distributing a three-minute video content on social media.

For instance, Tohoku Tourism Promotion Organization, which covers Aomori, Iwate, Miyagi, Akita, Yamagata and Fukushima Prefectures, distributed a promotional video "Autumn Colors in Tohoku, Japan 4K".

(https://www.youtube.com/watch?time_continue=142&v=ngNSWQIXFBc) on YouTube in December 2016. The number of views exceeded 27 million, and the video was mentioned worldwide. Tohoku regional authorities has completed many PR projects for foreign tourists so far. Of them, this can be considered as the most successful case in terms of cost efficiency, resulting in a video that is as short as 3 min and 25 s.

In the past few years, there have been quite active tourism PR efforts using online videos. Despite an increasing number of promotional videos uploaded, some countries and regions are not getting more foreign tourists or even losing them because the videos fail to present the tourist attractions effectively. These are the cases where the video productions led to deficits. Tourism PR videos are intended to make people want to visit the place, so we need to make that particular country or region look more attractive than they really are. In many cases, localities that have few tourist attractions or places of interest can be popular destinations thanks to well-presented promotional videos. Therefore, it is reasonable to assume that what sets the successful countries apart from unsuccessful ones is how well their PR videos are made, not the inherent problem of the localities.

In this research, we focus on the way a tourism promotional video should be because it is now the primary avenue of tourism PR, and thus is the factor separating success from failure in attracting tourists. We will specifically look into the techniques used in internet video promotion, which include the structure of the video content and the methods as well as highlights of content creation. We will specifically look into the techniques used in internet video promotion, which include the structure of the video content and the methods as well as highlights of content creation.

Of all the videos available that are created to promote inbound tourism, we compare and analyze in this research the high-profile videos that successfully attracted foreign tourists, and the ones that did not turn out to be so fruitful in this effort. We will then define more advanced techniques and ideas for promotion. We will also examine the techniques of creating videos for successful inbound tourism promotion, which should be used by travel agencies and local governments looking to attract foreign tourists.

3 The Characteristics of Promotional Videos Created by Countries with Growing Tourism Industries

In this chapter, we take three countries as typical cases of rapid growth in tourism industry, and discuss the characteristics of their tourism promotional videos.

3.1 Iceland

Iceland is one of the countries visited by a great number of foreign tourists in recent years. The number of visitors from abroad in 2018 was four times larger than in 2011 (Fig. 1). The highlights of Iceland are activities for visitors to natural attractions such as huge glaciers and volcanoes, which tourists can get up close.

This time, we analyze the YouTube channel videos by Inspired by Iceland, which promotes inbound tourism of Iceland. Inspired by Iceland has uploaded a total of 206 videos on YouTube since 2011. The videos that get relatively high number of views are from Iceland Academy series (Fig. 4). This series of videos covers different subjects such as "how to enjoy activities" and "things you should not do in Iceland". Most of the videos are catchy and are less than 1.5 min long, and the entire series is provided in English language.

3.2 The Netherlands

The Netherlands is another country where tourism industry is booming. The number of the foreign tourists almost doubled between 2010 and 2018 (Fig. 2). However, Amsterdam, the capital of the Netherlands, now has the problem of over tourism.

The Netherlands Bureau of Tourism and Congresses creates and distributes promotional videos, which can be viewed on their YouTube channel, Holland.com. Of the uploaded promotional videos, the most-viewed one is "Holland. The Original Cool" series (Fig. 4). In this series, a Dutch gentleman introduces the country's attractions other than the common Dutch stereotypes, which foreign people might have, such as windmills and tulips. These videos should help solve the over tourism problem by dispersing visitors away from popular tourist destinations.

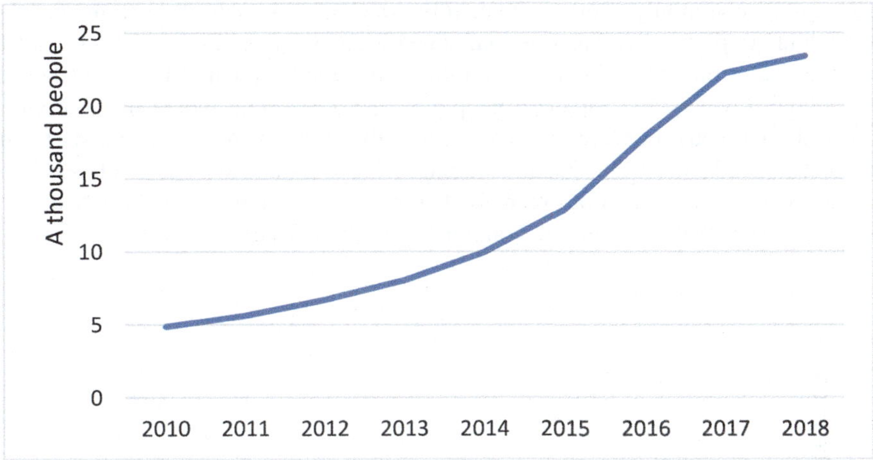

Fig. 1. Iceland's visitor arrivals

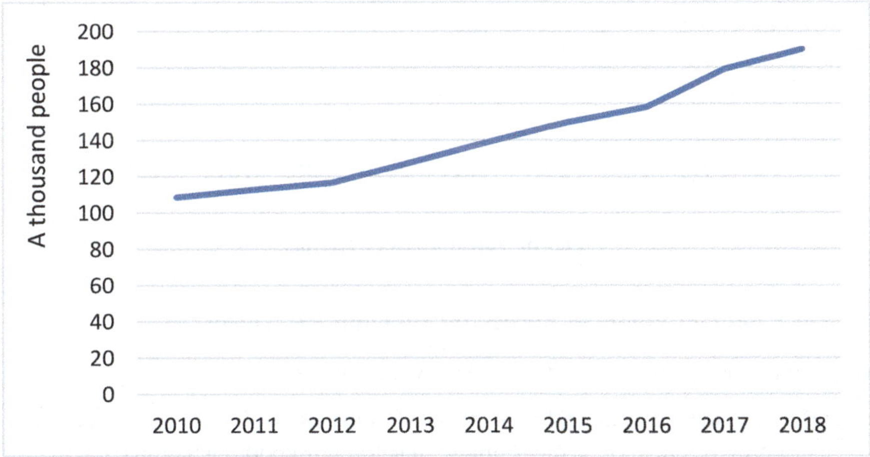

Fig. 2. Netherland's visitor arrivals

3.3 Japan

The number of foreign tourists to Japan has increased between 2011 and 2018 for 7 years in a row (Fig. 3). With major events like the Tokyo 2020 Summer Olympics and Osaka-Kansai World Expo set to take place, Japan expects to see a further increase in the number of international tourists.

Japan National Tourism Organization has uploaded 384 videos since 2015 on Visit Japan, the organization's YouTube channel. These include videos of various forms including 360°VR videos, those with titles translated into different languages, and videos featuring non-Japanese celebrities.

A feature commonly seen in frequently watched videos is the setting in which foreign tourists play a leading role and travel around Japan (Fig. 4). Also, visually-attractive, beautiful videos without narration are quite popular. Interaction between international travelers and Japanese people is also a recurrent theme in these videos. Japan National Tourism Organization is currently working to attract tourists to rural areas - an agenda item that Japan's inbound tourism industry must address. They particularly focus on the promotion of Tohoku region, and are making videos for tourists to discover the charms of Tohoku from different perspectives.

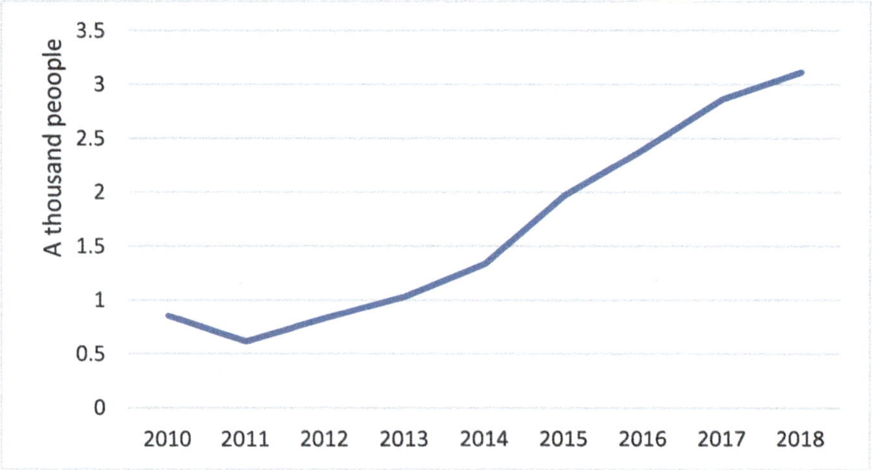

Fig. 3. Japan's visitor arrivals

3.4 The Common Points of Successful Inbound PR Videos

Here are the features of successful tourism videos of three countries.

① The video that appeals the country's charms, which foreign people do not know besides the typical image that foreign people generally have.
② The video that provides specific information about sightseeing spots and activities.
③ The video that enables the viewers to easily imagine traveling in the country by using a foreign tourist as a main character.
④ The promotion video of the local region in the country.
⑤ The video in which foreign tourist interacts with local people.

Regarding the three exampled countries' videos described in previous section, it was rare that the most popular sightseeing places are introduced as a main subject. The unique charms and everyday-scenes of the country are introduced. The unique charms and everyday-scenes of the country are introduced. Globally, tourists have already focused on popular sightseeing places, and there is a problem that those sightseeing places are overwhelmed by a large number of visitors because of the focus. It is considered that the is showing the charms of the local regions and getting viewers' attention with unique things is the best way to attract foreign tourists furthermore. Recently, tourists tend to spend money on "experience" instead of "things." To get attention of viewers and make them want to experience a different culture, it is effective to introduce the life of local people rather than information on the tourist destinations.

Fig. 4. The pictures of the PR videos

4 The Features of the Promotion Videos of the Countries with the Decreased Number of Foreign Tourists

In this chapter, we analyze the features of the videos of the countries where the inbound industry is not growing.

4.1 Panama

Panama is a small country in central America and it is a tourism nation, which has beautiful beaches, casinos and carnival as tourism resources. The rapid growth of the tourism industry has been found until 2014, but the number of foreign tourists decreased after 2015 despite that the tourism business is becoming more and more powerful in the world (Fig. 5). The inbound promotion of Panama has been conducted by Tourism Authority of Panama. Tourism Authority of Panama uploaded 110 videos on their YouTube channel, "Visit Panama."

The popular videos are short movies of a length of 15–16 s. The videos are not for the introduction of specific destinations, instead, the activities that people can enjoy in Panama such as "Nature", "Surf", "Fishing" are featured as theme. Both of English and Spanish versions are uploaded. Although almost all videos consist of beautiful sceneries, the tourism resources unique to Panama are not introduced. The common activities that people can also enjoy in other countries are focused in the videos.

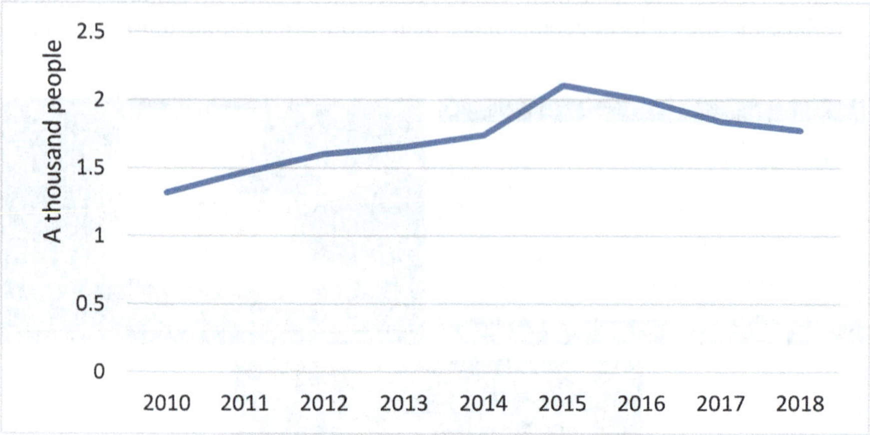

Fig. 5. Panama's visitor arrivals

4.2 Bolivarian Republic of Venezuela

The number of foreign tourists in Venezuela remarkably increased from 2011 to 2013. However, the number decreased for the fourth consecutive year after 2013 (Fig. 6). The one of the reasons for the decrease of foreign tourists is the least safe situation of Venezuela. Venezuela national tourism organization is in the charge of the inbound promotion of Venezuela. Venezuela national tourism organization uploaded 997 videos on their YouTube channel, "Ministerio de Turismo y Comercio Exterior".

Spanish is used as common language for all the videos of this channel. From this, we can figure out the target tourist of Venezuela is the people who speak Spanish. Although the beautiful beaches and nature are introduced in the promotion videos, these attractions look rather common in the world and "difference" from other countries are not emphasized. The positive point in respect of that is the use of Latin American music as background music.

4.3 The Common Points of the PR Videos Regarding the Cases of Declining Inbound Tourism

The following are the features of the typical incomplete video cases regarding two countries exampled above.

① They cannot provide proper amount of information because the length of videos is quite short, they are less than 15 s in most of the cases.
② The target is limited because the language used in the videos is not English.
③ There is no difference from other countries because the sceneries in the videos are not unique tourism resources.

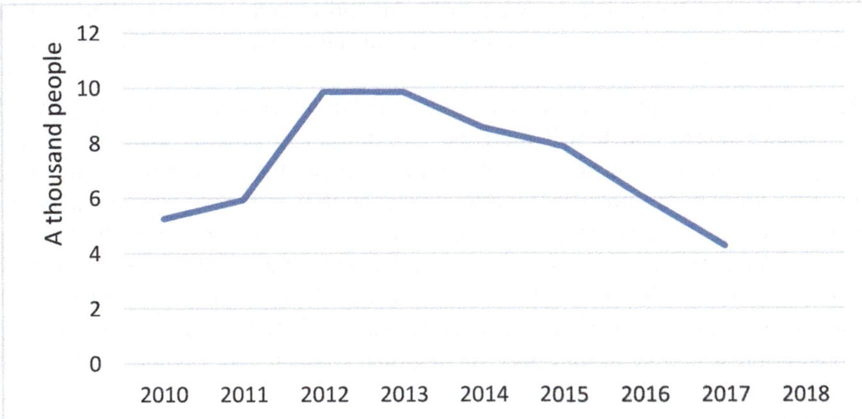

Fig. 6. Venezuela's visitor arrivals

④ The viewer cannot feel the cultures and a variety of lifestyles of the countries because almost all the videos focus on typical tourist destinations.

It is difficult for the viewers to recognize that the video is about Panama or Venezuela when they watch the applied video. The beautiful beaches and wonderful nature are introduced but these are the common popular tourism resources in the world. People can enjoy them in other countries as well. In short, it is the problem that what people can see and experience only in Panama or Venezuela are not introduced. To get more viewers for the videos, the production of video that supports a few or several languages is necessary. Furthermore, the scenes of tourists interacting with local people are rarely used in the videos of the two countries. With regards to Venezuela, there are people who perceive that the country is known as the least safe country in the world. To dispel this negative impression, showing scenes in which people are living in friendly atmosphere can be a strategy.

5 Consideration

The result of the comparison and analysis of the promotion videos from five countries, two effective promotion techniques were found.

The first technique is to introduce information such as the unique cultures, daily lives of local people and the traditional food adding to the popular information that foreign people already know. These days, the tendency of what the tourists spend their money on is gradually changing from "tangible goods" to "experience." People rather spend their money on experiences such as staying at a local guest house and participating in a traditional cooking class than spending on material things like souvenirs. People place their values on different cultural experiences. Showing various information in the promotion videos can give people the start to further research on something with which they were impressed. It is considered that the best length of the video is between 3 min and 4 min to convey a lot of information to the viewers without making them bored. Because they can watch the video of the length feeling like enjoying the music video.

The second technique is to deliver positive points that can cover the negative side of the country image and it will motivate the viewers to actually go to the country. Although Venezuela is known with the impression of the least safe country in the world, if the tourism resources that is attractive enough to make people want to go there in spite of the typical negative perception, the tourists would be more likely to come to Venezuela. As a strategy, it is not effective to introduce the popular tourism resources such as beautiful beaches and hiking trails, because many tourists would preferably choose other resort destinations such as Hawaii and Bali with safer image. The important point would be to introduce the regional characteristics unique to Venezuela for which the viewers can find the unforeseen needs.

6 Summary

With recent energetic tendency of the inbound tourism industry, various tourism promotion videos have been created all over the world. It goes without saying that not only PR videos are the cause for the current prosperity of the tourism industry. To enhance other various factors is also necessary. Especially, it is often the case that the first contact point is online PR videos when people look for or happen to know the tourism information on the Internet. It is also true that detailed information by texts is important, but the intuitiveness of videos can affect the decision making for tourist destinations. Nowadays, the tourism promotion by using PR videos is becoming prevailing. Upon the tendency, a number of tourism videos have been released, but the number of unsuccessful cases are increasing as well. As typical cases, almost all the videos produced so far just partially introduce famous destinations that everyone knows. The possibility of attracting new tourists is quite low.

To survive the tourist enticement competition that is intensified internationally, it is important to introduce the features such as experience and dipping into the local culture, of the presented country and make a lot of fascinating points for viewers by placing the scenes and information that cover various kinds of fields in one video.

Acknowledgement. This work was supported by JSPS KAKENHI Grant Number 17K00730.

References

1. UNWTO World Tourism Organization: International Tourism Highlights 2019 Edition
2. Kureha, M., Funck, C., Arima, T.: Examining the development of inbound tourism in Japan from geographical perspectives. In: Proceedings of the General Meeting of the Association of Japanese Geographers, vol. 2017a, no. 0, p. 100064 (2017)
3. Watanabe, J., Kirimura, T.: An analysis of temporal change of interest in a tourist spot; using SNS's Photograph data. In: Proceedings of the General Meeting of the Association of Japanese Geographers, vol. 2018s, no. 0, p. 000198 (2018)

4. Kawamura, Y.: A study on the technique-rhetorics and effects of commercial films. Cogn. Stud. **14**(3), 409–423 (2007). (in Japanese)
5. Iwasaki, M.: Image of city in promotional video of world cities. In: Summaries of Technical Papers of Annual Meeting, Architectural Institute of Japan EE, pp. 335–336 (2014)
6. Ministry of Land, Infrastructure, Transport and Tourism: The summary of the market of experience-based tourism contents (2019). (in Japanese)
7. ISLANDSSTOFA promote Iceland: Inspired by Iceland, 20 January 2019. https://inspiredbyiceland.com/?fbclid=IwAR1aljaLTkC4M7UDfCB4rguXWLGCrKUxfhiC_yDKhkl0q6RHZiye_k81vc0
8. Icelandic Tourist Board: Foreign visitors to Iceland 1949–2019
9. The Netherlands National Tourism Organization: Holland.com, 21 January 2020. https://www.nbtc.nl
10. CEIC Data: Netherlands NL: International Tourism: Number of Arrivals, 21 January 2020. https://www.ceicdata.com/en/netherlands/tourism-statistics/nl-international-tourism-number-of-arrivals
11. Japan National Tourism Organization: The statistical date of visitor arrivals, 01 February 2020. https://www.jnto.go.jp/jpn/index.html
12. Japan National Tourism Organization: Japan National Tourism Organization Web Site, 02 February 2020. https://www.japan.travel/en/
13. CEIC Data: Venezuela Tourism Statistics, 14 February 2020. https://www.ceicdata.com/en/venezuela/tourism-statistics
14. Tourism authority of Panama: Visit Panama, 04 February 2020. https://www.visitpanama.com
15. CEIC Data: Panama Tourism Statistics, 06 February 2020. https://www.ceicdata.com/en/panama/tourism-statistics
16. Ministerio de Turismo y Comercio Exterior: YouTube channel 「Ministerio de Turismo y Comercio Exterior [Spanish]」, 05 May 2020. https://www.youtube.com/user/minturvenezuela/videos

SNS Application with "Peeping Function" to Prevent Information Leakage on Social Media

Taishi Nemoto(✉) and Takayuki Fujimoto

Graduate School of Information Sciences and Arts, Toyo University, Tokyo, Japan
nemotouc@gmail.com, fujimoto@toyo.jp

Abstract. Negative perception people have about internet issues such as information leakage and online slander is deep-rooted. These incidents are caused by casual communication through media like SNS, rather than the Internet itself, which can be viewed by much more users than the author expects. One example is Internet flaming. In order to avoid such troubles, you should be aware that you are being watched online and simply stop posting what you do not want leaked. An ideal solution to this problem would be an application that allows you to peek at social media posts created by any users. In this paper, based on the fairy tale, The Little Match Girl, we devise a system to "peep" the posts only while the match is burning. The "peeper" purchases a match as a token of the right to keep it burning for tens of seconds, and to the "peeped", selling matches proves that they are free from a sense of guilt. The number of matches sold means the number of their followers on social media. The number of matches serves as the measure of popularity, and the matches are designed to bring benefit to both sides. In this research, we propose a new social media application that is designed to satisfy the peeped users' esteem needs, while allowing "peeping" as part of entertainment.

Keywords: Social media · SNS · Information · Smartphone · Messenger · Online communication

1 Introduction

Social networking service (SNS) as a communication tool has become popular in recent years. SNS and messenger applications enable an easier and faster communication than email. Using them is now an established way of communication, transforming our lifestyle. It is no exaggeration to say that they are the easiest and most accessible communication tools in use today. On the other hand, there are many problems regarding communication in modern society that derive from SNS. Being easier to use, SNS is causing more troubles than email. For example, cyberbullying via Internet often starts on SNS. Incidents of personal and confidential information leakage caused by SNS have also increased rapidly in recent years. The contributing factors of troubles with SNS may be its ease of use and shortness of text. Because the tool is so handy and accessible, we are apt to forget the fact that we are using existing communication

H. Selvaraj et al. (Eds.): ICSEng 2020, LNNS 182, pp. 324–331, 2021.
https://doi.org/10.1007/978-3-030-65796-3_31

methods just like letters and emails. It is easy for us to fall into the illusion that we are 'whispering' to our friends, making it difficult for us to be fully aware of the danger that written text stays and can be easily viewed by others. For this reason, not a few users keep, record and share on SNS the information that they should not leak. It is likely that an SNS message which was supposed to be sent to a specific person as a private message is forwarded or copied repeatedly and thus instantly spreads worldwide. With a group chat, the risk would increase dramatically.

We feel SNS is easy to use and accessible because we are under an illusion that a third party never gets access to the tool. Needless to say, that is theoretically impossible. Of all the existing means of communication, it is a system with rather weak security and poor confidentiality management that allows for human errors. A tool built without a sense of crisis over the possibility of a third party looking at the content has higher confidentiality risk. In that sense, in order to avoid today's SNS troubles, people should be acutely aware of such risk. Therefore, in this research, we design a system that helps users realize the risk of being watched by a third party. However, a method like monitoring system would require an administrator, which would discourage the use of the tool, and is not suitable for communication. Therefore, in the system proposed in this research, we achieve the situation - "a third party person watching you" - by incorporating such function as part of the entertainment.

2 Purpose

2.1 Purpose of Research

The fact that the interactions on SNS is "viewed by a third party" means a "risk" from the user's point of view while it can be "entertainment" from the viewer's. It is annoying to have one's personal communication seen by a third party while, as a third party, the act of peeping another person's communication can be fascinating. It will be more so when the communication is of a celebrity.

Based on this assumption, we design a system for avoiding SNS-related troubles while aiming at entertainment. This chapter describes the concept of the Match Girl Communication System designed in this study.

2.2 Design of Match Grill Communication System

The Match Girl Communication System is an SNS application with a function that allows users to peek at other people's communication on SNS without their knowledge. In other words, users must always be aware of the risk of being watched by someone when communicating. The Match Girl Communication System is based on the fairy tale, The Little Match Girl. In the story of "The Little Match Girl", the girl can see a dim vision of her dreams in the flame of the matches only while they are burning. Each match burns for about 25 s. "The Little Match Girl" was able to dream each dream for 25 s. People can barely see something well in just 25 s, but it can be long enough to focus on one particular point and watch it intently. Match Girl Communication System is based on these episodes.

With the System, purchasing a match allows you to peek at any private communication on SNS for only 25 s. Just like real matches, the flame dimly projects the image of the area just around the match on the screen. The image may not be vividly clear but should be legible (Fig. 1).

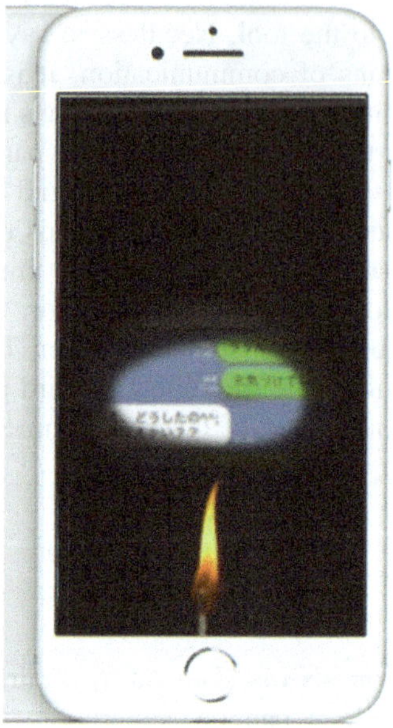

Fig. 1. Illustration of App

Matches that allow viewers to peek at other people's interactions are priced and sold by application users. For example, a celebrity could sell his/her matches for $100 or $1000 each. From unknown users or those with inactive accounts, no one would want to buy even for 1 cent. Of course, not a few users, even if they were not famous, would want to sell matches to those who want to peek at interactions of their friends and business colleagues.

However, in order to prevent peeping from going beyond entertainment and giving pressure on the peeped, the following restrictions are applied to "peeping".

(Restriction 1) A user can purchase up to 12 matches per account per year.
(Restriction 2) A user must allow an interval of one month or longer before viewing again the timeline once peeped.
(Restriction 3) A user whose interactions were peeped will not be informed of the details of peeping activity including who and when they were peeped.

The reason for limiting the number of matches to be purchased to 12 per year per account is that we assume viewers would feel an urge to try once a month. Also, from the viewpoint of the peeped, an activity that happens only once a month should be tolerable. An interval of more than one month makes it difficult for the peepers to collect detailed information by continuous watching because, in general, topics are expected to change significantly after one month. In addition, strategic plan for using the twelve matches and a sales promotion strategy for selling matches to an unspecified number of users are also important. Huge profits can be achieved.

On the other hand, all participating users face the risk of being watched by the third party. Therefore, they would refrain from exchanging personal information, confidential information, illegal exchanges or embarrassing information that they do not want the third party to know about, which can deter SNS-related troubles.

The Match Girl Communication System is a completely new entertainment SNS that enables "peeping" entertainment and monetization of the activity, while also integrating with them a deterrent of SNS-related troubles.

3 Precedent Research

We will discuss three related applications below. They have similarities in specifications, time limits, and self-exposing nature.

One similar research is on a communication application "peep", which uses blurry photos. The application's effect of gradually making blurred images clear when the screen is rubbed stirs users' desire to see the pictures. However, it is also a problem that the blurred image reveals itself when rubbed in that the users eventually get the whole picture. This means it ends up in an ordinary SNS with which users can get information effortlessly.

"Ephemeral SNS," where posts disappear after a certain period of time, is gaining popularity as a function that encourages casual posting. A typical example is Instagram Stories, a system in which short movies of a few seconds disappear automatically after 24 h. It can be described as a preferred SNS over other sites among users who cannot help watching its movies before they disappear.

Another example is "Omoide Nozokimado (a viewing window for memories)", which uses AR technology to share landscape pictures from the past and enable users to have their own images viewed. This makes it possible to leave the image a user have seen in the past in that same location. In this way, "peeping" is not necessarily a negative act.

4 Mechanism

In the application, a "match" is needed to peek at another person's timeline. Users buy matches and use them for the users whom they are interested in. Therefore, a user who makes people use a lot of matches becomes a popular user, but on the other hand, the people who purchase matches for negative purposes are also assumed. The sales of the matches will be split between the service provider and the user. Even if a certain user is peeped for negative purposes, the peeping is given back in the form of sales. For this mechanism, in "Match Girl Communication System", the more a user's content of SNS and messages that a user usually do not want to make public are seen by other users, the more he or she can improve the status and profit. From the different point of view, with this mechanism, all users are inevitable to have a sense of danger that "someone may be watching over me" by default, and the consciousness not to disseminate unnecessary information on SNS, specifically, the consciousness not to allow other users to peep something swill result in the improvement of the whole users' media literacy. In a real society, we always live by restraining expressions and remarks, thinking that 'someone may be overhearing us,' but this care tends to be forgot on SNS and messenger applications. Celebrities and public figures are often criticized for making slanderous and discriminatory statements on SNS that they do not make on the general media. With this system, it is thought to be possible to reduce the number of slander and personal information leakage on the Internet.

5 System Design

We incorporate the design image of the 'match-selling girl' into the system. Only while the match is burning, you can browse the timeline of the account that you want to see to a limited extent. The Match Girl Communication System adopts the motif of the vision in the match's flame, and for the reality, it also reproduces the 'match's features': the limited lit area, the appearance of the fire, and the fragility for which the fire disappears when any vibration is applied. If you move the match too much because you want to see wider range, the flame will be weaken fast and goes out.

The following specifications are implemented on the user screen of the peeper.

① While the match is burning, the surrounding area becomes bright and you can see the applied account user's interactions.
② The match is lit for about 25 s.
③ Moving the match will brighten another area, which you can look into.
④ If you move the match too much, the counting speed will be fast, and the time for which you can browse the specific area will be decreased.

The following is a screen image (Fig. 2).

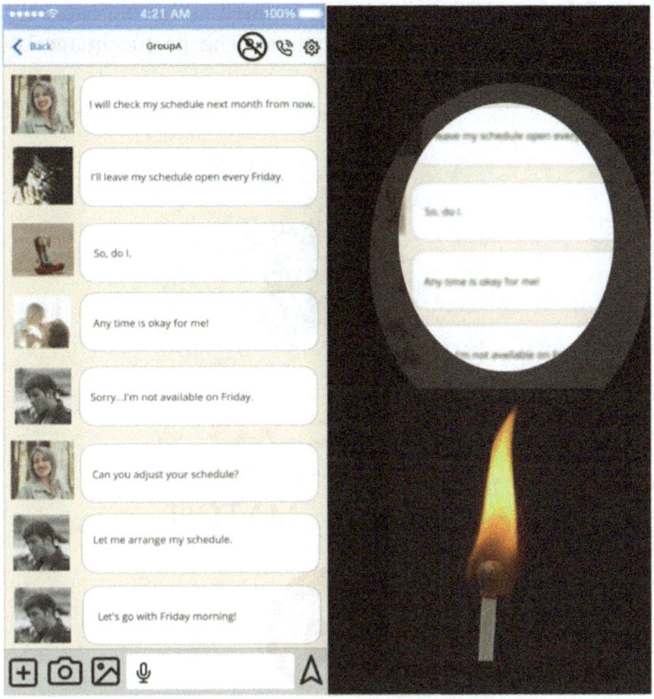

Fig. 2. Peeped screen and peeper screen

The lit area is narrow and blurry, but the content and atmosphere can be seen slightly. In fact, because each area is fragmented and it is necessary to purchase another match again to peek more. The purchasing power depends on how much you want to look into the areas.

There are four steps for how to use the system as the "peeper."

1. Select the user that you want to peep.
2. Purchase the matches required for peeping.
3. With match's flame, light the dark screen that appears randomly.
4. The screen that you have targeted for peeping will disappear in about 25 s, and then you will transit to the initial screen.

On the other hand, there are five steps for using the system as the peeper.

1. You need to agree that personal information could be leaked.
2. Set the maximum number of the matches that an other user can buy to peep your account's interaction.
3. Online communication with the registered users
4. Other users buy matches and you get looked into the conversation at random.
5. The number of the purchased matches is counted and recorded on the status screen of the peeped user.

The number of the purchased matches is counted and it indicates how many times the applied user have been peeped with interest. The profile image is shown below (Fig. 3).

Fig. 3. User profile image

6 Conclusion

Through this study, we have created a system that allows the users to peek by default from two viewpoints: 'demand for peeking' and 'prevention of personal information leakage.' Users who are looked into the SNS content are the users that attract a lot of users, and by focusing the action, 'peeking', we aimed to enhance the awareness of being watched. The design purpose is a mechanism upon which users do not post slanderous content or content that they do not want to get out to. Making good use of the people's peeking curiosity not only about celebrities but also about all those who are included in human relationship, we devised the system. The period in which people have wanted to connect with many people has gone, and now the use of just following people whom each user is interested in, is becoming widespread. The connection with which posts can be sometimes viewed when you want to catch up or peek at would be suitable for SNS communication. As a matter of course, some users would use the

'Match Girl Communication System' for frequent peeping, but the nature of the peeping under the condition that both of the "peeper" and the "peeped" interests coincide is very different from the sticky activities on regular SNS, which demand extremely frequent a target's reactions.

7 Future Work

As Future work, the implementation experiments are required. Aside from whether this system actually makes a business as entertainment, it is necessary to clarify a numerical value how many users actually become careful when posting on the SNS, namely, this application, which has the risk of personal information or conversations leakage by default. Moreover, it is required to prove that the number of the used matches works as a status or a popularity barometer. We will also confirm that the match's burning time specified by limited seconds is appropriate for peeping.

Acknowledgement. This work was supported by JSPS KAKENHI Grant Number 17K00730.

References

1. Ishii, K.: The "Strong-tied" SNS and "Weak-tied" SNS: a comparison regarding the disclosure of personal information and personal relationships. JSICR **29**(3), 25–36 (2011)
2. Niedzviecki, H.: The Peep Diaries: How We're Learning to Love Watching Ourselves and Our Neighbors. City Lights Publishers, San Francisco (2009)
3. Raiman, L., Antbring, R., Mahmood, A.: WhatsApp messenger as a tool to supplement medical education for medical students on clinical attachment. BMC Med. Educ. **17**(1), 1–9 (2017)
4. Schultz, F., Utz, S., Göritz, A.: Is the medium the message? Perceptions of and reactions to crisis communication via twitter, blogs and traditional media. Public Relat. Rev. **37**(1), 20–27 (2011)
5. Andersen, H.C.: The Little Match Girl. Dreamscape Media, LLC, Holland (2017)
6. Amâncio, M.: "Put it in your Story": Digital Storytelling in Instagram and Snapchat Stories (2017)
7. Fujimoto, T.: Ideology of AoD: analog on digital-operating digitized objects and experiences with analog-like approach. In: 7th International Congress on Advanced Applied Informatics (IIAI-AAI). IEEE (2018)

Proposal and Prototype of Video Playback Method Based on Walking Speed

Tianyu Zhu(✉), Takayuki Fujimoto, and Ziran Fan

Graduate School of Information Sciences and Arts,
Toyo University, Tokyo, Japan
cjtjl996@gmail.com, fujimoto@toyo.jp, swterc@gmail.com

Abstract. This paper proposes a new video viewing style by designing a video control application based on a user's 'actual walking.' Nowadays video players on smartphones are generally controlled by users' gesture operations. However, this is just the operation to play videos and users cannot interact with the video control reflecting other information. Through this research, we design a video player application controlled by speed that a user actually walks holding a smartphone, instead of the gesture operations on the screen.

Keywords: Player · iOS development · GPS sensor · Watching videos

1 Introduction

1.1 Purpose

Smartphones are devices that are mainly controlled by users' gestures operations such as swiping and tapping on the screens. It goes without saying that smartphones these days are equipped with various functions and they can be controlled by corresponding ways. For example, iPhone implements various sensors such as proximity sensor, ambient light sensor, gyroscope and accelerometer sensor and each enables different kind of control. However, at present, these sensors are rarely used to control the device. They are often used to detect information for location confirmation, pedometer, health management, game control and so on. For example, watching video is one of the main purpose of the smartphone use. While smartphones are used in various ways, regarding watching video, the user operations to control video playback are only limited to the gesture operations.

Therefore this research proposes a new video watching method controlled by the movements and the speed of the user, who walks holding the smartphone. The purpose of this research is not just to propose the new way to watch videos on smartphones. Utilizing walking speed to video playback control, this research ultimately demonstrate the possibility of new video representation, which is different from previous method of watching video: to gaze the screen standing/sitting still. Watching video is a "static entertainment" that seldom requires the subject's motions. This research undermines that assumption by proposing the video watching way in dynamic. Specifically, it

H. Selvaraj et al. (Eds.): ICSEng 2020, LNNS 182, pp. 332–341, 2021.
https://doi.org/10.1007/978-3-030-65796-3_32

presents the possibility of playback by the motions of the users who actually move their bodies. According to the concept, we design the application that enables sensory user experience based on his or her actual movements, and the experience is different from the interaction by gesture operations on the screen.

The significance of this research is indicated by the following procedures:

1. Examination of the video watching method that introduces actual motions
2. Designing content for reliving experience with physical sensation
3. Offering the new possibilities that represent the sensory experience in digital

Today, many kinds of content that represent activities in the real life, are provided. For example there are video works such as mountain climbing, temple viewing, scenic spot viewing. However, with those "static entertainment videos," user cannot experience the physical sensation of climbing the mountain, walking through the temple, or viewing the scenic spot. Their experience is just limited to viewing the screen. As a solution, this research aims at providing the video method that viewer can relive the experience that video content represents, with actual feeling of motions: walking. This idea is different from VR that changes the point of view, or, AR that merges videos or content and real space. It can be said that this conception is very meaningful and innovative in respect of making a sensory experience out of digital content.

1.2 Background

With the development of the computer's performance and technologies in recent years, now VR videos and 360-degree videos, which require high-speed processing are common in our life. For example, speaking of video content service, we can image YouTube. It has 1 billion users and enables everyone to easily deliver live-streaming. Other than the ease of use, it has the complete video controlling functions: adjusting the playback rate and video quality, auto-generated subtitles, and auto-translation by speech recognition. Recently, as represented by 8K display resolution, the quality of video has been actually increasing. User can also easily access the high-performance video player such as VLC player, which supports all kinds of devices and has various functions: adding or adjusting the subtitles, playing the 360-degree video, and controlling network stream. However, those video services and video player applications commonly use the conventional video playback way by default. Although they enables advanced performance, but they do not provide the new possibility of video expression. Even the 'play' button is not placed on the system proposed in this research. It can free users' hands from operations and it provides a new video watching experience by controlling the playback with user's walking actions instead of gesture operations on the screen (Figs. 1 and 2).

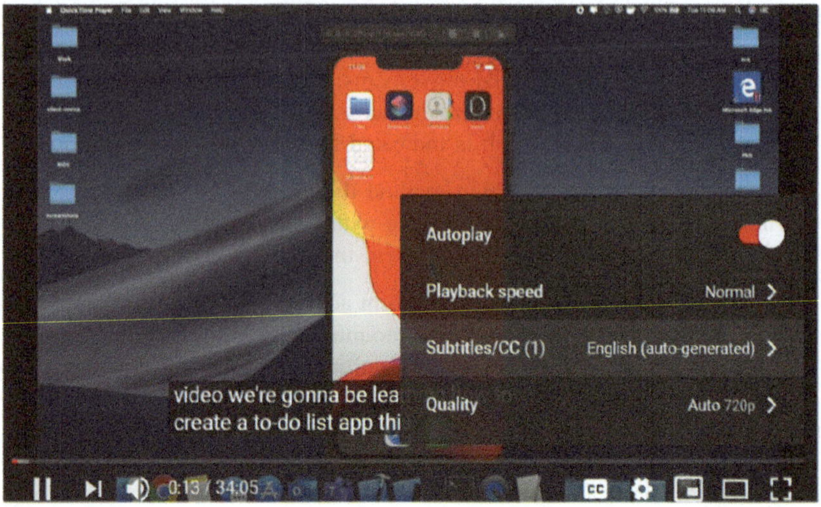

Fig. 1. Youtube

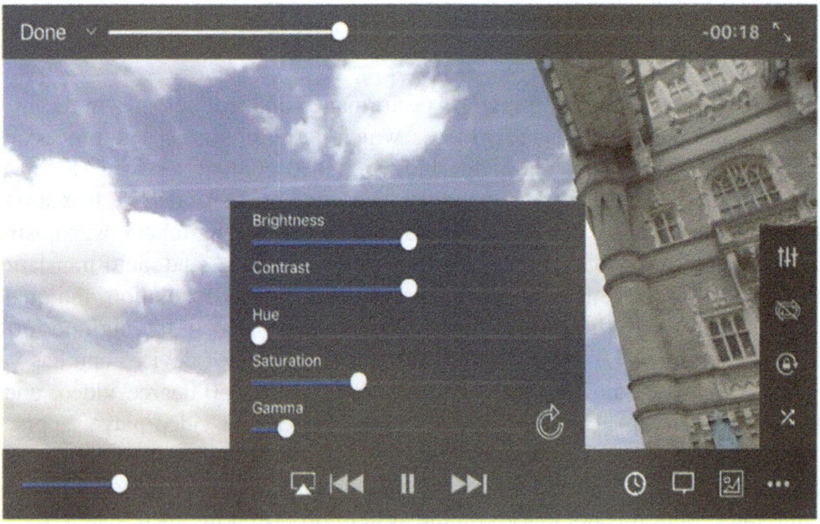

Fig. 2. VLCPlayer

2 Related Studies

There are various researches regarding video playback, but most of them are the methods to display or represent the videos such as VR videos and 360-degree videos. In recent years, 4D film in which the fourth element is added to 3D film has appeared. As the examples of 4D film systems, there are 4DX developed by CJ 4DPLEX company and D-box developed by D-Box company. Upon those systems, the seats in

Fig. 3. D-BOX

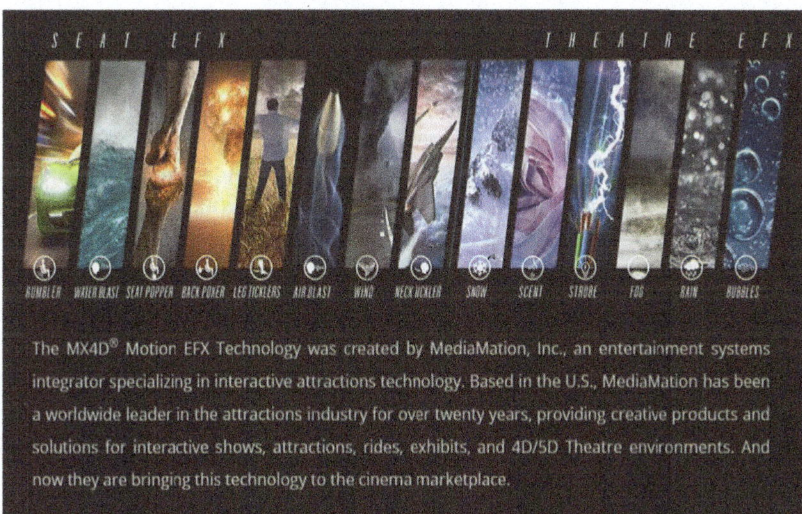

Fig. 4. MX4D

theater can be shaked with the film and voice. Audience can feel as if they were just like in the film. In addition, MX4D developed by MediaMation company does not only shake the seat, but also add scents, and other sensation such as water aerosol distribution as fog (Figs. 3 and 4).

The tries to add different sense besides visual and auditory sensation regarding video have been widespread since the 1980s. But these are limited to the entertainment in amusement parks and it is just one of the displaying methods that enable people to further enjoy the videos. Considering the concept of 4D video, it is easy to add the fourth element to 3D videos. However, because the variety of elements such as vibration, smell, water, and wind is limited to the factors that people can easily list, and as a result, 4D video's expression ability is also restricted.

'Dynamic video viewing' aimed at by this research is designed to be based on the user's action. The 'dynamic video viewing' is superior to 4D video viewing in respect of sensory experience. Because a user can link his or her own action in spontaneity to how the video is played. The proposed video application is different from 4D video system with which people are influenced by external elements such as vibration, smell, water and wind. On the other hand, the popularity of 4D film viewing indicates that people need new way to watch video and this need also links to our new design concept based on the actual actions and sensation.

Therefore this research proposes a new technique that can change the style of watching video by incorporating a new video control function; providing a new way of controlling the video playback, instead of just adding the fourth element. As far as the authors know, there is no other concept like our proposal and therefore this research has high novelty.

3 Video Playback Control by Walking Speed

3.1 Application Design

The proposed video player application was developed for iOS using Xcode. This application can detect user's walking speed to control the video playback. Specifically, it is based on the AVFoundation and CoreLocation framework; AVfoundation is used to control video playback and CoreLocation is used to detect the user's walking speed. Figure 5 shows the image of application use.

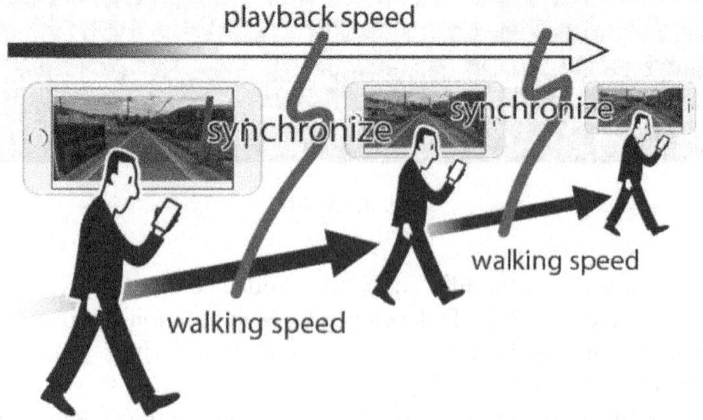

Fig. 5. How to use application

Figure 6 shows the execution screen of the application. As shown in Fig. 5, this application is not like a conventional media player. It does not have even the 'play button' and 'playback rate button.' The video playback rate automatically changes by user's walking speed. Also, when a user stops walking, the video will stop playing. If you walk faster than the provided video, it will be played at double-speed, and if you walk slowly, it will be played in slow-motion.

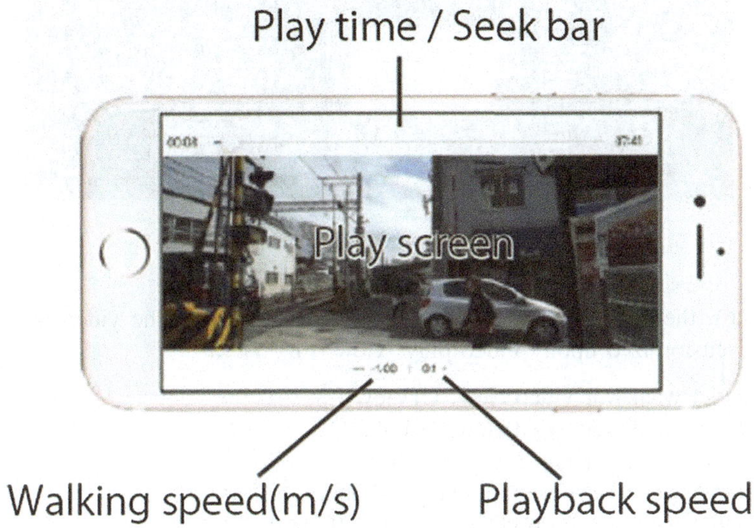

Fig. 6. Application screen

3.2 Process of Application Development

Mapping
Regarding this application, the UI widget outlets on the storyboard and code in Swift need to be connected for the following items.

Playback rate(label)
Walking speed(label)
Video play(view)
Current playback time(label)
Playback duration time(label)
Time slider(slider)
Time slider change(slider)

Regarding connect for the label, properties need to be declared. When data is acquired, properties are changed accordingly and the data will be shown on the screen. It is same to the slider, declaration of the properties is necessary at first. Slider's position changes by video playback, and therefore method also needs to be declared at

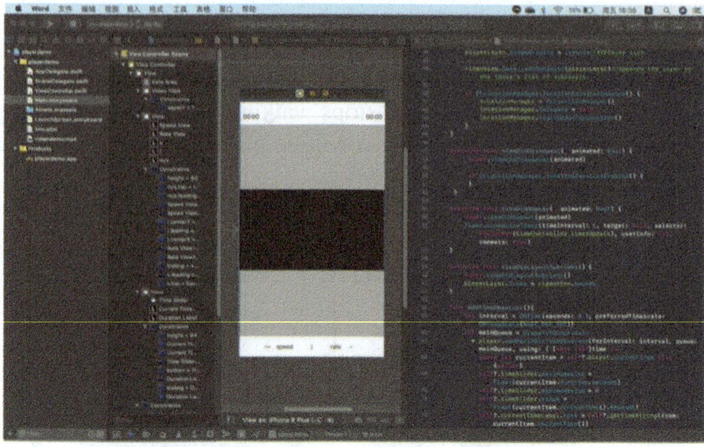

Fig. 7. Application design screen in Xcode

first. At last, the outlet created in storyboard is connected to the video that was initialized or customized upon 'Video play' view (Fig. 7).

```
@IBOutlet weak var videoView: UIView!
@IBOutlet weak var speedView: UILabel!
@IBOutlet weak var rateView: UILabel!
@IBOutlet weak var durationLabel: UILabel!
@IBOutlet weak var currentTimeLabel: UILabel!
@IBOutlet weak var timeSlider: UISlider!
@IBAction func sliderValueChanged(_ sender: UISlider).
```

Detection of Walking Speed

Here is the code to detect a user's walking speed. The initialization code is omitted.

```
func locationManager(_ manager: CLLocationManager, didUpdateLocations
locations: [CLLocation]) {
guard let newLocation = locations.last, //initialization
CLLocationCoordinate2DIsValid(newLocation.coordinate) else {
self.speedView.text = "Error"//when get failed throw error
return
}
self.speedView.text = "".appendingFormat("%.2f", newLoca-
tion.speed)//change the format to show in the view
speed = newLocation.speed// backup the data
}
```

To detect user's walking speed, we must use CoreLocation framework. CoreLocation framework is prepared and provided by Apple Inc. The framework operates location information related to various sensors built into the iPhone.

Before use this framework, it must request authorization and the system prompts the user to grant or deny the request. To set authorization, it must initialize the prompt in Info.plist which control privacy and add the notification sentence "Location When In Usage Description" To use the framework, various code is necessary to initialize that. For example, when a user clicks the application icon and before the application view shows, the location service will start. When the application is put into the background, the location service will stop. At last, walking speed is acquired according to the code and it is saved for backup.

Playback Rate Setting

Here is the code for playback rate setting.

```
Timer.scheduledTimer(timeInterval: 1, target: self, selector: #selector(View-
Controller.timerUpdate)      //add timer
@objc func timerUpdate(){       //timerUpdate function
self.transform()                //run transform function
player.rate = self.rate         //change the playback rate
rateView.text = "\(self.rate)" //show playback rate in view
}
func transform(){
switch speed {
case -1 : rate = 0.1
case 0 : rate = 0
case 0..<0.5 : rate = 0.5
case 0.5..<1 :rate = 0.75
case 1..<1.2 :rate = 1
case 1.2..<1.5:rate = 1.25
case 1.5..<2:rate = 1.5
case 2..<100:rate = 2
default :rate = 0
  }
}
```

Walking speed is updated every second, therefore you need to add a timer here to run the function every second. With the setting according to the above code, the function called 'timerUpate' works every second.

Regarding the timerUpdate function, specifically there are three functions to be run every second: (1) conversion from walking speed to playback rate (2) change of playback rate (3) indication of the playback rate in view.

Finally, we also refer to the code of conversion from walking speed to playback rate. The correspondence between walking speed and video playback rate is determined based on the general human walking speed of 1 to 1.2 m/s (Fig. 8).

Speed m/s	Play rate x
-1(default)	0.1
0	0
0~0.5	0.5
0.5~1	0.75
1~1.2	1
1.2~1.5	1.25
1.5~2	1.5
2~100	2
other	0

Fig. 8. Correspondence between walking speed and playback rate

4 Conclusion and Future Work

In this paper, we developed an application to control video playback by walking speed and considered a new expression of video entertainment. Since the spread of 3D videos, the video expression has been developed by updating the image quality. However it is still stagnating as a mean of expression. 4D film only improved the reality by the techniques, which has existed since 1980s. Nowadays, videos are basically controlled by 'play button', 'back button', 'fast forward button', and 'stop button.' It is also common to control playback rate by seek bar. There is no different way of controlling other than them basically.

Regarding the proposed system, a user can watch videos while walking and it is possible to control playback rate according to the walking speed. This brings a new possibility for future video expression.

As the future task, playback direction and its control are considered. Currently, playback forward controlled by walking speed is achieved. We will also develop 'reverse' function. For example, by user's walking backwards or changing to a certain direction, backward playback will be started.

Acknowledgment. This work was supported by Mitsubishi UFJ Charitable Trust Fund for Ohata Kinen Research.

References

1. Chen, S.E.: QuickTime VR – an image-based approach to virtual environment navigation. Apple Computer, Inc. (1995)
2. Ito, E., Fujita, K., Fujimoto, T.: A proposal of communication-type broadcasting system which can share pictures in real time. Hum. Comput. Interact. **142**(14), 1–6 (2011)
3. Miyawaki, T., Kikuchi, T.: Possibility of the video expression by 360-degrees video. Technical Report of the Institute of Image Information and Television Engineers, vol. 40, no. 11, pp. 165–167 (2016)
4. Apple Developer Documentation, CoreLocation. https://developer.apple.com/documentation/corelocation/cllocation
5. Apple Developer Documentation, Avfoundation. https://developer.apple.com/documentation/avfoundation
6. Fujimoto, T., Fujita, K.: Development of scale processing method to fuse background image and real time video. IEICE Technical Report, vol. 113, no. 473, pp. 311–314 (2014)
7. Nishida, M.: Why we should separate VR video and 360-degree video (2020). https://www.itmedia.co.jp/news/articles/2002/17/news032.html
8. Fujimoto, T., Ogawa, R.: The possibility of 3D-origmi system as a tourism promotion tool for Japan. J. Adv. Res. Soc. Sci. Hum. **3**(1), 1–10
9. Fan, Z., Fujimoto, T.: Proposal of a 'Disposable Camera' app providing sensory reality of analog operation on smartphone. In: 2nd International Conference on Interaction Design and Digital Creation/Computing (2019)
10. Oh, E., Lee, M., Lee, S.: How 4D effects cause different types of presence experience. In: Proceedings of the 10th International Conference on Virtual Reality Continuum and Its Applications in Industry, pp. 375–378.

A Research of the Utilization of an Educational Space as an Advertising Medium

Kanata Itoh(✉) and Takayuki Fujimoto

Graduate School of Information Sciences and Arts, Toyo University, Saitama, Japan
{s3B102000037, fujimoto}@toyo.jp

Abstract. In recent years, advertisements via the Internet have become widespread, and the balance of the 4-mainstream media as television, radio, magazines, and newspapers has been severely disrupted, which has led a major impact on the advertising industry. As "Boba tea movement" in the food service industry showed, such a suddenly global boom has been expanding in recent years due to the information diffusion by SNS. In the present SNS age, advertisements "diffused by private media" via SNS centered on young people have a great influence instead of traditional "mass media" such as television and magazines. This study focuses on the consumption activities linked to the youth's SNS diffusion power and examines the utilization of educational spaces such as universities as an advertising medium. The purpose of this research is to identify the educational space, such as "university lecture" which attracts these young people without exception and regards it as an "advertising medium", and to clarify its potential.

Keywords: Advertisements · SNS · Information · University lecture · Private media

1 Introduction

This study attempts to explore the possibility of utilizing an educational space as an advertising medium and clarify the merits and demerits along with the optimum solution in treating the classroom as an advertising medium.

2 Background

In recent years, advertisements via the Internet have become widespread, and the balance of the 4-mainstream media as television, radio, magazines, and newspapers has been severely disrupted which has led a major impact on the advertising industry. As the comparison of advertising costs of each medium shows; Internet advertising has maintained a growth rate more than 10% every year, and the advertising costs that were hundreds of billions of years ago are now have grown to 1.5 trillion yen, making it realistic to exceed TV advertising.

H. Selvaraj et al. (Eds.): ICSEng 2020, LNNS 182, pp. 342–347, 2021.
https://doi.org/10.1007/978-3-030-65796-3_33

As “Boba tea movement” in the food service industry showed, such a sudden global boom has been increasing in recent years due to the information diffusion by SNS. A lot of “Fashionable photos” taken by consumers are posted on SNS and attracted public attention that would be more influential than advertisements created by companies at high costs. In present SNS age, advertisements “diffused by private media” via SNS centered on young people have a great influence instead of traditional “mass media” such as television and magazines.

3 Purpose

This study focuses on the consumption activities linked to the youth’s SNS diffusion power and examines the utilization of educational spaces such as universities as an advertising medium. Needless to say, the educational space such as universities where young university students gather most of the day is also the space, and they pay attention to the information provided there. The purpose of this research is to identify the educational space, such as “university lecture” which attracts these young people without exception and regards it as “advertising medium”, and to clarify its potential.

Regarding the use of advertising media in the educational space, traditional or old methods such as “distributing leaflets during class” and “announcement by a business operator or an office worker who suddenly comes in during class” have been used. Therefore, the idea of using the educational space as an advertising medium is not new. In addition, various ways have been already conceived.

However, many conventional methods, albeit trivial, have been established by interrupting lessons, which is not only an ethical issue, but also “disconnection of lessons and advertisements”. It can be easily assumed that the interest of “classroom = advertising” will be lost at the moment of “discontinuation”.

Therefore, in this research, we examine a seamless “educational space-time advertising medium” through university lessons, which is premised on the use of presentation tools such as Microsoft PowerPoint. The medium of “classroom” used in this study is different from conventional mediums which are actively contacted by consumers, such as the 4 mainstream media and the Internet mentioned above. It is considered by applying the time during lectures which is with a high degree of attention and coerciveness. This way is thought as the same as the type of advertisement like outdoor advertisements and traffic advertisements.

Moreover, “classroom” is a medium that enhances the impression by increasing the number of contact points with students, like outdoor advertisements and traffic advertisements. Since advertisements are streamed during lectures, efficient and repeated contact can be achieved. Furthermore, the practical application is predicated on the particular age distribution data of users concentrated on the age to take university lessons.

4 Utilized Method of Advertisements in Educational Space

This chapter describes what kind of advertising can be effectively used in the educational space. In this study, the advertisements in educational space is apart from the advertisements outside the school hours and classrooms such as school broadcast, the solicitation of a driving school in front of the university co-op and travel guide for members of the university student association, etc. We aim to integrate advertisements and lessons using presentation tools like Microsoft PowerPoint.

4.1 Method to Play Advertisement Videos Before and After Lessons

Against the noise through preparation and the talks of students, playing advertisements before and after lessons would be expected as a lubricating oil that can smoothly start lessons. It is possible to play advertisement video through the lesson, however, it is concerned to be a frustration that the advertisements would encumber both professors and students when they concentrate on the lesson.

For example, playing advertisements in the half of the time of the lesson (generally 45 min of a 90 min lesson) could not be a practical way because it would be a conflict between advertising and class works such as test and writing tasks.

4.2 Overlay Advertisement Displayed During Lessons

The advertisement used in this case is not a video, but an indelible overlay advertisement that always displays the image of products or services and the descriptions of those which client want to advertise during the lesson.

Although the method mentioned in 4.1 is a short video which takes 30 s, students may think that they do not pay tuition fees for advertisements since the video occupies a part of lessons. On the other hand, overlay advertisements is always displayed during the lesson so that it does not interrupt the lesson. In this way, the concern mentioned above could be solved.

As Fig. 1 shows, advertisements are placed in the edge of a document with the picture and the description of products. The QR code would also be displayed so that students can access to Amazon to purchase them by scanning. However, the problem would be pointed out that it would be extremely rude to the professor who is giving a lecture when many students point their smartphones to the QR code which is not relevant to the lesson. Therefore, the advertisement would be selected as the product to be relevant to the lesson such as the reference documents or the book written by the professor.

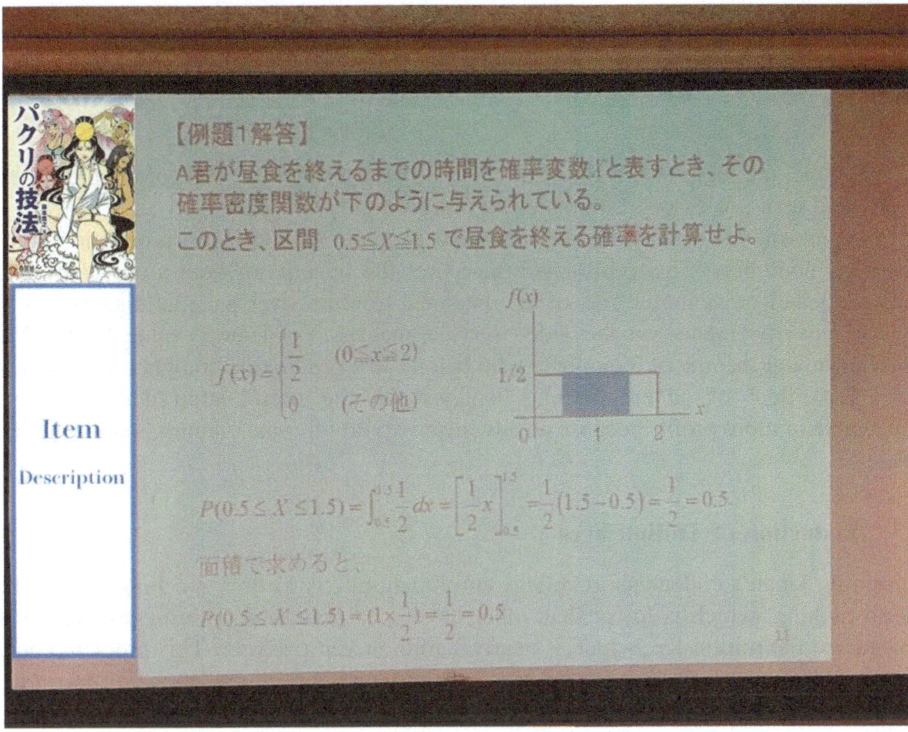

Fig. 1. Overlay advertisement

5 Profit Sharing

In either case (4.1) or (4.2) mentioned above, there is no merit for the professor to permit displaying the advertisement on the presentation documents during the lesson. Moreover, some of them would think that the advertisements may be an obstacle in teaching. Even if they have the choice to refuse the advertisements, what if we have some way of a win-win proposal?

5.1 Advertisements of the Book Written by Professor

As described in the previous chapter, displaying the products written by the teacher himself as an advertisement will directly lead to profit the professor. There is still room of consideration that the professor would be the only one client in this case. However, it would be successful in constructing the educational space as an advertisement medium without any discomfort.

On the other hand, this study depends on young people's purchasing possibility and information diffusion ability so that some new or epoch-making things would be effective for advertising. In this case, the economic effect could be unpredictable when the products are only targeted as technical books written by professors.

5.2 Profit Calculation

Advertiser → University → Professor

For example, the legitimate profit-sharing could be considered if 80% of the profits are distributed as a professor's share and 20% is as a university's.

This allocation is based on the precedent example of UUUM, a production company for Youtubers. This profit-sharing secures the benefits of both a university and professors and prevents the risk of interpersonal troubles over unequal distribution.

The advertisements via the video service such as YouTube is quite lively. The average annual income of Youtubers who belong to a production company is 8 million yen. The effect of advertisements displayed on the presentation documents are expectable in more profits because an advertiser would offer the commission at a higher price.

5.3 Reduction of Tuition Fees

Currently, Japan's college/short college enrollment rate is 61.46% reaching 32nd in the world ranking, which is lower than other developed countries. The biggest reason is considered the tuition fee, which is nearly 1 million yen per year. This problem could be solved by putting advertisements in a university and providing the advertisement fee into a university. Although the word of "profit" sounds impure in the educational situation, it would be an effective way to reduce the burden for students while maintaining the quality of education.

5.4 Efficient Time Allocation

The investments of advertisement would be changed according to the value and popularity of lessons. For example, highly rewarded investments would be provided for popular lessons by famous professors with many participants and less rewarded investments would be allocated to unpopular lessons with few participants. As a result, the efficient time allocation of lessons could be implemented as popular lessons can be placed in the afternoon class, which is with the high participation rate. On the other hand, unpopular lessons can be distributed on Saturday 5th period class, which is with the lowest participation rate.

6 Summary

We focused on the diffusion power and purchasing power by young people and sought to utilize educational space as a new advertising medium. As a result, this study showed the great potentiality. There are various subjects existing in the practical use such as the curtailment of lesson by advertisements and the ethical problems between the economic earnings and educational environment. However, reducing tuition fees and differentiating the lessons lead to the improvement of the quality of education and would bring the great advantages in the future. As the future work, we would like to enhance the methodology toward the practical application of study.

References

1. Fujita, H., Kuriyama, S., Nose, T., Kubo, S.: Study on the Effect of Advertising on the Internet. Trans. Jpn. Soc. Manage. Eng. **51**(6), 587–593 (2001)
2. Fujita, A.: Advertising model on the web. J. Jpn. Soc. Inf. Knowl. **16**(4), 23–32 (2006)
3. Amano, E.: In-school marketing of food and beverage companies in the United States: schools as 'markets' and children as 'consumers'. Kanto Gakuin Univ. Econ. Assoc. **251**, 35–49 (2012)
4. Kato, T., Tsuda, K.: Difference in purchasing behavior between web advertisement induction and spontaneous web search. In: Proceedings of the 22nd Pacific Asia Conference on Information Systems, pp. 95–98 (2018)
5. Wakae, M., Mori, R., Yoshii, H.: Current status of tracking in online advertising and its legal consideration. J. Inf. Commun. Policy **2**(2), 1–21 (2019)
6. Syouda, T.: Website as an ad and interactivity. Advertising Sci. **44**, 127–143 (2003)
7. Arai, T.: Role of internet advertising in the media planning. Aichi Gakuin Daigaku ronso Shogakukenkyu **46**(1), 67–81 (2005)
8. Sakata, T.: Instagram marketing strategy: an advertising communication for instagram user engagement. Takachiho Univ. Acad. Repository **51**(2), 1–33 (2016)
9. Wakimoto, K., Kawamoto, S., Zhang, P.: Keyword-based text generation for internet advertisement. In: Proceedings of the 34th Annual Conference of the Japanese Society for Artificial Intelligence, pp. 1–4 (2020)
10. Okada, M.: Takaeshi Kinoshita effects of transition on late of click in banner advertisement-on the effect of overlap and wipe. Yamaguchi Univ. Fac. Eng. Res. Rep. **62**(2), 33–40 (2012)

Examination of Problems and Improvements of Virtual Library Systems

Xuezhen Li(✉), Takayuki Fujimoto, and Ziran Fan

Graduate School of Information Sciences and Arts, Toyo University, Tokyo, Japan
505livehouse@gmail.com, fujimoto@toyo.jp, swterc@gmail.com

Abstract. In recent years, online service in public libraries have been making progress. In addition, there are services to use VR to guide library facilities. All these online services increase the convenience of information retrieval for users. However, there are many cases to lose functionality that libraries originally hold. In this paper, we point out the problems and risks of online libraries, and aim at developing a new form of virtual library based on the shape of traditional open-stack libraries.

Keywords: Online library · Visual tour · Visual library · Information design · Knowledge spaces

1 Introduction

In the 16th century, with the advent of modern public libraries, anyone could read books for free and gain knowledge and information easily. In recent years, with the development of the Internet, the spread of e-books, and the popularization of online libraries, it becomes possible for anyone to use the library without being restricted to the environment. Many university libraries and public libraries not only provide search services, but they are also be able to rent and browse books using database without having to go to the library. Knowledge and information are becoming more and more democratized and popularized.

Online libraries are expected to become more and more popular in the future, but on the other hand, the need for traditional libraries which have bookshelves will not decline. The main reason is that online libraries and traditional libraries have completely different functions and roles which we called “knowledge spaces”.

In this paper, we clarify the differences between online libraries and traditional libraries, and reexamine the conventional functions of online libraries so far, and investigate the possibilities of new virtual libraries that combine the advantages of both.

H. Selvaraj et al. (Eds.): ICSEng 2020, LNNS 182, pp. 348–357, 2021.
https://doi.org/10.1007/978-3-030-65796-3_34

2 The Differences Between Online Libraries and Traditional Libraries

In this chapter we describe the differences between online libraries and traditional libraries.

2.1 Features and Functionality of Online Libraries

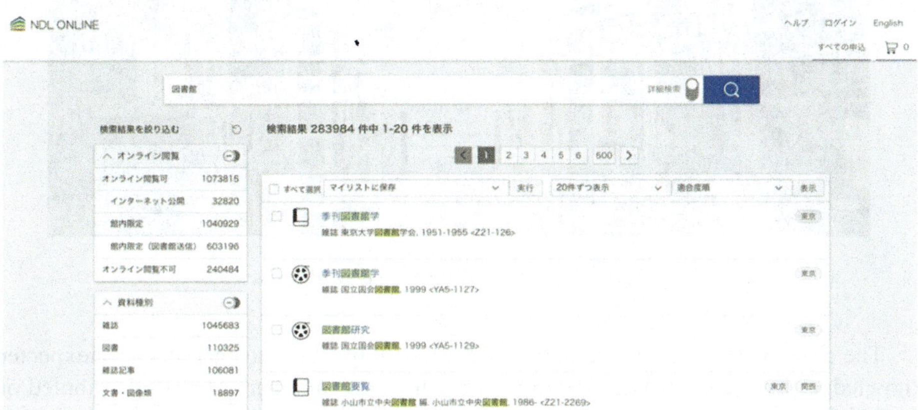

Fig. 1. Search page of NDL

First, online libraries and library databases currently provided place importance on searching for book information.

When you enter a keyword using an online system or information terminal, the information of the book related to the keyword stored by the library is listed (Fig. 1).

This method is very efficient if search for known books, or you have clearly topics and keywords. The troublesome process of searching on a bookshelf by relying on call numbers can be omitted.

However, if you do not clarify the keywords of the book you want to find in advance, you will not be able to search them. Unless the user enters a keyword the database cannot be used.

For users who don't clarify the keywords, the library's vast database is like a "keyless treasure chest" and can be wasted. In other words, an online library is a mechanism that can enhance functionality by relying on user's knowledge and searching skills, which means it maybe not suitable for general readers.

2.2 Features and Functionality of Traditional Libraries

Traditional libraries can't search for information as easily as online libraries. It is general to wander through books arranged in fields and next reach books which you search in a "seeing while walking" mode. Users have to get books in a much less efficient way than online libraries.

Traditional libraries are significantly less functional than online libraries in terms of information retrieval and book detection, however they are capable of characteristic searching features that online libraries do not have (Fig. 2).

Fig. 2. Traditional libraries' bookshelves

The characteristic search method of a traditional library is to encounter "unexpected knowledge" by "seeing while walking" through a large number of books distributed on bookshelves. In the case of online search, searching keyword is clear, and only related information is presented as a list.

However, books arranged on bookshelves are only roughly allocated for their "field" or "area", the genre and within the genre, books that you did not assume according to search keywords, or books, authors, keywords, etc. and that you did not anticipated that related can be encountered. Of course, as you go to the bookshelf you are aiming for, you walk through a relevant bookshelf, so unexpected books may "come into sight". Such "unexpected encounter with knowledge" and "discovery of unexpected relevance" are the distinctive search methods that are not found in online libraries. Database-based online libraries can improve the efficiency of searching for book information, but at the same time, they can break the relevance between books or knowledge, which can narrow users' visual fields. With this, it is hard to say that users can fully use the original functions of libraries with a huge amount of knowledge.

Also, most online libraries up to now cannot feel the physical characteristics of books by only using letters. Books are not just about textual information. Design as an object, such as the color of a cover, font size, typeface, book size and thickness, is also an important factor when readers make the choice, but such options are often lost in online libraries.

Table 1. The Comparison between online library and traditional library

	Online library	**Traditional library**
Functionality	Very efficient	Ineffective
Purpose of use	Search for specific books	Search for specific books + 「browse」
How to use	Search for information based on keywords	Walking and seeing
How to encounter knowledge	Inevitably	Randomness
Elements of expression in books	Text information only	Physical visual elements of books
Experience of use	Mechanical way	Have fun with encountering knowledge

As shown in Table 1, as the result of comparing online libraries and traditional libraries, online libraries are called "libraries", but the difference between them is significant. It is not possible to encounter unexpected books or knowledge, and the realization of efficient search can only be done when users have basic knowledge to clarify search keyword. Rather than reading search results in same font, touching real books and feeling their thickness and weight enhances the user's satisfaction. Compared to traditional libraries where looking for a book can be enjoyable, online libraries are more than search engines.

3 Related Research

In this chapter we introduce virtual libraries and their studies related to this study.

In recent years, many museums, libraries, and other facilities have virtual tours or such services. But many of the so-called virtual tours are made of videos or movies, and just can be watched it or have a simple information display function. Additionally, most of the movements in the library are just to watch introduction texts in each area.

3.1 Case Study of Virtual Libraries

In 1988, Mamoru Sato created the first visualization virtual library, "SungWoKung". The virtual library "SungWoKung" uses CG animation technology to reproduce the library as a geometric space (Fig. 3).

Fig. 3. The entrance of "SungWoKung"

Users can achieve open access search as if they were walking around the actual library (Fig. 4).

Fig. 4. Users get books from bookshelf of "SungWoKung"

In 1995, NEC's Masaki Hara and others researchers developed a virtual library system that allows users who do not have a clear search purpose to freely explore the library by walking around.

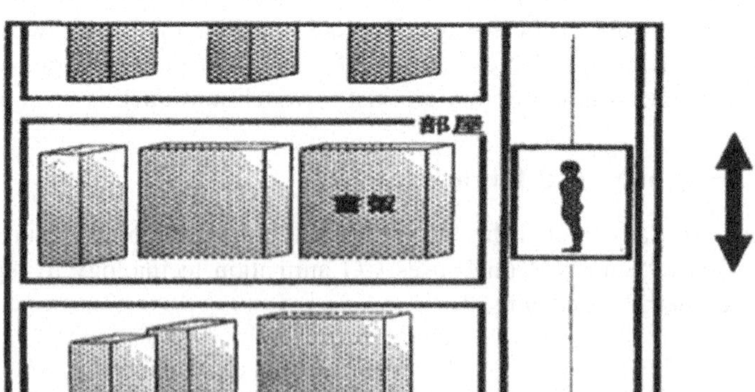

Fig. 5. Re-enactment of taking the elevator and moving between reading rooms

In order to enhance the user's actual feelings, all of the movements between reading rooms are carried out by elevators, and the degree of perfection as an initial stage in a virtual library is high (Fig. 5).

Kyushu University developed a related bibliography display system in the form of virtual bookshelves. By improving the appearance according with CG to use a photograph of the spine cover, it enhanced the reality of a virtual library.

Fig. 6. Virtual search page of "the document of Japanese economic history database".

Osaka City University Academic Information Center has developed a system that provides two patterns, keywords search and virtual search, when creating a database of its collected materials. Users can look at the entire bookshelves and select books on bookshelves (Fig. 6).

3.2 King's College Library' UA-PTC Ottenheimer Libraries[1]

Few existing libraries have what can be so-called "virtual tour" in a strict sense. Examples include King's College Library in the United Kingdom, the UA-PTC Ottenheimer Libraries, and so on. These virtual tours increase the sense of immersion by allowing users to choose what they want to browse in units of rooms and bookshelves.

For example, the virtual tour in King's College Library allows users to take a look at the map of an entire building and choose the area you want to visit. In each area, the description is displayed by hovering over, it is possible to get detailed information (Fig. 7). However, there are many problems that have not been solved even in these cases. For example, many virtual tours introduce libraries as famous places and public facilities. It is not the main purpose to virtually experience the original functions of the library, like the searching and browsing of books and the state of display of books. Users also can't walk freely to find books like "see while walk" as in a real library.

[1] https://www.uaptc.edu/vr/library.

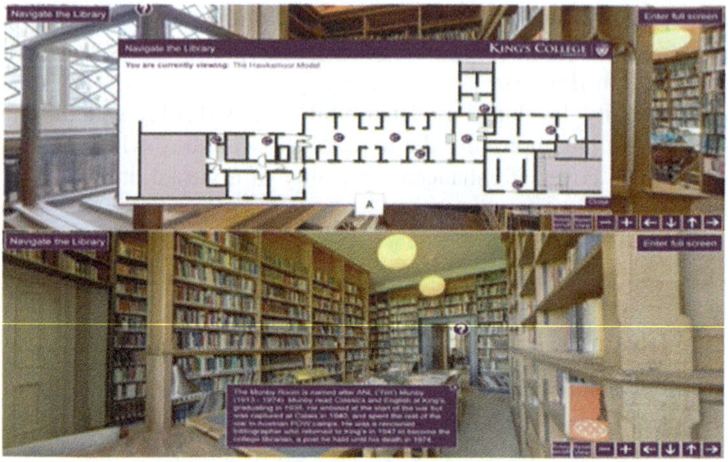

Fig. 7. Visual tour of King's College Library

3.3 Conventional Problems of Virtual Library So Far

With the advancement of technology, virtual libraries have actively adopted CG animation, images, photos, etc. in order to enhance the sense of reality. In the virtual space, not only the books themself, but also the library where bookshelves and reading rooms are included, and the expressive power is increasing more and more.

With the development of computer performance, the expressiveness and reality of virtual libraries have increased, but there is no significant change in the way of usage. Lastly it is limited to "virtually viewing libraries as content" by operating the mouse, There is no practical virtual libraries equipped with the functionality of "searching, browsing, and reading books" which are provided by traditional libraries.

4 Purpose of Research

In recent years, online libraries have become more sophisticated as search engines, and the convenience of information retrieval has increased significantly. However, this method has risk of filtering "unexpected knowledge" and "overlooked knowledge" as "extra information" and losing the chance to "encounter related information" in libraries. Therefore, the purpose of this research is to design a new virtual library based on the shape of open-stack libraries after pointing out the problems and risks of online libraries so far.

Therefore, in this research, in order to reproduce the knowledge space originally possessed by the library, based on the problems of existing virtual libraries, we construct the physically digital library space rather than "SungWoKung" or virtual bookshelves developed by Kyushu University. In addition, in order to demonstrate the function of the library as "a place of encountering knowledge", rather than a so-called virtual tour of sightseeing spots, we adopt a method of looking for a book while walking in the "space" of the visualized library. Finally, in order to realize the characteristic of traditional libraries, “encounter knowledge by accident”, we will devise a unique virtual library construction mechanism in this study.

First, in order to digitally construct the physical space of library, we use images of an actual library taken by a 360° camera not by a conventional CG animation method. In addition, in order to reflect the actual situation that the display of books in library changes when new books arrive, the image of the library will be updated once a month to purchase in a real daily world, which cannot be realized by CG. Next, we propose a function to “walk through” the visual library by linking the actual movement of “walking” with 3D videos.

The virtual library proposed in this study has a mechanism to arrange and display books by various elements, as contents, authors, publication place, etc., instead of sticking to the Japanese decimal classification method. Users can overlook at the entire field by looking at the “bookshelves” like in an actual library. Also, as a virtual library, books can be freely displayed and arranged in various ways, it can be expected to clarify the relationship between books that are difficult to discover in actual libraries. Of course, even if the book you want to browse has not been decided in advance, by “looking around” the virtual library, as in actual libraries, without being limited by pre-knowledge, users can explore books freely and intuitively.

Particularly, in this research, we propose the function to display past books and bookshelves in the library, and name it as “Knowledge Ruins”. Humans have cognitive mechanism that senses that the coming time as the future and back time as the past time.

In this research, we propose the concept of knowledge mapping in space-time by constructing a system that can view library’s books and bookshelves in the past. This function is based on the accumulation of actual image data and reflect the spatiality of time in library’s book display.

The factor that users have an influence on choosing books in the library do not come from the search result of books, but also from a wide variety of factors. That point lacks in today’s online library. Therefore, the purpose of this study on a virtual library is designed to convey the space-time sensibility of the actual libraries to the user while reflecting the book-search results by various elements and to enhance a real experience. We aim for novelty beyond the functions that existing online libraries cannot cover.

5 Conditional Functions for Virtual Libraries that Give Users Real Feelings

Various virtual libraries and library systems have been developed so far, but there are no systems or services that fully satisfy both “feelings of using an actual library” and “functions of book-searching” for users. Since online, virtualization, and systemization of information and knowledge are inevitable in future libraries, not only knowledge and information but also the feelings in reality should be provided to users, which will increase convenience and opportunities to contact with knowledge. These attempts are important.

It is important to try to enhance convenience and opportunities to contact knowledge by giving users not only knowledge and information, but also usefulness. Based on this discussion so far, this chapter gives a concrete overview of the conditional functions for realizing a virtual library that we are currently developing.

5.1 Conditional Functions① Function of Rearrange Books in Various Styles

Existing libraries have many physical restrictions, making it difficult to freely and easily rearrange books and bookshelves. In addition, in online libraries and library databases, the corresponding information that relates to the search keywords is only displayed as a list, and it is difficult to change the display except for a simple method such as ascending/descending orders.

In this paper, taking into account the convenience of the user's purpose of use, it is essential to have a function to rearrange bookshelves and books in various styles. In a general library, bookshelves are divided by "fields", among which the arrangement of books is organized by categories, themes, and authors. In the system we consider, the standard of the organization can be arbitrarily determined. For example, it can be considered variously depending on users' needs, such as "orders by the highest number of rentals" and "orders by the smallest number of pages".

5.2 Conditional Functions② Function of Reproducing Changes in Bookshelves for a Long Term

Libraries' bookshelves and displays vary along with time and eras, change, needs of users and social conditions. Therefore, it is no exaggeration to say that the arrangement of the bookshelves and the books of the library represents the knowledge situation and the trend of that time. Therefore, in the system that we are developing, we are considering to furnish the function of restoring the arrangement of bookshelves and books that allows to go back to the real situation of a selected era. This enables users to feel the knowledge situation and trend of the time.

5.3 Conditional Functions③ Function of "Seeing While Walking" Through the Virtual Library

In the past, virtual libraries are generally separated from the system for traveling around library facilities and the system for searching and viewing book information. Therefore, the former is for "guidance of the facility", and "search for information and knowledge" which is the original role of the library was not satisfied. Therefore, we plan to adopt a video control application based on "real walking operation". This function allows users to browse the images of bookshelves in a virtual library while walking. In addition, rather than the screen operation by a mouse or a hand, the action of actually walking is reflected in the movement within a virtual library. In particular, by controlling the playback speed of bookshelves image according to walking speed, it can control the movements between bookshelves. Therefore, it is possible to feel like being in an actual library even if it is not in an actual library.

5.4 Conditional Functions④ Function that Allows the Users to Move Around in 360°

This function is associated with condition 2. In a virtual library we are developing, all contents use the images taken by a 360-degree camera. As a result, it is possible to reproduce the feeling of "being surrounded by bookshelves and books" when visiting an actual library.

6 Summary and Future Works

In this paper, we reviewed the current status of virtual libraries and online libraries, examined the problems and issues, and discussed the possibility of a virtual library with a high utility and novelty. Libraries are not a place just to store and borrow books. They are facilities with diverse possibilities that can lead us to the search for new fields and to improve cultural life. With the availability of online libraries, the convenience for users in terms of information retrieval is increasing, but it is hard to say that other functions and roles of libraries are satisfied.

The future works are to reexamine the method of expressing the relationship between books and knowledge. The shape of bookshelves gave a hint about how to visualize the relationship. Based on this, beyond just imitating traditional libraries, we consider a knowledge mapping system that can be realized only in the Internet era in order to make one's position clearly in the vast knowledge world.

References

1. Campbell, J.W.P.: The Library: A World History. University of Chicago Press, Chicago (2013)
2. Clark, W.J.: The Care of Books. Sagwan Press, New York (2015)
3. Crary, J.: Suspensions of Perception. The MIT Press, Cambridge (2001)
4. Howsam, L.: Old Books and New Histories. University of Toronto Press, Scholarly Publishing Division (2006)
5. Yates, A.F.: The Art of Memory. University of Chicago Press, Chicago (2001)
6. Mamoru, S.: Electoronic Libraries "SungWoKung" which support retrievals of books using presentation of CG pictures of libraries, IPSJ SIG Technical reports (FI), pp.1–8 (1991)
7. Hara, M., Kamiya, T., Miyai, H.: Development of digital library system using 3D CG. In: Proceedings of the 52th National Convention of IPSJ, pp. 365–366 (1996)
8. Miyagawa, T., Omori, Y., Ikeda, D.: Automation of back cover image generation in virtual bookshelf. Digit. Libr. **30**. 25–38 (2006)
9. Nishikawa, H.: Development of virtual library system: its concept and retrieval method based on categories. IPSJ-DPS **1997**(35), 177–182 (1997)
10. Shibayama, M.: Digital archives and virtual library for historical materials. IPSJ SIG Comput. Humanit. 109–116 (2001)
11. Kamimura, H.: Support of Information access and utilization for trend information ecosystem. In: JSAI2015, pp. 1–2 (2015)
12. Fujimoto, T.: The imagination of information design. Gendai-Sugakusha (2005)

Enhanced Sign Changes for SEMG Onset Detection

Mateusz Magda(✉), Jerzy Świątek, and Piotr Wyrostek

Faculty of Computer Science and Management,
Department of Computer Science and Systems Engineering,
Wroclaw University of Science and Technology, Wroclaw, Poland
{mateusz.magda, jerzy.swiatek}@pwr.edu.pl,
246749@student.pwr.edu.pl

Abstract. Electrical activity of human muscles is a valuable source of information for the analysis of different types of human movements. SEMG (Surface Electromyography) is a technology that allows us to observe it. Information extracted from sEMG signals is already being used in many areas like medicine, sport or robotics. For some applications it is crucial to detect the exact moment of a muscle activation. Pointing the precise onset basing solely on sEMG signals is a non-trivial task. Low SNR (Signal to Noise) ratio, crosstalk from surrounding muscles and its ontogenetic nature are only some of the problems to be listed. This paper presents an improved version of automatic sEMG onset estimator called Sign Changes. It ensembles its basic version with the AGLR algorithm – a model approach. Taking the advantage of reliability and accuracy provided by both algorithms we achieve better estimations. A comprehensive comparison is conducted to test the quality of presented solution. It occurs to be both more robust and precise than its previous version.

Keywords: EMG · sEMG · Onset · Changepoint · Detection · Activity · Estimation · Sign changes

1 Introduction

SEMG (Surface Electromyography) is a technology that allows for capturing an electrical signal that accompanies a muscle activity. By processing it, one can conclude information about human movement. This is why it is deployed in diagnosis of neuromuscular diseases [1], prosthetics, robotics [2], professional sport [3], rehabilitation [4], and many other areas.

EMG (Electromyography) differs from sEMG on the place where signal is being captured. The first technique measures it directly in the muscle tissues and thus requires a proper skin preparation. It consists of degreasing, disinfecting, depilating and piercing the skin layer. The latter one offers noninvasive recordings. It measures the signal directly on the skin surface and requires no preparation nor medical certification to conduct the examination. SEMG might be called the successor of EMG technology since it was invented later and has certain advantages over its initial version. It is much more convenient to use sEMG and that is why it became more popular. Unfortunately,

H. Selvaraj et al. (Eds.): ICSEng 2020, LNNS 182, pp. 358–366, 2021.
https://doi.org/10.1007/978-3-030-65796-3_35

there are several shortcomings of this technology. Low SNR (Signal to Noise Ratio) which is a consequence of observing the signal not directly but through several layers of skin. Lack of skin preparation causes additional noise which is hard to predict. Crosstalk caused by activity of surrounding muscles [5]. Depending on which muscle and where exactly is being observed, interference produced by surrounding muscles can be present in the observed sEMG signal.

One of the motivations for undertaking this research and improving the accuracy of existing solutions is the experimental use of sEMG signals. Onset detection algorithms are applied to calculate the EMD (Electromechanical Delay) and determine the asymmetry of muscles responsible for stabilizing a spinal column [3]. Since the measured value varies between 30 ms and 100 ms it requires considerably high accuracy.

2 Onset Detection Algorithms

Presented in this paper enhanced version of sEMG onset estimator consists of two algorithms combined to work together. Its initial version, called Sign Changes, proved high level of reliability and accuracy [6, 7]. Results achieved by this method are more reliable than widely known reference solutions. For further improvements an ensemble of algorithms was suggested. Adapting the idea of two stage onset estimator proposed in [8] we prepared a combination of two different approaches to muscle activity onset estimation.

2.1 Sign Changes

The first algorithm is based on the fact that sEMG signal changes its sign more frequently in case of resting muscle state. This combined with the information about greater amplitudes leads us to the factor illustrated on Fig. 1.

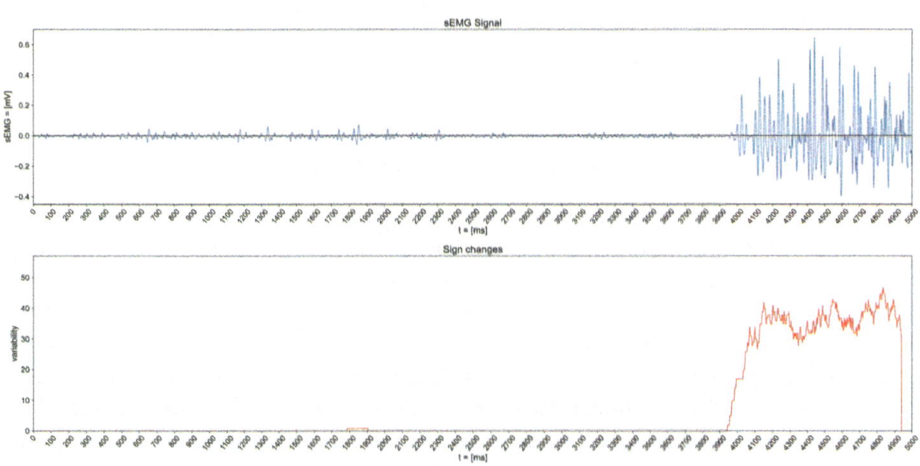

Fig. 1. SEMG signal and the corresponding value of sign changes algorithm

We can observe how well it is correlated to the sEMG activity. This solution is very robust but lacks in high precision. It is used to point a range of samples in which the muscle begins its activity and pass it to the second stage of detection.

2.2 AGLR

The second part of presented solution is a model based approach to onset detection called AGLR in step version proposed by Staude et al. [9, 10]. It deploys log-likelihood ratio test and threshold mechanism. Consecutive samples have its likelihood calculated. The first which exceeds a predefined threshold value is selected and proposed as the final onset estimation. This algorithm proved relatively high level of accuracy. However, the testing dataset used in this research was simulated by a mathematical model illustrated on Fig. 2.

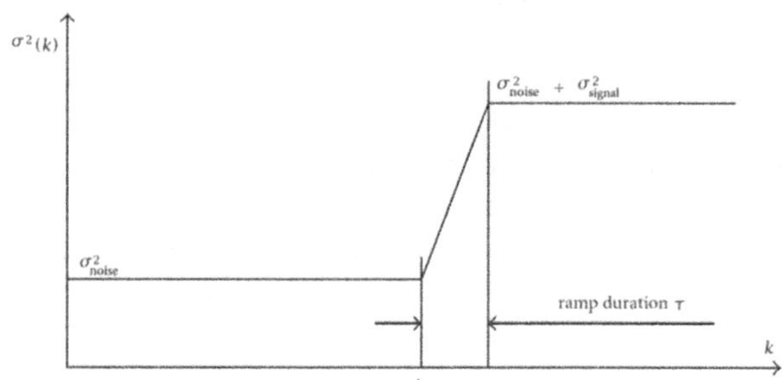

(a) Power envelope of modeled SEMG signals. After a period of background activity the SEMG power linearly rises to the active state within τ. Parameters were systematically varied during simulation.

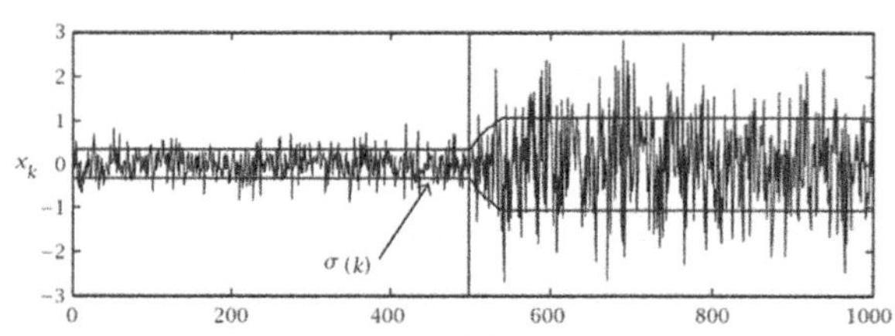

(b) Sample trace of simulated SEMG.

Fig. 2. Simple model of sEMG data [10]

It consists of three states: noise, transition, and the muscle activity represented by corresponding variance levels. It led to false conclusions since it did not reflect all complexity of the problem. Testing AGLR algorithm on a real world sEMG signals shows that it only works good on signals that match this pattern. Any signal crosstalk or disturbances which occur before the muscle activation state will very likely reflect in worse accuracy and premature indications. However, this solution serves well when working in a narrowed range of a signal, and as such it was deployed in our two-stage onset estimator.

3 Comparison

In order to test and relate the quality of presented solution to already existing one, a comprehensive comparison has been conducted. Since proposed onset detection method is an enhanced version of previously presented Sign Changes algorithm combined with AGLR step these three methods were analyzed and compared.

3.1 Criteria

To be able to assess the estimators we have to predefine the criteria. Equation (1) defines X_i which will be later referred as an error. It is the difference between estimator's indication and the predefined point of reference.

$$X_i = \hat{t}_i - t_i \tag{1}$$

Note that this value can be both positive and negative. It is risky to calculate its mean value since opposite values of −30 ms and 30 ms would diminish themselves to 0 ms. The only valuable information we can conclude from observing the mean error is whether our solution is biased or not. Skewed distribution of errors can tell us how far the results are from the benchmark. As an alternative, we can observe the mean absolute value of error, presented in (2)

$$\frac{1}{N}\sum\nolimits_{i=1}^{N}|X_i|, \mathit{where}\ |X| = \begin{cases} X, & \text{for } X \geq 0 \\ -X, & \text{for } X < 0 \end{cases} \tag{2}$$

The last of basic criteria we need to make a proper comparison is the variance. It informs us how different the results are from each other. Its unbiased approximation with Bessel's correction [11] is presented in (3)

$$\sigma^2 = \frac{1}{N-1}\sum\nolimits_{i=1}^{N}(X_i - \bar{X})^2 \tag{3}$$

3.2 Implementation and Execution

Control flow and data structures used to implement the two compared solutions were unified. Execution environment, operating system, and library versions were identical. The process was executed in isolation on a Dell Server (T5610, 2x Xeon 12 cores/ 24 threads, 64 GB RAM, SSD V-Nand, Quadro K4200). The input and output data structures were customized in order to work in the same comparison framework.

3.3 Parameter Optimization

Each onset detection algorithm takes one or more hyperparameters. The results of its work strongly depend on how well these hyperparameters will be suited. In order to provide equal conditions for all compared solutions we applied Tree-structured Parzen Estimator algorithm available in Optuna library for Python.

Criteria defined below serve as a fitness function which inform the algorithm on how well the considered set of hyperparameters perform. This way we ensure that each compared solution was run under optimal conditions, with well suited parameters and thus presented results are reliable. The cost function of estimations used for the sake of parameter optimization is described as f_1. Criteria presented in (4, 5) define the accuracy as value of absolute mean error. Using multi objective optimization algorithm, first introduced to the area of sEMG onset detection in [12], we can easily add a second criterion (f2) and optimize the number of outliers or another criterion, according to requirements.

$$f_1\left(x_i^j\right) = \frac{1}{i * j}\sum\nolimits_{i=1}^{n}\sum\nolimits_{j=1}^{n} g_1(x_i^j) \tag{4}$$

$$g_1(x_i^j) = \begin{cases} \left|f\left(x_i^j\right) - \hat{t}_i^j\right|, \textit{ if } \left|f\left(x_i^j\right) - \hat{t}_i^j\right| \leq 5000 \\ 5000, \textit{otherwise} \end{cases} \tag{5}$$

, where:

$$f_1 \rightarrow \text{cost function}$$

$$\hat{t}_i^j \rightarrow \text{onset in the } i^{th} \text{ signal, } j^{th} \text{ electrode}$$

$$f\left(x_i^j\right) \rightarrow \text{detected onset } i^{th} \text{ signal, } j^{th} \text{ electrode}$$

$$5000 \rightarrow \text{signal length, expressed in [ms]}$$

3.4 Data

Both algorithms were tested on a real world sEMG dataset collected at the University School of Physical Education in Wrocław, Poland, which was briefly described in [3]. Data used in this research was collected in 26 trials resulting in 312 signals, 172 of which were marked as valid and were further processed. Selected sEMG signals come from observation of abdominal muscles. They were recorded with the frequency of 1000 Hz on 6 electrodes with 2 additional channels, one serving as a baseline and another, measuring a muscle torque, as a reference value. Each signal had a predefined onset marked by a specialist, what by physicians is still considered to be the best point of reference while assessing automatic estimators.

Comparing onset detection algorithms on a real sEMG datasets is an important aspect. Using mathematical model of a signal to test solutions can jeopardize the validity of a whole research. When statistically based estimators are tested on a statistically modeled data the results are only as good as these two processes match.

4 Results

After adjusting all algorithms to data and preparing them to work in a common execution environment they were compared. In order to conduct a proper comparison, and for the sake of further analysis, four criteria were selected: mean error (bias), absolute mean error, variance, and the number of outliers (errors larger than 500 ms). The results are divided and presented in two sections. Table 1 shows full results without any rejections. At the cost of larger mean errors, the automatic optimization algorithm found presets which for both Sign Changes algorithms provided onsets for all signals. In case of AGLR algorithm there were 3 cases in which no indication was made. These cases were excluded from the results statistics and it must be stressed. For some application areas no indication is worse than a large error.

Table 1. Results of onset estimations, no exclusions.

	AGLR	Sign changes	Enhanced sign changes
$\bar{X}_{[ms]}$	−81.00	−48.48	−8.19
$\lvert\bar{X}\rvert_{[ms]}$	102.52	79.05	38.07
σ^2	210876.59	9924.91	8223.06
No onset	3	0	0

Table 2 shows the set of narrowed results, with outliers excluded. What means that all estimations in this summary were in the proximity of | 500 ms | from the real onset.

Table 2. Results of onset estimations, outliers excluded.

	AGLR	Sign changes	Enhanced sign changes
$\bar{X}_{[ms]}$	−5.25	−42.84	−1.62
$\lvert\bar{X}\rvert_{[ms]}$	27.49	73.65	31.74
σ^2	2010.88	5955.84	2808.71
$\lvert\bar{X}\rvert_{> 500\ ms}$	4	1	1

Table 3 shows the results of cross validated statistical significance test. The overall p-value for hypothesis that all the result groups come from the same distribution is 9.63 $* 10^{-15}$, what means that we can reject it with very high level of confidence. It clearly shows that there are significant differences between sets of results achieved by Sign Changes and AGLR (p-value $3.23 * 10^{-11}$). The same for Sign Changes and Enhanced Sign Changes (p-value $5.06 * 10^{-13}$).

Table 3. Kruskal-Wallis statistical significance test

P-value (overall, $9.63 * 10^{-15}$)	AGLR	Sign changes	Enhanced sign changes
AGLR	1	$3.23 * 10^{-11}$	0.62
Sign changes	$3.23 * 10^{-11}$	1	$5.06 * 10^{-13}$
Enhanced sign changes	0.62	$5.06 * 10^{-13}$	1

However, the p-value of 0.62 for the two sets of results achieved by AGLR and Enhanced Sign Changes means we cannot reject the hypothesis that their results come from the same distribution. Since enhanced Sign Changes has implemented AGLR as its inner function, almost 71% of indications are the same. More than 29% of indications that differ these two algorithms stand for almost 3 times better results achieved by Enhanced Sign Changes in general case. Even though most results match, the remaining part is significantly different.

Figure 3 shows an example of sEMG signal with relatively low SNR. In addition, we can see a slow transition between the rest and active states of a muscle.

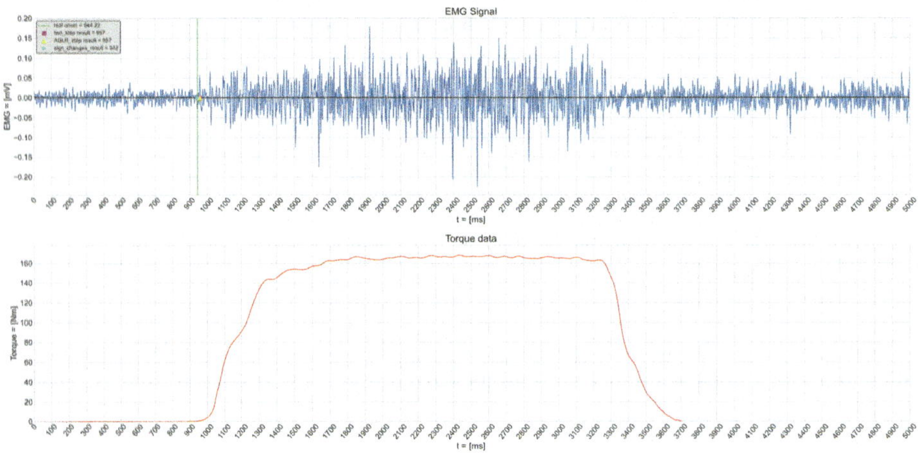

Fig. 3. sEMG with low SNR

This kind of signal shall be difficult to process. Nevertheless, all compared algorithms managed to point the onset accurately with absolute value of error within (0, 20) ms range. It confirms that all algorithms perform well thanks to a proper hyperparameter optimization.

However, AGLR algorithm base strongly on the resting variance which it assumes to occur at the beginning of a signal. Figure 3 show an example of signal where the disturbance is pure noise and is regularly distributed throughout its whole length. On Fig. 4 we can observe another type of disturbance – unevenly distributed fragments with larger variance. It is caused by the signal crosstalk coming from surrounding muscles and for AGLR algorithm – it is very difficult to distinguish from real signal. Enhanced version of Sign Changes algorithm, thanks to the two staged detection process, managed to point the exact onset of a muscle activity.

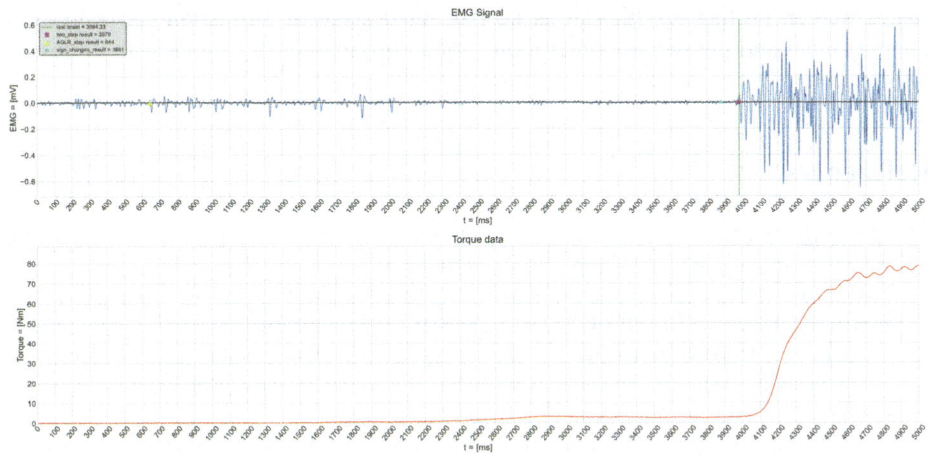

Fig. 4. sEMG difficult case – an example of signal crosstalk

5 Conclusions

Results Analysis. Looking at the results we can draw many conclusions. Comparing the two methods of automatic onset detection Sign Changes and Enhanced Sign Changes we can clearly say that adding the second stage of detection helps estimate the muscle activity more precisely. Lower mean error (−8.19 ms vs −48.48 ms) and its mean absolute value (38.07 ms vs 79.05 ms). For the narrowed results the differences are even greater: lower mean error (−1.62 ms vs −42.84 ms), absolute value of mean error (31.74 ms vs 73.65 ms) and two times lower variance which means the results are more cohesive.

Relating the two Sign Changes algorithms to AGLR we can state that the latter achieved significantly worse results in general: −81.00 ms mean error and 102.52 ms of its mean absolute value. It is caused by premature onset detections of AGLR as we saw on Fig. 4. It presents relatively low mean absolute value of error for narrowed results 27.49 ms (rejecting errors > 500 ms). Analyzing its results, we must stress that in 3 cases AGLR did not point any onset which for some application areas may be unacceptable. It might influence the results – these 3 signals for which the algorithm could not take any decision might occur its worse indications if considered in final statistics. Bias is present in all examined cases what means there is a space for further improvement.

According to the computational cost and the time of execution the Enhanced Sign Changes occurred to be 4 times faster than the standard AGLR method. Applying Sign Changes as initial indicator, the algorithm calculates log-likelihood test only for narrowed sample range.

Possible Improvements. Applying filters to the field of precise sEMG onset detection is risky because they can easily skew the signal by more than 10 ms and the estimations will no longer be precise. However, having an estimator that consists of two separate stages we can profit from filtering at no risk. First stage of estimation can work on a filtered signal and the second, precise one, on an original sEMG signal with no loss of information.

References

1. Sisk, J.E.: Electromyography. In: Longe, J.L. (ed.) The Gale Encyclopedia of Nursing and Allied Health, pp. 925–927, Gale, Detroit (2006
2. Artemiadis, P.K., Kyriakopoulos, K.J.: EMG-based control of a robot arm using low-dimensional embeddings. IEEE Trans. Robot. **26**, 393–398 (2010)
3. Szpala, A., Rutkowska-Kucharska, A., Drapała, J., Brzostowski, K., Zawadzki, J.: Asymmetry of electromechanical delay (EMD) and torque in the muscles stabilizing spinal column. Acta Bioeng. Biomech. **12**, 11–18 (2010)
4. Sarasola-Sanz, A., Irastorza-Landa, N., López-Larraz, E., Bibián, C., Helmhold, F., Broetz, D., Birbaumer, N., Ramos-Murguialday, A.: A hybrid brain-machine interface based on EEG and EMG activity for the motor rehabilitation of stroke patients. In: 2017 International Conference on Rehabilitation Robotics (ICORR), pp. 895–900 (2017)
5. Mezzarane, R.A., Kohn, A.F.: A method to estimate EMG crosstalk between two muscles based on the silent period following an H-reflex. Med. Eng. Phys. **31**, 1331–1336 (2009)
6. Magda, M.: EMG onset detection – development and comparison of algorithms
7. Magda, M.: A Novel Approach to Onset Detection in EMG. In: Nauki medyczne i nauki o zdrowiu. T. 1, pp. 98–103. MN2015, Poznan, Poland (2015)
8. Drapala, J., Brzostowski, K., Szpala, A., Rutkowska-Kucharska, A.: Two stage EMG onset detection method. Arch. Control Sci. **22**, 427–440 (2012)
9. Staude, G.H.: Precise onset detection of human motor responses using a whitening filter and the log-likelihood-ratio test. IEEE Trans. Biomed. Eng. **48**, 1292–1305 (2001). https://doi.org/10.1109/10.959325
10. Staude, G., Flachenecker, C., Daumer, M., Wolf, W.: Onset detection in surface electromyographic signals: a systematic comparison of methods. EURASIP J. Appl. Signal Process. **2001**, 67–81 (2001)
11. Reichmann, W. I.: Use and Abuse of Statistics. Methuen & Co. Ltd. Publishers, London (1961). 336 S., Preis 22 s 6 d net - Heinisch - 1964 - Biometrische Zeitschrift - Wiley Online Library
12. Magda, M., Martinez-Alvarez, A., Cuenca-Asensi, S.: MOOGA parameter optimization for onset detection in EMG signals. In: New Trends in Image Analysis and Processing – ICIAP 2017, pp. 171–180. Springer, Cham (2017)

Special Session: Optimal, Robust, and Adaptive Control Systems

Optimal Energy Resources Scheduling of Hybrid Diesel/Battery Ships in Shallow Waters

Mohsen Banaei[1], Jalil Boudjadar[2(✉)], and Mohammad-Hassan Khooban[2]

[1] Department of Engineering, Ferdowsi University of Mashhad, Mashhad, Iran
[2] Department of Engineering, Aarhus University, Aarhus, Denmark
jalil@eng.au.dk

Abstract. In this paper, optimal energy resources scheduling for a hybrid diesel/battery ship that sails in shallow water is studied. Sailing in shallow waters increases the impact of water depth on the ship's resistance. Changing the ship's resistance affects the speed, consumed fuel and consequently the operation cost. This paper proposes a non-linear model to find the optimal operation of diesel generator and battery bank for supplying the ships' engines considering the impact of variable water depth of the traveling route on the ship's operation cost. To this end, first, different types of resistance are calculated for the understudy real case ship. Then the impact of the water depth on the resistance is investigated, and finally, the optimization problem is formulated and solved using a Genetic Algorithm. Simulation results show the efficiency of the proposed scheduling method in decreasing the total operation cost while considering the different operational constraints.

Keywords: Real-time scheduling · Energy management · Cost optimization

1 Introduction

The utilization of diesel generators is known as one of the common techniques to generate power in marine vessels [1]. Energy storage systems can also be used in parallel with the diesel generators to cover part of the fast electricity demand fluctuations that the diesel generators cannot secure. Moreover, charging the batteries via cold-ironing at harbors and use their energy during the sailing can help in reducing the greenhouse gas emission caused by diesel generators [2]. Optimal energy resources management is an important issue for cost-effective and reliable operation of ships. This problem has been widely studied in the literature so far. Studies [3–5] present energy management strategies for fuel cell/battery hybrid zero-emission ships considering different assumptions. The work in [6] applies a non-linear method for deriving a control strategy of a ship with only a hybrid energy storage system as the energy resource. The authors of [2] tackle the energy management problem of an all-electric ship using dynamic programming. In [7], model predictive control is adopted with integrated perturbation analysis and sequential programming algorithms to solve

H. Selvaraj et al. (Eds.): ICSEng 2020, LNNS 182, pp. 369–378, 2021.
https://doi.org/10.1007/978-3-030-65796-3_36

the optimal scheduling problem of a ship with hybrid batteries and ultra-capacitors. In [1], optimal energy resources scheduling of a ship with diesel generators, photovoltaic panels, batteries, and cold ironing is solved. The work in [8] studies the energy management of a hybrid diesel//battery ship alongside with a high efficient combined cooling and heat power plant.

To the best of our knowledge, most of the studies in the marine vessels' energy management are performed assuming sailing in deep waters and ignore the impact of water depth on the efficiency of the engines. In shallow waters, the ship's resistance is affected by the water depth [9]. Thus, the efficiency of the engines at different portions of the traveling route will be different and to find the optimal power scheduling of the ship during the sailing, the changes in the water depth should be included in the behavior model. In this paper, the optimal energy management of a hybrid diesel/battery ship that sails in the shallow waters is studied. Main contributions of this paper are as follows:

1. Formulating the main resistance coefficients for an understudy case ship.
2. Proposing a non-linear model for energy management of the hybrid ship in shallow waters by considering different operation constraints.
3. Proposing an optimization process that calculates the optimal speed leading to minimal cost for each water depth of a given traveling route and vessel.

The rest of the paper is organized as follows. In Sect. 2 the problem is defined. In Sect. 3 ship's resistance is formulated. The relation between fuel consumption and engines power is determined in Sect. 4. An optimization method is proposed in Sect. 5. Numerical results are shown in Sect. 6. Finally, Sect. 7 concludes the paper.

2 Problem Definition

A hybrid diesel/battery ship that sails on a specific route in shallow water is considered. The ship information is presented in Table 1. The rated power of the diesel generator is 1000 kW. Batteries capacity is 120 kW. The depth of the water at different points of the traveling route is known. Following the water depth in different locations, the ships' resistance for sailing varies. Thus, the consumed fuel for reaching a specific speed in different locations will be different. The goal of this paper is to adjust the speed of the ship in different portions of the traveling route such that the minimal operation cost is obtained.

Table 1. Ship information

ρ_s	S	L	ϑ	g	∇	B	T	β	LCB
1.005	490.77	49.830	1.188e−6	9.8	485.5	11.297	1.920	0.7874	−4.696

3 Ship's Resistance Formulation

The relation between the speed and engines' power is as below [10]:

$$P_x^e = R_x V_x \tag{1}$$

P_x^e is the engine's power (kW), R_x is the resistance, and V_x is the speed (m/s) at portion x of the traveling route. Ship's resistance is formulated as below:

$$R_x = C_x^T \frac{\rho_s}{2} V_x^2 S \tag{2}$$

C_x^T is the total resistance coefficient, ρ_s is the mass density, and S is the wetted surface of the ship. C_x^T is calculated as below [10]:

$$C_x^F = C_x^F + C_x^R + C_x^A + C_x^{AA} + C_x^{AS} \tag{3}$$

C_F represents the frictional resistance coefficient, C_x^R is the residual resistance coefficient, C_x^A is the incremental resistance coefficient, C_x^{AA} is the air resistance coefficient, and C_x^{AS} is the steering resistance coefficient. The frictional resistance coefficient is formulated as below [10]:

$$C_x^F = 0.075\left(log_{10}^{\frac{V_x L}{\vartheta}} - 2\right)^{-2} \tag{4}$$

L is the length of the waterline and ϑ is the coefficient of kinematic viscosity. C_x^R is obtained as a function of the Froude number using the curves that are drawn for standard sizes of the ships [10]. The Froude number is as below:

$$F_{n,x} = V_x / \sqrt{gL} \tag{5}$$

g is the standard gravity. In order to know which curve should be used for finding the relation between C_x^R and $F_{n,x}$ two below factors should be calculated.

$$\varphi = \nabla / LBT\beta \quad \mu = \mathrm{L}/\nabla^{1/3} \tag{6}$$

∇ is the volumetric displacement, B and T are the breadth and draught of the ship, respectively, and β represents the mid-ship section area coefficient. Based on the obtained values of φ and μ for the understudy ship, the proper curve that represents the relation between C_x^R and $F_{n,x}$ is obtained using the diagrams in [10]. This curve can be used for the ships in which $B/T = 2.5$. In the case that this ratio is different, the value of C_x^R needs to be updated as follows [10]:

$$C_x^R = C_x^{R\left(\frac{B}{T}=2.5\right)} + 0.16 \times 10^{-3} \times \left(\frac{B}{T} - 2,5\right) \tag{7}$$

The real samples that are extracted for the curves are presented in Fig. 1. A fourth-degree polynomial function is fitted with these data samples using cftool in Matlab software as below:

$$10^3 C_x^R = 0.00942 \times V_x^4 - 0.1827 \times V_x^3 + 1.413 \times V_x^2 - 4.927 \times V_x + 6.75 \tag{8}$$

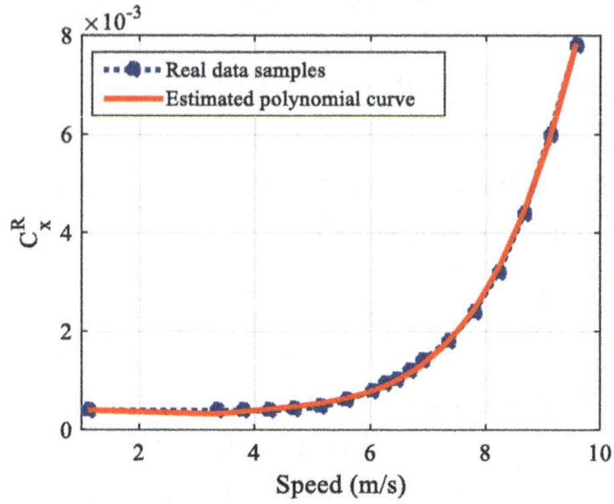

Fig. 1. Estimating the C_x^R curve

The incremental resistance coefficient C_x^A depends on the length of the waterline and for understudy ship is equal to 0.4×10^{-3}. In this paper, we consider that $C_x^{AA} = C_x^{AS} = 0$.

Now, the impact of water depth on the resistance is formulated. Figure 2(a) compares the ship's resistance for shallow waters and deep waters as a function of depth Froude number. The depth Froude number for water depth h_x at portion x of the traveling route is defined as below:

$$Fr_{h_x} = V_x / \sqrt{gh_x} \tag{9}$$

Experiments show that for speeds around $V_x^c = \sqrt{gh_x}$ the resistance increases significantly. There is no mathematical formulation that presents the relation between resistance and water depth. However, based on the information in Fig. 2(a), an approximated formulation is proposed to model this relation:

$$\gamma_x(V_x) = 1.01 - 0.01V_x + \frac{10}{\sqrt{2\pi g h_x}} \exp\left(-\frac{V_x - \sqrt{gh_x}}{\left(\sqrt{2\pi g h_x}/5\right)^2} \right) \tag{10}$$

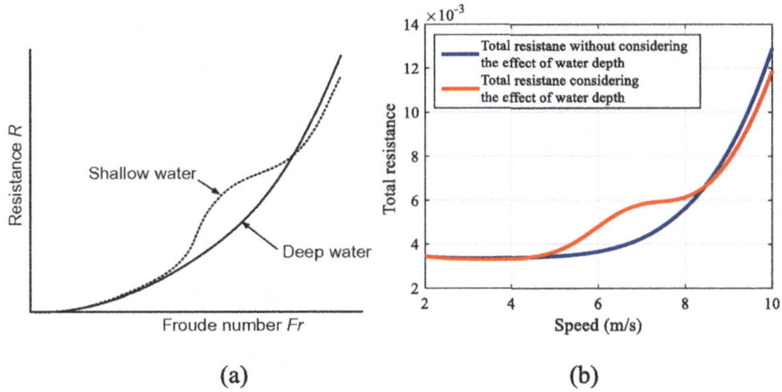

Fig. 2. Comparing the ship's resistance for deep and shallow waters, a) experiment results in [9] b) proposed formula for estimation

The ship's resistance after considering the impact of the water depth is obtained from the following formulation:

$$R_x^C = \gamma_x(V_x) \times R_x \tag{11}$$

Figure 2(b) compares the total resistance with and without considering the shallow water impact on the resistance for 4.5 m water depth using Eq. (10).

4 The Relation Between Fuel Consumption and Power

We use the formulation below to find the mass of fuel consumption for a specific output power of the diesel generator at portion x of the traveling route [11]:

$$FC_x^{DG} = SFOC \times P_x^{DG} \tag{12}$$

P_x^{DG} is the output power of the diesel generator (kW) and FC_x^{DG} is the fuel consumption (g/h). $SFOC$ represents the Specific fuel oil consumption (g/kWh). The SFOC curve for a diesel generator is depicted in Fig. 3 [11].

As shown in Fig. 3, SFOC depends also on P_x^{DG}. This curve is also estimated with a fourth-degree polynomial function. Considering that SFOC in 100% output power of the power generator is 211 g/kWh [11], we obtain the following formulation for SFOC using cftool in Matlab software:

$$SFOC\left(P_x^{DG}\right) = 2.04 \times 10^{-10} \times P_x^{DG^4} - 6.88 \times 10^{-7} \times P_x^{DG^3} + 0.87 \times 10^{-3} \times P_x^{DG^2} - 0.474 \times P_x^{DG} + 300.8 \tag{13}$$

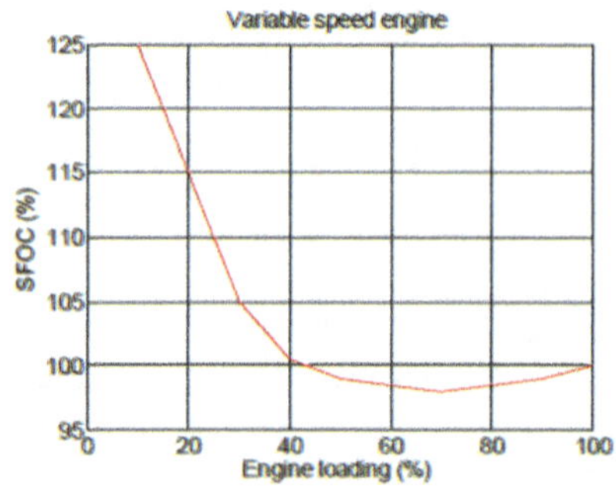

Fig. 3. SFOC curve for a diesel generator

5 Optimization Process

The sailing route is divided into several portions, each of which has a different water depth. The length of each portion x is D_x. Our goal is to find the optimal value for ship's speed i.e., V_x at each portion that minimize the total operation cost. This will estimate what would be the post state (cost) if a given speed is used [12]. So, V_x is the decision making variable. Since the ship is a hybrid diesel/battery ship the cost of both fuel and battery should be considered in the optimization. We define the objective function as follows:

$$\min Cost = \sum_{x \in X} C^F FC_x^{DG} t_x + C^{CH}\left(B_0^s - B_{n_x}^s\right) + \sum_{x \in X} C^{BD} B_x^{Dch} t_x \tag{14}$$

X is the set of all portions. t_x is the time point of leaving route portion x (h). C^F is the fuel price (\$/g), C^{CH} is the cost of charging the batteries at harbor before starting the journey, and C^{BD} is a cost that refers to batteries degradation due to discharging. B_x^{Dch} is the discharged power of batteries in portion x. n_x is the number of traveling route portions. B_0^s and $B_{n_x}^s$ are batteries state of charge (SOC) at the beginning and the end of the journey, respectively. The first term in (14) represents the total fuel consumption cost. FC_x^{DG} in (14) is obtained as a function of P_x^{DG} using (12) and (13). The second term is the cost of charging the batteries at harbor before starting the journey. The third term is a cost that is imposed due to batteries degradation by discharging [5]. The following constraints are considered in the optimization process:

Diesel generator's constraint:

$$0 \leq P_x^{DG} \leq \overline{P^{DG}} \tag{15}$$

$$\left(P_x^{DG} - P_{x-1}^{DG}\right)/t_x \leq \overline{R} \tag{16}$$

$\overline{P^{DG}}$ is the rated power of the diesel generator. $\overline{R}$ is the power ramp rate of the diesel generator (kW/h), and t_x is formulated as follows:

$$t_x = D_x / \left(3.6 \times 10^3 V_x\right) \tag{17}$$

Engines' power consumption constraint:

$$0 \leq P_x^e \leq \overline{P^e} \tag{18}$$

P_x^e is obtained by (1) considering (2)–(11). $\overline{P^e}$ is the rated power of the engines. Batteries operation constraints are as follows [13]:

$$\underline{SOC} \leq B_x^s \leq \overline{SOC} \tag{19}$$

$$B_{x-1}^s + e^{Ch} B_x^{Ch} t_x \leq \overline{SOC} \tag{20}$$

$$B_{x-1}^s - B_x^{Dch} t_x / e^{Dch} \geq \underline{SOC} \tag{21}$$

$$B_x^s = B_{x-1}^s + e^{Ch} B_x^{Ch} t_x - B_x^{Dch} t_x / e^{Dch} \tag{22}$$

$\overline{SOC}$ and $\underline{SOC}$ are upper and lower operation bounds of the batteries, respectively. B_x^s is the batteries SOC at portion x. B_x^{Ch} is the charged power in batteries at portion x. e^{Ch} and e^{Dch} are charging and discharging efficiencies. Ship's acceleration constraint is presented as below:

$$|V_x - V_{x-1}| / \left(3.6 \times 10^3 t_x\right) \leq \overline{a} \tag{23}$$

$\overline{a}$ is the maximum allowable acceleration (m/s^2). Traveling time constraint is:

$$\underline{T} \leq \sum\nolimits_{x \in X} t_x \leq \overline{T} \tag{24}$$

Fuel consumption constraint is formulated as below:

$$\sum\nolimits_{x \in X} FC_x^{DG} t_x \leq \overline{FC} \tag{25}$$

$\overline{FC}$ is the maximum allowable fuel consumption for the journey. This constraint can be considered for either the limitation of the fuel tank or greenhouse gas emission. Power flow constraint is presented as follows:

$$P_x^{DG} = P_x^e + B_x^{Ch} - B_x^{Dch} \tag{26}$$

The proposed formulation (14)–(26) is a non-linear optimization problem. Genetic Algorithm tool in Matlab software is used to solve this optimization.

6 Numerical Results

This section presents simulation results. Relevant ship information is presented in Sect. 2 and Table 1. Upper and lower bounds for the SOC of batteries is assumed to be 80% and 20% of the batteries capacity, respectively. Fuel cost, Battery charging cost and the cost related to the battery degradation is 0.0037 \$/g, 0.16 \$/kWh, and 0.045 \$/kWh, respectively [1]. Lower and upper limits of the traveling time are 40 and 50 min, respectively. The power ramp rate of the diesel generator is 0.720 kW/h [8]. The acceleration for the ship is also limited to 0.0083 m/s^2. Batteries charging and discharging efficiencies are 0.85 and 1 [4]. The length of each portion of the route and the respective water depth are presented in Fig. 4. Simulation results are presented in Fig. 5. In Fig. 5(a), the ship's speed is presented in knots, where one knot is equivalent to 0.5144 m/s. The speed at each portion of the route is different based on the water

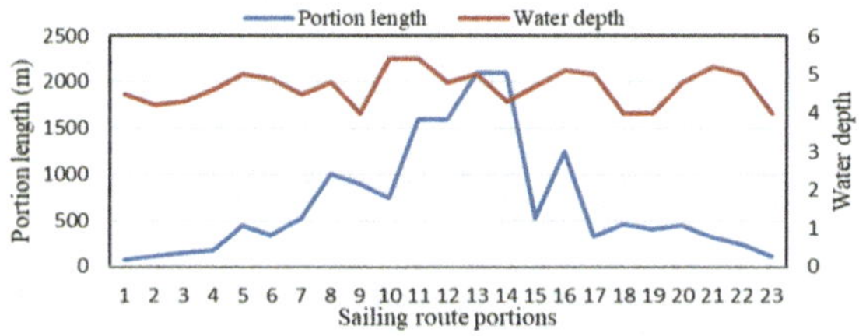

Fig. 4. Sailing route portions' length and water depth

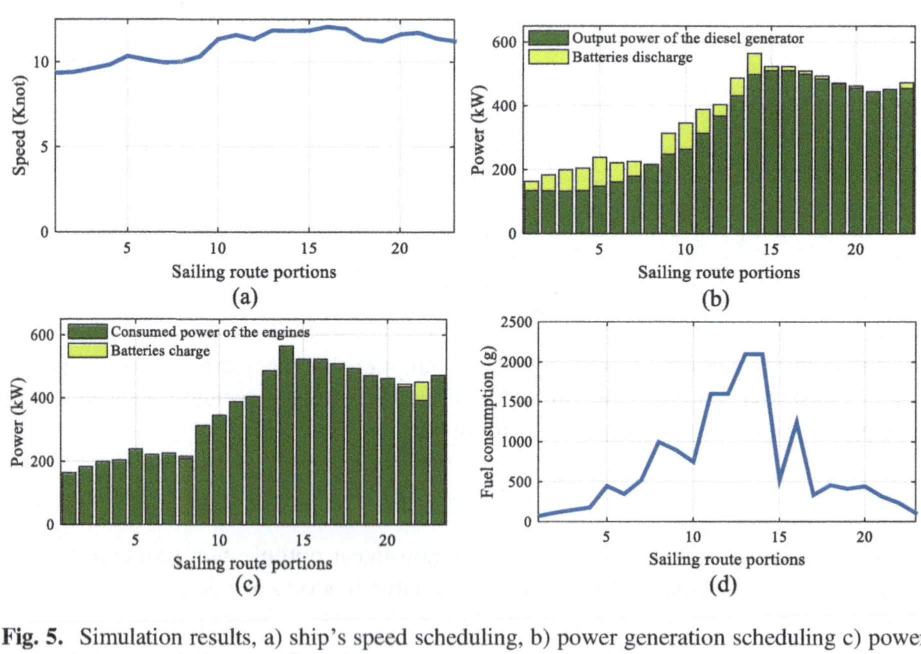

Fig. 5. Simulation results, a) ship's speed scheduling, b) power generation scheduling c) power consumption scheduling and d) fuel consumption of the diesel generator.

depth and operation constraints. It can be seen that the speed variation has been kept smooth as much as possible which makes it practical, and provides a comfortable journey to the passengers. Power generation and consumption flow in the ship are presented in Fig. 4(b) and Fig. 5(c), respectively. As shown in Fig. 5(b) the output power of the diesel generator is smooth and fast power fluctuations caused by the engines are supplied by the batteries. The fuel consumption process is illustrated in Fig. 5(d), which is dependent on both power generation by the diesel generator and the length of each route portion. According to Fig. 4, portions 13 and 14 have the longest distances and consequently high power consumption. The total operation cost obtained is 230 $, where 96.5% of that cost is related to the fuel consumption cost as the main source of energy. Total fuel consumption is 60 kg which is equal to the maximum limit of fuel consumption, and the total traveling time is 46 min.

7 Conclusions

In this paper, the optimal energy management of a hybrid diesel/battery ship that sails in shallow waters is studied. Sailing in shallow waters relates the ship's resistance to the water depth and this affects the efficiency of engines and consequently the total operation cost. In order to study these impacts, first, the ship resistance is formulated. Then the impact of the water depth on the resistance is mathematically quantified, and finally, a non-linear optimization problem is proposed to find the optimal speed of the ship in different portions of a traveling route. Simulation results show the efficiency of the proposed method in achieving a speed scheduling program that leads to a smooth and stable traveling for the ship while satisfying the operation and time table constraints. Moreover, the simulation results highlight the role of batteries in reducing the power fluctuations of the diesel generator caused by the engines.

References

1. Tang, R., Li, X., Lai, J.: A novel optimal energy-management strategy for a maritime hybrid energy system based on large-scale global optimization. Appl. Energy **228**, 254–264 (2018)
2. Kanellos, F.D.: Optimal power management with GHG emissions limitation in all-electric ship power systems comprising energy storage systems. IEEE Trans. Power Syst. **29**(1), 330–339 (2014)
3. Khooban, M.H., Vafamand, N., Boudjadar, J.: Tracking control for hydrogen fuel cell systems in zero-emission ferry ships. Complexity **20** (2019)
4. Letafat, A., Rafiei, M., Sheikh, M., Afshari-Igder, M., Banaei, M., Boudjadar, J., Khooban, M.H.: Simultaneous energy management and optimal components sizing of a zero emission ferry boat. J. Energy Storage **28**, 101215 (2020)
5. Khooban, M.H., Gheisarnejad, M., Farsizadeh, H., Masoudian, A., Boudjadar, J.: A new intelligent hybrid control approach for DC–DC converters in zero-emission ferry ships. IEEE Trans. Power Electron. **35–6**, 5832–5841 (2019)
6. Balsamo, F., Capasso, C., Miccione, G., Veneri, O.: Hybrid storage system control strategy for all-electric powered ships. Energy Procedia **126**, 1083–1090 (2017)

7. Hou, J., Song, Z., Park, H., Hofmann, H., Sun, J.: Implementation and evaluation of real-time model predictive control for load fluctuations mitigation in all-electric ship propulsion systems. Appl. Energy **230**, 62–77 (2018)
8. Zhengmao, L., Yan, X., Sidun, F., Yu, W.: Multi-objective coordinated energy dispatch and voyage scheduling for a multi-energy cruising ship. In: Proceedings of I&CPS Conference, Calgary, AB, Canada, pp. 1–8 (2019)
9. Molland, A.F., Wilson, P.A., Taunton, D.J., Chandraprabha, S., Ghani, P.A.: Resistance and wash measurements on a series of high speed displacement monohull and catamaran forms in shallow water. Trans. R. Inst. Naval Architect. **146**, 19–38 (2004)
10. Kristensen, H.O., Lutzen, M.: Prediction of resistance and propulsion power of ships. Technical university of Denmark, Project no. 2010-56, Emissionsbeslutningsstesystem Work Package 2, Report no. 04 (2013)
11. Yuan, L.C.W., Tjahjowidodo, T., Lee, G.S.G., Chan, R., Adnanes, A.K.: Equivalent consumption minimization strategy for hybrid all-electric tugboats to optimize fuel savings. In: Proceedings of ACC Conference, pp. 6803–6808 (2016)
12. Bodeveix, J.P., Boudjadar, A., Filali, M.: An alternative definition for timed automata composition. In: Proceedings of ATVA 2011. LNCS, vol. 6996, pp. 105–119. Springer (2011)
13. Banaei, M., Rezaee, B.: Fuzzy scheduling of a non-isolated micro-grid with renewable resources. Renew. Energy **123**, 67–78 (2018)

Linear Quadratic Regulator for Helicopter Model with a Prescribed Degree of Stability

Kaushik Raj[1(✉)], Santosh Kumar Choudhary[2], and Venkatesan Muthukumar[1]

[1] Department of Electrical and Computer Engineering, University of Nevada Las Vegas, Las Vegas, NV 89154, USA
raj@unlv.nevada.edu, venkatesan.muthukumar@unlv.edu

[2] Manipal Institute of Technology, Manipal Academy of Higher Education Manipal, Karnataka 576104, India
santosh.kumar@manipal.edu

Abstract. The 3-degree of freedom (DOF) helicopter is a benchmark aerodynamical laboratory model having non-linear characteristics and open-loop unstable dynamics that make the control of such system for either posture stabilization or trajectory tracking a challenging task for the control community. This article investigates a linear quadratic regulator (LQR) based optimal solution for the position and attitude control problem of 3-DOF helicopter system with a prescribed degree of stability to achieve desired attitude (elevation and roll angles) and position (travel rate). Numerical simulations are demonstrated to verify the effectiveness of the proposed control scheme.

Keywords: Attitude control · Position control · 3-DOF helicopter · Optimal control · Linear-quadratic regulator · Dynamic modeling

1 Introduction

Helicopters have become indispensable for the aviation sector due to their ease of take-offs, landings, and maneuverability. The helicopters are controlled broadly by linear non-linear design schemes. The design of the flight control system especially for the 3-DOF helicopter system can be regrouped into (i)Model-based method and (ii) Method based on experience. The linear–quadratic–regulator (LQR), linear– quadratic–Gaussian (LQG), and robust sliding mode control, etc. fall into the former group and proportional–integral–derivative (PID), fuzzy control, and neural network control into the latter.

In this article, the 3-DOF helicopter model produced by Googol Technology Ltd., Kowloon, Hong Kong is used in the experiment is a prototype model of laboratory helicopter (depiction in Fig. 1). Both propellers are driven by two DC motors mounted at both extreme ends of a rectangular frame. As depicted in Fig. 1. The prototype helicopter model frame is hanged from the one-end of the long arm, the other-end of the arm supports a balance block. The arm can roll across an elevation axis and swivel across a

H. Selvaraj et al. (Eds.): ICSEng 2020, LNNS 182, pp. 379–389, 2021.
https://doi.org/10.1007/978-3-030-65796-3_37

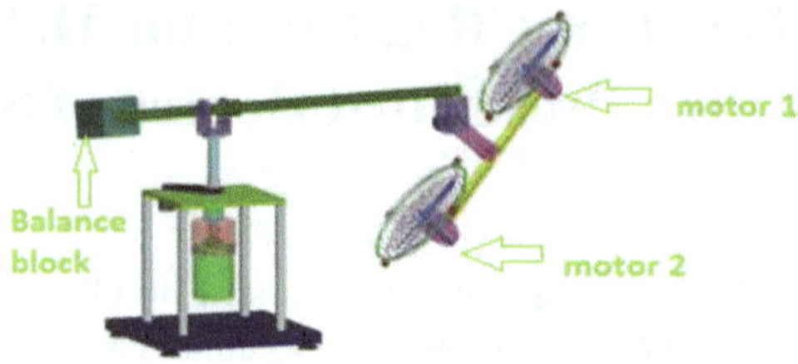

Fig. 1. 3-DOF laboratory helicopter model [5]

travel axis. The elevation of the body can be achieved by passing positive voltage across either motor. When the body rolls, the thrust vectors result in the travel of the body. The counterweight is provided for the balance which helps reduce the power requirements of the motors. The open-loop stimulation of the 3-DOF helicopter model reveals the inherently unstable dynamics. Model encounters uncertainty owing to external disturbances such as wind-gust, and coupling among roll angle and travel rate dynamics. In this scenario, control becomes a very complicated and challenging task. Over the years researchers have tried to develop various control schemes for the prototype 3-DOF helicopter model. In [10], a $\mathscr{H}_\infty$ based control robust controller was designed with limitations on restraining the effects of external disturbances. In [16], the essential motion in the travel axis was neglected and the goal was tracking the elevation and roll angle by using robust attitude control. A Linear–quadratic regulator (LQR) based controller is also designed in [7] under input and state constraints without discussing the prescribed degree of stability. In [9] researchers have developed a rigid body dynamical model and used an input-output linearization scheme whereas the article uses trajectory tracking of the two degrees-of-freedom. A unique reinforcement. A unique reinforcement learning design scheme has been used in the article [12]. A newly introduced negative imaginary control also applied to 3I DOF helicopter model in [3]. The article [3] discussed about $\mathscr{H}_\infty$ loopshaping based robust controller synthesis for prototype helicopter model. In article [15] a model based development of micro air vehicle altitude control using ground equipment for stable control in unpredictable flight-test environment is discussed. Acceleration feedback based hierarchical control scheme for robust trajectory control of a quadrotor drone in presence of external disturbances is demonstrated in [17]. The article provides an optimal solution for position and tracking control problem using a new Linear–quadratic regulator (LQR) with a prescribed degree of stability for the 3-DOF helicopter system. The article is limited to the digital simulation result using MATLAB and does not delve into the real-time interface with helicopter. The prototype helicopter model hangs from the one-end of the long-arm and the other-end of the arm supports a balance block. The arm can roll across an elevation-axis and swivel across a travel-axis. The elevation of the body is achieved by passing positive voltage across either motors. When body rolls, the thrust vectors culminate in the travel of the body. The counterweight is provided for the balance which helps reduce the power requirements of the motors.

The article provides optimal solution for the position and tracking control using a new Linear–quadratic regulator (LQR) with prescribed degree of stability for the

3-DOF helicopter system. The article is limited only to simulation results in MATLAB and does not delve into the real-time interfacing with helicopter. There are a number of other approaches that might benefit scientific community under various forms of present work [4, 13, 14]. The remainder of the article is categorized as following: Sect. 2 deals with brief formulation of control objectives and the mathematical modeling of a 3-DOF helicopter. The optimal controller design methodology in the Sect. 3 with a prescribed degree of stability is illustrated and the Sect. 4 is dedicated to verifying the effectiveness of the methodology through the numerical simulation results for controller design of helicopter.

2 Problem Formulation

2.1 Dynamical Model of the Laboratory Helicopter

The 3 DOF laboratory helicopter models are benchmark problems researched extensively within the academic and commercial circles. The flight behavior can be understood by simulating the mathematical model of the helicopter. Due to the cost constraint and complex design issues, evaluation of the system is quintessential to ensure the desired control objectives as required before implementation in the real-time.

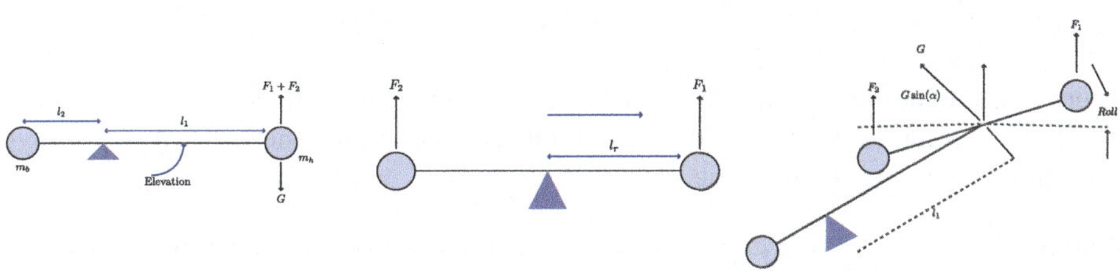

(a) Elevation axis schematic diagram

(b) Roll axis schematic diagram

(c) The schematic representation of travel-axis

Fig. 2. Schematic diagram of 3-DOF helicopter model

Consider the elevation axis schematic diagram as shown in Fig. 2a. The elevation axis dynamics of a 3-DOF helicopter system can be modeled as:

$$J_1 \frac{d^2\theta}{dt^2} = l_1(F_1 + F_2) - l_1 G = K_c l_1 (V_1 + V_2) - T_g$$

After ignoring the gravity torque disturbance,

$$J_1 \frac{d^2\theta}{dt^2} = K_c l_1 V_s \tag{1}$$

where θ is the elevation angle, system's moment of inertia $J_1 = m_h l_1^2 + m_b l_2^2$ is about elevation axis, m_b is the mass of the counter-weight blocks, m_h is propeller motor total-mass, voltages applied are V_1 and V_2 to the front and back motors which generates a force F_1 and F_2, motor-propeller force constant is K_c , pivot point to the propeller motor distance is l_1, pivot point to the counter weight blocks distance is l_2, effective gravitational-torque T_g is due to the mass differential G about the elevation-axis. If the force generated by the right motor is lower than the force generated by the left motor, the body of the prototype helicopter overturns clockwise. This phenomenon is considered as roll axis (see Fig. 2b) dynamics of the prototype 3-DOF helicopter system and it can be modeled as:

$$J_2\frac{d^2\phi}{dt^2} = l_r(F_1 - F_2) = l_rK_c(V_1 - V_2)$$
$$\Rightarrow J_2\frac{d^2\phi}{dt^2} = K_cl_rV_d \tag{2}$$

where ϕ is roll angle, J_2 = moment of inertia of the helicopter system about the roll axis, distance between the roll-axis to either motors are l_r. The horizontal-component of G causes a torque about that the travel-axis when the roll axis is tilted and overturned, which ensues in an acceleration along the travel-axis direction. Assuming that the helicopter body has rolled up atleast by the angle ϕ as depicted in Fig. 2c, then travel rate motion of the 3-DOF helicopter prototype model can be described as:

$$J_3\frac{d^2\psi}{dt^2} = -K_p\sin(\phi)l_1 \approx G\sin(\phi)l_1 \tag{3}$$

where J_3 is the moment of inertia of the helicopter model along the travel-axis direction, ψ is the travel angle, K_p is the force essential for maintaining the helicopter prototype in flight and is approximately the mass differential G

Remark 1. *When the roll angle value ϕ is equal to zero, then there will not be any force transmission along the travel-axis direction. A negative rolling motion produces an acceleration in the positive travel-axis direction and vice-versa.*

The simplified dynamic mathematical model of 3 DOF prototype helicopter depends on the assumptions listed below:

Assumption 1. *To validate the linear approximation of $3-DOF$ prototype helicopter, the angles θ, ϕ and ψ are assumed to be sufficiently small.*

Assumption 2. *The frictional forces and coupling-dynamics are ignored.*

Based on aforementioned assumptions, the prototype helicopter dynamical model established by (1), (2) and (3) is now effectively reduced in the equations shown below:

$$\ddot{\theta} = \frac{K_cl_1}{J_1}u_1 \tag{4}$$
$$\ddot{\phi} = \frac{K_cl_r}{J_2}u_2 \tag{5}$$
$$\ddot{\psi} = \frac{Gl_1}{J_3}\phi = \frac{T_g}{J_3}\phi \tag{6}$$

where control inputs u_1 and u_2 are given by

$$\begin{cases} u_1 = \frac{V_s+V_d}{2} \\ u_2 = \frac{V_s-V_d}{2} \end{cases} \tag{7}$$

Remark 2. *The u_1 and u_2 are the input voltages for motor 1 and motor 2 respectively.*

If $x = \begin{bmatrix} \theta \\ \phi \\ \dot{\theta} \\ \dot{\phi} \\ \dot{\psi} \end{bmatrix}$ and $y = \begin{bmatrix} \theta \\ \phi \\ \dot{\psi} \end{bmatrix}$ are the state variables and outputs of the helicopter system respectively, then a state-space realization of system described by (4), (5) and (6) can be constructed as follows:

$$\begin{cases} \dot{x} = \begin{bmatrix} \dot{\theta} \\ \dot{\phi} \\ \ddot{\theta} \\ \ddot{\phi} \\ \ddot{\psi} \end{bmatrix} = \begin{bmatrix} 0 & 0 & 1 & 0 & 0 \\ 0 & 0 & 0 & 1 & 0 \\ 0 & 0 & 0 & 0 & 0 \\ 0 & 0 & 0 & 0 & 0 \\ 0 & \frac{-T_g}{J_3} & 0 & 0 & 0 \end{bmatrix} \begin{bmatrix} \theta \\ \phi \\ \dot{\theta} \\ \dot{\phi} \\ \dot{\psi} \end{bmatrix} + \begin{bmatrix} 0 & 0 \\ 0 & 0 \\ \frac{K_c l_1}{J_1} & \frac{K_c l_1}{J_1} \\ \frac{K_c l_r}{J_2} & -\frac{K_c l_r}{J_2} \\ 0 & 0 \end{bmatrix} u \\ y = \begin{bmatrix} \theta \\ \phi \\ \dot{\psi} \end{bmatrix} = \begin{bmatrix} 1 & 0 & 0 & 0 & 0 \\ 0 & 1 & 0 & 0 & 0 \\ 0 & 0 & 0 & 0 & 1 \end{bmatrix} \begin{bmatrix} \theta \\ \phi \\ \dot{\theta} \\ \dot{\phi} \\ \dot{\psi} \end{bmatrix} \end{cases} \tag{8}$$

2.2 Control Objective

The article attempts to find the optimal solution for the position control (travel rate) and attitude control problem (elevation and roll angle) of 3-DOF prototype helicopter model via LQR with a prescribed degree of stability. The design objective for the controller can be stated as: finding the control input $u(t) \in [0, \infty)$ to obtain controller gain value K, which is the state feedback gain matrix, that provides optimal control law $u(t) = -Kx(t)$ to drive the system defined by (8) from perturbed state (non-zero) to nominal state (zero) so that cost function (performance index) (10) is optimized.

3 Controller Design Methodology

The design methodology of the controller is reproduced for improving the readability of the article. The elaborate materials are also in the research papers [1,2,6,8,11]. Consider a generalized linear time-invariant (LTI) model defined as:

$$\begin{cases} \dot{x} = Ax(t) + Bu(t); \;\; x(t = t_0) = x(0) \\ y = Cx + Du \end{cases} \tag{9}$$

If all the poles of the closed-loop system are placed on the left-side of $s = -\alpha$ in the controller design on the s-plane, where $\alpha > 0$, this is known as the *degree-of-stability of* α. The new cost function (performance measure) for LQR problems can be defined as

$$J = \frac{1}{2}\int_{t_0}^{\infty} e^{2\alpha t}\left[x^T(t)Qx(t) + u^T(t)Ru(t)\right] dt \tag{10}$$

Here, Q and R corresponds to the constant symmetric positive definite matrices and it is assumed that the pair $[A+\alpha I,\ B]$ is completely stabilizable
Now, the task remains to obtain the optimal control that minimizes the cost function (performance measure) in (10) based on the equation for dynamical-constraint in (9). Considering the transformation shown below

$$\hat{x}(t) = e^{\alpha t}x(t);\ \hat{u}(t) = e^{\alpha t}u(t) \tag{11}$$

Then, using the transformation scheme produced in the Eq. (11), system can easily be transformed into

$$\begin{aligned}\dot{\hat{x}}(t) = \frac{d}{dt}\left\{e^{\alpha t}x(t)\right\} &= \alpha e^{\alpha t}x(t) + e^{\alpha t}\dot{x}(t)\\ &= \alpha\hat{x}(t) + e^{\alpha t}\left[Ax(t) + Bu(t)\right]\\ \dot{\hat{x}}(t) &= (A+\alpha I)\hat{x}(t) + B\hat{u}(t)\end{aligned} \tag{12}$$

It is to be noted that there exists a relationship of $\hat{x}(t_0) = e^{\alpha t_0}x(t_0)$ in the initial condition of the modified systems (12) and the initial conditions for the original system (9) and in a scenario, when $t_0 = 0$, initial conditions for both original and modified systems are the same. By applying the transformation (11), the original performance index (10) can be reformulated as

$$J = \frac{1}{2}\int_{t_0}^{\infty}\left[\hat{x}^T(t)Q\hat{x}(t) + \hat{u}^T(t)R\hat{u}(t)\right] dt \tag{13}$$

For the minimization of modified system shown in the (12) and (13), an optimal control law can be defined as

$$\hat{u}^*(t) = -R^{-1}B^TP\hat{x}^*(t) = -K\hat{x}^*(t) \tag{14}$$

where, gain matrix $K = R^{-1}B^TP$ and the positive definite symmetric matrix P is the solution of the algebraic Riccati equation

$$P(A+\alpha I) + \left(A^T + \alpha I\right)P - PBR^{-1}B^TP + Q = 0 \tag{15}$$

By applying control law (14) in the modified system (12), the feedback system transforms into

$$\dot{\hat{x}}^*(t) = \left(A + \alpha I - BR^{-1}B^TP\right)\hat{x}^*(t) \tag{16}$$

Now, these results can simply be applied in the original system. Therefore, using the transformation (11) in (14), the state feedback control of the original system (9) becomes

$$\begin{aligned}u^*(t) = e^{-\alpha t}\hat{u}^*(t) &= -e^{-\alpha t}R^{-1}B^TPe^{\alpha t}x^*(t)\\ &= -R^{-1}B^TPx^*(t) = -Kx^*(t)\end{aligned} \tag{17}$$

However, the optimal control of the desired (original) (17) exhibits a similar layout to the optimal control of the modified system (14). Moreover, the cost function (performance index) for original or modified system is same and it is given by

$$\begin{aligned} \hat{J}^* &= \frac{1}{2}\hat{x}^{*T}(t_0)P\hat{x}^*(t_0) \\ J^* &= \frac{1}{2}e^{2\alpha t_0}x^{*T}(t_0)Px^*(t_0) \end{aligned} \tag{18}$$

One can observe that the feedback system (16) has eigenvalues with real parts less than $-\alpha$. This means that x^* tends to zero at least as fast as $e^{-\alpha t}$. Hence, one can conclude that the feedback system (16) has a degree of stability of at least α. The schematic diagram of the feedback control system is displayed in Fig. 3a as well.

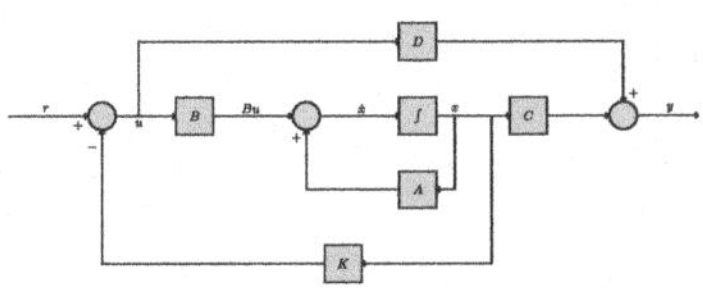

(a) The closed-loop optimal control system representation

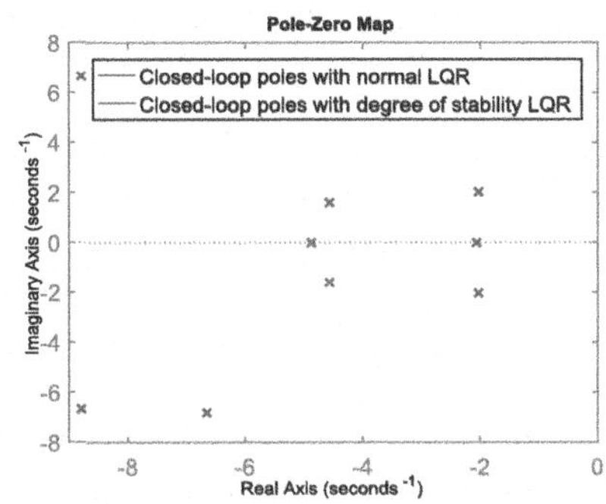

(b) Poles location of 3-DOF helicopter feedback system

Fig. 3. 3-DOF helicopter feedback control system

3.1 Weighting Matrix Selection

In the linear quadratic regulation (LQR) scheme, the weighting matrix selection is although an indispensable but dreadful task to achieve. The weighting matrices Q and R are chosen on trial and error basis to satisfy the expected performance criterion. Hence, the responses vary depending upon the different Q and R values. In a certain range, if Q is larger the system becomes more robust to disturbance and the settling time becomes shorter. However, there is no straight-forward method to choose weighting matrices and it is generally done via an iterative simulation. In this article, the Q and R weighting matrices are chosen as follows:

$$\begin{cases} Q = \rho C^T C \\ R = I \end{cases} \tag{19}$$

The increment in ρ suppresses the control cost and provides quicker response and vice-versa. Hence, It could be concluded as
Cheap control, $\rho \rightarrow \infty$: The control law stabilises and increases the rate of the system response.
Expensive control, $\rho \rightarrow 0$: Control law can only stabilises the system, effect on response isn't as fast as the cheap control scheme.

4 Numerical Simulation and Analysis of Results

The analysis of results and numerical simulation framework for a 3-DOF helicopter system includes posture stabilization (elevation and roll angle) and position tracking (travel rate), specified by desired elevation (θ) angle, yaw (ϕ) angle and travel rate (ψ). Now, the weighting matrices are chosen using (19) as mentioned in the Sect. 3.

$$Q = C^T C = \begin{bmatrix} 1 & 0 & 0 & 0 & 0 \\ 0 & 1 & 0 & 0 & 0 \\ 0 & 0 & 0 & 0 & 0 \\ 0 & 0 & 0 & 0 & 0 \\ 0 & 0 & 0 & 0 & 1 \end{bmatrix}$$

$$R = \begin{bmatrix} 1 & 0 \\ 0 & 1 \end{bmatrix}$$

The non-zero or perturbed states of initial condition of 3-DOF helicopter systems are considered as

$$x_0 = \begin{bmatrix} 1 & 1 & 0 & 0 & 1 \end{bmatrix}^T$$

Now the state feedback gain matrix K can be calculated and obtained by `MATLAB` code for gain `K=lqr(A,B,Q,R)`. Then the following control gain matrix is computed:

$$K = \begin{bmatrix} 0.7071 & 0.9225 & 0.3486 & 0.1201 & -0.7071 \\ 0.7071 & -0.9225 & 0.3486 & -0.1201 & 0.7071 \end{bmatrix} \tag{20}$$

In addition, the eigen values (closed-loop poles) of the system are placed at $-2.0286 \pm 2.0286i, -6.6486 \pm 6.8070i, -2.0644$. For acquiring all the poles of the system far from the imaginary axis to obtain a degree of stability of at least 2, the `MATLAB` code for gain `K=lqr(A+2× eye(size(A)),B,Q,R)` shall be employed to get desired solution in such scenarios. The state feedback controller gain matrix K is computed for driving the helicopter model from perturbed state (non-zero state) to nominal state (original zero-state) via optimal control law (14) and the matrix K is obtained as

$$K = \begin{bmatrix} 2.0111 & 1.6222 & 0.7843 & 0.1757 & -2.2446 \\ 2.0111 & -1.6222 & 0.7843 & -0.1757 & 2.2446 \end{bmatrix} \tag{21}$$

The desired location for closed-loop poles are depicted in Fig. 3b. The closed-loop system simulation is carried out to observe the performance of the position and tracking control of a 3-DOF helicopter. The initial state or perturbed state of the 3-DOF helicopter system is considered as $x(0) = [1\ 1\ 0\ 0\ 1]^T$. The Fig. 4a, Fig. 4b and Fig. 4c show

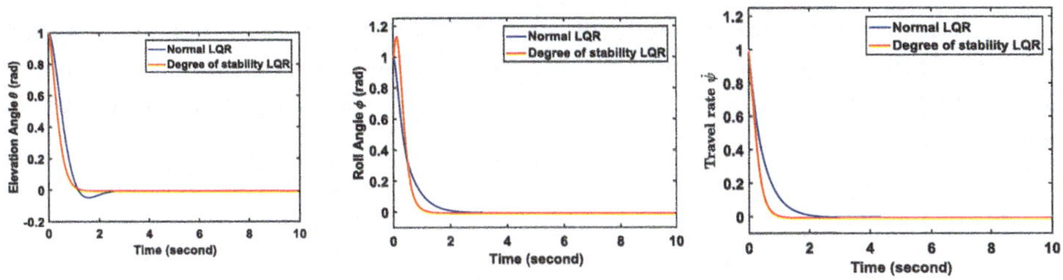

(a) Controlled state response of helicopter elevation axis

(b) Controlled state response of helicopter roll axis

(c) Controlled state response of helicopter travel axis

Fig. 4. States response of 3-DOF helicopter system

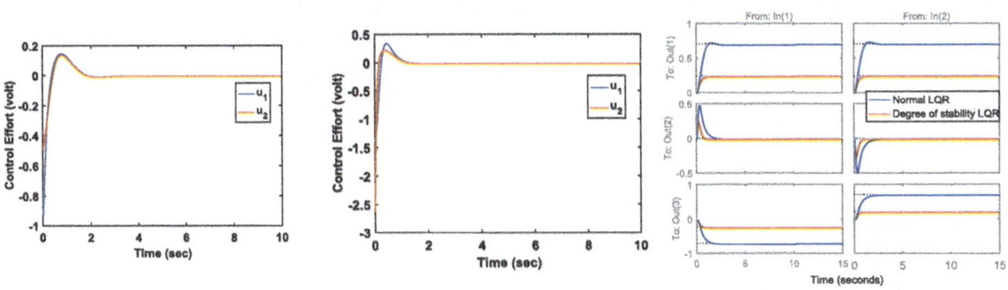

(a) Inputs for normal LQR

(b) Inputs for the degree of stability LQR

(c) The Closed-loop output response of a 3-DOF helicopter System

Fig. 5. Input/output of 3-DOF helicopter model

that controlled states for elevation, roll and travel axis of helicopter respectively and it is observed that both feedback controller gain matrices in Eqs. (20) and (21) successfully drive the helicopter from perturbed state to the nominal state using optimal control law (14).

The figures (Fig. 5a) and Fig. 5b demonstrate the control actions to stabilize helicopter by using normal Linear–quadratic regulator (LQR) and degree of stability Linear–quadratic regulator (LQR) techniques respectively and it is observed that the degree of stability LQR stabilises helicopter faster than the normal LQR. Thus, it can be concluded from the Fig. 4a, Fig. 4b, Fig. 4c, Fig. 5a and Fig. 5b that the LQR with prescribed degree of stability provides superior result than the normal LQR controller.

Finally, the authors check the design by plotting the step responses of the closed-loop system. It is recommended to always plot responses of the test signals such as the step responses to ensure that the transient response of the system is within the tolerance limit. The step response plots are the easiest way to check system stability. The plots depicted in Fig. 5c illustrate the tracking curve of reference input with control matrix (20) and (21). There are six plots in this figure that show the step response from each input to each output channel. These plots clearly show that the system is stable. Moreover, the optimal state feedback control for the prototype helicopter system with a prescribed degree of stability provides reasonable system performance. It enables the desired trajectory tracking, exhibits almost nil overshoot, and attain shorter settling time.

5 Conclusion

This article provides an intelligible mechanism for the optimal controller design of a 3-DOF helicopter system through a linear-quadratic regulator (LQR) with a prescribed degree of stability. The article first presents the dynamic mathematical equations together with the state-space model of 3 DOF helicopter and then, an optimal design approach is briefly described. The numerical simulation results of the article successfully demonstrate the optimal response of position and attitude control problem of a 3 DOF helicopter system with a prescribed degree of stability. Moreover, the robustness of optimal feedback control 3 DOF prototype helicopter model with a prescribed degree of stability is comparatively superior with regard to short settling-time, insignificant overshoot, and stable tracking.

References

1. Choudhary, S.K.: Design and analysis of an optimal orbit control for a communication satellite. Int. J. Commun. **9**(1), 28–35 (2015)
2. Choudhary, S.K.: Optimal feedback control of twin rotor MIMO system with a prescribed degree of stability. Int. J. Intell. Unmanned Syst. **4**(4), 226–238 (2016)
3. Choudhary, S.K.: Negative imaginary feedback control for a 3-DOF helicopter system. In: 2019 8th International Conference on Modeling Simulation and Applied Optimization (ICMSAO), pp. 1–6. IEEE (2019)
4. Choudhary, S.K., Raj, K., Muthukumar, V.: Optimal trajectory design and analysis for soft landing on the moon from lunar parking orbits. Int. J. Space Sci. Eng. **5**(4), 321–338 (2019)
5. Googol Technology Ltd., 3 DOF Helicopter Experiment manual, Kowloon, Hon Kong (2009)
6. Hespanha, J.P.: Linear Systems Theory. Princeton University Press, Princeton (2009)
7. Kiefer, T., Graichen, K., Kugi, A.: Trajectory tracking of a 3 DOF laboratory helicopter under input and state constraints. IEEE Trans. Control Syst. Technol. **18**(4), 944–952 (2010)
8. Kirk, D.E.: Optimal Control Theory: An Introduction. Courier Corporation, Chelmsford (2012)
9. Koo, T.J., Ma, Y., Sastry, S.: Nonlinear control of a helicopter based unmanned aerial vehicle model. IEEE Trans. Control Syst. Technol. (2001). https://people.eecs.berkeley.edu/~sastry/pubs/SastryAllJuly07/SastryALL/2002/KooNonlinear2002.pdf
10. Meza-Sánchez, I.M., Aguilar, L.T., Shiriaev, A., Freidovich, L., Orlov, Y.: Periodic motion planning and nonlinear $\mathscr{H}_\infty$ tracking control of a 3 DOF underactuated helicopter. Int. J. Syst. Sci. **42**(5), 829–838 (2011)
11. Naidu, D.S.: Optimal Control Systems. CRC Press, Boca Raton (2002)
12. Ng, A.Y., Coates, A., Diel, M., Ganapathi, V., Schulte, J., Tse, B., Berger, E., Liang, E.: Autonomous inverted helicopter flight via reinforcement learning. In: Ang, M.H., Khatib, O. (eds.) Experimental Robotics IX, pp. 363–372. Springer, Heidelberg (2006)
13. Saidi, A., Naceri, F., Vaidyanathan, S.: A new robust adaptive fuzzy synergetic control for nonlinear systems with an application to an inverted pendulum. Int. J. Model. Ident. Control **33**(2), 89–97 (2019)

14. Swain, S.K., Sain, D., Mishra, S.K., Ghosh, S.: Real time implementation of fractional order PID controllers for a magnetic levitation plant. AEU Int. J. Electron. Commun. **78**, 141–156 (2017)
15. Wright, S.: Model-based development of MAV altitude control via ground-based equipment. Int. J. Model. Identif. Control **28**(4), 361–369 (2017)
16. Yu, Y., Zhong, Y.: Robust attitude control of a 3 DOF helicopter with multi-operation points. J. Syst. Sci. Complexity **22**(2), 207–219 (2009)
17. Zaki, H., Alcan, G., Unel, M.: Robust trajectory control of an unmanned aerial vehicle using acceleration feedback. Int. J. Mechatron. Manuf. Syst. **12**(3–4), 298–317 (2019)

A Safety-Driven Cost Optimization for the Real-Time Operation of a Hybrid Energy System

Jalil Boudjadar(✉) and Mohammad Hassan Khooban

Department of Engineering, Aarhus University Denmark, Aarhus, Denmark
jalil@eng.au.dk

Abstract. Thinking-green consciousness is pushing international authorities to impose new regulations about the use of energy and CO2 emissions. Many hybrid energy solutions using renewable energies have been proposed and deployed. Designing real-time controllers that perform efficient and safe energy management is a challenging task due strong safety requirements, dynamic operation cost and real-time power demands. In this paper, we consider a hybrid combination of Fuel Cells (FC) and batteries sub-systems to aliment a marine vessel. Our contribution is to design a safety-driven efficient energy management control to dynamically schedule the energy sources to supply the real-time power requests so that 1) we maintain the system safety by not overloading or heating up an energy source; 2) reduce the operation cost by considering the cheapest energy source in a real-time manner. We use the symbolic model checker Uppaal to verify the safety properties. To demonstrate the efficiency of our energy scheduling, we analyze an actual case study.

1 Introduction

Due to the increasing energy demand, different authorities have dictated many regulations and standards to reduce energy utilization and pollutant emissions [11]. Few renewable energy technologies have been, therefore, emerged.

Although it is promising, Fuel cell (FC) is still regarded as a fledgling technology for marine applications [15,32], and usually paired with reversible energy storage devices such as batteries and ultracapacitors to create hybrid energy supply solutions, shortly known by FC-hybrid power systems [10]. FC-based systems have very strict safety requirements due to the existence of hazardous phenomena such explosion of hydrogen tanks and batteries, melting of fuel cells, storage risks, etc. The complexity and criticality of such systems make the underlying real-time control complex as it has to consider different constraints related to safety and reliability [1], operation cost [32] and performance [7]. To comply with the safety requirements, FC-hybrid energy systems are subject to a rigorous verification and validation prior to deployment [32].

The safety and energy efficiency of FC-based systems have been studied for different application areas [3,7,15,19,28] such as road vehicles [2], buses [21]

H. Selvaraj et al. (Eds.): ICSEng 2020, LNNS 182, pp. 390–400, 2021.
https://doi.org/10.1007/978-3-030-65796-3_38

and trams [33], however only few attempts consider both metrics together when designing controllers for safety-critical applications. Making safety as the main driving property in the design of safety-critical systems may lead to expensive operation cost and inefficient utilization of the energy resources. This paper introduces a safe controller for cost-effective real-time scheduling of the energy sources of a hybrid fuel cell-based power system (HFC). Our real-time controller focuses on finding a compromise between safety and cost to supply the real-time energy requests from propulsion motors so that the operation cost is reduced as low as the safety permits. Runtime scheduling decisions are made with respect to the actual system state such as devices temperature and available energy reserves. We use Uppaal model checker to examine the safety properties.

The rest of the paper is organized as follows: Sect. 3 presents the HFC systems architecture and the mathematical modeling of operation and temperature dynamics. Section 4 describes our safety-driven energy scheduling algorithm. Section 5 demonstrates our formal safety verification and cost effectiveness analysis on a case study. Section 6 concludes the paper.

2 Related Work

The safety and energy scheduling efficiency of FC-based systems have been studied for different application areas [6,7,19,20,28] such as road vehicles [2], buses [21] and trams [33]. However, only few attempts focused on the safety analysis of FC for marine applications [27,30].

The authors of [6,7] introduce a scheduler for the FC-based hybrid vehicle power systems. The control algorithm is coupled with an optimization process to find the efficient injection rate so that minimizing the gap between FC supply and the load demand. However, the main driving optimization factor relies on performance only which may jeopardize the system safety. Compared to that, our energy scheduling algorithm guarantees cost optimal operation while maintaining the system safety.

The authors of [12,13] propose a scheduling process for multi-stack FC systems to improve the reliability and lifetime of individual FC units. A fuel cell is chosen to supply the load energy if the cost related to its activity is the smallest possible. Making cost as the decisive factor may overload the cheap resource units, thus contributes in the temperature increase and violates safety.

Bigdeli et al. [5] propose a scheduling technique for a FC hybrid electric power generation system. The collaborative scheduling, where different energy sources operate simultaneously, is achieved using performance as the main criterion. This might lead to safety violation of the mostly active energy supply units.

The authors of [18] studied the scheduling of energy supply for FC hybrid electric vehicles with the objective to minimize the overall cost of hydrogen consumption. The collaborative scheduling of the energy sources was performed by considering availability, performance and state of charge constraints.

Our paper is a design and validation of an efficient real-time scheduler, as a decision making process, of the energy sources of the hybrid fuel cell-based energy

(HFC) system. It focuses on finding a compromise between safety, performance and cost so that one maintains the system safe while reducing the operation cost as low as the safety permits.

3 System Architecture and Modeling

The system we consider is formed by a composition two subsystems [4]: FC-based and battery-based. The fuel cell-based subsystem consists of three Proton-Exchange Membrane FC units having different capacity (300 kwh, 300 kwh, 100 kwh). These units can operate collaboratively and exclusively to supply energy. The battery-based system is formed by two lithium batteries (200 kwh each) operating in similar way to FC units. The two energy subsystems have different safety and performance characteristics as well as cost. The control architecture of our HFC system is given as a hierarchical scheduling system. As depicted in Fig. 1, an individual controller is dedicated to operate each of the two energy subsystems locally. Moreover, a top level controller is considered as the main energy management system coordinating the different subsystems. Each subsystem can interact via 2 ways: send alarm events to the main controller and receive commands from the main controller through command bus, or inject energy to the DC network. The battery-based system can also receive energy from DC bus.

The top level controller H-Ctrl receives power requests from the load (propulsion engines) as signals via port R, real-time cost for the operation of both battery and FC via port C, and the internal configuration of each subsystem such as the amount of energy left in the storage units (battery charge level, volume of hydrogen left in FC tanks) and temperature of the different energy sources. Based on the different real-time inputs, the top-level controller calculates a scheduling decision and mediates the control to the controller of the energy subsystems identified to be suitable to deliver the requested power. The local controller decision depends again on the internal state of each energy unit, and the amount of energy to deliver.

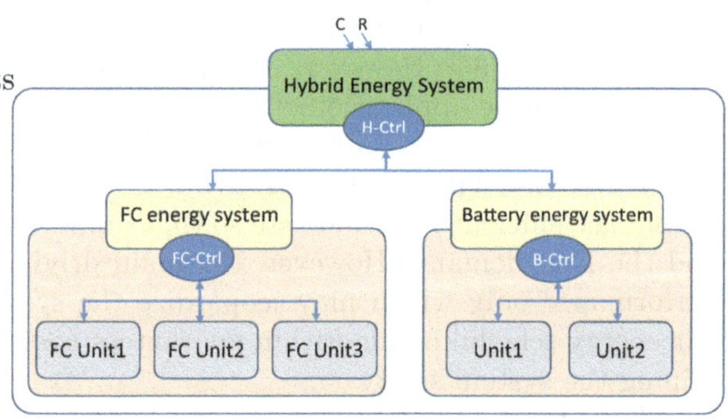

Fig. 1. Control architecture of a marine HFC

To model the system dynamics, we show how the temperature of energy sources is affected by their operation over time. Then, we model how the reserve of an energy source drains during operation. Finally, we show how the battery level is updated following a charging operation.

We define a power request $R = \langle e, d \rangle$ to be the energy supply rate e and the time duration d for which the supply holds. This is in fact one way to capture the load requests according to the propulsion force, speed and torque [16].

- We use F_i, with $i \in [1, 2, 3]$, to represent FC units and B_j, with $j \in [1, 2]$, for battery units.
- $Temp(X, t)$ is a function that returns the temperature of a energy source unit X at a given time point t.
- $H(t)$ returns the actual hydrogen volume at a given time point t.
- $SoC(B_j, t)$ returns the charge level of a given battery unit B_j at a given time point t.
- $U(X, t)$ returns the unit price of a given energy source at time point t.

3.1 Temperature Dynamics

We use thermal models to compute the temperature of battery and FC units following their real-time operations.

FC Temperature. The fuel cell temperature dynamics depends mainly on FC voltage level and the duration of each voltage level supply [24]. For a stable voltage and current density, which is the case here, FC temperature increases linearly with the power injection time [22]. Assuming the temperature of a FC unit F_i at time t is $Temp(F_i, t) = x$. The empirical analysis made in [22] leads to a simplified estimation of the temperature of F_i after satisfying a power request

$$\langle e, d\rangle \text{ as follows: } Temp(F_i, t + d) = Temp(F_i, t) + 2.58 * d \quad (1)$$

FC temperature can be reduced using a cooling system. For the sake of simplicity, we consider a static cooling coefficient k which means that the cooling system requires constantly the same current density e_c.

Battery Temperature. We calculate the real-time temperature of battery units following the charging/discharging operations using the thermal model introduced in [23]. The best temperature of batteries to operate with high performance is between 25 °C and 40 °C. Operating with a high temperature degrades the performance of a battery due to the internal resistance [23].

The temperature of a battery unit increases linearly when charging for a request $R = \langle e, d\rangle$ as follows [14]:

$$Temp(B_i, t + d) = Temp(B_i, t) + (d * 0, 162) \quad (2)$$

The battery temperature increases much faster during discharge mode with an exponential rate [25]. The higher discharge is, the faster the temperature increase will be. This is due to the rapid increase in cell impedance near the end of cell capacity. The battery temperature increase during discharge is given by:

$$Temp(F_i, t + d) = Temp(F_i, t) + d^{9/10} \quad (3)$$

3.2 Energy Injection

Injecting energy from a battery or a FC unit into the DC network, to satisfy a power request, decreases the energy reserve available in that storage device. Such a decrease depends on how much the current density supply is, and for how long the supply lasts. For the FC-based subsystem, to satisfy a power request $R = \langle e, d \rangle$ the selected FC unit needs to consume a hydrogen volume v calculated as follows [17]:

$$v(R) = (e * d/60 - 73.331)/55.944 * 10^3 \tag{4}$$

$v(R)$ is given in terms of liters. For the sake of simplicity, we ignore the dynamic performance of FC where the same volume of hydrogen can generate different energy amounts under different circumstances. Draining a battery with a current density e over a time duration $[t, t']$, namely an energy request $R = \langle e, [t, t'] \rangle$, decreases its charge with the following amount [19,23]:

$$drain(B_j, R) = \int_t^{t'} e \, d\pi \tag{5}$$

Thus the battery state of charge (SoC) at time point t' will be as follows [19]:

$$SoC(B_j, t') = SoC(B_j, t) - \frac{drain(B_j, (\langle e, [t, t'] \rangle))}{C_N^j} \tag{6}$$

C_N^j is the nominal capacity provided by the manufacturer for battery B_j.

3.3 Battery Charging

A battery charges if its SoC level is below the max SoC threshold. In a similar way to discharging, charging a battery increases its SoC following the current density and the charging duration. Thus, we define the charged amount $charge(B_j, R)$ as follows:

$$charge(B_j, R) = -1 * drain(B_j, R) \tag{7}$$

We omit the cooling down operation and the underlying energy consumption as it can simply be captured as a power request from the cooling system, which has to be added to the current power request of the top level controller.

4 Cost-Effective Scheduling Control

In this section we propose a real-time scheduler to arbitrate the two energy subsystems following the energy demands. The energy scheduling algorithm runs every time the controller H_Ctrl receives a new energy request from propulsion motors, or an alarm from one of the subsystems. The alarm signals can be:

- FC unit heating up to a threshold.
- Battery unit heating up to a threshold.
- Hydrogen tank of the FC-based systems running out.
- Battery unit reaching minimum or maximum storage level.

An alarm is issued by a local controller, FC_system or B_system, and communicated to the main controller H_Ctrl where a new scheduling is re-calculated.

To satisfy a power request, H_Ctrl controller checks first the availability of the requested energy in both subsystems. If such an energy amount is available in both resources, a safety check is performed to make sure that if a subsystem is used to deliver the requested energy it will not violate its safety properties such as maximum temperature and minimum reserve. If both subsystems are able to safely satisfy such a request, a cost analysis is carried out to choose the less expensive subsystem. In case neither of the subsystems is able to satisfy the power request individually, a collaborative scheduling will be triggered.

4.1 Energy Reserve Availability Check

This analysis stage is about to check whether the remaining energy reserve of a subsystem is sufficient to satisfy a given power request. In fact, for each demand we perform an analysis of the post-supply state that would be obtained if the demand request is satisfied [8]. Given a power request $R = \langle e, [t, t'] \rangle$, we first calculate the hydrogen volume $v(R)$ (Eq. 4) needed to produce the energy requested in R. If the required hydrogen volume $v(R)$ is larger than the actual hydrogen volume $H(t)$, at time instant t when R is issued, then FC-based subsystem is a candidate to satisfy R. We define the candidate function as follows:

$$Candidat(FC\_Ctrl, R) = \begin{cases} True \;\; If \;\; H(t) > v(R) \\ False \; Otherwise \end{cases}$$

Similarly, we define the function $Candidat(B\_Ctrl, R)$ for battery subsystem by:

$$\begin{cases} True \;\; If \; \exists B_j \mid drain(B_j, R) \leq C_N^j * (SoC(B_j, t) - Min\_SoC_j) \\ \qquad\quad Or \; drain(B_1, \langle e_1, [t, t'] \rangle) \leq C_N^1 * (SoC(B_1, t) - Min\_SoC_1) \\ \qquad\qquad \wedge \, drain(B_2, \langle e_2, [t, t'] \rangle) \leq C_N^2 * (SoC(B_2, t) - Min\_SoC_2) \\ \qquad\qquad \wedge \, e_1 + e_2 = e \\ False \; Otherwise \end{cases}$$

where $R_1 = \langle e_1, [t, t'] \rangle$, $R_2 = \langle e_2, [t, t'] \rangle$ and Min_SoC_j is the minimal state of charge of battery B_j used to prevent against over-discharge situations. In fact, we check whether there is a battery unit that is able to provide the totality of the requested amount by comparing the state of charge and draining that the request would make. If neither of the two battery units is able to provide the requested energy individually, we consider collaborative scheduling [9,26] where both battery units supply together two sub-requests, respectively R_1 and R_2 such that $e = e_1 + e_2$.

4.2 Safety Check

This analysis stage is about analyzing whether a candidate energy unit would violate its maximum temperature in case it is used to satisfy an energy request. To such an end, we use Eq. 2 and Eq. 3 to calculate what would be the temperature after satisfying a request. If the maximum temperature of a candidate energy source unit is not violated then such a device will proceed further for the cost analysis, otherwise the energy source unit will be discarded for supplying the totality of the requested energy and a collaborative scheduling has to be considered. Formally, we define a function $Safe(F_i, R)$ to check the safety of a FC unit as follows:

$$Safe(F_i, R) = \begin{cases} True \;\; If \; Temp(F_i, t') < Max\_Temp_i \\ False \; Otherwise \end{cases}$$

Accordingly, the fuel cell safety function $Safe(FC\_Ctrl, R)$ is given by:

$$\begin{cases} True \;\; If \; \exists \; i \mid Safe(F_i, R) \; Or \\ \qquad\quad \exists \; i \; i' \mid Safe(F_i, \langle e_1, [t, t'] \rangle) \; \wedge \; Safe(F_{i'}, \langle e_2, [t, t'] \rangle) \wedge e = e_1 + e_2 \\ False \; Otherwise \end{cases}$$

Similarly, we define the temperature safety property of a battery unit regarding a power request as follows:

$$Safe(B_j, R) = \begin{cases} True \;\; If \; Temp(B_j, t') < Max\_Temp_j \\ False \; Otherwise \end{cases}$$

Thus, battery subsystem is safe for a request R if either one of both units safely satisfy R. We define accordingly the safety function $Safe(B\_Ctrl, R)$ of such a subsystem as follows:

$$\begin{cases} True \;\; If \; \exists \; j \mid Safe(B_j, R) \\ \qquad\quad Or \; Safe(B_1, \langle e_1, [t, t'] \rangle) \wedge Safe(B_2, \langle e_2, [t, t'] \rangle) \wedge e = e_1 + e_2 \\ False \; Otherwise \end{cases}$$

The individual contribution percentages must satisfy both safety and availability check. In case both subsystems succeed in the safety check, we consider the related cost operation.

4.3 Cost Calculation

The cost analysis amounts to calculate the operation cost of each energy subsystem when satisfying a given energy request, using real time energy rates. We define the cost of FC_Ctrl system for a request $R = \langle e, [t, t'] \rangle$ by the volume of hydrogen to consume multiplied by the unit price of hydrogen $U(\mathcal{H}_2, t)$: $Cost(FC\_Ctrl, R) = v(R) * U(\mathcal{H}_2, t)$.

In a similar way, we calculate the budget of using battery-based system by the amount of energy to inject multiplied by the actual unit price of charging $U(\mathcal{E}, t)$: $Cost(B\_Ctrl, R) = drain(B_1, R) * U(\mathcal{E}, t)$

Since our batteries are identical, we use the unit price of battery B_1 no matter of which battery unit is selected by B_Ctrl. In case both energy subsystems have the same operation cost, we prioritize to use the FC-based subsystem due to its fast tanking operation.

4.4 Collaborative Scheduling

As mentioned earlier, if none of the energy subsystems passes successfully the availability and safety check to serve a power request individually we consider a collaborative scheduling where both subsystems contribute with different rates. The individual contributions of each subsystem must be checked for availability and safety [29].

Collaborative scheduling [9] is achieved by a binary process carried out in an iterative way following Dichotomy method [31]. For a given power request R, the start point of the iterative process consists in checking whether each subsystem is able to supply $R/2$. In case one of the subsystems fails the ability check, we reduce its contribution by half, i.e. to $R/4$, and so on until we find a contribution rate x the given subsystem is able to secure while the other subsystem is able to provide $R-x$ safely. Otherwise, when both subsystems are able each to provide $R/2$ the iterative process considers increasing the contribution of the resource having cheaper cost. We keep increasing the contribution x of the cheap subsystem on each iteration, by 50% of the value added on the last iteration, until it becomes not able to secure the contribution rate x. The final value of x, accordingly $R-x$ for the other subsystem, is most likely the contribution rate leading to optimal operation cost.

5 Formal Verification of Safety

To perform a formal verification of the safety properties, we mechanize the system model in Uppaal. In fact, we use symbolic model checking to examine safety properties whereas statistical model checking (SMC) is used for quantitative analysis. We consider four different safety properties given as follows:

$$S_1 \doteq \forall\ i\ t\ \ Temp(F_i,t) \leq Max\_Temp_i; S_2 \doteq \forall\ j\ t\ \ Temp(B_j,t) \leq Max\_Temp_j$$
$$S_3 \doteq \forall\ j\ t\ \ SoC(B_j,t) \leq Max\_SoC_j; S_4 \doteq \forall\ j\ t\ \ SoC(B_j,t) \geq Min\_SoC_j$$

Each safety property S_i is examined individually using Uppaal symbolic model checking as follows: $\forall[]\ S_i$. The following Uppaal SMC queries are used respectively to perform quantitative analysis of the batteries SoC and operation cost: `E[time≤1e6; 1000] (max:SoC`$_j$`)`; `E[time≤1e6; 1000] (max:cost)`. Each query specifies that SMC runs 1000 simulations, each of which last for one million time unit. To show the applicability and efficiency of our algorithm to schedule and control the HFC system, we have analyzed an actual marine case study system provided by our industrial partner Hvide Sande Ship Yard Co with real data of an actual 24-h journey: energy requests, real energy unit price changes, refueling/charging. Our formal analysis concludes that none of the safety properties is violated for the given journey data. A probability distribution of the state of charge (SoC) of battery unit B_1 is depicted in Fig. 2. One can see that our energy management algorithm maintains SoC_1 between the accepted bounds 15% and 90%, corresponding to 60 kwh and 360 kwh of the maximum capacity 400 kwh.

The analysis outcomes of the case study show a trade-off between the energy unit prices and the subsystems temperature. This is in fact due to a resource having a cheaper cost is most likely to be used with an extremely high contribution rate, which escalates its temperature.

6 Conclusion

This paper presents a safety-driven cost-effective controller to schedule the energy sources of a hybrid energy solution formed by a set of batteries and fuel cell units. The scheduling algorithm operates the different energy supply units following real-time demand, safety constraints and operation cost.

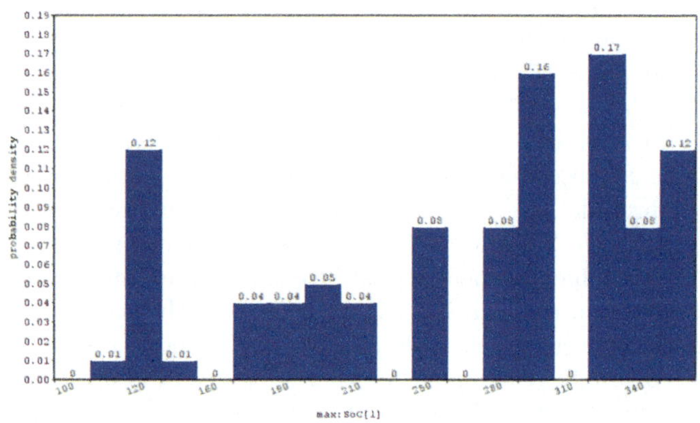

Fig. 2. Probability distribution of SoC_1 values

The proposed controller performs energy availability check availability followed by a safety check and cost calculation. An optimal scheduling is identified to be the cheapest safe energy supply configuration. We conducted a formal verification of the safety using Uppaal model checker where our algorithm proved to be capable to secure the system viability and safety while reducing the operation cost. We examined the proposed controller on an actual case study.

References

1. Alkaner, S., Zhou, P.: A comparative study on life cycle analysis of molten carbon fuel cells and diesel engines for marine application. J. Power Sources **158**(1), 188–199 (2006)
2. Andari, W., Ghozzi, S., Allagui, H., Mami, A.: Optimization of hydrogen consumption for fuel cell hybrid vehicle. Indian J. Sci. Technol. **11**(2), 1–9 (2018)
3. Banaei, M., Rafiei, M., Boudjadar, J., Khooban, M.: A comparative analysis of optimal operation scenarios in hybrid emission-free ferry ships. IEEE Trans. Transp. Electr. **6**(1), 318–333 (2020)

4. Bendjedia, B., Rizoug, N., Boukhnifer, M., Bouchafaa, F.: Hybrid fuel cell/battery source sizing and energy management for automotive applications. IFAC-PapersOnLine **50**(1), 4745 – 4750, 2017. 20th IFAC World Congress
5. Bigdeli, N.: Optimal management of hybrid PV/fuel cell/battery power system: a comparison of optimal hybrid approaches. Renew. Sustain. Energy Rev. **42**, 377–393 (2015)
6. Bizon, N.: Energy optimization of fuel cell system by using global extremum seeking algorithm. Appl. Energy **206**, 458–474 (2017)
7. Bizon, N.: Real-time optimization strategy for fuel cell hybrid power sources with load-following control of the fuel or air flow. Energy Convers. Manage. **157**, 13–27 (2018)
8. Bodeveix, J.-P., Boudjadar, A., Filali, M.: An alternative definition for timed automata composition. In: Bultan, T., Hsiung, P.-A. (eds.) Automated Technology for Verification and Analysis, pp. 105–119 (2011)
9. Boudjadar, J., Ramanathan, S., Easwaran, A., Nyman, U.: Combining task-level and system-level scheduling modes for mixed criticality systems. In: 23rd International Symposium on Distributed Simulation and Real Time Applications (2019)
10. Das, V., Padmanaban, S., Venkitusamy, K., Selvamuthukumaran, R., Blaabjerg, F., Siano, P.: Recent advances and challenges of fuel cell based power system architectures and control - a review. Renew. Sustain. Energy Rev. **73**, 10–18 (2017)
11. Eyring, V., Köhler, H.W., Lauer, A., Lemper, B.: Emissions from international shipping: 2. Impact of future technologies on scenarios until 2050. J. Geophys. Res. Atmos. **110**(D17), 1–18 (2005)
12. Herr, N., Nicod, J., Varnier, C.: Prognostics-based scheduling in a distributed platform: model, complexity and resolution. In: 2014 IEEE International Conference on Automation Science and Engineering (CASE), pp. 1054–1059 (2014)
13. Herr, N., Nicod, J.-M., Varnier, C., Jardin, L., Sorrentino, A., Hissel, D., Péra, M.-C.: Decision process to manage useful life of multi-stacks fuel cell systems under service constraint. Renew. Energy **105**, 590–600 (2017)
14. Hoffart, F.: Fast, high efficiency, standalone NiMH/NiCd battery charging. In: Dobkin, B., Hamburger, J. (eds.) Analog Circ. Des., pp. 429–430. Newnes, Oxford (2015)
15. Khooban, M.H., Gheisarnejad, M., Farsizadeh, H., Masoudian, A., Boudjadar, J.: A new intelligent hybrid control approach for DC DC converters in zero-emission ferry ships. IEEE Trans. Power Electron. **35**, 5832–5841 (2020)
16. Mapelli, F.L., Tarsitano, D.: Modeling of full electric and hybrid electric vehicles. In: Stevic, Z. (ed.) New Generation of Electric Vehicles, Chapter 7. IntechOpen, Rijeka (2012)
17. Minnehan, J.J., Pratt, J.W.: Practical application limits of fuel cells and batteries for zero emission vessels. Sandia National Lab (2017)
18. Neffati, A., Guemri, M., Caux, S., Fadel, M.: Energy management strategies for multi source systems. Electr. Power Syst. Res. **102**, 42–49 (2013)
19. Odeim, F., Roes, J., Heinzel, A.: Power management optimization of an experimental fuel cell/battery/supercapacitor hybrid system. Energies **8**(7), 6302–6327 (2015)
20. Olatomiwa, L., Mekhilef, S., Ismail, M., Moghavvemi, M.: Energy management strategies in hybrid renewable energy systems: a review. Renew. Sustain. Energy Rev. **62**, 821–835 (2016)

21. Peng, J., He, H., Xiong, R.: Rule based energy management strategy for a series–parallel plug-in hybrid electric bus optimized by dynamic programming. Appl. Energy **185**, 1633–1643 (2017). Clean, Efficient and Affordable Energy for a Sustainable Future
22. Rabbani, A., Rokni, M.: Dynamic characteristics of an automotive fuel cell system for transitory load changes. Sustain. Energy Technol. Assess. **1**, 34–43 (2013)
23. Samba, A.: Battery electrical vehicles-analysis of thermal modelling and thermal management, PhD thesis (2016)
24. Strahl, S., Costa-Castelló, R.: Temperature control of open-cathode PEM fuel cells. IFAC-PapersOnLine **50**(1), 11088–11093 (2017). 20th IFAC World Congress
25. Taheri, P., Bahrami, M., Fraser, S.: Temperature rise in prismatic polymer lithium-ion batteries: an analytic approach. SAE Int. J. Passeng. Cars Electron. Electric. Syst. **5**, 164 – 176 (2012)
26. Toor, A., ul Islam, S., Sohail, N., Akhunzada, A., Boudjadar, J., Khattak, H.A., Din, I.U., Rodrigues, J.J.: Energy and performance aware fog computing: a case of DVFS and green renewable energy. Future Gener. Comput. Syst. **101**, pp. 1112–1121 (2019)
27. Tronstad, T., Astrand, H.H., Haugom, G.P., Langfeldt, L.: Study on the use of fuel cells in shipping. EMSA European Maritime Safety Agency, pp. 1–108 (2017)
28. Vafamand, N., Boudjadar, J., Khooban, M.: Model predictive energy management in hybrid ferry grids. In: International Conference on Power and Energy Systems Engineering CPESE (2019)
29. Vafamand, N., Khooban, M.H., Dragičević, T., Blaabjerg, F., Boudjadar, J.: Robust non-fragile fuzzy control of uncertain dc microgrids feeding constant power loads. IEEE Trans. Power Electron. **34**(11), 11300–11308 (2019)
30. van Biert, L., Godjevac, M., Visser, K., Aravind, P.: A review of fuel cell systems for maritime applications. J. Power Sources **327**, 345–364 (2016)
31. Villa, T., Kam, T., Brayton, R.K., Sangiovanni-Vincentelli, A.L.: Synthesis of Finite State Machines: Logic Optimization (2011)
32. Wang, J.: Barriers of scaling-up fuel cells: cost, durability and reliability. Energy **80**, 509–521 (2015)
33. Zhang, W., Li, J., Xu, L., Ouyang, M.: Optimization for a fuel cell/battery/capacity tram with equivalent consumption minimization strategy. Energy Convers. Manage. **134**, 59–69 (2017)

Special Session: Software Engineering

Context Driven Software Development

Arunkumar Khannur(✉)

Christ University, Bengaluru, India
arunkumar.khannur@res.christuniversity.com

Abstract. The Context-Driven Software Development (CDSD) is a novel software development approach with an ability to thrive upon challenges of 21st century digital and disruptive technologies by using its innovative practices and implementation prowess. CDSD is a coherent set of multidisciplinary innovative and best practices like context-aware and self-adaptive system modelling, human-computer interaction, quality engineering, software development-testing-and continuous deployment frameworks, open-source tools-technology-and end-to-end automation, software governance, engaging stakeholders, adaptive solutioning, design thinking, and group creativity. Implementation prowess of CDSD approach stems from its three unique characteristics, namely, its principles, Contextualize-Build-Validate-Evolve (CBVE) product development element, and iterative and lean CDSD life cycle with Profiling, Contextualizing, Modelling, Transforming, and Deploying phases with in-process and phase-end Governance and Compliances. CDSD approach helps to address issues like complexity, software ageing, risks related to internal and external ecosystem, user diversity, and process-related issues including cost, documentation, and delay.

Keywords: Context-aware software · Self-adaptive software · Design thinking · Adaptive Problem Solving · Quality engineering · Software governance

1 Introduction

The value of software to users and organizations arises from its actual behavior in a particular context. Context-Driven Software Development (CDSD) approach is a novel software development approach for building quality software to implement this. CDSD brings in much needed major shift and value addition in software development. It uses evolved practices to deal with the modern-day disruptions. It enables the technology-enabled digital society to make quick progression towards Mark Weiser's vision of ubiquitous or pervasive computing [1] whose essence is the creation of an environment saturated with computing and communication capabilities, yet gracefully integrated with human users [2].

CDSD is based on context and modelling. Its characteristics and capabilities are derived as a conglomeration of theories and innovative multidisciplinary practices drawn from domains like context-aware and self-adaptive system modelling, human-computer interaction, quality engineering, software development-testing-and continuous deployment frameworks, open-source tools-technologies, end-to-end automation,

H. Selvaraj et al. (Eds.): ICSEng 2020, LNNS 182, pp. 403–412, 2021.
https://doi.org/10.1007/978-3-030-65796-3_39

software governance, engaging stakeholders, adaptive solutioning, design thinking, and group creativity. CDSD approach helps to address issues like complexity, software ageing, risks related to internal and external ecosystem, user diversity, and process-related issues including cost, documentation, and delay. The characteristics of CDSD approach deliver software with relevance that increasingly satisfies user and customer requirements, make modern-day organizations to create organizational strategic advantages and value in terms of building applications, operational efficiency, decision making, automation, and competitive edge in business. Following sections describe the CDSD Approach and its three characteristics, namely, CDSD principles, Contextualize-Build-Validate-Evolve (CBVE) product development element, and CDSD life cycle with the phases.

2 Context-Driven Software Development (CDSD) Approach

The power CDSD approach comes from its three unique characteristics, namely, its principles, Contextualize-Build-Validate-Evolve (CBVE) product development element, and CDSD life cycle (CDSDLC) with the phases with a set of tasks.

2.1 CDSD Principles

CDSD Principles provide the required basis and novelty to make the CDSD approach significantly different and unique. The principles include: focus on contextualizing, be adaptive, engage stakeholders, strive for solutioning using design thinking and context modelling, generate business value, engineer quality using lean philosophy, manage risks over defects and continuously deploy, end-to-end automation, and respect governance, compliance, and ethics.

Focus on Contextualizing
Software is becoming more and more user-centric with increasing priority for behavior of the system in a highly complex technical environment. To develop such a system, there is a need to define the context, making the system to be context-aware, and building an ability in the system so that interactions between users and the system will be effective. This requires the facility to capture explicit acts of communications from users, making communication effective by using the context to interpret and respond to these explicit acts, and interoperability through co-existence and collaboration among computers-content-communication-and connectivity to produce required functional and non-functional characteristics.

Be Adaptive
Modern-day software development shall be adaptive to thrive upon challenges by accommodating changing requirements and operational context. To achieve this, software development has to be seen as adaptive technical problem-solving and adaptive capacity shall be built through Adaptivity, Adaptive Leadership, and Adaptive Problem Solving [3]. Adaptivity is a strategy to inculcate the ability or tendency to adapt to different situations to survive and grow. Adaptive leadership helps in learning to control the context through experiments [3] by mobilizing and engaging

stakeholders with open mind towards a diversity of viewpoints to generate a large number of options. This brings in the ability to software development to deal with situations driven with an idea but with no clear definition of a problem, no well-defined problem-solving approach, and aligning to stakeholders. The entire exercise involves learning about: the problem-method-solutioning, implementing with solution-oriented goal driven mindset, and evolving on the move.

Engage Stakeholders

Stakeholders are individuals or organizations with direct or indirect influence on the requirements of the system [6]. Quality software development involves stakeholder engagement, evolving solutions by using their creative minds, and balancing the varying and conflicting expectations of the diversified stakeholders. Stakeholder engagement [7] is the process with a set of adaptive actions to balance stakeholder expectations. To address adaptive challenges it is very crucial to engage stakeholders to be conservative and progressive, exhibit experimentation mindset, build software by diversifying risks, and significantly displace-reregulate-rearrange some old practices [3]. The team with all stakeholders in it learns and practices a balanced approach of using the best possible previous good practices and know-how, and anchor change in values, competencies, and strategic orientation in each engagement. All this can be achieved using the process of engaging stakeholders with stages that include: understanding stakeholders, mobilizing stakeholders, and creating environment with creative dialogues.

In Understanding Stakeholders stage helps to reveal the stakeholders, their expectations, and the relationship among them. This is followed by Mobilizing Stakeholders stage to adapt to - situations and context, systems, and changes so as to collaborate to co-create and evolve value. In this regard, agile approach with group thinking and brainstorming will be of use. This requires transformation of every team member to be an evolved individual by transforming him from mere worker to involved team player to engaged intrapreneur. It is an ongoing process of creating eco-system to enagage stakeholders in dialogue and decision-making that supports experimentation with sharing project information, understanding stakeholder concerns, and building collaboration between stakeholders and stakeholders-implementers [4, 5].

Strive for Solutioning Using Design Thinking and Context Modelling

Software development shall focus on solutioning. Team attitude shall be built to exhibit solutioning mindset by keeping the end in mind, by being open for criticism, and respect diversified views of others to deliver the best. Solutioning excels with the use of design thinking for context and self-adaptive system modelling. Design Thinking [8] can be thought of as a discipline that uses designer's sensibility to context and methods to match the needs of stakeholders with what is technologically feasible and what is a viable business strategy to generate customer value, market, and strategic competitive edge. Design thinking starts with curiosity about any mystery or an idea more than observations and progresses by actively looking for new data points, challenge accepted explanations, and inferring possible new worlds [29]. It, according to Charles Sanders Peirce, believes in "logical leaps of the mind" far beyond what is possible at the moment by going beyond deductive and abductive logic and with ability to derive "inferences to the best explanation". The design shall fulfil certain requirements [8], just to mention: first requirement, the design must be matched to what is technologically feasible, and the second requirement, the innovation must make business sense. This guarantees implementability with business viable value addition. Design thinking involving creativity driven experimentation where we progress from delving into mysteries to moving through the knowledge funnel to identifying patterns to arriving at heuristics to algorithm [29]. In this, delving into mysteries is the most expensive activity. To generate immense value, combine design thinking with context and self-adaptive system modelling.

Generate Business Value

Every software shall provide a solution with business value. Software shall improve efficiency and effectiveness to optimally use resources, improve productivity, reduce waste, and generate capability to build strategic advantage for the organization. Timing and sequencing become crucial and critical to success and survival [3]. This requires framing key questions and issues to identify adaptive challenges to decide on direction; disclosing threats from business, technical, and operational environments; skakeholder engagement; diversified thinking and exposing conflicts; and challenge norms of let challenges be challenges. To facilitate this, software development team members shall inculcate observe-interpret-and intervene process [3].

Engineer Quality Using Lean Philosophy

Software system and subsystems shall be built by quality engineering and by using lean philosophy. Quality Engineering (QE) is preferred over Quality Control (QC) and Quality Assurance (QA) and it aims at building functionality, reliability, efficiency, usability, maintainability, and portability characteristics and related sub-characteristics based on ISO 25000 standard [33] and Dromy's model [32]. Lean Philosophy is used by incorporating Contextualize-Build-Validate-Evolve (CBVE) product element where the product is built based on the principle of building a little, validating it, and evolving over many iterations by going to field with active stakeholder participation.

Manage Risks over Defects and Continuous Deployment

Risk management is very crucial in software development [10]. Risk is a futuristic and potential concern that may occur at a particular time and cause an adverse effect. Every software faces risks. However how much importance and consideration shall be given to these risks during software development shall be decided by Risk Exposure Factor (REF) which is the product of (Probability of Occurrence $\times$ Frequency of Occurrence $\times$ Timeliness). Based on the context of the software and different elements that influence software and its operation, risks shall be identified, REF shall be calculated and used as the basis in prioritising risks, and strategy to address these risks shall be decided upon. In CDSD, software developments and related testing shall move away from scope bound, on-time, within budget, zero-defect mindset towards managing risks and performing risk based testing. Releasing the product with defects is a permissible practice here and is accompanied by risk management and continuous deployment with a small release cycles with automated framework for frequent build and release. This requires proper change management, configuration management, build and release strategy, continuous deployment, and end-to-end automation.

End-to-End Automation
One of the essential factors and principles of CDSD is End-to-End Automation. Appropriate automation framework shall be opted for with open-source tools for each phase. Also, automation of software management related life cycles like change management, documentation, configuration management, build and release also shall be implemented. Framework like DevOps [11] which supports automation of development and operation with a host of open-source tools may act as a high potential candidate for consideration.

Respect Governance, Compliance, and Ethics
The governance of software projects is defined as the complex process that is responsible for the control of project scope, progress, and continuous commitment of developers [34]. Governance is recognized to have a key role in enabling software project success [35]. Good Software Development Governance (SDG) tends to be focused on achieving strategic goals. SDG has three main concerns [35, 36], namely, manage value [10], develop flexibly, and manage risks. It helps a software organization realize its goals, leading to business benefits and creating value. It also helps mitigate risks and improves team effectiveness by enabling good communication and effective measurement and control. All this requires, effective Governance-Risk-Compliance in place with carefully incorporated governance and compliance structures with appropriate levels of ethical practices, policy, code of conduct, and training.

2.2 Contextualize-Build-Validate-Evolve (CBVE)

The CDSD life cycle for software development is based on lean product development principles and its phases are implemented using phases and small iterative Contextualize-Build-Validate-Evolve (CBVE) product development element. This CBVE Cycle has inbuilt ability to build and simultaneously validate with stakeholders and governance requirements.

2.3 CDSD Life Cycle

Lehman's laws of software evolution [40] state that the system must be continually adapted else it becomes progressively less satisfactory in use and the functional capability of systems must be continually increased to maintain user satisfaction over the system lifetime. This demands for evolving software to adapt to changing user needs and context in which the software is being used. Also, software development started to understand the value of building context-aware (pervasive computing) [1, 2] software which makes provision for incorporating timely changes and individual changes. In summary, self-adaptive and context-aware software development gained grounds. Self-adaptive software systems are software systems that can modify their behavior and/or structure in response to their perception of the environment and the system itself, and their goals [41]. CDSD Life Cycle considers all of these and aims at building self-adaptive and context-aware systems by using phases, namely, Profiling, Contextualizing, Modelling, Transforming, and Deploying. Each phase is implemented by engaging mobilized stakeholders and use of CBVE with support of the concepts of lean philosophy, design thinking, and quality engineering to propose an adaptive model with a viable solution. Each of these phases is explained below.

Profiling
Profiling Phase involves understanding the requirements of stakeholders and performing User Profiling, Business Value Profiling, Quality Profiling, Risk Profiling and using them as input to define Requirement Profiling. Based on Requirements Profiling, context profiling shall by derived by retrieving contextual information, followed by continuous evaluation of sensed/ inferred contextual information, and finally context adaptation through adaptation mechanisms. More details on this are available in [30].

Contextualizing
Contextualization understands that building and maintaining modern software systems requires developers to perform various tasks [9] with the help of models arrived at by context modeling. Context modelling is the task of representing the environment's states and situations as computational structures, to allow the software to process context information and react to changes in the environment. The effectiveness context modelling depends on tasks that include defining context of building software, adaptive context recommendations, creating complex context models using mental models, and finally organizing context.

As a first step in contextualizing is defining context. The context [22] is defined as any information that can be used to characterize the situation of an entity (i.e. whether a person, place or object) that are considered relevant to the interaction between a user and an application. Effective contextualizing depends upon various dimensions that include stakeholders, business, operational environment, quality, risks, behavior, adaptivity, trustworthiness, and sustainability.

As a second step, adaptive context recommendations shall be made by understanding and capturing the context [31] to provide relevant information about software development to developers. This makes developers to focus on creative tasks, how to execute a specific procedural task or what they should do next.

As a third step, mental models shall be defined to (re)establish their context of the software development which should culminate into context modelling. Based on the context model, model-centric development is initiated in which high-level, causally connected views of software and their application domain are available at run-time which ensure systematic transformation of the model to the software product [28]. It is significant to implement the solution at service level, functional level, logical cluster level, implementation level. Furthermore, such systems must be context-aware to control the scope of adaptations and changes. In this regard, proper development framework, technical platform, and programming language shall be used. Also, end-to-end automation across all phases of the development life cycle will be very crucial.

In the fourth step, defining and organization of context-awareness is carried out [22] to propose on Using Context and Adapting to Context to provide relevant information and/or services to the user, where relevancy depends on the user's task. Using Context considers the ability of computing devices to detect and sense, interpret and respond to aspects of a user's context [23–25]. Adapting to Context considers the ability of an application to dynamically change or adapt its behavior based on the context of the application and the user [26, 27].

The entire exercise of contextualizing is complex and challenging. Business Process Management (BPM) [12, 13] and Process Management System (PMS) will be handy in this exercise. BPM focuses on organizing context, uses techniques and tools to design, enact, control, and analyze business processes involving humans, organizations, documents and other sources of information [14], and finally define business processes. The business processes are the key instruments for visualizing and organizing the activities of a product or service and their interrelationships [14]. Following this, by using PMS the process models are defined by modelling activities and the relationships by considering three dimensions [15], namely, the control-flows, the data perspectives, and the resource perspective. Proper exception handling strategies shall be devised to act on exception that may be reported in the control-flow, data and/or resource models. Contextualizing shall consider different dimensions that include stakeholders, business, quality, behaviour, adaptivity, trustworthiness, sustainability and carry-out activities related to feasibility analysis, product persona creation, and value proposition.

Modelling

Modelling Phase considers Context, Requirement Profiling, and Process Models as input and by using model-driven engineering (MDE) defines models. Models are first-class entities that aim to develop, maintain and evolve software by performing model transformations to ensure software adaptation for both functional and non-functional goals. [19–21, 38]. Models are also the primary means by which developers and other systems understand, interact with, configure and modify the runtime behavior of software [37]. These models result in design and implementation of the adaptive system which can adapt their behavior to withstand unpredictable, uncertain, and continuously changing execution environments [38]. Developing such a model shall have context-aware and adaptive features and shall be supported by domain-specific modelling language to build models that are dynamic and include [39] abstractions of runtime phenomena, automate runtime adaptation, analyze running software systems, controller with a specific technique [38], and feedback loops [39]. Quality of models is achieved by having some characteristics like modularity, contextual alignment, clarity, completeness, unambiguity, and verifiability.

Transforming

Transformation of a source models into a target models shall be automatic and shall be based on a set of transformation rules that describe on how one or more constructs in the source model can be transformed into one or more constructs in the target model. In this, a taxonomy proposed by Mens et al. for the possible transformations may be of help [16]. This model proposes a different type of transformations, namely, endogenous Ex: the reverse engineering, the migration (Ex: from one platform and language to another), exogenous transformations (Ex: refactoring of a software), horizontal (Ex: refactoring of software to improve maintainability characteristic) and vertical transformations (Ex: refining of a specification into a full-fledged implementation), Syntactical and Semantical transformations by the means of successive refinement steps add more concrete details [17, 18].

Deploying: Continuous Deployment

Automated Continuous Deployment (CD) shall be done with the help of an appropriate automation framework for build and release and continuous deployment. In the selected automation framework, the currently operational software will be replaced by new software at a defined frequency to build and deploy new software. To implement CD, tasks that include- defining change management, and build and release related policy and procedure, defining CD strategy, arriving at build and release strategy, choosing tools for building and release, developing required utilities, defining back-up and release strategies, planning configuration audit, establishing archival strategy and also, implementation of these shall be ensured.

3 Conclusion

In the 21st century, digital world with disruptive technological eco-system is progressing towards ubiquitous or pervasive computing [1] faster than predicted. The nature of software and systems, their capabilities, and applications that would emerge will depend on altogether new characteristics and these shall operate on very complex but connected technical environment with requirements for functional and non-functional characteristics, adaptability, sustainability, and availability with entire innovation approaches in human-computer interaction. This situation created a motivation for seeking a new software development approach. CDSD is an approach that is an attempt in this regard and is based on Context-Aware and self-adaptive system modelling. It uses multidisciplinary innovative practices and implementation using three unique characteristics, namely, its principles, Contextualize-Build-Validate-Evolve (CBVE) product development element, and iterative and lean CDSD life cycle with Profiling, Contextualizing, Modelling, Transforming, and Deploying phases with in-process and phase-end Governance and Compliances. There is a scope in abundance for future work on developing techniques and methods to implement, human-computer interactions, defining frameworks – tools - and automation, and new research on developing self-adaptive related controls and feedbacks.

References

1. Weiser, M.: The computer for the 21st century. Sci. Am. **265**, 94–104 (1991)
2. Satyanarayanan, M.: Pervasive computing: vision and challenges. Pers. Comm. **8**, 10–17 (2001)
3. Heifetz, R., Grashow, A., Marty Linsky, M.: The Practice of Adaptive Leadership. Harvard Business Review Press, Boston (2009)
4. Ben-Chaim, Y, Levy, M., Irit, H., Eitan, F., Alla, B.: Engaging stakeholders. In: Globally Distributed Software Development Processes. MCIS 2010 Proceeding (2010)
5. Ahmad, S.J., O'Regan, N., Ghobadian, A.: Leadership, decision making and internal stakeholder engagement. Int. J. Manage. Decis. Making **6**(3/4), 345–358 (2005)
6. Kotonya, G., Sommerville, I.: Requirements Engineering: Processes and Techniques. Requirements Engineering: Processes and Techniques. Wiley, New York (1998)

7. Partridge, K., Jackson, C., Wheeler, D., Zohar, A.: The Stakeholder Engagement Manual Volume 1: The Guide to Practitioners Perspectives on Stakeholder Engagement, Stakeholder Research Associates Canada Inc. (2005)
8. Brown, T.: Design Thinking. Harv. Bus. Rev. **86**, 84 (2008)
9. Bradley, N., Fritz, T., Holmes, R.: Context-aware conversational developer assistants. In: 2018 IEEE/ACM 40th International Conference on Software Engineering (ICSE), Gothenburg, 2018, pp. 993–1003 (2018)
10. Boehm, B.W.: Software risk management: principles and practices. IEEE Softw. **8**, 32–41 (1991)
11. Ebert, C., Gallardo, G., Hernantes, J., Serrano, N.: DevOps. IEEE Softw. **33**(3), 94–100 (2019)
12. Duma, M., La Rosa, M., Mendling, J., Reijers, H.A.: Fundamentals of Business Process Management. Springer, Heidelberg (2013)
13. Weske, M.: Business Process Management: Concepts, Languages, Architectures. Springer, Heidelberg (2007)
14. Di Ciccio, C., Andrea Marrella, A., Alessandro Russo, A.: Knowledge-intensive processes characteristics, requirements and analysis of contemporary approaches. J. Data Seman. **4**, 29–57 (2015)
15. La Rosa, M., Dumas, M., Hofstede, A.H.M., Mendling, J.: Configurable multi-perspective business process models. Inf. Syst. **36**(2), 313–340 (2011)
16. Mens, T., Czarnercki, K., Van Gorp, P.: A taxonomy of model transformation. In: Bezivin, J., Heckel, R. (Hrsg) Language Engineering for Model-Driven Software Development. Dagstuhl, Germany 2005 (2005)
17. Wirth, N.: Program development by stepwise refinement. ACM 14 (1971)
18. Back, R.J., Von Wright, J.: Refinement Calculus. Springer, Heidelberg (1998)
19. Mukerji, J., Miller, J.: Model Driven Architecture, July 2001. http://www.omg.org/cgibin/doc?ormsc/2001-07-01
20. Atkinson, C., Kühne, T.: Model-driven development: a metamodelling foundation. IEEE Softw. **20**(5), 36–41 (2003)
21. Frankel, D.: Model Driven Architecture: Applying MDA to Enterprise Computing. OMG Press (2003)
22. Dey, A.K., Abbowd, G.D.: Towards a better understanding of context-awareness. In: Proceedings of the Workshop on the What, Who, Where and How of Context-Awareness, CHI2000 Conference on Human Factors in Computer System. ACM Press, New York (2000)
23. Hull, R., Neaves, P., Bedford-Roberts, J.: Towards Situated Computing. In: 1st International Symposium on Wearable Computers, pp. 146–153 (1997)
24. Pascoe, J.: Adding generic contextual capabilities to wearable computers. In: 2nd International Symposium on Wearable Computers, pp. 92–99 (1998)
25. Pascoe, J., Ryan, N.S., Morse, D.R.: Human-computer-giraffe interaction – HCI in the field. In: Workshop on Human-Computer Interaction with Mobile Devices (1998)
26. Cheng, B.H.C., de Lemos, Giese R.H., Inverardi, P., Magee, J.: Software engineering for self-adaptive systems. In: Dagstuhl Seminar Proceedings, Schloss Dagstuhl, Germany (2008)
27. Cheng, B.H.C., Atlee, J.M.: Research directions in requirements engineering. In: Future of Software Engineering 2007 (FOSE 2007), pp. 285–303. IEEE Computer Society, Minneapolis, May 2007
28. Nierstrasz, O., Denker, M., Renggli, L.: Model-centric, context-aware software adaptation. In: Cheng, B., de Lemos, R., Giese, H., Inverardi, P., Magee, J. (eds.) Software Engineering for Self-Adaptive Systems, vol. 5525, pp. 128–145. Springer, Heidelberg, (2009)

29. Martin, R.: The Design of Business – Why Design Thinking is the Next Competitive Advantage. Harvard Business Press, Boston (2009)
30. Sindico, A., Vincenzo Grassi, V.: Model driven development of context-aware software systems. In: International Workshop on Context-Oriented Programming, COP 2009, July 2009, pp. 1–5 (2009). Article No. 7
31. Melo, G., Alencar, P., Cowan, D.: Context-Augmented Software Development Projects: Literature Review and Preliminary Framework, arXiv: 1910.08167v1 [cs.SE] (2019)
32. Dromey, R.G.: A model for software product quality. IEEE Trans. Softw. Eng. **21**(2), 146–162 (1995)
33. Estdale, J., Georgiadou E.: Applying the ISO/IEC 25010 quality models to software product. In: Larrucea X., Santamaria I., O'Connor R., Messnarz R. (eds) Systems, Software and Services Process Improvement. EuroSPI 2018. Communications in Computer and Information Science, vol. 896. Springer, Cham (2018)
34. Renz, P.S.: Project Governance: Implementing Corporate Governance and Business Ethics in Nonprofit Organizations. Physica-Verl (2007)
35. Van Genuchten, M.: Why is software late? An empirical study of reasons for delay in software development. IEEE Trans. Software Eng. **17**(6), 582–590 (1991)
36. Biffl, S., Aurum, A., Boehm, B., Erdogmus, H., Paul Grünbacher, P.: Value-Based Software Engineering. Springer, Heidelberg (2005)
37. Chulani, S., Williams, C., Yaeli, A.: Software Development Governance and its Concerns, SDG2008. ACM (2008)
38. Filieri, A., Maggio, M.: Control strategies for self-adaptive software systems. ACM Trans. Auton. Adaptive Syst. **11**, 1–31 (2017)
39. Vogel, T.: Model-Driven Engineering of Self-Adaptive Software, UCT CS Colloquium, South Africa (2015)
40. Lehman, M.M., Belady, L.A.: Program Evolution: Processes of Software Change. Academic Press, Cambridge (1985)
41. Cáma, J., de Lemos, R., Ghezzi, C., Lopes, A.: Assurances for Self-Adaptive Systems: Principles, Models, and Techniques. Springer, Heidelberg (2013)

A Framework for the Semiotic Quality of User Stories

Pankaj Kamthan[1(✉)] and Nazlie Shahmir[2]

[1] Concordia University, Montreal, Canada
pankaj.kamthan@concordia.ca
[2] Canadian Pacific Railway Limited, Calgary, Canada
nazlie_shahmir@cpr.ca

Abstract. The use of user stories for expressing software requirements, in agile software projects and beyond, has only increased over the years. However, the results on the quality of user stories have been mixed. This paper proposes a semiotics-based, technology-and-tool-independent, semi-formal framework, comprising interrelated conceptual (meta-)models that provide an understanding to the concept of user story, user story context, user story quality, and violations of user story quality. In doing so, it elicits the unique nature of user story, highlights the challenges in modeling and addressing user story quality, presents the results of a preliminary survey of students and professionals on the use of user story, and, through various real-world examples, illustrates violations of user story quality.

Keywords: Agile methodology · Conceptual modeling · Continuous requirements engineering · Interactive system · Requirements debt · Standard

1 Introduction

It has been known for past several years that, irrespective of the software development methodology being used, *requirements engineering* (RE) is one of the earliest and most crucial phases in software development [1]. In recent years, the user stories are being transitioned to and adopted increasingly by academia and industry in their software projects for expressing requirements in a scenario-oriented RE approach to interactive software systems [2–4].

There are several reasons for paying special attention to the quality of (agile) software requirements, in general, and user stories in particular [5]. The quality of design, implementation, and testing, as well as customer and user acceptance, depends intimately on the quality of software requirements [6]. It has been reported that unacceptable quality of (agile) software requirements is among the common root causes of software project failures [7]. Finally, practices for ensuring quality of software requirements are correlated directly with an organization's process maturity.

The rest of the paper is organized as follows. In Sect. 2, necessary background is provided and related work is discussed. The specifics of the construction of elements of the framework, along with associated challenges, are presented in Sect. 3. In Sect. 4,

H. Selvaraj et al. (Eds.): ICSEng 2020, LNNS 182, pp. 413–422, 2021.
https://doi.org/10.1007/978-3-030-65796-3_40

directions for future research are outlined. Finally, in Sect. 5, concluding remarks are given.

2 Background and Related Work

2.1 Evolution of Requirements Engineering and User Stories

The conventional RE, as per IEEE Standard 830 and its successor the ISO/IEC/IEEE 29148 Standard, has evolved through multiple states, including agile RE, the current state being continuous RE, as part of the movement towards continuous software engineering [8]. There are a number of similarities between conventional RE and continuous RE, but there are also notable differences, as shown in Table 1.

Table 1. A comparison of conventional and continuous RE.

Criterion	Conventional RE	Continuous RE
People	▪ Small-scale collaboration (Local in SDLC) ▪ Feedback is late ▪ Feedback is not among the Reasons for modification of requirements	▪ Large-Scale collaboration (Global in SDLC) ▪ Feedback is early ▪ Feedback is among the primary reasons for modification of requirements
Project	▪ Idealistic ▪ Rigid philosophy ▪ Plan driven ▪ Outlook is long-term ▪ Requirements: Constant; Resources and time: Variable	▪ Realistic ▪ Agile philosophy ▪ Value driven ▪ Outlook is short-term ▪ Requirements: Variable; Resources and time: Constant
Process	▪ Predictive in nature ▪ Oblivious to environment (Market, Stakeholders) ▪ Requirements process is more linear, less nonlinear ▪ Shift-right: Tested late	▪ Adaptive in nature ▪ Sensitive to environment (Market, Stakeholders) ▪ Requirements process is more nonlinear, less linear ▪ Shift-left: Tested early
Product	▪ Targets an entire software system ▪ Preference for formality in expression/representation ▪ Aims to be complete	▪ Targets a part of software system (MVP or MMP) ▪ Preference for informality in expression/representation ▪ Aims to be just enough

2.2 Intrinsic Nature of User Stories and Its Implications

- **Abstractness.** A user story is about problem, and, by definition, a problem is abstract. It is not always easy, especially for non-technical stakeholders, to be able to reduce the *cognitive distance* between problem and solution. To be able to relate a user story to its concretization requires, apart from creativity, education and experience in the problem domain as well as in RE.

- **Futuredness.** A user story is about 'delayed gratification'. The scope as well as (positive or negative) effects of a user story are not always immediate, and may only be visible during later phases of the software development process. In particular, the value provided by a user story may not be apparent until it appears in an executable product.
- **Interrelatedness.** A user story is usually not completely isolated, and may, for some reason, be related to other user stories. For example, this is the case when a collection of user stories represents an *emergent property*, such as safety or security. As the number of user stories *scales*, it becomes ever more challenging to check for certain quality attributes, such as consistency and (horizontal) traceability.
- **Human-Centeredness.** A user story is a result, in part, of human endeavor, involving empathy, ingenuity, and sociability. Furthermore, as implied by Table 2, the user story process involves a conversation among different types of stakeholders, and cannot be automated entirely. It has been reported in empirical studies that inadequate human-to-human communication is one of the primary causes of errors in (agile) software requirements [9].

2.3 Previous Work on User Story Quality

The study of quality of software requirements has a long and interesting history [1, 10]. The attention on the quality of user stories is relatively recent, and, in that regard, there have been a number of proposals identified by mnemonic acronyms: C3 (Card, Conversation, Confirmation) [11], DEPT (Discrete, Estimatable, Prioritized, Testable) [3], INVEST (Independent, Negotiable, Valuable, Estimatable, Small, Testable) [2, 12], QUS (Quality User Story) [13], and ARVF (Agile Requirements Verification Framework) [14].

However, these initiatives are limited by one or more of the following: there appears to be no theoretical basis for modeling quality, the quality attributes as a collective appears arbitrary, the quality attributes are not all at the same level of abstraction, and the means of addressing the quality attributes are not considered.

3 The Construction of Elements of a Framework for the Semiotic Quality of User Stories

3.1 A Conceptual Meta-Model for User Story

Figure 1 shows a UML Class Diagram that represents a conceptual meta-model for user story.

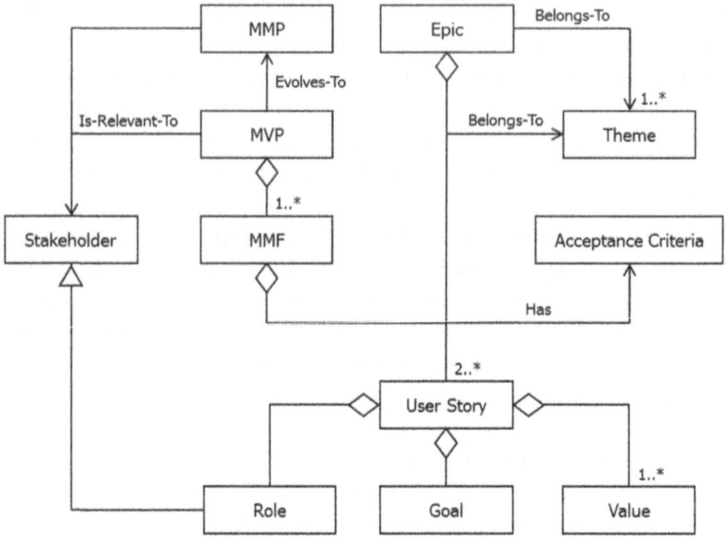

Fig. 1. A conceptual meta-model in UML for user story.

This meta-model is from two different perspectives, namely non-technical (managerial or business) and technical (development), as described in the following:

- **Conceptual Meta-Model for User Story: Non-Technical Perspective.** A *minimum marketable feature* (MMF), which is an atomic feature that creates market value in some way, is an aggregation of multiple user stories. A *minimum viable product* (MVP), which is a not necessarily deployable release 1, aimed primarily towards internal stakeholders, consists of at least one MMF. An MVP could, over time, evolve as a deployable release ≥ 2, aimed primarily towards external stakeholders, to become a *minimum marketable product* (MMP) [15].
- **Conceptual Meta-Model for User Story: Technical Perspective.** A user story is an aggregation of role, goal, and value. A role is a kind of stakeholder. An epic is an aggregation of multiple user stories [2]. An epic and a user story both belong to some theme, implicitly or explicitly. A user story is associated with *acceptance criteria*, which is usually expressed as a collection of tests.

3.2 A Conceptual Meta-Model for User Story Quality

Figure 2 shows a conceptual meta-model for user story semiotic quality. It is anticipated that such a model could be useful for authoring and reviewing user stories. For the sake of this paper, the purpose of software project artifacts is communication to human or machine, and *semiotics* is the study of languages for communication.

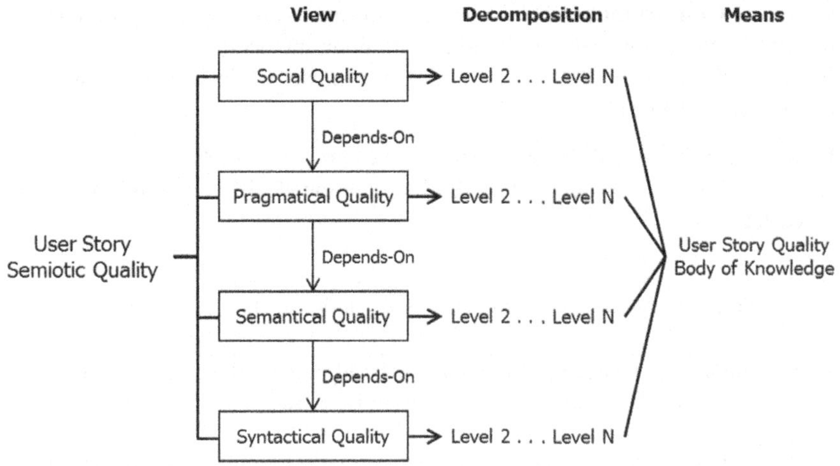

Fig. 2. A conceptual meta-model for user story semiotic quality.

The notion of semiotic quality in and of itself is at a too high a level to be useful, and therefore it needs to be *decomposed* into multiple, low levels to be meaningfully practical. (The decomposition relies on the software engineering principles of *separation of concerns* by views and *abstraction*.) This decomposition lends a *hierarchical structure* to the conceptual meta-model.

In Fig. 2, the Level 1 decomposition is from a linguistic perspective, and the Level $\geq$ 2 decomposition is from a continuous RE perspective. At Level 1, the views of interest are *View 1: Syntactical Quality*, *View 2: Semantical Quality*, *View 3: Pragmatical Quality*, and *View 4: Social Quality*.

The Level 1 can, if necessary, be refined further into sublevels, and, in some cases, doing so is necessary. For example, role, goal, or value of View 1, or ambiguity and consistency of View 2, could each be decomposed further. For operationalization, the quality attributes at the 'bottom' level of such a hierarchy are associated with entities of *body of knowledge on user story quality* (USQBOK). The mapping between the set of quality attributes at the bottom levels and the set of such entities is, in general, many-to-many.

3.3 Salient Characteristics of User Story Quality Views

- **Nature of Relationship.** The views are mutually orthogonal but interdependent for their existence. In particular, View 4 depends on View 3, View 3 depends on View 2, and View 2 depends on View 1.
- **Nature of Knowledge.** The type of knowledge varies upon traversal from View 1 to View 4. The knowledge of View 1 is explicit, while that of View 4 is a combination of tacit, implicit, and explicit knowledge.
- **Nature of Agent.** The type of agent varies upon traversal from View 1 to View 4. In particular, View 1 is machine-oriented, while View 4 is human-oriented.

- **Nature of Conformance.** The scale of conformance moves from discreteness and linearity to continuousness and nonlinearity upon traversal from View 1 to View 4. The conformance to View 1 is discrete and linear, while that to View 4 is continuous and nonlinear.

Table 2 provides a collection of user stories from a variety of different problem domains with violations of one or more aforementioned views of semiotic quality, as discussed later.

Table 2. A collection of examples of user stories of 'low' semiotic quality.

US_1. A student can register for any course offered in the Fall semester so that he or she can work towards completing the program he or she is in.
US_2. A surgeon can click on the `Surgery Schedule` button on the XYZ tablet computer to check which patients in the database are due for surgery on a particular day of the month.
US_3. A patron can download the restaurant menu as quickly as possible to see which food items to order.
US_4. An attendee can register for the conference, reserve a flight, and/or book a hotel room as early as possible.
US_5. A customer can use the membership card to receive a discount on any item in the store.
US_6. A customer can use the membership card to receive a discount on any item that is not already on sale in the store.

View 1: Syntactical Quality of a User Story. The syntactical quality is a contract between a user story and its underlying expression language. The prevalent expression language is a natural language, such as English, used in a controlled manner. A user story is typically expressed as a statement that contains one abstract role, one concrete goal, and (at least) one concrete value, in that order [2, 16].

It is possible to violate syntactical quality in different ways. For example, if a user story is a compound statement, such as when it has several roles or several goals, then that is a violation of syntactical quality. For another example, a user story that is not associated with any role, does not have any goal, or does not provide any value to the role in question, then that is a violation of syntactical quality.

View 2: Semantical Quality of a User Story. The semantical quality is a contract between a user story and its underlying problem domain. (For the sake of this paper, the "semantics" is lexical semantics.)

It is possible to violate semantical quality in different ways. For example, if a user story is indeterminate (does not represent an established reality of the problem domain), is incorrect (contradicts an established reality of the problem domain, or includes specifics of the solution domain, namely design or implementation), is ambiguous, is vague, or is redundant, then that is a violation of semantical quality. For a specific example, consider US_1. If a student can register for a course only after satisfying certain conditions, such as prerequisites, then that is a violation of semantical quality. Furthermore, US_2 is oriented towards the solution domain, US_3 is vague, US_4 has coordination ambiguity, and US_5 and US_6 are mutually inconsistent.

View 3: Pragmatical Quality of a User Story. The pragmatical quality is a contract between a user story and its stakeholders.

It is possible to violate pragmatical quality in different ways. For example, if a user story is not identifiable, cannot be read (easily), does not provide value to customers and/or users during any iteration, has not been estimated or cannot be estimated properly, has not been prioritized or has been prioritized incorrectly, cannot be traced, cannot be completed in a single iteration, cannot be modified (easily), or cannot be tested (easily), then that is a violation of pragmatical quality. For a specific example, using one of its interpretations, US_4 is rather large to be completed in a single iteration. (It is an epic, not a user story, and therefore needs to be split into multiple user stories such that its subset could lead to a MMF.)

View 4: Social Quality of a User Story. The social quality is a contract among the stakeholders of the user story.

It is possible to violate social quality in different ways. For example, advancing a user story for development without negotiation or consensus among the stakeholders is a violation of social quality. This could occur if a user story is decided by producers only, ignoring the needs of the customers and/or users, or is based solely on introspection and/or speculation.

3.4 Means for User Story Semiotic Quality

The user story resources include the (experiential) USQBOK, including metrics, and user story tools, such as ScopeMaster, among other possibilities. The entities of USQBOK could be classified as either *process-oriented* (about user story process) or *product-oriented* (about user story). In each of these categories, the USQBOK could be classified as either *positive* ("things to do") or *negative* ("things to not do"). For example, guidelines [2, 15] and patterns are positive, while smells [2] and anti-patterns are negative.

3.5 Scope of the Framework

For considerations of space, the proposed framework does not discuss the stakeholders of a user story; the specifics of role, goal, and value of a user story; the quality of acceptance criteria of a user story [17]; the severity of the violations of a user story semiotic quality; the tools that support user story semiotic quality; or the ROI of the user story semiotic quality.

3.6 A Survey on User Stories

To better understand the difficulties experienced by those in academia and industry in satisfying certain attributes of the semiotic quality of a user story, a small-scale survey was conducted during Fall 2019 and Winter 2020. The respondees were from Canada, and consisted of graduate students in software engineering programs and professionals in software-intensive organizations. The survey had 14 items, each based on a 5-point Likert Scale ranging from Strongly Disagree (1) to Strongly Agree (5). Figure 3 shows the results of 20 complete responses.

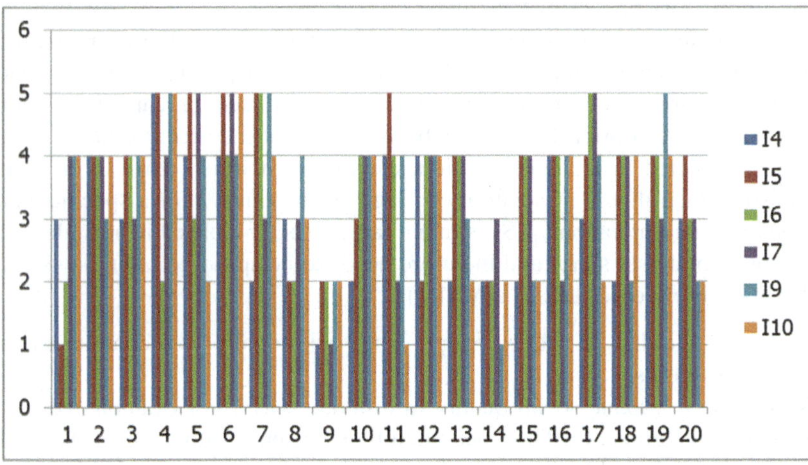

Fig. 3. The distribution of selected responses from a user story survey.

It can be concluded from Fig. 3 that a majority of respondees found identifying value of a user story (I4), separating problem from solution (I5), ensuring unambiguity of a user story (I6), ensuring consistency among user stories (I7), estimating a user story (I9), and ensuring that a user story can be completed in a single iteration (I10) to be relatively difficult.

4 Directions for Future Research

4.1 Ontological Representation of User Story Knowledge

In recent years, it has been shown that the ontological engineering of software requirements knowledge can aid communication; guide elicitation by problem domain knowledge representation; provide a formal classification; reduce ambiguity, inconsistency, incompleteness, and redundancy; and help towards evolvability, interoperability, and traceability [18]. The user story process is knowledge-intensive. In that regard, ontological representation of user story knowledge is of research interest. For example, the UML Class Diagram of Fig. 1 could be expressed as an OWL DL ontology [19], and from its individuals that include assertions of the form `Subject-Predicate-Object` (for role, goal, and value, respectively) desirable logical inferences could be deduced using a query language, such as SPARQL.

4.2 Causal Analysis of Violations of User Story Semiotic Quality

There can be a number of underlying causes of violations of a user story semiotic quality [9]: scarcity of resources, inadequate education in the problem domain or USQBOK, insufficient access to customer needs or preferences, slip (mental error), short-term expediency due to the pressure of time-to-market, lack of foresight (short-sightedness), presence of cognitive biases, resorting to logical fallacies, or deficiency of soft skills necessary for interviewing or negotiating. Indeed, knowing the origins (or root causes) of such violations could be useful for a *preventative* approach towards the semiotic quality of a user story, and is therefore also of research interest.

5 Conclusion

Even though the nature of RE has changed over the years, the classical and perennial question of the quality of software requirements remains as pertinent today as it was yesterday, regardless of the software development methodology being used or the RE process being followed.

In answering this question, it is important that the approach towards understanding, analyzing, and improving the quality of user stories be not only systematic but also realistic, as this paper has attempted to show. In particular, the quality of user stories needs to be *forethought*, not afterthought, as the user story process is not a time-independent, reversible process.

Furthermore, it should be underscored that, although any commitment to quality entails cost that needs to be rationalized, disregarding the violations of user story semiotic quality could contribute to the accrual of *user story debt*, a kind of requirements debt [20], from which it may be difficult for the stakeholders to recover, if at all, before the software system is delivered and is in operation.

Acknowledgement. The authors are grateful to CCE for a Professional Development Grant, and respondees of the survey for their interest and participation.

References

1. Boehm, B.W.: Verifying and validating software requirements and design specifications. IEEE Softw. **1**(1), 75–88 (1984)
2. Cohn, M.: User Stories Applied: For Agile Software Development. Addison-Wesley, Boston (2004)
3. Alexander, I., Maiden, N.: Scenarios, Stories, Use Cases through the Systems Development Life-Cycle. Wiley, New York (2004)
4. Leffingwell, D.: Agile Software Requirements: Lean Requirements Practices for Teams, Programs, and the Enterprise. Addison-Wesley, Boston (2011)

5. Medeiros, J., Vasconcelos, A., Silva, C., Goulão, M.: Quality of software requirements specification in agile projects: a cross-case analysis of six companies. J. Syst. Softw. **142**, 171–194 (2018)
6. Wiegers, K., Beatty, J.: Software Requirements, 3rd edn. Microsoft Press, Redmond (2013)
7. Curcio, K., Navarro, T., Malucelli, A., Reinehr, S.: Requirements engineering: a systematic mapping study in agile software development. J. Syst. Softw. **139**, 32–50 (2018)
8. Bosch, J. (ed.): Continuous Software Engineering. Springer, Cham (2014)
9. Hu, W., Carver, J.C., Anu, V., Walia, G.S., Bradshaw, G.L.: Using human error information for error prevention. Empirical Softw. Eng. **23**, 3768–3800 (2018)
10. Bell, T.E., Thayer, T.A.: Software requirements: are they really a problem? In: The Second International Conference on Software Engineering (ICSE 1976), San Francisco, USA, 13–15 October 1976
11. Jeffries, R.: Essential XP: Card, Conversation, and Confirmation. XP Magazine (2001)
12. Wake, W.C.: Extreme Programming Explored. Addison-Wesley, Boston (2002)
13. Lucassen, G., Dalpiaz, F., van der Werf, J.M.E.M., Brinkkemper, S.: Improving agile requirements: the quality user story framework and tool. Require. Eng. **21**(3), 383–403 (2016)
14. Heck, P., Zaidman, A.: A systematic literature review on quality criteria for agile requirements specifications. Softw. Q. J. **26**, 127–160 (2018)
15. Aschauer, B., Hruschka, P., Lauenroth, K., Meuten, M., Rogers, G.: Handbook of RE@Agile According to the IREB Standard. International Requirements Engineering Board (2019)
16. Wautelet, Y., Heng, S., Kolp, M., Mirbel, I.: Unifying and extending user story models. In; The Twenty Sixth International Conference on Advanced Information Systems Engineering (CAiSE 2014), Thessaloniki, Greece, 16–20 June 2014
17. Oliveira, G., Marczak, S., Moralles, C.: How to evaluate BDD scenarios' quality? In: The XXXIII Brazilian Symposium on Software Engineering (SBES 2019), Salvador, Brazil, 23–27 September 2019
18. Dermeval, D., Vilela, J., Bittencourt, I.I., Castro, J., Isotani, S., Brito, P., Silva, A.: Applications of ontologies in requirements engineering: a systematic review of the literature. Require. Eng. **21**(4), 405–437 (2016)
19. Murtazina, M., Avdeenko, T.: An ontology-based approach to the agile requirements engineering. In: The Twelfth International Andrei P. Ershov Informatics Conference on Perspectives of System Informatics (PSI 2019), Novosibirsk, Russia, 2–5 July 2019
20. Lenarduzzi, V., Fucci, D.: Towards a holistic definition of requirements debt. In; The 2019 ACM/IEEE International Symposium on Empirical Software Engineering and Measurement (ESEM 2019), Porto de Galinhas, Brazil, 19–20 September 2019

On Marketable User Stories for Customer Satisfaction

Pankaj Kamthan[1(✉)] and Nazlie Shahmir[2]

[1] Concordia University, Montreal, Canada
pankaj.kamthan@concordia.ca
[2] Canadian Pacific Railway Limited, Calgary, Canada
nazlie_shahmir@cpr.ca

Abstract. The profitability, reputability, and sustainability of any organization depend intrinsically on customer satisfaction. In the context of an agile project, this paper proposes a set of conceptual models that form an input to a customer-centered, market-oriented, user story engineering process aiming to provide a continual positive customer experience. In doing so, organizational constraints pertaining to economics of the user story engineering process are considered.

Keywords: Agile methodology · Axiology · Conceptual modeling · Requirements engineering · Software consumerization · User experience

1 Introduction

The consumerization of software has radically changed how software, as a commercial *product*, is viewed, perceived, managed, developed, tested, promoted, released, distributed, sold, deployed, used, reproduced, and maintained. In recent years, many organizations of all sizes have adopted agile methodologies for their software development-related initiatives [1]. The viability of any commercial organization rests on *customer satisfaction* [2]. It has been shown that customer satisfaction is a multi-dimensional concept contingent on many factors [3], including continual *positive customer experience*.

It is acknowledged broadly that principles and practices of in one of the earliest stages in development, namely *requirements engineering*, have a major influence on the direction of software development and outcome of a software project [4]. The purpose of this paper is to provide a starting point for augmenting 'conventional' agile RE with knowledge derived and distilled from developmental psychology, human-computer interaction, and experiential marketing in order to provide a positive customer experience.

The rest of the paper is organized as follows. In Sect. 2, background is provided and related work is highlighted. The conceptual models and, based on these models, a customer-centered and market-oriented process, necessary for user story engineering, are introduced in Sect. 3. In Sect. 4, directions for future research are outlined. Finally, in Sect. 5, concluding remarks are given.

H. Selvaraj et al. (Eds.): ICSEng 2020, LNNS 182, pp. 423–432, 2021.
https://doi.org/10.1007/978-3-030-65796-3_41

2 Background and Related Work

2.1 Agile Methodologies and Customer Experience

The agile methodologies are characterized by the *Agile Manifesto*. The Agile Manifesto includes principles and practices that prompt software development methodologies to satisfy the needs of small-and-medium-sized enterprises (SMEs).

In the evolution of the agile methodologies, there have been a number of paradigmatic stages. The interest in this paper is in the integration of agile methodologies and user-centered design methodologies with similar goals, such as *user experience design* (UXD) [5].

Figure 1 shows the intrinsic relationship between customer experience and related relevant concepts from UXD. It can be concluded that user experience is necessary, but not sufficient, for customer experience. For example, customer experience can involve aspects beyond the use of the product, such as those related to replacing the product or its parts, or returning the product and refunding the customer.

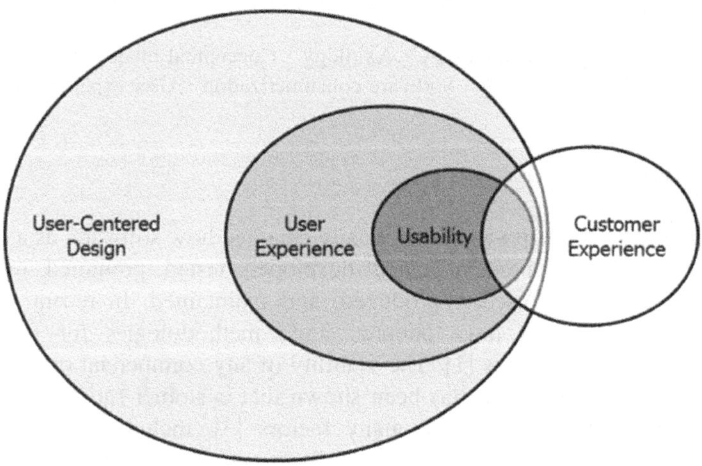

Fig. 1. The customer experience in context.

A *customer* is an individual who pays for software, and a *user* is an individual who for some purpose interacts with software. Therefore, in general, a user may not be a customer, and, conversely, a customer may not be a user. However, in this paper, for the sake of simplicity, user and customer are used synonymously, unless otherwise noted.

2.2 Agile Requirements

In agile methodologies, software requirements are usually expressed as either *user stories* or lightweight *use cases* [4]. The interest in this paper is in user stories [6], which are supported natively by many agile methodologies, including Extreme

Programming (XP), Scrum, and their extensions, such as BizDevOps and continuous requirements engineering; by organizational initiatives, such as the *Guide to the Business Analysis Body of Knowledge* (BABOK); and by the ISO/IEC/IEEE 26515 Standard.

A user story statement in controlled natural language text is usually structured as: A role can goal to value. For example, "A customer can make a payment using a credit card to purchase a shopping item." is a user story statement for a shopping system. For practical purposes, a user story statement is usually annotated with metainformation elements, such a priority, estimate, and acceptance criteria.

2.3 Marketable User Stories

Figure 2 shows a user story as part of different spatial and temporal configurations, each of which is described subsequently.

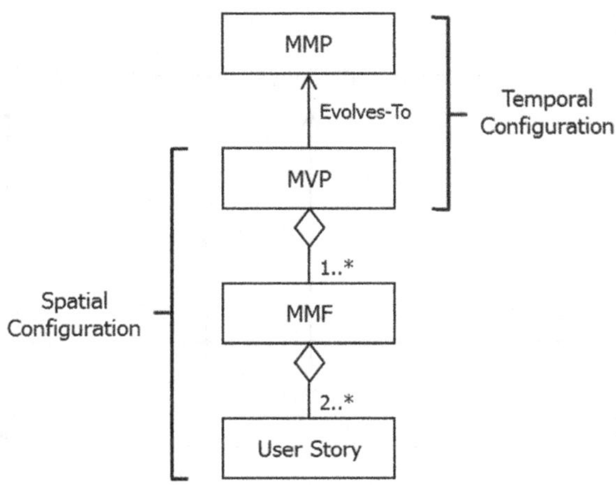

Fig. 2. An abstract hierarchy comprising MMP, MVP, MMF, and user stories.

A *feature* is a collection of capabilities of a software system that provides value to a user. A *minimal marketable feature* (MMF) is an atomic unit of software value creation [7]. An MMF creates market value (for the organization) in one or more of the following ways: competitive differentiation, revenue generation, cost saving, brand projection, and enhanced loyalty. In an agile project, an MMF is developed in a single iteration of the underlying agile process, and is composed of multiple user stories. From the perspective of release engineering, a *minimum viable product* (MVP), which is a not necessarily deployable release 1 and aimed primarily towards internal stakeholders, consists of at least one MMF. An MVP could, over time, evolve as a deployable release ≥ 2, aimed primarily towards external stakeholders, to become a *minimum marketable product* (MMP) [8].

3 A Framework for Customer-Centered, Marked-Oriented, User Story Engineering

3.1 Conceptual Models for a User Story

A *model* is a simplification of a thing for some purpose. In software engineering-related modeling, a thing can be almost anything, including a concept; a model is usually conceptual; the purpose of a model is usually reducing the size and/or complexity of a thing for the sake of understanding that thing; and a model is usually created through abstraction of a thing.

Figure 3 shows a compendium of preliminary, interrelated, conceptual models in relation to a user story statement.

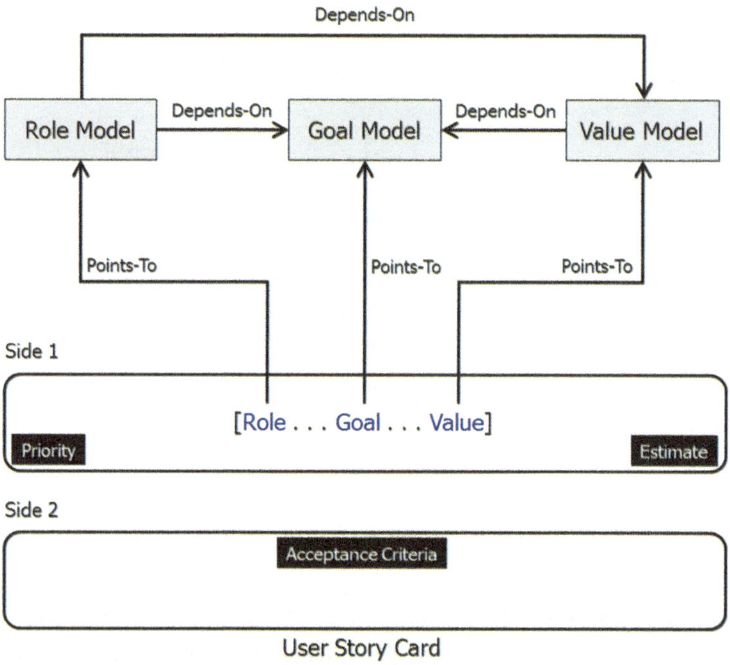

Fig. 3. An abstract user story derives knowledge from conceptual models for its description.

These conceptual models, each of which is described subsequently, are a requisite for appropriate user story engineering. This is because of a number of reasons, including that these models identify and separate relevant concerns, encapsulate thought processes, lend an understanding to and provide rationale for some of the activities during user story engineering, are an outcome of preparation necessary for user story engineering, can serve as sources during requirements traceability, and, being independent of any technique, technology, or tool, can be reused in similar contexts if necessary.

The development of these conceptual models requires a suitable combination of *ethnographic techniques* (such as interviews, observations, and surveys) and *ideation techniques* (such as brainstorming, mind mapping, and sketching).

These conceptual models could be complemented by a *glossary* to facilitate communication among stakeholders, create a common ground for terminological reference, and reduce the potential for ambiguity.

Finally, if deemed necessary, the conceptual models as well as other agile project artifacts could be placed on a suitable Social Web-based version control system, such as Wiki, to foster collaboration, support evolution, and expedite dissemination.

3.1.1 Role Model

This model provides an understanding of the characteristics and behavior of a typical user (playing a particular role with respect to the software system) [9]. There can be different kinds of role models. For example, a *persona* (an archetypical user of a software system) or a *proto-persona*, structured according to the ISO/IEC/IEEE 26515 Standard, is a candidate for the desired model [10]. In general, the mapping between the set of people and the set of roles is many-to-many.

3.1.2 Goal Model

This model provides an understanding of the goals of a typical user for using the software system. The construction of an appropriate goal model is especially crucial as both the role model and the value model depend on it. There can be different kinds of goal models. For example, the following organization of goals can be used for the desired model [11]: an *experience goal* (is at a visceral level and is about how a user wants to *feel* while using the system), an *end goal* (is at a behavioral level and is about what a user wants *to do* with the system, and a *life goal* (is at a reflective level and is about who a user wants *to be* by using the system).

3.1.3 Value Model

This model provides an understanding of the values that could be gained by the use of the software system. There can be different kinds of value models. For example, one possible model is a model for *consumption values* [12], which is aligned with the *Maslow's Hierarchy of Needs* and provides the following classification of values: *conditional* (say, value associated with a COVID-19 tracing application), *emotional* (say, value associated with a computer game), *epistemic* (say, value associated with the knowledge of a new shopping portal), *functional* (say, value associated with a statistical computing software), and *social* (say, value associated with an electronic social network).

3.2 Elements of Customer-Centered, Market-Oriented, User Story Engineering

3.2.1 A Customer-Centered, Market-Oriented, User Story Engineering Process

Figure 4 shows a single iteration of the desired customer-centered, market-oriented, user story engineering process. The solid lines indicate the transitions between steps, and the dashed lines suggest that the feedback is optional.

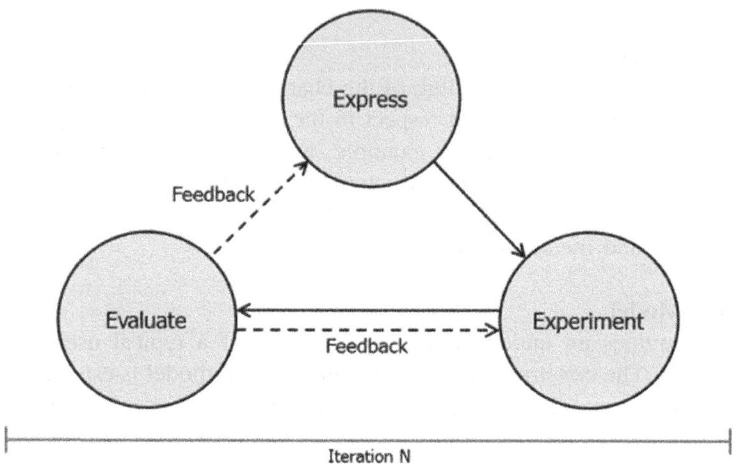

Fig. 4. The basic steps of a customer-centered, market-oriented, user story engineering process.

The steps of the aforementioned process can be explained as follows:

1. *Express* the user stories for all the MMFs aimed for a given iteration of an agile process. This would typically involve authoring, reviewing, and publishing the user stories. The knowledge from the aforementioned conceptual models is relevant to this step. The collection of user stories could be organized visually using techniques such as *user story mapping*, or, more generally, *customer journey mapping*. An economic challenge in this step is the accrual of *user story debt*.
2. *Experiment* with the MMFs by developing a high-fidelity UXD prototype. The goal of such a prototype should not only be to *maximize user experience*, but also to *minimize the cost of interaction* for the users. For example, such cost can manifest itself in form of inconvenience, unnecessary cognitive complexity, loss of information, unanticipated latency, or (excessive) time and effort spent in recovering from mistakes. This requires *empathy* of designers with the users [13], and an alignment between the *mental models* of designers and users [14]. It also requires adequate support in the prototype for error handling and helpful error messaging. To realize that, the prototype could rely on established marketing design knowledge, such as principles of persuasion [15], device knowledge, and information and interaction design knowledge, such as patterns [16]. An economic challenge in this step is the cost of representativeness of the prototype.

3. *Evaluate* the prototype for experience and satisfaction (by involving actual users, for example, through crowdsourcing [17] and survey, respectively). This could lead to feedback about the *product* (which could prompt a refinement of the user stories and/or the prototype), as well as about the *process* (which could prompt an improvement of the user story engineering process). An economic challenge in this step is the cost of testing, including regression testing, and the cost of *user story analytics*, including analysis of data extracted from the feedback.

The previous steps could be repeated to create an MVP, and, eventually, an MMP.

3.2.2 Economics of Customer-Centered, Market-Oriented, User Story Engineering

It is crucial that the value an MMP provides to an organization does not conflict with the value it provides to the (potential) customers. Indeed, for the agile software development to be successful, the aim of the business model should be to *maximize* the intersection of the two types of aforementioned values, namely the "Win-Win" value, as shown in Fig. 5. This can be done by applying *design thinking* [18] and *lean principles* [19] suitably, and, at the same time, avoiding *dark patterns* [20] that aim for exploitability rather than usability, and reducing the potential for *negative uses* [21] that emanate from the violations of, say, privacy, safety, and/or security.

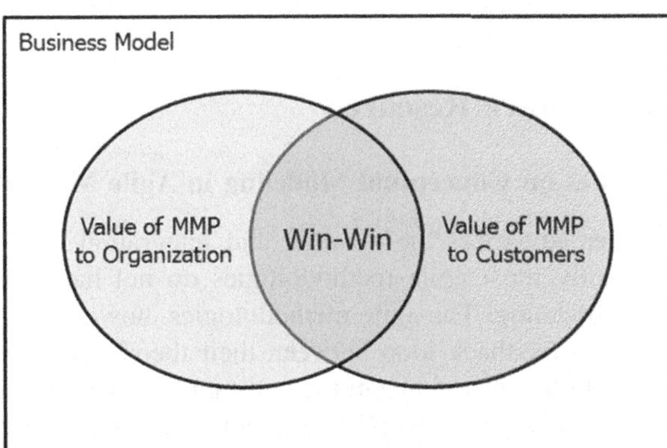

Fig. 5. The "Win-Win" value proposition.

The "Win-Win" value is neither automatic, nor free-of-cost. Indeed, a successful realization of "Win-Win" value is subject to trade-offs between certain organizational constraints pertaining to economics.

Figure 6 shows an equilateral triangle, representing indispensable, minimal, and (for the sake of simplicity) equally-significant constraints of a commercial agile project. It can be concluded from an examination of this triangle that it is necessary to make compromises among 'high value', 'low cost', and 'short time-to-market', and that it is usually possible to achieve only one out of three.

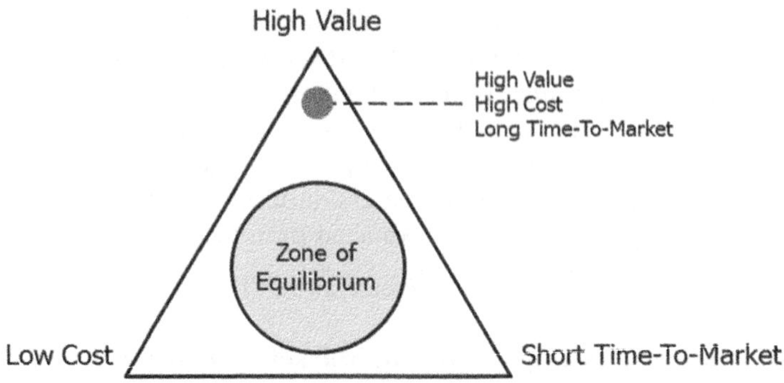

Fig. 6. The triangle of organizational constraints for a commercial agile project.

For example, an MMP aimed at providing 'high value' needs to be of 'high quality', and such an MMP, in turn, tends to take longer to develop and costs more; conversely, it is possible to shorten the initial development time, but that comes at the expense of reducing the value provided by the delivered MMP, assuming cost is kept almost constant.

Thus, practical considerations necessitate *equilibrium* among the constraints (and that, evidently, requires moderation of each of the constraints). To provide some flexibility, however, this equilibrium is not represented as a point but rather as a region.

4 Directions for Future Research

4.1 Empirical Studies on Conceptual Modeling in Agile Methodologies

Even though the paper advocates the creation and integration of a number of conceptual models, currently most agile methodologies do not have native support for extensive conceptual modeling. The agile methodologies, however, continue to evolve because of the continual feedback loop between their theory and practice. Therefore, conducting empirical studies in organizations with appropriate agile process maturity levels for eliciting the technical as well as non-technical challenges, analyzing associated risks, and using *earned value management* (EVM) as well as other metrics for calculating the *return on investment* (ROI) in the constructions of the conceptual models, is of research interest.

4.2 User Stories for Affective Values

The attention in this paper is almost exclusively on *utilitarian values*. To provide a positive user experience, it is also important to attend to *hedonic values* [11], such as enjoyment and engagement. The type of enjoyment in browsing a shopping system, however, can be different from playing a computer game, and the type of engagement in an interaction with a rental car reservation system is different from that of a banking system. It has been observed that due consideration of the hedonic values in software development may not only help the software satisfy the customers better, it may also help the organization gain competitive advantage by being able to market the product better [22]. Therefore, investigating the degree to which customers place significance on what kinds of hedonic values in their interaction with different types of software systems, and the implications of it for user story engineering, is of research interest.

4.3 Market-Oriented User Story Education

There are certain characteristics inherent to the *nature* of user stories (namely, abstractness, futuredness, and insipidness) that already pose a challenge in teaching user stories. Furthermore, in a real world context, non-technical concerns often constrain technical concerns. However, the students in 'conventional' software requirements engineering education are usually not exposed to realities of software-as-a-business, to intricacies of marketing aspects of a software product, or to the operational environment of a software product. Therefore, introducing such topics in software requirements engineering-related courses, and exploring their impact on the motivation and performance of students, is of research interest.

5 Conclusion

A market-oriented approach to software development opens new vistas for customers, but also brings new challenges for the software project team to be overcome. An agile customer-centered, market-oriented, methodology holds considerable promise for an organization committed to software development-related activities. If the industrial surveys are any sign, this trend is likely to continue for the foreseeable future.

A successful realization of prospective benefits of such an endeavor, however, requires the organization to be explicitly sensitive to the needs, values, and preferences of its customers, and be able to manage the trade-offs involved. It also requires customer satisfaction to be forethought rather than afterthought. This is possible by equipping and executing user story engineering with conceptual models that encapsulate knowledge related to customers, as this paper has attempted to show.

Acknowledgement. The authors are grateful to CCE for a Professional Development Grant.

References

1. Brunetti, G., Feld, T., Heuser, L., Schnitter, J., Webel, C.: Future Business Software: Current Trends in Business Software Development. Springer International Publishing (2014)
2. Fečiková, I.: An index method for measurement of customer satisfaction. TQM Mag. **16**(1), 57–66 (2004)
3. Klaus, P.: Measuring customer experience: how to develop and execute the most profitable customer experience strategies. Palgrave Macmillan, London (2015)
4. Wiegers, K., Beatty, J.: Software Requirements, Third Edition. Microsoft Press, Redmond (2013)

5. Brown, D.D.: Agile User Experience Design: A Practitioner's Guide to Making It Work. Morgan Kaufmann, Burlington (2013)
6. Cohn, M.: User Stories Applied: For Agile Software Development. Addison-Wesley, Boston (2004)
7. Denne, M., Cleland-Huang, J.: Software by Numbers: Low-Risk, High-Return Development. Prentice-Hall, Upper Saddle River (2003)
8. Aschauer, B., Hruschka, P., Lauenroth, K., Meuten, M., Rogers, G.: Handbook of RE@Agile According to the IREB Standard. International Requirements Engineering Board (2019)
9. Wendel, S.: Designing for Behavior Change: Applying Psychology and Behavioral Economics. O'Reilly Media, Sebastopol (2014)
10. Cooper, A., Reimann, R., Cronin, D., Noessel, C.: About Face: The Essentials of Interaction Design. Wiley, Hoboken (2014)
11. Norman, D.A.: The Design of Everyday Things. Basic Books, New York (2013)
12. Sheth, J.N., Newman, B.I., Gross, B.L.: Why we buy what we buy: a theory of consumption values. J. Bus. Res. **22**(2), 159–170 (1991)
13. Kamthan, P., Shahmir, N.: Effective user stories are affective. In: The Eleventh International Conference on Ubiquitous Computing and Ambient Intelligence (UCAmI 2017), 7–10 November 2017, Philadelphia, USA (2017)
14. Young, I.: Mental Models: Aligning Design Strategy with Human Behavior. Rosenfeld Media, New York (2008)
15. Cialdini, R.B.: Influence: The Psychology of Persuasion. HarperCollins, New York (2007)
16. Tidwell, J., Brewer, C., Valencia, A.: Designing Interfaces: Patterns for Effective Interaction Design, Third Edition. O'Reilly Media, Sebastopol (2020)
17. LaToza, T.D., van der Hoek, A.: Crowdsourcing in software engineering: models, motivations, and challenges. IEEE Softw. **33**(1), 74–80 (2016)
18. Maedche, A., Botzenhardt, A., Neer, L.: Software for People: Fundamentals, Trends and Best Practices. Springer-Verlag, Heidelberg (2012)
19. Harris, M.D.S.: The Business Value of Software. CRC Press, Boca Raton (2018)
20. Gray, C.M., Kou, Y., Battles, B., Hoggatt, J., Toombs, A.L.: The dark (patterns) side of UX design. In: The 2018 CHI Conference on Human Factors in Computing Systems (CHI 2018), 21–26 April 2018, Montreal, Canada (2018)
21. Kamthan, P., Shahmir, N.: A characterization of negative user stories. In: The Twenty Eighth International Conference on Software Engineering and Knowledge Engineering (SEKE 2016), 1–3 July 2016, Redwood City, USA (2016)
22. Agarwal, A., Meyer, A.: Beyond usability: evaluating emotional response as an integral part of the user experience. In: The ACM CHI 2009 Conference on Human Factors in Computing Systems (CHI 2009), 4–9 April 2009, Boston, USA (2009)

Special Session: Engineering of Cyber-Physical Systems

Hierarchical Versioning to Increase Compatibility in Signal-Oriented Vehicle Networks

Andreas Vetter[1](✉) and Eric Sax[2]

[1] Mercedes-Benz AG, Networking and Software Technologies, Böblingen, Germany
andreas.a.vetter@daimler.com
[2] Karlsruhe Institute of Technology (KIT), Institute for Information Processing Technologies (ITIV), Karlsruhe, Germany
eric.sax@kit.edu

Abstract. As the importance of electronics and software in automobiles grows, software maturity becomes an increasing challenge. To avoid the delays of production starts as well as recalls, new development processes are needed. To increase the testing abilities the compatibility between different development releases of electronic control units must be improved. To achieve this, we propose "hierarchical versioning", a novel versioning scheme and set of rules how to handle compatible and incompatible modifications, using the hierarchical structure of vehicle networks. Data of the AUTOSAR (AUtomotive Open Software ARchitecture) based development history of a current model series automobile has been analyzed to gauge the applicability of hierarchical versioning. We found a worst case bandwidth increase of about 30% for intermittent development releases when using the scheme to introduce complete backward compatibility for an additional release cycle. The bandwidth of the final release is not affected. During this analysis we also noticed that less than 3% of changes cause incompatibilities in more than 100 communication connectors, while most of the changes only affect less than 10 communication connectors. This may further aide in increasing compatibility between releases and reduce the added bandwidth.

Keywords: Embedded communication · Signal-oriented architecture · Versioning · Change management · AUTOSAR

1 Introduction

A major part of today's automotive engineering is the design of the electric and electronic architecture (E/E-architecture) and the software running on its electronic control units (ECU). Until 2025 the development costs for embedded software are expected to make up for 40% of the vehicle development costs [8]. By this time, the over the air updates are expected between annually and monthly [9]. Challenges in electronic and software design delay starts

H. Selvaraj et al. (Eds.): ICSEng 2020, LNNS 182, pp. 435–444, 2021.
https://doi.org/10.1007/978-3-030-65796-3_42

of new model series or cause callbacks [1,11,13]. With the increasing amount of electronics and software, those events will happen more frequently unless development processes are modified towards greater software maturity.

1.1 Automotive Network Development and Compatibility

Part of the E/E-architecture is the vehicle network. The architecture consists of ECUs and automotive field buses to connect them to each other. To meet the real time requirements of systems like airbags or the electronic stability program (ESP) the networks are statically designed [6]. This means that all the messages sent and received by all ECUs are planned during design time. A message may contain one or more signals. A signal is a digitized data element, i.e. a temperature, which would have formerly been transmitted through its own wire. An E/E-architecture, based on the static routing of signals between ECUs, is called a signal-oriented architecture.

The information about signals and their routes between ECUs is stored in a central database called communication matrix (CM) [12]. The CM-database also contains all the information required to derive the actual representation on a physical bus for the abstract signals. This includes for example the data types and value ranges of signals, addresses of frames and the positions of signals within frames (when multiple signals are transmitted within the same frame).

The development of a vehicle network consists of the initial creation of a communication matrix and iterative enhancements. To ensure a common interpretation of bus messages all ECUs have to be based on the same version of the CM. During the development phase of a model series, there are defined CM release dates, synchronized to the rest of the E/E development (i.e. each ECU software release is based on a CM release). Between those points the CM is modified to correct errors or to add additional features.

With up to 150 ECUs in a top class vehicle [10]), compatibility between different development versions becomes a significant issue. If not every single ECU is delivered on time, the testing has to be stalled until the last ECU of the current release has arrived. This reduces the time in which test results can be integrated into the next release and the overall software maturity. As an alternative, the missing ECUs can be substituted with their previous versions, if the old interface configuration is compatible to the current CM release.

Currently the compatibility of single ECUs to different communication matrix releases is surveyed manually. The CM history is checked for changes to the interface of an ECU in question. If there are any, they are inspected for aspects that would interfere with the operation of the unchanged ECU in the new vehicle.

An example for a change that may be present, but would not prohibit a prototype vehicle from using the previous version of an ECU, is the transmission of a new signal representing a new event (i.e. a button press) not present in previous releases. A prototype vehicle could be built with the old ECU and everything but the reaction to the button press could be tested.

1.2 Aim and Structure of this Paper

The goal of this paper is to define a methodology that meets the following three requirements:

Rq1 The compatibility between an ECU and a vehicle, based on different CM releases, can automatically be detected.
Rq2 The compatibility between consecutive development releases is optimized.
Rq3 Legacy signals and interfaces are removed from the CM in a well defined, predictable way.

In Sect. 2 existing methods and technology is presented, on which the new methodology can build on. Section 3 describes the proposed scheme. An analysis of the CM databases of a current model series proving the applicability of the concept is presented in Sect. 4. Section 5 goes through possible threats to the validity of the concept and comes to the conclusion. An outlook to future work is given in Sect. 6.

2 State of the Art

2.1 AUTOSAR

Communication and ECU Software design in the automotive industry is standardized in the Automotive Open Software Architecture (AUTOSAR) by a consortium consisting of major car manufacturers, suppliers, software- and hardware-vendors. AUTOSAR provides a standardized software stack, communication middleware, a metamodel to describe communication systems and an exchange format ("AUTOSAR XML") for the transmission and storage of models based on the AUTOSAR metamodel, between car manufacturers and suppliers [7].

Vehicle network design is based on the AUTOSAR metamodel. It defines, beside other things, how an abstract piece of information communicated between two pieces of software is mapped to actual field buses like the controller area network (CAN), the local interface network (LIN) or FlexRay [4]. Each communication medium has a way to group and transport digital information. AUTOSAR defines this abstractly as a protocol data unit (PDU), independent of the actual technology. A PDU can contain one or more signals. The abstract nature of a PDU allows it to be transmitted through different parts of the vehicle network without a dependency on a specific technology. The PDUs are mapped to frames which are part of the physical buses (which are called "Communication Cluster" in AUTOSAR). The Interfaces connecting ECUs to buses are called "Communication Connectors" in AUTOSAR.

The communication matrix relates not only signals to sending and receiving ECUs but also contains the information which PDUs are sent and received over which communication connectors and which signals are within those PDUs. This information is exported as an AUTOSAR XML file called "ECU extract". The extract is used during code generation, to configure the COM stack (the part of the AUTOSAR stack responsible for managing communication).

The behavior of the COM stack is standardized. The compatibility of a modified interface depends on the handling of PDUs with unexpected properties. Wherever possible (i.e. where the available buffer space allows it) Gateways will forward longer PDUs unchanged [3]. Receiving ECUs will discard the additional bytes at the end of a PDU, which is longer than expected. The data fitting into the expected length is used as planned and the mismatch of sizes is tolerated [2]. An ECU only uses PDUs with a known ID so that unexpected (i.e. new) PDUs are automatically ignored [5]. Therefore additional PDUs are also a backward compatible extension.

2.2 Semantic Versioning

Having to either update all packages of a software project, if a package on which they depend changed, or assuming compatibility without verification, is undesirable. To avoid such situations Tom Preston-Werner proposed the "semantic versioning" scheme to assign and increment version numbers [14]. It groups modifications of the versioned software into three categories:

Major *Breaking Changes* are the largest possible modifications where the new version of a package is *not compatible* with software that was based on the old one.

Minor *Additional features* added to an API without modifying the existing ones is a *backward compatible* change that allows the new version of the package to operate in combination with software designed for an older version.

Patch *Bug fixes and corrections* are modifications to a software that do not influence the API at all and are *fully compatible*.

According to [14] a version consists of three dot-separated numbers each of which is linked to one of the three change types. The first number references the Major Version, the middle one the Minor Version and the last one annotates the Patch Version. When the Minor Version is incremented the Patch Version is reset to 0 and when the Major Version is incremented both Minor and Patch Version are reset. The numbers may be followed by a hyphen and a label (e.g. "alpha", "beta" or "final").

3 Hierarchical Versioning

The AUTOSAR based communication design as shown in Subsect. 2.1 can be modeled as a hierarchical structure with the bus and its interface ("Communication Cluster"/"Communication Connector") at the top, the PDU in the middle and the signal at the bottom. The compatibility between an ECU and a vehicle network depends on this structure. As shown there are different mechanisms allowing for backward compatibility on the individual hierarchy levels. With fully compatible, backward compatible and breaking changes semantic versioning is applicable. This is a way to identify compatibility issues between development releases and would satisfy **Rq1**. To also meet **Rq2** and **Rq3** we propose a methodology based on the following rules:

HV1 Semantic versioning is used for each independent object in the hierarchy of a communication connector.
HV2 When the minor or patch version of an object is incremented, the minor or patch version of the parent object is also incremented by one. (A minor increment resets the patch version, as defined in [14].)
HV3 If there are multiple child objects which changed in minor or patch version between two releases, the patch version of the parent object is still only incremented by one.
HV4 If the major version of an object has to be incremented the element is copied. The existing (compatible) version stays unchanged and at the same position. The breaking change is applied to the copy, which is appended to the end of the parent object. This step shall be called "copy-changing" an entity. This is a backward compatible change of the parent object and results in a minor version increment.
HV5 Each entity has a guaranteed minimum lifetime, in which the given major version will be available. The lifetime is given in a relative form, e.g. as a number of how many future releases the major version of this object will be available. Within the lifetime of an entity, only minor and patch changes are allowed, while major modifications have to be implemented by copychanging.
HV6 Once an object has been copychanged, the guaranteed minimum lifetime turns into a lifetime countdown.
HV7 At the end of the lifetime of an entity, it is removed from its parent object. This in turn causes a breaking change of the parent object
HV8 Breaking changes at the end of the lifetime of an entity are possible without copychange.

Rules **HV1–HV3** work toward meeting **Rq1** in enabling to automatically deduce from a version number if an object is compatible or not. This works without iterating through all entities, because due to **HV2** the versions of parent objects are automatically adjusted when their child objects are changed. **HV3** prevents the minor and patch versions of entities on higher hierarchy levels from incrementing unnecessarily fast.

Rule **HV4** causes changes to be backwards compatible, optimizing the compatibility between development releases as demanded by **Rq2**.

Rules **HV5–HV8** define a way to orderly remove legacy elements from the system, fulfilling **Rq3**. This is important, to balance out the bandwidth increase caused by **HV4**. The lifetime guarantees given by **HV5** allows to plan features depending on a certain communication connector version. **HV6** lets entities exist for arbitrary amounts of time, as long as no major version changes occur. Despite this duration being unpredictable, dependent software always has the minimum guaranteed lifetime to adapt to a major version change, as soon as it occurs. **HV7** frees up bandwidth from objects that aren't used anymore. **HV8** is to prevent the copychanging of legacy elements.

3.1 Example

As an example for the practical application of the rules HV1-HV8 we look at *Communication Connector A* consisting of the *PDUs L,M and N*. We also assume that the *PDU L* consists of the *Signals X and Y*. Three releases of this hierarchy are pictured in Fig. 1.

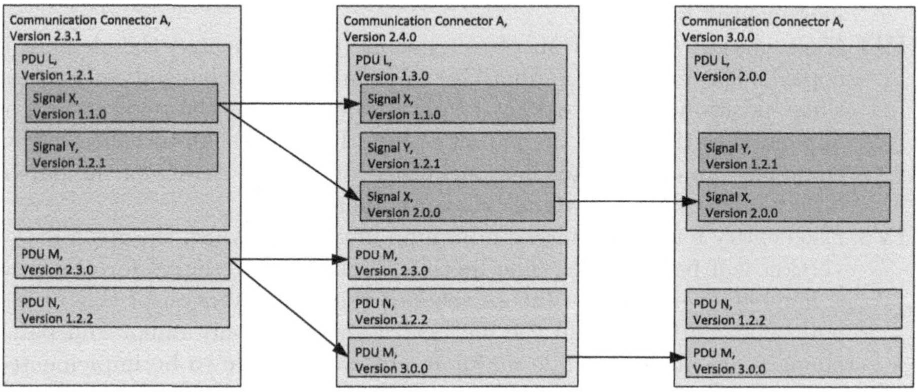

Fig. 1. An example of multiple development releases of Communication Connector A

The first, pictured release of *Communication Connector A* is version 2.3.1. An incompatible modification requires *Signal X, Version 1.1.0* to be copychanged to *Signal X, Version 2.0.0* as specified by **HV4**. This Major Version change within *PDU L* causes a minor version increment from 1.2.1 to 1.3.0 as specified by **HV4**. Following **HV2**, the minor version increment of *PDU L* causes a minor version increment from version 2.3.1 to 2.4.0 for the *Communication Connector A*. Within the same release cycle a breaking change to *PDU M* is resolved by a copychange to version 3.0.0. Due to **HV3** the Minor Version of *Communication Connector A* is only incremented once, despite the presence of two backward compatible changes to it. With a larger *PDU L* and two different versions of *PDU M* the second, pictured release requires more bandwidth than its predecessor.

The copychanges to *Signal X* and *PDU M* started the lifetime countdowns for their legacy versions following **HV6**. The third release in Fig. 1 shows the *Communication Connector A* after the end of those lifetimes. Without the legacy versions the bandwidth used by *Communication Connector A* is as low again as before the changes. The major versions of *Communication Connector A* and *PDU L* have been incremented, as the removal of objects is a breaking change.

This example shows how hierarchical versioning can enhance the compatibility between development releases. Although the copychanging mechanism increases the bandwidth usage of intermediary releases the end product is not affected by this.

4 Applicability

To analyze the applicability of hierarchical versioning to real world vehicle development we looked at the CM-development of a current model series of Mercedes-Benz with more than 45.000 Signals, 6.000 PDUs and 600 communication connectors.[1] By inspecting the changes between 25 development releases we answer how common breaking changes are. This allows to estimate the bandwidth increase caused when introducing a backward compatibility of one release cycle by copychanging. The analysis is done recursively for the three hierarchy levels (communication connector, PDU and signal).

In the CM-database are version numbers for the whole model (the development releases) as well as for the elements in the hierarchy below. The version numbers are incremented whenever any change has been made to an object. To get information about such a change of an entity, its child elements have to be inspected as well. The objects were analyzed by a python script using a generic algorithm, that is applied to all levels of the hierarchy. This way all changes of the hierarchy are recursively analyzed. To control for fluctuations in the parent levels, the numbers of PDU changes has been divided by the numbers of communication connector changes per release and the same for signal and PDU changes.

The results are plotted in Fig. 2. Towards the end of the development phase there are less changes as the vehicle design matures. Due to the fact that even a single modified signal causes a PDU to be marked as changed and a single modified PDU causes a communication connector to be marked as changed, the number of modified communication connectors (and therefore incompatible ECUs) stays large even in late development stages. In most cases there are two or less signal modifications per PDU and three or less PDU modifications per communication connector. With an average of 10 PDUs per communication connector this leads to a 30% bandwidth increase during the design phase, if all the changes between two releases are implemented through copychanging. Inspecting changed signals in PDUs yields similar results.

With a 30% bandwidth increase, during development time and under the worst case assumption of all PDU changes being breaking ones, hierarchical versioning could have been applied to the shown development process. In reality this number will be lower, because a part of the PDU changes will be non-breaking due to signal modifications implemented as copychanges. The final product will not be affected at all due to rules **HV6–HV8**.

[1] Gateway ECUs have multiple communication connectors and a communication matrix contains all the variants (e.g. the low-, mid- and high-end entertainment system) of all ECUs. Therefore the number of communication connectors in the CM is larger than the number of ECUs in an actual vehicle.

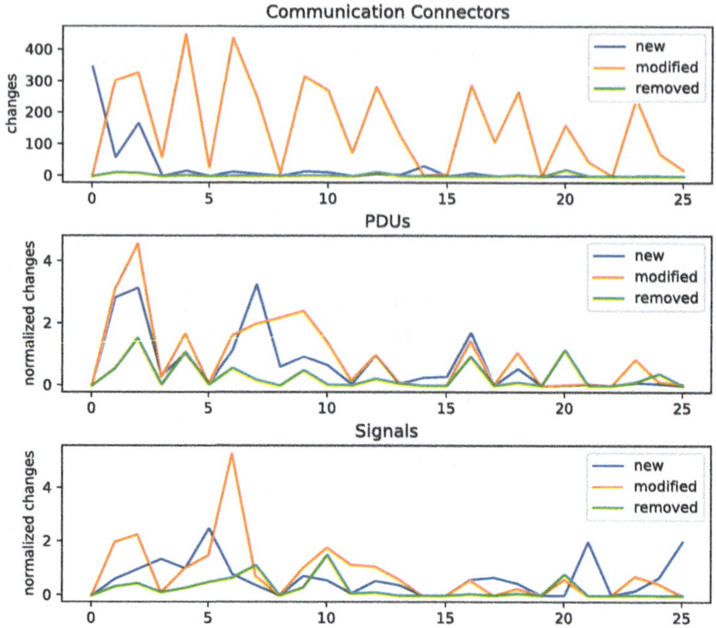

Fig. 2. Absolute changes to communication connectors and normalized changes to PDUs and signals

The analyses have shown a tendency that few PDU modifications cause a large number of changed communication connectors. To further inspect this, histograms have been plotted showing how many communication Connectors have been affected by each modified PDU (ordered from most significant to least significant PDU change). The plot is shown in Fig. 3. The first three percent of each histogram have been marked in red. Most PDU modifications affect less than ten communication connectors. Within the top three percent, a single modified PDU causes changes to more than 100 communication connectors. This indicates that even a partial implementation of hierarchical versioning, applying only to these "superchangers", could improve the compatibility between releases.

5 Threats to Validity and Conclusion

Hierarchical Versioning relies on the ability to extend an existing data structure without breaking te compatibility. This is currently not a fixed part of the AUTOSAR standard, but a side effect of its resilience features. New versions of the standard could break the copychange mechanism. To prevent this, hierarchical versioning and copychanging could become part of the standard itself.

A potential show stopper for the proposed concept is the increase in bandwidth requirements. The estimations in Sect. 4 show a bandwidth increase of 30% for full backward compatibility one release into the past. Having more

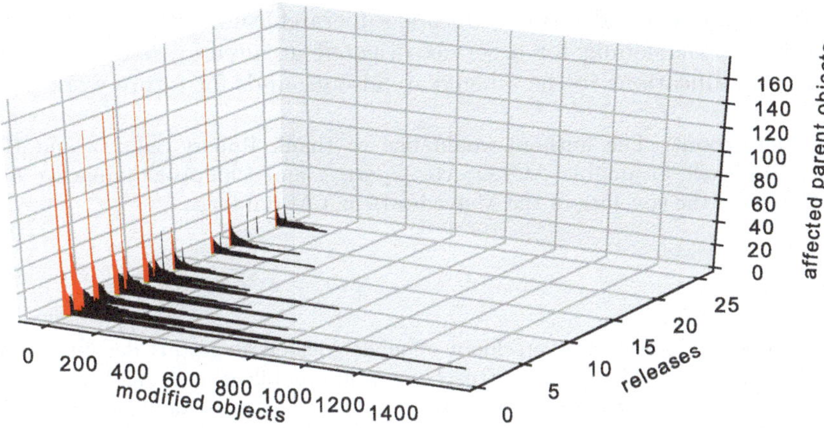

Fig. 3. The influence of modified PDUs on communication connectors

changes per release or a backward compatibility for more than one past release will increase this number, possibly into impractical ranges. Therefore the relation between the frequency of breaking changes and the guaranteed minimum lifetimes is crucial. An indicator for a bad ratio would be objects with more than two major versions.

With Hierarchical Versioning, a scheme has been proposed, which increases backward compatibility while at the same time offering a well defined process to remove legacy elements. An analysis has shown that the proposed concept would have been feasible in the network development of a current automobile model series. Our concept would have increased the bandwidth required by 30% or less for intermittent develop releases while leaving the bandwidth required by the final release unaffected. The authors suggest applying it to future development projects to allow for easier prototype construction to increase testing abilities and software maturity.

6 Future Work

In the future we plan further analyses about the bandwidth requirements of Hierarchical Versioning. Especially the specific impact of different minimum lifetime guarantees and frequencies of breaking changes. Also of interest is the possibility to use different minimum lifetime guarantees on different hierarchy levels. This investigation should yield a set of rules that help to chose the ideal parameters for a development process based on Hierarchical Versioning.

The phenomenon of the "superchangers" also warrants further investigation, possibly leading to other best practice rules for network design.

Currently the automotive industry is transitioning from signal-oriented communication schemes [15], as presented in this paper, towards service-oriented communication. A service consisting of multiple methods which can consist of multiple parameters and return values is a hierarchical structure, too. The use of hierarchical versioning for a service-oriented architecture and the possibly necessary modifications to the scheme are also grounds for future work.

Acknowledgment. The authors would like to thank Robert Sakretz and Holger Schnürer for their invaluable AUTOSAR expertise and Philipp Schumacher for his great help with the specifics of the Mercedes-Benz CM-database.

References

1. Software-Probleme: Audi verschiebt offenbar Auslieferungsstart des e-tron. Automobilwoche (21102018). https://www.automobilwoche.de/article/20181021/NACHRICHTEN/181029997/software-probleme-audi-verschiebt-offenbar-auslieferungsstart-des-e-tron
2. AUTOSAR: General Specification of Transformers (28112019)
3. AUTOSAR: Requirements on Gateway (28112019)
4. AUTOSAR: Requirements on System Template (28112019)
5. AUTOSAR: Specification of CAN Interface (28112019)
6. Çela, A., Gaid, M.B., Li, X.G., Niculescu, S.I.: Optimal design of distributed control and embedded systems. Springer Science & Business Media (2013)
7. Durisic, D.: Automotive software architectures. In: Staron, M. (ed.) AUTOSAR Standard, pp. 81–116. Springer International Publishing, Cham (2017)
8. Meissner, F., Shirokinskiy, K., Alexander, M.: Computer on wheels: disruption in automotive electronics and semiconductors. https://www.rolandberger.com/de/Publications/Das-Auto-wird-zu-einem-Computer-auf-R
9. Guissouma, H., Klare, H., Sax, E., Burger, E.: An empirical study on the current and future challenges of automotive software release and configuration management. In: Bures, T., Angelis, L. (eds.) SEAA 2018, Piscataway, NJ, pp. 298–305. IEEE, New York (2018). https://doi.org/10.1109/SEAA.2018.00056
10. Priyatham, J.: Sales of automotive electronic control units soar as cars go electric and add driver-assist features (2019). https://technology.ihs.com/614010/sales-of-automotive-electronic-control-units-soar-as-cars-go-electric-and-add-driver-assist-features
11. Menzel, S.: VW will die Software-Probleme rasch lösen: ID.3 kommt im Spätsommer. Handelsblatt (10062020). https://www.handelsblatt.com/unternehmen/industrie/volkswagen-vw-will-die-software-probleme-rasch-loesen-id-3-kommt-im-spaetsommer/25904772.html
12. Streichert, T., Traub, M.: Elektrik/Elektronik-Architekturen im Kraftfahrzeug. Springer, Heidelberg (2012). https://doi.org/10.1007/978-3-642-25478-9
13. Maier, T.: Volvo-Rückruf: Probleme beim automatischen Bremssystem. Auto Service Praxis (16032020). https://www.autoservicepraxis.de/rueckrufe/artikel/volvo-rueckruf-probleme-beim-automatischen-bremssystem-2581604
14. Preston-Werner, T.: Semantic Versioning 2.0.0. https://semver.org/spec/v2.0.0.html
15. Traub, M., Maier, A., Barbehön, K.L.: Future automotive architecture and the impact of IT trends. IEEE Softw. **34**(3), 27–32 (2017). https://doi.org/10.1109/MS.2017.69

Towards Data-Driven Digital Twins for Smart Manufacturing

Deena P. Francis[1(✉)], Sanja Lazarova-Molnar[2], and Nader Mohamed[3]

[1] DTU Compute, Technical University of Denmark, Anker Engelunds Vej 1, 2800 Kgs. Lyngby, Denmark
deena.francis@gmail.com

[2] Mærsk Mc-Kinney Møller Institute, University of Southern Denmark, Campusvej 55, 5230 Odense, Denmark

[3] Department of Computer Science, Information Systems and Engineering, California University of Pennsylvania, California, PA 15419, USA

Abstract. The adoption of a digital twin for a smart factory offers several advantages, such as improved production and reduced costs, and energy consumption. Due to the growing demands of the market, factories have adopted the reconfigurable manufacturing paradigm, wherein the structure of the factory is constantly changing. This situation presents a unique challenge to traditional modeling and simulation approaches. To deal with this scenario, we propose a generic data-driven framework for automated construction of digital twins for smart factories. The novel aspects of our proposed framework include a pure data-driven approach incorporating machine learning and process mining techniques, and continuous model improvement and validation.

Keywords: Data-driven · Digital twin · Simulation · Smart factory · Reconfigurable manufacturing · Machine learning · Process mining

1 Introduction

The term digital twin, as first introduced by Grieves in a talk in 2003 at the University of Michigan, is a term that has been endowed with varied definitions over the years [1]. A quick study into these varied definitions reveals certain distinct commonalities among them. The first one has to do with the fact that a "real world entity" is being "digitally modeled" for various purposes. The second commonality among the definitions has to do with the "real-time" nature of the digital model. This refers to the need for updating the digital model as soon as changes in the real entities are observed, resulting in a high-fidelity replica of it [2,3]. The real differences in the definitions arise from the requirement of a digital twin to "interact" with the physical entity. This interaction is described by some works as a "two-way data flow" [4], resulting in changes in both the digital and real entities. This connectivity, which is enabled by technology, such as the Internet of Things (IoT), is considered to be an important part of a digital

H. Selvaraj et al. (Eds.): ICSEng 2020, LNNS 182, pp. 445–454, 2021.
https://doi.org/10.1007/978-3-030-65796-3_43

twin [3]. In the interest of improving clarity and reaching a consensus regarding what we mean by a digital twin in this work, we define the following constituent parts: (1) Real-world entity, such as a machine, a group of machines, or an entire factory; (2) Digital model, consisting of: (a) algorithms, machine learning and software, and (b) connectivity components, i.e. IoT systems; and (3) Data that has been or is being generated by the real-world entity. This third component is a crucial part of a digital twin, which impacts how well the goals of digital modeling are achieved, as we discuss in Sect. 2.2.

The development of digital twins can be seen as a natural extension of traditional simulation modeling coupled with increased data availability, connectivity and evolving needs of end users. Like simulation modeling [5], a digital twin helps to understand, monitor and experiment with complex physical systems. In addition, digital twins use the knowledge extracted from the simulation to provide feedback to the physical systems with the objective of controlling some aspect of it in order to optimize some end-user given parameters. There are several uses for a digital twin in areas ranging from automotive industry to healthcare. In this paper, we consider the case of a data-driven digital twin modeling for smart manufacturing. As such, we have noted the following relevant challenges: (1) *Entire product life cycle management*: issues related to product lifecycle in a factory starting from the product design to sale can be handled by specialized digital twins [3,6]. (2) *Reconfigurable manufacturing*: in this scenario, the configuration and structure of the factory changes rapidly in accordance with changing product demand [7]. This type of manufacturing scenario particularly benefits with the employment of digital twins [8–10]. (3) *Remanufacturing*: it refers to the process of converting old and used products to new ones. Digital twins have been deployed in waste electrical and electronic equipment (WEEE) recovery scenarios [11,12].

This paper is organized as follows. In Sect. 2, we start with an overview of the problems of traditional simulation modeling, an introduction to digital twins for smart factories and the data-driven simulation modeling paradigm for smart factories. In Sect. 3, we describe the proposed framework for data-driven digital twins for smart factories, detailing the key constituent elements. In Sect. 4, the challenges of the proposed framework are discussed, and in Sect. 5, the summary and general outlook of future developments are described.

2 Background

A digital twin is considered to be among the technologies that help to realize the promises of Industry 4.0, which is a term associated with the incorporation of smart technology such as Internet of Things (IoT), Big Data Analytics and Smart Sensors in the traditional manufacturing industries. Some of the challenges associated with today's manufacturing systems are as follows.

- *Frequently changing demands*: the needs of the end-users and market are continuously changing. This results in demands with respect to the quantity and types of products. In such situations, custom-made models become obsolete

as soon as they are developed. Consequently, there has been a shift in the manufacturing paradigm from traditional to customizable, flexible and reconfigurable systems.
- *Need for real-time monitoring*: due to the high frequency of changes within manufacturing systems, it becomes essential to monitor these changes in real-time to make informed decisions that could impact the attainment of their performance goals.
- *Complexity of the manufacturing system*: due to the paradigm shift to reconfigurable systems, the layout and working of the whole manufacturing system becomes extremely complex. In addition to the challenge of correctly modeling such systems, control and optimization of these systems to meet the demands of the end-users becomes another challenge.

The challenges mentioned above motivate the use of digital twins in manufacturing systems. Modeling of complex systems, control, optimization and real-time monitoring can be, in theory, handled by the digital twin. There are, however, certain aspects that are currently ignored. A detailed description of these aspects of digital twins will be explored in Sect. 2.3 after going through the relevant background.

2.1 Challenges with Traditional Simulation Models

Simulation modeling has been considered a great tool to evaluate the design, operation, and control of complex systems. It offers a number of advantages, such as the ability to model causal relationships between events, provision for predictive, planning and controlling capabilities for the system and ability to validate models with collected data. These models, however, often make idealistic assumptions about the systems they are representing [13]. Typically, discrete-event simulation models are used for modeling complex systems, and they are manually designed by experts who select abstractions appropriate to model the systems of interest, resulting in custom simulation models. This method of simulation model generation is not suitable for the reconfigurable manufacturing systems because these systems undergo rapid changes in their physical and software structure, in an attempt to match changing demands in the market (e.g.. as typical for mass customization [14]). Hence, custom-made models are rendered obsolete soon after their development, and it becomes necessary to develop new models or manually update existing ones. It is easy to see that this scenario is not ideal in a fast-changing industry, such as manufacturing.

2.2 Data-Driven Simulation Modeling

Simulation models that are developed and parametrized by data are referred to as data-driven simulation models [15]. Some of the advantages of adopting a data-driven approach, as opposed to a representation or intuition-driven approach, are listed as follows:

- *Closeness to the real system*: when compared to traditional methods, data-driven models tend to be more accurate to the actual systems being modeled. This is because data encapsulates systems' behaviors and characteristics, and it can reflect phenomena that cannot be foreseen.
- *Ability to take advantages in developments in Machine Learning (ML) and Artificial Intelligence (AI)*: the use of data provides opportunities to use advancements in ML and AI to gain deeper understanding of systems' behaviors and make more informed decisions.

Because of the above-mentioned advantages, and several successful adoptions of data-driven approaches, there has been a shift in focus from traditional simulation modeling to more data-driven approaches. The underlying motivations for this shift have been brought about by the advancements in data generation and storage, algorithms and ever-changing needs of the manufacturing market. Recently, there have been a number of approaches, such as knowledge-based [15], spatio-temporal-data based [16], ad-hoc [17], multimodal-data and digital-twin-based [18] and data-driven remanufacturing models [19]. Previous research in data-driven simulation modeling suffers from the following drawbacks.

- *Lack of dominantly data-driven model generation*: Most of the existing approaches do not fall under the category of pure automatic simulation modelling with the exception of [8,9]. These works [15–19] also do not consider the whole Product Life Cycle (PLC) of a typical reconfigurable smart factory. The PLC is of interest because it encompasses all aspects of a manufacturing system.
- *Industry 4.0*: The real-time synchronization between models and real-world entities has become possible via the enabling technologies of Industry 4.0. Previous works in data-driven modeling are not specifically designed to accommodate the Industry 4.0 settings in which modern factories are designed.
- *Data heterogeneity*: A smart factory has a number of sources of data such as logs, images, videos, etc., which are often collected via Internet of Things (IoT) systems. Exploiting such wide varieties of data requires implementation of integration methods. Most of the existing works do not consider heterogeneous data obtained from various sources within the smart factory.
- *Data validation*: Data collected within factories often contains noise and missing values. Volume and frequency of collection of such data vary considerably, and they are often corrupted by noise, introduced by the factory environment and human errors. Performing validation checks on the data is of foremost importance because the quality of modeling depends heavily on the quality of the data. Most of the previous works do not perform data validation, which is crucial in ensuring correctness of the model.
- *Reconfigurable systems*: In order to deal with the demands of the market to support mass customization, the adoption of reconfigurable manufacturing systems have taken place. Ideas for simulation modeling described in previous methods are not suitable for frequently changing and challenging scenarios such as reconfigurable manufacturing systems.

2.3 Digital Twins for Smart Factories

Smart factory is a manufacturing solution that provides flexible and adaptive processes that solve problems in production facilities with dynamic and rapidly changing boundary conditions. Smart factories need digital twins to model and control the typically complex manufacturing systems. The benefits of adopting digital twins are the following: ability to monitor real-time status of machines, improved reliability, possibility for predictive maintenance, customized services to stakeholders, etc. [20]. IoT devices continuously collect data from machines and factory environment, which are then communicated to various systems in the factory that use this data for scheduling, maintenance, logistics and decision making. The attainment of an intelligent and data-driven decision-making within the smart manufacturing environment to improve profits and sustainability is the ultimate goal of Industry 4.0.

3 Data-Driven Digital Twins for Smart Factories

The huge and varied amounts of data, generated by the smart factory environment, is the driving force behind a data-driven approach to digital twins. The data, or the knowledge that can be derived from it, reflects the state of the system. The underlying workings of the complex dynamical systems is embedded in the data. Hence, it is from exploiting such data that we can hope to achieve true high-fidelity models of factories. Deriving insights from data is typically carried out using ML and process mining. The different types of data that are generated within the smart factory can be used to help develop our model or form informed decisions by utilizing specialized ML algorithms. For example, time series data related to machine operation that are generated within the factory can be used by specialized time series models to detect or predict presence of faults.

3.1 Key Elements of Data-Driven Digital Twin

We have identified the following key elements that enable a data-driven digital twin for a smart factory.

- **Data collection.** The first task to be performed before collecting the data is entity identification. This step involves locating and collecting data about the entities within the real-world system that is being modeled. On completion of this task, the data that is continuously being generated by the identified entities within the factory's manufacturing systems are stored within a database. The key sources of data are as follows: sensors, end-user inputs, needs, and the types of data that are being generated include diagnostic data, logs and traces and market/end-user data.
- **Data validation.** The main problems associated with data include: Noise, missing values, multiple modalities: image, text, structured, audio and huge sizes. Data preprocessing is a key task that needs to be performed to ensure validity of the data. Some of the preprocessing tasks include imputation and

integration. Imputation refers to the task of filling in the missing values using statistical or ML approaches and integration refers to the process of converting multi-modal data in the form of images, logs, audio etc. into usable and unified format, that can be used by the knowledge extraction step.

- **Knowledge extraction.** The task of knowledge extraction from validated data obtained from the previous step consists of the following sub-tasks.
  (a) *Event detection and labeling*: Identifying simulation goal-relevant events that take place in a smart factory is an important part of data-driven simulation modeling. Events are key elements that drive the development of discrete-event simulation models, and the ability to automatically detect events that occur within the smart factory determines the effectiveness of the simulation modeling. Detection of relevant events, such as faults, is considered to be an important part of reliability analysis of manufacturing systems. Previous research on smart manufacturing has already identified measuring and ensuring reliability in smart factories as a complex and non-trivial task [21,22]. Automatically detecting the occurrence of certain events that signal the occurrence of faults is a crucial task that impacts the responsiveness and accuracy of the smart factory model. This task could be considered a form of anomaly detection, which has seen a number of developments in the past few years [23,24].
  (b) *Process discovery*: Discovering the relevant processes that occur within a smart factory is beneficial to the knowledge extraction part of a smart factory because processes are considered to relay important information about the working of the factory. The event logs that have been collected as part of the data collection part is the input to process mining algorithms, which discover the main processes automatically from the data. In this context, a formal description of the process flow in the system is derived using a modeling language or a process specification language, such as Petri Nets [25]. Such a language helps to convert the raw processes in the event logs to formal specifications of these processes. Using the formal specifications, process discovery and associated conformance checking can be done. Process discovery aids in understanding parameters of the discovered processes such as reliability, speed, delay, duration etc.
- **Automatic/semi-automatic simulation modeling.** The development of a unified data-driven model is the next key component of the proposed scheme. The input to this step is the events and processes that have been identified in the previous steps. Once relevant events and processes have been discovered, the simulation model can be constructed, which can happen with some degree of human intervention for the first time, and, subsequently, the initial model can be updated automatically to reflect the changes in the smart factory. For this, links between the model and the smart factory data streams need to be clearly identified. Furthermore, adequate algorithms for model extraction and model updates will need to be developed to support the automatic/semi-automatic simulation modeling processes.

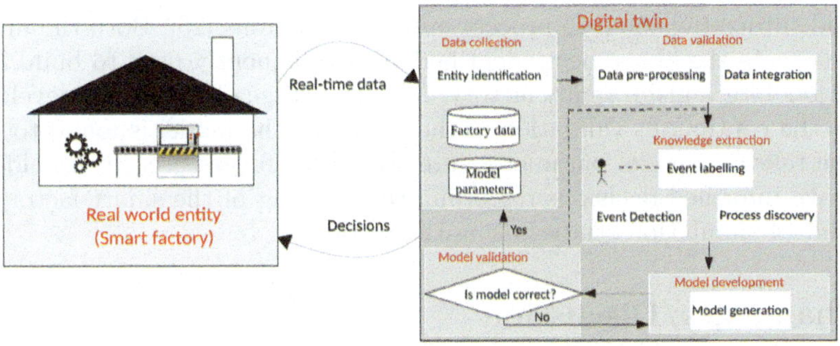

Fig. 1. Framework for the proposed data-driven digital twin for smart factory

- **Continuous validation of model.** The data-driven simulation modeling approach requires rigorous model validation processes to construct high-fidelity models that assist in correct and timely decision-making. Since the data is being continuously collected in real-time, an opportunity to continuously validate the simulation models is clear. The complete and intense reliance on data to perform model validation necessitates that data itself is validated rigorously, such as to ensure that the models are derived from high-quality data. This is a challenge that will need to be solved to enable the framework for data-driven simulation modeling.

3.2 Framework for Data-Driven Digital Twins

We propose a framework for data-driven digital twin for smart factory, shown in Fig. 1. The smart factory, which is the real-world entity that is being modeled continuously produces data through the IoT devices and sensors. This data is the starting point of the data-driven modeling approach. The data collection processes include relevant entity identification and data storage in databases. Entity identification includes explicitly listing the relevant entities such as production systems, control technologies etc. within the smart factory. One interesting approach to utilize to collect and understand manufacturing data is the Haystack standard [26]. This standard is an open source initiative that offers standard semantic data models with the target of simplifying the extract value from the data generated by different IoT devices including devices in manufacturing facilities. The next logical step is data validation by means of general data cleaning or preprocessing and integration. The data implicitly captures relevant events occurring in the factory. In order to make these events more explicit and to aid the model development stage, event labeling is performed semi-automatically, i.e. by human intervention. A number of relevant events are identified and labeled manually by experts or stakeholders with specialized knowledge of the smart factory. In addition, ML models are used to further detect events automatically using such labeled data. Process discovery is also done using the generated event logs, and subjecting them to process mining algorithms [25]. Both the processes and the events that have been identified are subsequently used to build a simulation model for a given smart factory. A crucial stage of the model development stage is the continuous validation of the model. If the model is found to be correct, the relevant model parameters are stored for future use. Stakeholders can then make informed decisions regarding the working of the smart factory using the result of simulation studies using the model.

4 Challenges/Discussion

Implementing a framework for data-driven digital twins in smart factories is not trivial and it has a number of associated challenges. We have identified the following main challenges:

- Striking the right balance between human intervention and pure data-driven approach: using a data-driven approach that combines human intervention at the right levels and automation, a good balance that ensures interpretability and efficiency can be achieved.
- Decision making in real-time: the digital twin can assist stakeholders make the correct decisions. The timeliness needs should be matched to the criticality and weight of the associated decisions, which should further be reflected in the granularity of the data that is being collected. Apparently, safety related decisions in safety-critical systems would need higher timeliness than production related decisions.
- Evolving goals: the ultimate goal of smart manufacturing systems is to ensure steady and profitable production rates. This goal also includes aspects like ensuring reliability of the manufacturing systems, reducing material wastes and energy consumption, etc. Besides these, from an environmental perspective, reducing greenhouse gas emissions is another goal that is and will be an important aspect that is to be considered. Building simulation models to reflect, update and serve such varying and evolving goals is a huge challenge that needs to be considered.

The presented framework, however, also presents new opportunities, listed as follows:

- Development of high-fidelity digital twins: due to the data-driven aspect of the proposed framework, the model tends to be an accurate representation of the current behavior of the data and processes within the smart factory. Moreover, the continuous validation of data and model ensures that this accuracy is maintained consistently.
- Improved understanding of processes and decision-making: the proposed approach relies on process discovery through the event logs in addition to other data-driven components. This improves the understanding about the process flows in the system and therefore the aids in decision making. By utilizing the process flows in the system, the stakeholders can make relevant decisions about the future of the system as well as the digital twin.

- Increased reliability: with continuous validation of data and model in place, our proposed approach can ensure reliable working of the manufacturing system.
- The availability of data-driven digital twin can offer a good support for the concept of agile manufacturing which enables manufacturing enterprises to respond rapidly to market changes while still maintaining good quality and controlling costs.

5 Summary and Outlook

We studied the needs and opportunities to enable data-driven digital twins in the domain of smart manufacturing. Building digital twins in real-time, based on data collected from IoT devices in smart factories will serve a number of purposes, i.e. we will have robust models that are always up to date and support more flexible and reconfigurable factory layouts; it will, furthermore provide an opportunity to have integrated ongoing model validation, etc. However, there are still significant advances that need to be made to move towards data-driven digital twins. Some of the steps might never be fully automated and a human intervention will always be necessary to a certain degree. In this manner, the question is how to minimize and mainstream the human intervention. In attempt to address the needs of data-driven digital twin development, we have developed a high-level conceptual framework too. We anticipate that this data-driven simulation modeling will gain more attention as more mechanism for ensuring data quality become available. We also discuss the associated challenges and opportunities.

References

1. Grieves, M.: Digital twin: manufacturing excellence through virtual factory replication. White Pap. **1**, 1–7 (2014)
2. Kritzinger, W., Karner, M., Traar, G., Henjes, J., Sihn, W.: Digital twin in manufacturing: a categorical literature review and classification. IFAC-PapersOnLine **51**(11), 1016–1022 (2018)
3. Tao, F., Zhang, H., Liu, A., Nee, A.Y.: Digital twin in industry: state-of-the-art. IEEE Trans. Industr. Inf. **15**(4), 2405–2415 (2018)
4. Söderberg, R., Wärmefjord, K., Carlson, J.S., Lindkvist, L.: Toward a digital twin for real-time geometry assurance in individualized production. CIRP Ann. **66**(1), 137–140 (2017)
5. Law, A.M., Kelton, W.D., Kelton, W.D.: Simulation Modeling and Analysis, vol. 3. McGraw-Hill, New York (2000)
6. Tao, F., Zhang, M.: Digital twin shop-floor: a new shop-floor paradigm towards smart manufacturing. IEEE Access **5**, 20418–20427 (2017)
7. Koren, Y., Gu, X., Guo, W.: Reconfigurable manufacturing systems: principles, design, and future trends. Front. Mech. Eng. **13**(2), 121–136 (2018)
8. Radziwon, A., Bilberg, A., Bogers, M., Madsen, E.S.: The smart factory: exploring adaptive and flexible manufacturing solutions. Procedia Eng. **69**, 1184–1190 (2014)

9. Gola, A., Świć, A.: Simulation based analysis of reconfigurable manufacturing system configurations. In: Applied Mechanics and Materials, vol. 844, pp. 50–59. Trans Tech Publ (2016)
10. Zhang, C., Xu, W., Liu, J., Liu, Z., Zhou, Z., Pham, D.T.: A reconfigurable modeling approach for digital twin-based manufacturing system. Procedia CIRP **83**, 118–125 (2019)
11. Wang, X.V., Wang, L.: Digital twin-based WEEE recycling recovery, and remanufacturing in the background of industry 4.0. Int. J. Prod. Res. **57**(12), 3892–3902 (2019)
12. Yang, S., MR, A.R., Kaminski, J., Pepin, H.: Opportunities for industry 4.0 to support remanufacturing. Appl. Sci. **8**(7), 1177 (2018)
13. Kim, B.S., Kang, B.G., Choi, S.H., Kim, T.G.: Data modeling versus simulation modeling in the big data era: case study of a greenhouse control system. Simulation **93**(7), 579–594 (2017)
14. Qi, Y., Mao, Z., Zhang, M., Guo, H.: Manufacturing practices and servitization: the role of mass customization and product innovation capabilities. Int. J. Prod. Econ. **107747** (2020)
15. Lattner, A.D., Bogon, T., Lorion, Y., Timm, I.J.: A knowledge-based approach to automated simulation model adaptation. In: Proceedings of the 2010 Spring Simulation Multiconference, pp. 1–8 (2010)
16. Charpentier, P., Véjar, A.: From spatio-temporal data to manufacturing system model. J. Control Autom. Electr. Syst. **25**(5), 557–565 (2014)
17. Rodič, B., Kanduč, T.: Optimisation of a complex manufacturing process using discrete event simulation and a novel heuristic algorithm. Int. J. Math. Models Methods Appl. Sci. **9**, 320–329 (2015)
18. Uhlemann, T.H.J., Lehmann, C., Steinhilper, R.: The digital twin: realizing the cyber-physical production system for industry 4.0. Procedia CIRP **61**, 335–340 (2017)
19. Goodall, P., Sharpe, R., West, A.: A data-driven simulation to support remanufacturing operations. Comput. Ind. **105**, 48–60 (2019)
20. Rasheed, A., San, O., Kvamsdal, T.: Digital twin: values, challenges and enablers from a modeling perspective. IEEE Access **8**, 21980–22012 (2020)
21. Lazarova-Molnar, S., Mohamed, N.: Reliability assessment in the context of industry 4.0: data as a game changer. Procedia Comput. Sci. **151**, 691–698 (2019)
22. Ali Farsi, M., Zio, E.: Industry 4.0: Some challenges and opportunities for reliability engineering
23. Chandola, V., Banerjee, A., Kumar, V.: Anomaly detection: a survey. ACM Comput. Surv. (CSUR) **41**(3), 1–58 (2009)
24. Francis, D.P., Raimond, K.: A random fourier features based streaming algorithm for anomaly detection in large datasets. In: Rajsingh, E., Veerasamy, J., Alavi, A., Peter, J. (eds.) Advances in Big Data and Cloud Computing, vol. 645, pp. 209–217. Springer, Singapore (2018)
25. Van Der Aalst, W.: Process mining. Commun. ACM **55**(8), 76–83 (2012)
26. Haystack: Haystack, project haystack. https://project-haystack.org/ (2020). Accessed 1 August 2020

A KNN Approach for Blockchain Based Electronic Health Record Analysis

Sazia Parvin[1(✉)], Sonia Farhana Nimmy[2], Sitalakshmi Venkatraman[1], Sa'ed Abed[3], and Amjad Gawanmeh[4]

[1] Melbourne Polytechnic, Melbourne, VIC, Australia
{saziaparvin,SitaVenkat}@melbournepolytechnic.edu.au

[2] University of New South Wales, Canberra, Australia
snimmy.1984@gmail.com

[3] Computer Engineering Department, College of Engineering and Petroleum, Kuwait University, Kuwait City, Kuwait
s.abed@ku.edu.kw

[4] Department of Electrical Engineering, College of Engineering and IT, University of Dubai, Dubai, UAE
amjad.gawanmeh@ud.ac.ae

Abstract. Blockchain plays a significant role to connect end users with utmost security and reliability as it is well-suited to handle critical privacy issues in medical and business data sharing. It is considered as one of the best cutting edge technology in providing a suitable framework for sharing critical medical information across various experts globally. In this paper, an integrated method is designed to combine Blockchain with data analysis for an automated and secure sharing of medical datasets among healthcare experts in improving their patient diagnosis and advice. The first key component of our framework involves classification of a large number of medical records containing hundreds of variables or features using linear discriminant analysis (LDA) and K-nearest neighbours (KNN) algorithm. Using LDA, highly related and significant features are grouped for sharing them with medical experts in order to quickly and accurately reach a decision. The next key component of our framework addresses the challenges involved in sending the processed information to the targeted persons by introducing a dynamic Blockchain discovery process. For this, we develop an automated Blockchain discovery system that repeatedly checks the nodes in the Blockchain to transfer the information in order to ensure robust security. Further, we evaluate the effectiveness of KNN algorithm with and without LDA in making accurate decisions. A rigorous experimental analysis confirms that LDA based KNN algorithm is preferable in this context.

1 Introduction

Application of technology has a great influence on disease diagnosis and treats by enlarging medical knowledge and explaining the conditions diagnosis as disease entities. The expenditure of healthcare service is repeatedly becoming higher

H. Selvaraj et al. (Eds.): ICSEng 2020, LNNS 182, pp. 455–464, 2021.
https://doi.org/10.1007/978-3-030-65796-3_44

due to insufficient support from the system and dependence on different intermediate factors [20]. In addition, several challenges have emerged due to the integration of healthcare systems with other paradigms such as IoT [16] and mobile technologies [2], and hence prominent solutions are required including secure authentication mechanisms [14,15].

The technological concept of Electronic Health Records (EHRs) provides the physiological, biochemical, bimolecular and morphological inventions that are used to define diseases as well as involve treating patients with medicative, protective, reformative and pacifying nursing. In EHRs, patient data is exceptionally fragmented and entities are interconnected under excessive regulatory outlines in a complex healthcare system [5]. Recently, the medical research focus is to serve the patients in a smooth, cost effective way and also provide top-level security for both parties and organizations that store sensitive information. EHRs have imposed the sensitivity to the paraclinical symbols and records. Only a secure and distributed technology can help to organize and manage this information in a robust way. Blockchain can be a unique data structure to store and access these types of sensitive data.

Blockchain technology has been introduced in every aspect of service and security and has been expanding rapidly worldwide. It implements a decentralized and distributed platform without any central authority [3,23]. In particular, cryptographic features of Blockchain provide secure and trustworthy transactions during data transmission. Blockchain technology is an innovative technology which can support dealing with the challenges of disease diagnosis and treating by adopting decentralized trust [4]. In healthcare, Blockchain technology has enormous plausibility due to the patient-centric approach that connects distinct systems and implement the accuracy of EHRs [21]. In addition the work in [1] proposed a framework to improve the quality of healthcare, lower the cost of delivery, and enhance medical records management.

The data of EHRs are gathered from various scattered sources and referred to as silos in healthcare sector. The challenge in using and access data silo is the authentic control and multi-level security. Therefore, patient data is not immediately accessible at emergency medical service [13]. The data frames are not organized like a one-stop service and as well as not developed using the same protocol. That's why sometimes the data become meaningless and incomplete to the third party [18]. Moreover, legal and ethical issues are complex and related to place and law [9]. Apart from these, illegal access, alter, and tempering patient data is the huge challenge for healthcare sector. EHRs handle and access massive amount of data from different sources and parties. Secure and trustworthy access and transactions always a big challenge for researchers.

The data frames created by Blockchain are like one-stop database service [19]. All the frames are homogenous and cannot be altered or deleted. All the parties participating the data frames have to authorise the operation to alter or delete data [12]. This feature of Blockchain imposes properties like data integrity and inception which is used to create solutions to prevent drug duplication and medical frauds. In addition, it can be used to prevent suspicious results and delete or temper clinical trials. From financial point of view, this technology could save billions of the clinical and pharmaceutical industries. Blockchain has been introducing every aspect of service and security and has been expanding rapidly. In healthcare, Blockchain technology has enormous plausibility due to the patient-centric approach that connects distinct systems and implement the accuracy of EHRs.

2 Related Work

The reliability of healthcare information systems has been a concern in the research community [7]. Griggs et al. [8] proposed the integrated WBANs with Blockchain smart deals for a secured real-time patient monitoring and medical interference system. The system introduces the combination of Blockchain to perform smart deals to estimate data collected by a IoT healthcare devices based on adapted threshold standards. This is tested to solve the logging transfer problem of data transactions in an IoT healthcare structure.

Zhang et al. [22] offered a securely healthcare system while sharing data among PSN nodes using two protocols. The first one developed through an IEEE802.15.6 enhanced version to offer protected links with unbalanced computational requests for portable devices and sensor nodes. The second protocol is implemented using Blockchain to share health data among PSN nodes. Ichikawa et al. [10] introduced a tamper-resistant based framework for mobile health system using Blockchain to ensure reliable records. The goal was to implement a mobile health system for cognitive behavioral treatment for insomnia using a smartphone application. Esposito et al. [6] proposed healthcare security and privacy data system by integrating various data structure in the Blockchain. To overcome the overhead complexity of storing huge data in the Blockchain, the infrastructure uses only the data hash as transaction.

Rahman et al. [17] offered a perceptive dyslexia analytics solution using decentralized large data repository to store and share with healthcare communities based on Blockchain. Remote multimedia health data is acquisitioned during dyslexia testing and stored in a decentralized data repository shared for advance clinical research and statistical analysis. Finally, Jo et al. [11] examined structural health monitoring by using both IoT and Blockchain technology. The authors designed an innovative scheme where locally centralized and globally decentralized distribution is activated by dividing them into edge and core network. Hence, it enlarges the performance of the Blockchain system.

3 KNN Based Approach for EHRs

In this section, we propose a framework that integrates a KNN approach with Blockchain based data storage and access for a secure analysis of electronic health records. An illustration of the integrated workflow for a secure medical data access is shown in Fig. 1. Two different types of information are collected for electronic health record. The first one is patient's personal information that includes patient name, patient age, patient id and address. The second part of the data that represents patients' medical history. In this section, we collect the information about diseases, respected medication, period of medication, update condition after completed the medication. All sample data was collected from Sir Salimullah Medical College Hospital, Bangladesh.

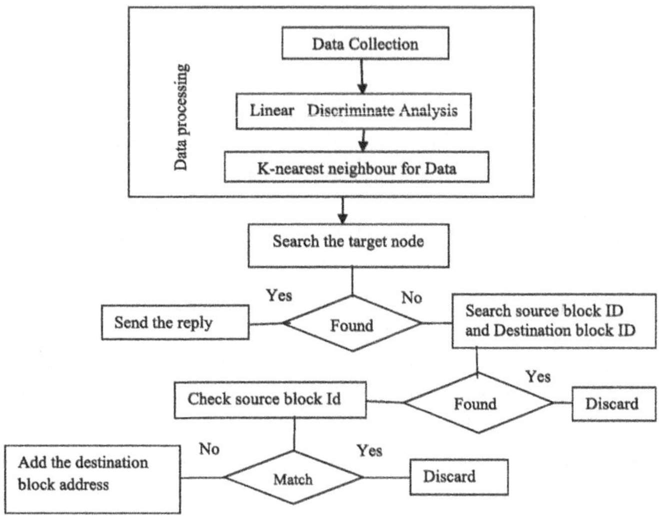

Fig. 1. Integrated workflow for Blockchain based medical data access

First data processed is conducted using two different types of classifier to process the collected data. After collecting the data, Linear Discriminate Analysis (LDA) algorithm is applied to separate the data in two different binary groups. Then, another algorithm called K-Nearest Neighbours (KNN) is applied to pick the data from the source data set. KNN algorithm finds the similarities among the data set in the same group. KNN algorithm is an ML algorithm which is used to handle both classification and regression predictive problems. It uses feature similarity to calculate the values of new data points, which specify that the new data point will be assigned a value based on how closely it matches the points in the training set as indicated in the following steps:

1. Calculate all the distances for test patients.
2. Find out the largest distance for the testing features.
3. Find the closely similar features.

LDA is a dimension reducing approach in a dataset to preserve as much information as possible. It can be splitting up into the following steps:

1. Collect the healthcare data of the patients.
2. Calculate co-variance of the patients' features.

3. Find out the eigenvalues of all features in an order from covariance.
4. Use he highest eigenvalues to identify patient's top features.
5. Finally, classify the collected medical records.

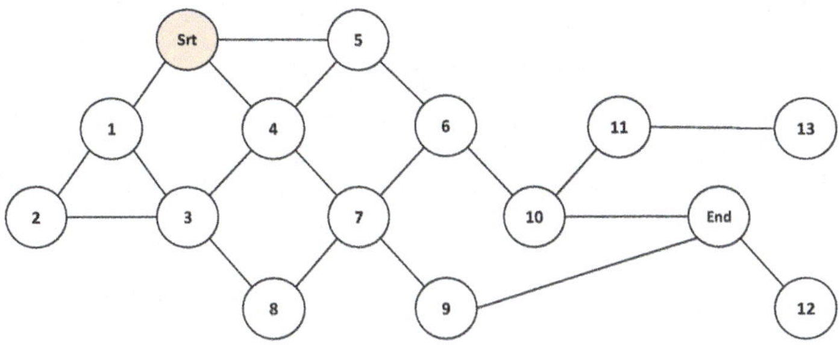

Fig. 2. Example of Blockchain based interaction from source Srt to destination End.

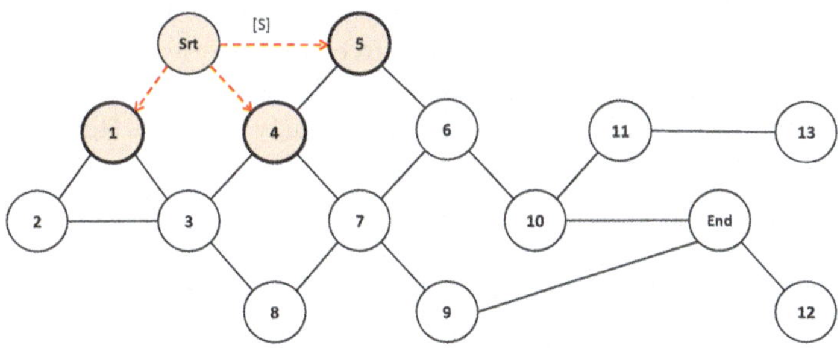

Fig. 3. Broadcasting messages in Blockchain based network

In the next stage Blockchain Road Discovery (BRD) is a reciprocal process that dynamically establishes relations among blocks. It is developed for use in multi-hop Blockchain network. Using BRD, the Blockchain network is self-organizing and mechanical. Network nodes interact to interchange and update signals to allow relationships over multiple hops between nodes. As Blockchain nodes in the network move about or join or leave the network, and as interconnections among blocks such as sources of interference change, all road finding data is automatically determined and maintained by the BRD process. Since the number or arrangement of intermediate block hops is required to reach any destination block nodes which may change at any time. The resulting Blockchain network structure may be quite large and can change rapidly at any time. BRD process allows block nodes to automatically discover a source road across multiple Blockchain network hops to any destination in the global Blockchain network.

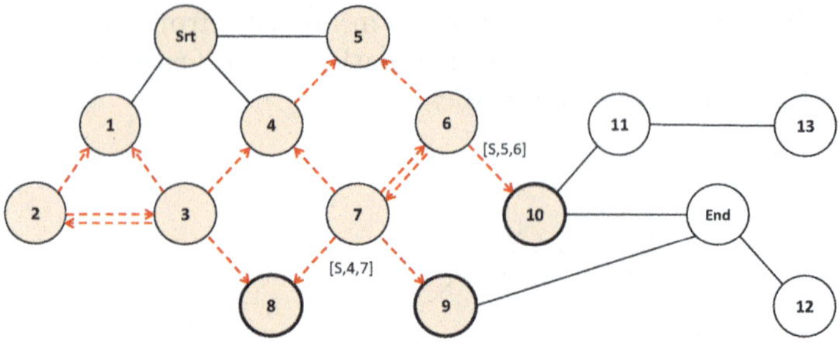

Fig. 4. Message propagation using IRR in Blockchain network

Information exchange among blocks sent then carries in Blockchain network where information keeps the track with sequence.

A Blockchain network may have huge number of blocks as nodes. In a Blockchain network comprises thousands of block nodes, BRD is the process where a source node Srt interacts to a destination node End and creates a path form source to destination. Basically BRD is used when a source node or any block node wants to make a relationship with any other block node that is a destination node. A Blockchain node first keeps the track of the destination node by placing information of destination in source node header and determines path cost. Figure 2 shows an example of a network where Srt will find a perfect source road by searching its Road Cache of road previously learned. However, if it fails, BRD automatically finds a new road to destination End.

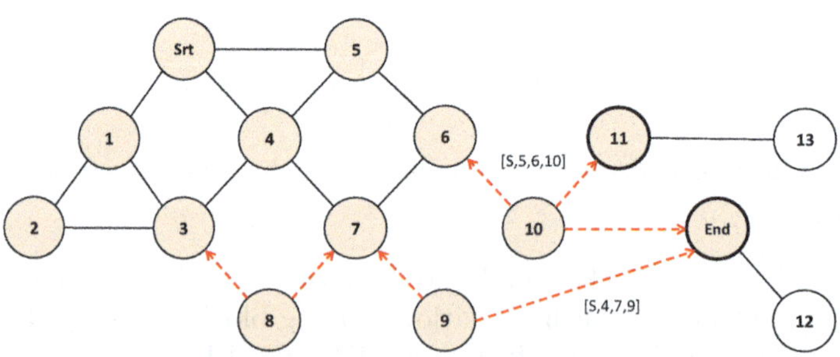

Fig. 5. Both of block 9 and block 10 send IRR to destination block End

To start the Road Discovery, Srt initiates Road Request Information (IRR) to its neighbour nodes. All network nodes accept this interaction-initiated request in the range of Srt and every RREQ message determines the initiator and target of the Road Discovery, and also contains a distinct request id, determined by the initiator of the request. Consequently, each IRR maintains a record list that contains the address of each internal node through which this fixed copy of the

IRR message has been forwarded as shown in Fig. 3, where the highlighted arrow indicates a broadcast message.

The process is continuous until the propagation is reached to destination node as shown in Fig. 4. Each source IRR contains source node id, current positions of source node, destination id, last known sequence number and broadcast id. Besides, initiate a timer to get wait reply. Intermediate nodes check the source id and broadcast id of IRR. If the requesting node already gets the response, it will discard the response. If it is not done, the intermediate node sets up a reverse road for the source node and increases the IRR hope count. Finally, the IRR is broadcast to all of its neighbours as demonstrated in Fig. 5.

When another node receives an IRR and if it is the target of the road discovery, it returns a Road Reply (RREP) message to the starting node of road discovery, which is indicated in the highlighted arrow in Fig. 6. Whenever starting node receives a RREP, it caches this road in its Road Cache (RC) for transferring information to the destination as shown in Fig. 6.

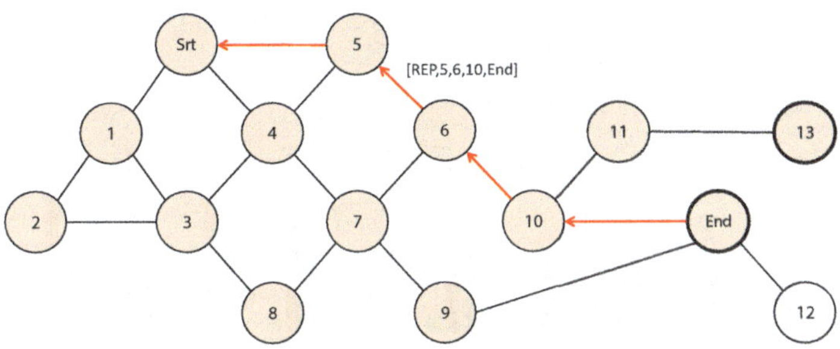

Fig. 6. Road reply (RREP) to the source block (Srt) from the destination block (End)

4 Experimental Results

Experimental analysis is organized using 10,000 medical records with 120 Blockchain nodes. Blockchain nodes are the end user computers connected with the network. LDA removes the redundant data from collected medical samples. The significant impact of LDA reduces the error rate to send the data to targeted nodes and reduce the time. KNN is imposed after LDA. The fundamental benefits of KNN is that helps to find the distances among the associated network as shown in Fig. 7. On the other hand, if we do not use LDA, there is large amount of error rate with large time consuming for KNN as demonstrated in Fig. 8.

As shown in Fig. 6, it is evident that for 1000 medical datasets and 120 destination end nodes, LDA based KNN is (21.33-10.25/21.33 * 100) 51.95% faster. Similarly, the error rate is 77.77% (9-2 = 7/9 * 100 = 77.77%) less. As data set is increased, LDA based KNN outperforms only KNN findings. For 2000, medical data records, LDA centric KNN is 60.67% faster and the error rate is 82.7% better than that of KNN less LDA. The significant of better findings are continuous for LDA centric KNN. Finally, for 10,00 medical records, the differences

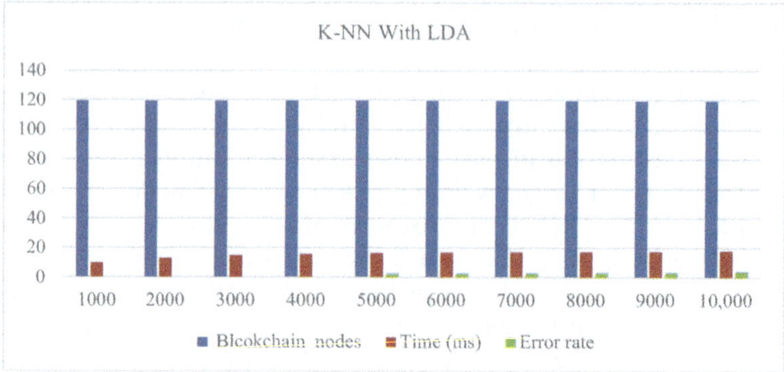

Fig. 7. Impact of LDA with KNN

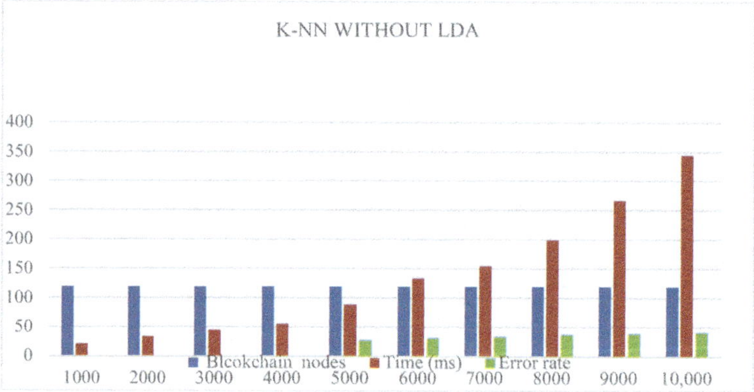

Fig. 8. LDA less KNN outcomes

are much higher than that of LDA less KNN and the resultant outcomes are about 95% faster with 89% error free.

5 Conclusion

Blockchain-based automated medical data regulation opens numerous opportunity to solve critical medical cases with wide range of expert support. This paper proposed a two stage dynamic Blockchain-based data processing structure as a highly secure system to connect a large number of medical experts from home and abroad for sharing electronic medical records. In the first stage, all medical records were processed to find very similar features or medical symptoms for a patient. LDA and KNN algorithms were used to process the primary medical records and generate highly correlated features. Next, dynamic Blockchain modelling was adopted to share information to the desired medical experts. To test our system, we considered the case of a patient getting advice or support from an expert or in special case, a board of five to six members. Our experimental results using 120 Blockchain nodes to transfer information securely demonstrated significant outcomes of the proposed framework. However, there is much scope to extend the current framework to include sharing of opinions among hundreds of experts. For future work, more than 1000 nodes with deep learning approaches will be explored to evaluate our proposed framework's performance.

References

1. Al-Karaki, J.N., Gawanmeh, A., Ayache, M., Mashaleh, A.: DASS-CARE: a decentralized, accessible, scalable, and secure healthcare framework using blockchain. In: 2019 15th International Wireless Communications & Mobile Computing Conference (IWCMC), pp. 330–335. IEEE (2019)
2. Al-Muhtadi, J., Shahzad, B., Saleem, K., Jameel, W., Orgun, M.A.: Cybersecurity and privacy issues for socially integrated mobile healthcare applications operating in a multi-cloud environment. Health Inform. J. **25**(2), 315–329 (2019)
3. Borkowski, M., Sigwart, M., Frauenthaler, P., Hukkinen, T., Schulte, S.: DeXTT: deterministic cross-blockchain token transfers. IEEE Access **7**, 111030–111042 (2019)
4. Danzi, P., Kalør, A.E., Stefanović, Č., Popovski, P.: Delay and communication tradeoffs for blockchain systems with lightweight IoT clients. IEEE Internet Things J. **6**(2), 2354–2365 (2019)
5. Downey, L.X., Bauchot, F., Röling, J.: Blockchain for business value: a contract and work flow management to reduce disputes pilot project. IEEE Eng. Manage. Rev. **46**(4), 86–93 (2018)
6. Esposito, C., De Santis, A., Tortora, G., Chang, H., Choo, K.K.R.: Blockchain: a panacea for healthcare cloud-based data security and privacy? IEEE Cloud Comput. **5**(1), 31–37 (2018)
7. Gawanmeh, A., Al-Hamadi, H., Al-Qutayri, M., Chin, S.K., Saleem, K.: Reliability analysis of healthcare information systems: state of the art and future directions. In: 2015 17th International Conference on E-health Networking, Application & Services (HealthCom), pp. 68–74. IEEE (2015)
8. Griggs, K.N., Ossipova, O., Kohlios, C.P., Baccarini, A.N., Howson, E.A., Hayajneh, T.: Healthcare blockchain system using smart contracts for secure automated remote patient monitoring. J. Med. Syst. **42**(7), 130 (2018)
9. Huang, D., Ma, X., Zhang, S.: Performance analysis of the raft consensus algorithm for private blockchains. IEEE Trans. Syst. Man Cybern. Syst. **50**(1), 172–181 (2019)
10. Ichikawa, D., Kashiyama, M., Ueno, T.: Tamper-resistant mobile health using blockchain technology. JMIR mHealt uHealth **5**(7), e111 (2017)
11. Jo, B.W., Khan, R.M.A., Lee, Y.S.: Hybrid blockchain and internet-of-things network for underground structure health monitoring. Sensors **18**(12), 4268 (2018)
12. Liu, D., Alahmadi, A., Ni, J., Lin, X., Shen, X.: Anonymous reputation system for IIoT-enabled retail marketing atop PoS blockchain. IEEE Trans. Industr. Inf. **15**(6), 3527–3537 (2019)
13. Liu, Y., Yu, F.R., Li, X., Ji, H., Leung, V.C.: Decentralized resource allocation for video transcoding and delivery in blockchain-based system with mobile edge computing. IEEE Trans. Veh. Technol. **68**(11), 11169–11185 (2019)

14. Parvin, S., Gawanmeh, A., Venkatraman, S., Alwadi, A., Al-Karaki, J.N.: Efficient lightweight mechanism for node authentication in WBSN. In: 2018 Advances in Science and Engineering Technology International Conferences (ASET), pp. 1–6. IEEE (2018)
15. Parvin, S., Gawanmeh, A., Venkatraman, S., Alwadi, A., Al-Karaki, J.N., Yoo, P.D.: A trust-based authentication framework for security of WPAN using network slicing. Int. J. Electr. Comput. Eng. **11**(2) (2020)
16. Pavithran, D., Shaalan, K., Al-Karaki, J.N., Gawanmeh, A.: Towards building a blockchain framework for IoT. Cluster Comput. **23**, 1–15 (2020)
17. Rahman, M.A., Hassanain, E., Rashid, M.M., Barnes, S.J., Hossain, M.S.: Spatial blockchain-based secure mass screening framework for children with dyslexia. IEEE Access **6**, 61876–61885 (2018)
18. Sun, Y., Zhang, L., Feng, G., Yang, B., Cao, B., Imran, M.A.: Blockchain-enabled wireless internet of things: performance analysis and optimal communication node deployment. IEEE Internet Things J. **6**(3), 5791–5802 (2019)
19. Wang, B., Dabbaghjamanesh, M., Kavousi-Fard, A., Mehraeen, S.: Cybersecurity enhancement of power trading within the networked microgrids based on blockchain and directed acyclic graph approach. IEEE Trans. Ind. Appl. **55**(6), 7300–7309 (2019)
20. Wang, J., Wang, S., Guo, J., Yanchang, D., Cheng, S., Li, X.: A summary of research on blockchain in the field of intellectual property. Procedia Comput. Sci. **147**, 191–197 (2019)
21. Wang, S., Ouyang, L., Yuan, Y., Ni, X., Han, X., Wang, F.-Y.: Blockchain-enabled smart contracts: architecture, applications, and future trends. IEEE Trans. Syst. Man. Cybern. Syst. **49**(11), 2266–2277 (2019)
22. Zhang, J., Xue, N., Huang, X.: A secure system for pervasive social network-based healthcare. IEEE Access **4**, 9239–9250 (2016)
23. Zhaofeng, M., Xiaochang, W., Jain, D.K., Khan, H., Hongmin, G., Zhen, W.: A blockchain-based trusted data management scheme in edge computing. IEEE Trans. Industr. Inf. **16**(3), 2013–2021 (2019)

Author Index

H. Selvaraj et al. (Eds.): ICSEng 2020, LNNS 182, pp. 465–466, 2021.
https://doi.org/10.1007/978-3-030-65796-3

The manufacturer's authorised representative in the EU is Springer Nature Customer Service Centre GmbH, Europaplatz 3, 69115 Heidelberg, Germany. If you have any concerns regarding our products, please contact ProductSafety@springernature.com

Printed and bound by CPI Group (UK) Ltd, Croydon, CR0 4YY
30/09/2026
02225495-0013